RHEOPHYSICS

The Deformation and Flow of Matter

Why is it necessary to strike while the iron is hot? What makes a good liquid crystal display? Why is rubber so elastic? These questions can be answered through rheophysics, using the mechanics of continuous media at the macroscopic scale, and statistical mechanics and the physics of defects at the microscopic level.

This book addresses problems involving the flow of matter, covering the main aspects of the mechanical response of fluids and solids to applied stress or strain. It includes the hydrodynamics of ordinary liquids, the elasticity and plasticity of solids, and the rheology of complex fluids such as suspensions, polymers and liquid crystals. Dislocations are described thoroughly, and special attention is given to instabilities.

Concepts and physical properties are illustrated by numerous experiments, historical anecdotes, and applications to aeronautics, metallurgy and geophysics, making this a valuable reference for researchers and graduate students in physics, engineering and materials science.

PATRICK OSWALD is Director of Research with CNRS at the Physics Laboratory at the École Normale Supérieure, University of Lyon, France. His research spans many areas in soft condensed matter, including thermotropic and lyotropic liquid crystals, crystal growth, defects, plasticity and rheology.

RHEOPHYSICS

The Deformation and Flow of Matter

PATRICK OSWALD

CNRS, University of Lyon, France

Translated by Doru Constantin

CAMBRIDGE
UNIVERSITY PRESS

University Printing House, Cambridge CB2 8BS, United Kingdom

Cambridge University Press is part of the University of Cambridge.

It furthers the University's mission by disseminating knowledge in the pursuit of education, learning and research at the highest international levels of excellence.

www.cambridge.org
Information on this title: www.cambridge.org/9781107439528

Original edition: Rhéophysique © Éditions Belin – Paris, 2005

Ouvrage publié avec le concours du Ministère français chargé de la culture – Centre national du livre

Published with the assistance of the French Ministry of Culture – National Book Centre.

First published 2009
First paperback edition 2014

A catalogue record for this publication is available from the British Library

ISBN 978-0-521-88362-7 Hardback
ISBN 978-1-107-43952-8 Paperback

To my wife Jocelyne,
and to my daughters Séverine and Magalie

Contents

Foreword

Should anyone still believe in a clear demarcation between fundamental science and applied science – that one is noble and the other laborious, that one moves in the higher spheres of universalism and the other makes do with down-to-earth utilitarianism – then reading this book will make them think again.

We are invited to travel through the border country where the world of ideas, principles and theories is not separated from but meets and mixes with the world of phenomena, empirical laws and concrete objects. Few topics lend themselves quite so readily to such cross-fertilisation as does rheophysics in the way Patrick Oswald presents it for us here.

Already in his famous book on *dislocations*, Jacques Friedel had accustomed us to bringing together the most classic and often the most ancient observations and knowledge – those of metallurgists (blacksmiths, founders, jewellers, wiredrawers and the like) or of potters, or even of flint cutters – with the science of continuous matter, then the science of crystal states and finally with the physics of solids. This gave rise to a whole new outlook for practitioners and engineers; they began to understand, at least in outline, the deformational behaviour of the materials they had erstwhile worked, processed and deformed quite empirically. The extreme complexity of these behaviours was reduced, using plain geometrical concepts (first at macroscopic and then at atomic scale), to insightful models that made interpretation possible. And so, as Jean Perrin put it, something complicated and visible was replaced by something simple and invisible, before the invisible appeared before our very eyes with the advent of the electron microscope in the late 1950s and its stupendous further development.

Here, in the same spirit, Patrick Oswald addresses the vast and important topic of *rheophysics* – a neologism created from *rheology*, the science of deformation and material flow. The topic is vast as it covers virtually all condensed matter. It is important because all matter is subjected to some force, be it gravity only, and so tends to deform, whether we like it (when we want to shape it) or not (when we want it to remain stable).

This treatise begins with an overview of materials in their extreme diversity and then with refreshers on continuum mechanics and hydrodynamics of simple liquids. The stage being set, a detailed analysis is presented of the various behaviours – elastic, elastodynamic, plastic, viscoelastic, hydrodynamic, etc. – that may occur when matter is subjected to forces. Wherever possible the major laws of macroscopic behaviour are associated with

xi

atomic or molecular models, the profusion of which is tempered by a constant endeavour to synthetise and to generalise.

Amid this wealth of data, hypotheses, laws, equations, experiments, formulas, criteria and materials (from metal alloys to rocks, from colloidal crystals to polymers, from suspensions to emulsions, from liquid crystals to rubbers, from micelles to glass, etc.) a firm line had to be kept. This was made possible by the global vision Patrick Oswald has acquired of a domain in which he is one of foremost experimental and theoretical specialists. He provides us here with a dense and thorough treatise, where experimental illustrations and practical examples justify and corroborate mathematical developments.

This book, which is to be lauded for the effort put into unifying the notation in a handy table and for the sheer quality of its presentation, will prove very useful in academic and industrial laboratories alike. May it also help students who, even if they will probably not read it from cover to cover, will discover a thousand and one topics and exercises, and will come to realise that between the mathematician's desk, the physicist's laboratory and the engineer's workshop there should be no barriers but, on the contrary, constant, supportive and stimulating feedback.

Yves Quéré
Emeritus Professor at the École Polytechnique

Preface

Rheophysics is the study of the deformation and flow of matter in all its states (gaseous, liquid, solid, and even glassy or mesomorphic[1]). Thus defined, the topic is extremely vast, comprising the hydrodynamics of simple liquids, the elasticity and plasticity of solids and the rheology of complex fluids exhibiting a behaviour intermediate between those of liquids and of solids.

There are two different (and yet complementary) approaches to rheophysics.

The *macroscopic* approach attempts to describe the behaviour of a a material over *scales much larger than that of its microstructure*. One then employs the concepts and techniques of the *mechanics of continuous media*. This theory, which stems directly from the fundamental laws of macroscopic physics (Newton's laws, the first and second principles of thermodynamics), can describe and quantify the average structure of flows[2] provided the *constitutive laws* of the material are known. Most important among them is that relating *stress* and *strain*. We recall that this law is often established in a *phenomenological* manner, relying on symmetry arguments and using the results of rheological measurements performed on model systems.

The second approach is *microscopic*. Its goal is to *identify the mechanisms of material deformation at the scale of its microstructure*. Though it depends strongly on the type of material under consideration, it remains nevertheless fundamental, as it brings into play extremely general mechanisms and concepts (such as *defects*). It also has great practical importance, since understanding the relation between the macroscopic constitutive law of a material and its microstructure enables us to act more effectively on its rheological behaviour. Note that microscopic theories require elementary notions of statistical mechanics (a brief reminder of which is given in the chapter on polymers) and of the physics of defects, to be considered in detail in the chapters on solids and on liquid crystals.

The goal of this book is thus to present rheophysics as generally as possible, striving to maintain the appropriate balance between these two approaches, at the experimental level as well as at the theoretical one. This is obviously a very difficult endeavour, considering the wide variety of materials and their rheological behaviour, impossible to exhaust in a

---

[1] Liquid crystals, whose structures are intermediate between liquids and solids, represent the mesomorphic states of matter. Note that the term mesomorphic stems from the Greek words 'meso' signifying intermediate and 'morphe', meaning form (of intermediate form).

[2] Namely, 'smoothed' over a scale larger than that of the microstructure, but still smaller than the sample size.

work of this size. For that reason, we had to make (sometimes difficult) choices between several possible topics, to the point of omitting certain important phenomena,[3] such as the turbulence of fluids, the rheology of glasses or the flow of dry granular media.

As to the matter of the book, it is presented in detail in the table of contents. Let us only point out that it is structured in eight chapters, grouped in four distinct parts.

Thus, the first two chapters, making up the first part, contain a general discussion of materials and their rheological behaviour, as well as a brief overview of the mechanics of continuous media. They are not indispensable at a first reading, but it is advisable to have a good grasp of the material here before tackling the remainder of the book.

Next comes the chapter on the hydrodynamics of simple liquids at small and large Reynolds numbers. Alongside the usual fundamental topics (lift force on wings, vortices, viscous drag on a moving body, etc.), this chapter deals with subjects seldom treated elsewhere, such as the spreading of a liquid on a rotating plate ('spin-coating'), microorganism propulsion or the Saffman–Taylor instability in porous media. This chapter, aimed primarily at hydrodynamicists, also contains an instrumental section on rheometers and their use, a paragraph on the microscopic theories of the viscosity of gases and liquids, as well as an appendix on shock waves, and more generally on discontinuity surfaces in fluid mechanics. It makes up the second part of the book.

One finds then two chapters on solids. The first one deals with linear elasticity in the static and dynamic regimes (waves, in particular). It discusses several classical topics of elasticity (the flexion of a plate, the torsion of a beam and its buckling instability under compression) alongside less common applications, such as seismic waves, aeroelastic instabilities, or the peculiar elasticity of colloidal crystals. The following chapter treats the problems of plasticity and brittle fracture, as well as the physics of defects, a subject both modern and fundamental, which only came into its own right after the Second World War, although it encompasses much more than the mere area of rheophysics. These two chapters, representing the third part of this book, should be of interest not only for physicists and metallurgists and but also for geologists.

We are left with the last three chapters on complex materials, two of them being dedicated to isotropic viscoelastic fluids and solids (polymers and dispersions, essentially), while the last one deals with liquid crystals, whose behaviour is clearly at the crossroads between hydrodynamics and metallurgy. These three chapters make up the fourth and final part of the book. It is addressed to readers with an interest in soft matter, whether physicists, mechanicists or engineers.

In conclusion, let us point out that some important references to fundamental papers or to recent work are cited at the end of each chapter, complementing the (in no way exhaustive) list of reference books given at the end of the book. Note that research papers are indicated in the following way: (Frank, 1954) (first author followed by the publication year of the paper), while the books are referenced by a number: [33], for instance. A list of the notation employed is also given at the end of the book. Finally, the equations of hydrodynamics and elasticity are written in cylindrical and spherical coordinates at the end of Chapters 3 and 4.

---

[3] Not completely understood to date.

Acknowledgments

This work appeared as a result of several lectures given between 1983 and 1993 at the Écoles Normales Supérieures de Fontenay-aux-Roses and Saint-Cloud, then, when they merged in 1987, at the École Normale Supérieure de Lyon. At the time, it was a revision course of the mechanics curriculum for physics students, including some elements of fluid hydrodynamics and solid elasticity. It was only in 2002 that I volunteered to present a lecture course in rheophysics to the department of material sciences of the École Normale Supérieure de Lyon, a proposal that was immediately supported by the directors of studies at that time, Professor E. Charlaix and Professor B. Berge, whom I thank once again for their confidence. It was while teaching this course that I started writing the book, taking advantage of the numerous questions and suggestions of the students to improve it along the way. I also benefited from the precious help of J.-F. Palierne, who advised me in my reading choices and who explained several of the more subtle points of polymer viscoelasticity. I should also acknowledge the outstanding part played by Doru Constantin who, while doing his Ph.D. under my supervision, took the initiative of studying the rheology of surfactant solutions, giving me no option but to get up to date on this topic. I would like to thank my early and assiduous readers, Doru Constantin (once again), Patrick Cordier, Éric Freyssingeas and Michel Saint-Jean, for their very constructive remarks and suggestions. Special thanks go to Professor Yves Quéré, member of the French Academy of Sciences, for whom I hold the greatest admiration and who gave me the honour of reading the manuscript and of writing the Foreword. It is a pleasure to thank my friend Pawel Pieranski, who never stopped encouraging me during the three years of work on the present book. I am equally grateful to my editors, C. Counillon at Belin and S. Capelin at Cambridge University Press, who made the English edition of this work possible. I would also like to express all my gratitude to Doru Constantin, who skilfully translated this book into English.

I will end with the most important: my family. Writing a book requires a lot of time, mainly taken on that which I should have dedicated to family (on weekends, in particular). That is why I would like to thank my wife Jocelyne, and my two daughters Séverine and Magalie, for their unfailing support; without them, this book would have never been written.

1

General points on materials and their rheophysical behaviour

The term 'rheology' was invented by E. C. Bingham in 1928.[1] In the beginning, it encompassed an extremely wide topic, dealing with the *study of the deformation and flow of matter*. This definition was finally adopted one year later, at the creation of the American Society of Rheology[2] (for a review of the history of rheology, see (Doraiswamy, 2002).

Nevertheless, this term became more 'specialised' with time and only applies today to fluids exhibiting behaviour intermediate between that of ordinary liquids and that of perfectly elastic solids. For this reason, we employ the term 'rheophysics' to define the science of deformation and flow in the general sense defined by Bingham, reserving the term 'rheology' for the study of complex fluids, as in modern use.

Schematically, matter can be found in either of two states, entailing extremely different rheological (or rheophysical) properties:[3]

The solid state This is the case for a block of metal at room temperature, which will conserve its shape indefinitely in the absence of applied forces. This holds, up to a small deformation, invisible to the naked eye, even under the influence of moderate forces, such as gravity. This deformation is *reversible*, since it disappears once the stress is removed. One speaks in this case of *elastic deformation*. *Elasticity* is one of the defining properties of solids. For most of them, it is of *enthalpic* origin, stemming from the variation in energy of bonds between atoms or molecules upon deformation of the sample. To this concept one associates that of *elastic modulus*, quantifying the resistance of the material to a small deformation (generally below 10^{-3}, for ordinary solids). It is worth remembering that the elastic response of a solid is often *anisotropic*, meaning that it depends on the direction of the applied forces with respect to the crystal axes. This anisotropy reflects the *crystal symmetry* of the material.

Experiment also shows that a distinction must be made between *'hard'* solids and *'soft'* solids.

For the first category, including *metals* and *minerals*, the elastic moduli are of the order of 10^{11} Pa. *Dissipation* is negligible in these materials in the elastic deformation regime up

---

[1] The word is coined from the Greek words '$\rho\acute{\varepsilon}\omega$' and '$\lambda\acute{o}\gamma o\sigma$' (pronounced rheos and logos) meaning 'to flow' and 'study', respectively.

[2] Counting among its founders M. Reiner, E. C. Bingham and G. S. Blair.

[3] At this point, one should note that we group the liquid, gas and liquid crystalline states under the denomination 'fluid state', a simplification justified as far as rheology is concerned, as we will show throughout this work.

to very high frequencies, in the gigahertz range. For this reason, sound experiences very little attenuation and can propagate over very large distances (thousands of kilometres) within them.

The second category have much smaller elastic moduli and dissipative effects are much more important. *Elastomers* (such as natural rubber), and also *polymer gels* and *colloidal crystals*, belong to this category of materials. Thus, colloidal crystals exhibit very low elastic moduli, typically a few tens of pascals, while for elastomers moduli vary between 10^5 and 10^8 Pa; gels (very often swollen by a solvent) have intermediate values. Dissipative effects also appear at very low frequencies, from the hertz to the kilohertz as the case may be: for this reason, waves are strongly attenuated in these solids, termed *viscoelastic*.

Elastomers and polymer gels are also peculiar in that they can sustain huge reversible deformations (sometimes by more than 100% in relative value). The reason is that their elasticity is of an *entropic* nature (instead of enthalpic, as in ordinary or colloidal crystals), originating in the loss of configuration entropy of the constituent polymer chains under elongation or compression.

We end this presentation of solids by noting that they can be in *crystalline* or *amorphous* form. In crystals,[4] the elastic properties are anisotropic, the deformation depending on the direction of the applied forces. Amorphous solids, on the other hand, exhibit isotropic elasticity. Finally, each type of solid has a *yield stress* (or *elastic limit*) above which it breaks or starts to *creep* (that is, to 'flow') *irreversibly*. One speaks of *rupture* in the first case, and of *plastic deformation* in the second one. In this latter regime, the material behaves more like a fluid, the other state of matter to be discussed in the following.

The fluid state   This is the case for gases and simple or complex liquids, which always end (after a certain time) by taking the shape of their container. The time taken is longer for a more *viscous* material, an intuitive concept that we will return to in more detail. Most fluids are *isotropic*, but there are also some that are *anisotropic*.

This latter type, also termed *liquid crystals*, are intermediate between liquids and crystals. Discovered at the beginning of the twentieth century by G. Friedel, they behave as solids in certain directions (along which they exhibit elasticity), and as liquids in the complementary directions (viscous behaviour). These *anisotropic* materials are thus *viscoelastic*, since their elastic or viscous behaviour depends on the direction of the applied stress.

There are also *isotropic viscoelastic fluids*, whose elastic or viscous behaviour depends on the *time scale* over which the deformation is applied. More precisely, these fluids respond elastically at short times, while they behave as viscous fluids over long times.

We emphasise that the behaviour of the viscoelastic solids mentioned above (elastomers, gels and colloidal crystals) is the exact opposite; their response being viscous over short times and elastic at long times (provided, however, their elastic limit is not exceeded).

---

[4] More precisely, in monocrystals.

In practice, the characteristic time separating these two regimes is called the *viscoelastic relaxation time*.[5] Obviously, this time varies enormously among materials, and can take values from 10^{-12} s (in the case of water, which will hence behave as an ordinary viscous fluid under normal conditions) and several hundred years (in the case of the Earth's mantle, behaving as a solid over the human lifetime and as a very viscous fluid over geological time scales).

These two examples show that the concepts of viscous or viscoelastic fluids, and more generally the distinction between solids and fluids, are only *relative*, crucially depending on the time scale over which one watches the material evolve.

After these comments, let us detail the contents of this chapter. We will begin by recalling the limiting behaviours one can encounter, which will lead us to defining the concepts of Hookean elastic solid and Newtonian viscous fluid (Section 1.1). We will then see that a solid can undergo plastic deformation and that a fluid can exhibit non-Newtonian (Section 1.2) and thixotropic (Section 1.3) behaviour. Finally, we will discuss the concepts of yield stress fluid (Section 1.4), and isotropic viscoelastic fluid and solid (Sections 1.5 and 1.6), ending with liquid crystals (Section 1.7).

1.1 The Hookean elastic solid and the Newtonian viscous fluid

As we have just shown, matter can be found in various forms. The two extreme cases are the elastic solid, described by Hooke's law (1678), and the viscous fluid, whose behaviour follows Newton's law (published in the *Principia* in 1687).

1.1.1 Hooke's law

This constitutive equation describes the behaviour of a perfect elastic solid.[6] It reads:

$$\sigma = G\gamma, \tag{1.1}$$

where σ is the applied stress (force per unit surface), $\gamma = u/d$ the strain (Fig. 1.1) and G the elastic modulus (the shear modulus in the case of Fig. 1.1). A proper definition of the concept of stress, fundamental in continuum mechanics, will be given in the next chapter. Since the stress is measured in Pa (the unit of pressure in the SI system) or in dyn/cm^2 in the CGS system (1 Pa = 10 dyn/cm^2) while the strain is dimensionless, the elastic modulus is also measured in Pa (or in dyn/cm^2).

The order of magnitude of the elastic modulus varies enormously, from 10^{10} or 10^{11} Pa for a metal or a rock, to a few tens of pascals for a colloidal crystal or a very dilute polymer gel in a solvent.

---

[5] This definition applies equally well to fluids and viscoelastic solids.
[6] It also applies to viscoelastic liquids 'at short times' and to viscoelastic solids 'at long times' (see Sections 1.5 and 1.6).

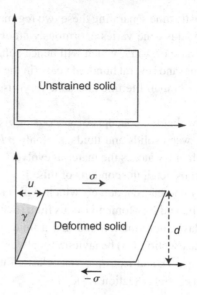

Fig. 1.1 Shear deformation of an elastic solid under an applied stress σ.

The theory of elasticity, dealing with the study of Hookean elastic solids, will be discussed in Chapter 4.[7]

Hooke's law states that the strain γ is proportional to the applied stress σ. Hence, if the stress is removed, the strain vanishes and the body recovers its initial shape, meaning that it keeps the memory of its previous state.

The concept of elasticity is thus intimately related to that of memory, the Hookean elastic solid representing the limiting case of a body endowed with *infinite memory*.[8]

It is noteworthy that Hooke's law makes the implicit assumption that the material *responds instantaneously* to mechanical action (since the law contains no characteristic time).

1.1.2 Newton's law

This constitutive law applies to viscous fluids.[9] It reads:

$$\sigma = \eta\dot{\gamma}, \tag{1.2}$$

where σ is the applied stress, $\dot{\gamma} = v/d$ the velocity gradient (or shear rate defined in Fig. 1.2) and η the *dynamic shear viscosity*. Viscous fluids obeying this law are termed

---

[7] In that chapter we will also study the response of a solid to deformation modes other than shear, namely elongation, flexion and torsion.

[8] The concept of 'memory' will exhibit its full importance when we consider the case of viscoelastic materials (see Sections 1.5 and 1.6).

[9] It also applies to viscoelastic liquids in the 'long time' limit, and to viscoelastic solids 'at short times' (see Sections 1.5 and 1.6).

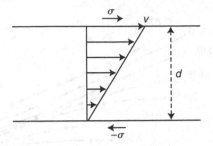

Fig. 1.2 Fluid under constant shear: v is the velocity of the upper plate with respect to the lower one and d is the thickness of the fluid layer.

Newtonian (this definition will be further refined in Section 1.5). Since $\dot{\gamma}$ is in units of s^{-1}, the viscosity is measured in pascal seconds, Pa s (also known as poiseuille in the SI system) or in poise in the CGS system (1 Pa s = 10 P).

At room temperature, water has a viscosity of 10^{-3} Pa s. Silicone oils, on the other hand, can exhibit viscosities up to several million Pa s, showing that this parameter can have hugely different values (without even mentioning the Earth's mantle, where the viscosity approaches 10^{21} Pa s).[10]

Let us also emphasise that *the Newtonian fluid responds instantaneously to shear* (since there is no characteristic time in Newton's law, just as there is none in Hooke's law) and *has no memory*, in contrast with the elastic solid. Indeed, after a stress σ is applied during a time t, the fluid undergoes a *permanent and irreversible deformation*, $\gamma = \sigma t/\eta$ and never again recovers its initial state after the force is removed.

The theory of hydrodynamics, describing the flow of Newtonian viscous fluids, will be treated in Chapter 3.

Let us now see in what way material behaviour differs from these ideal models. To start with, we will consider the case of the plastic solid, followed by that of the non-Newtonian viscous fluid.

1.2 The plastic solid and the non-Newtonian viscous fluid

For a long time (from the seventeenth century until the beginning of the nineteenth century), physicists contented themselves with classifying materials either in the family of Hookean elastic solids, or in that of Newtonian viscous fluids. It was only in the nineteenth century that they started having doubts.

In 1835, W. Weber (Weber, 1835), who was studying silk thread, noted that it is not perfectly elastic, and that it keeps on elongating slowly under the action of a load, even a moderate one. When the load is removed, the thread contracts but never recovers its initial length. Thus, under traction it undergoes a *permanent irreversible deformation*, also

---

[10] Other values will be given in Chapter 3, where we will also discuss the physical parameters that influence viscosity (temperature and pressure).

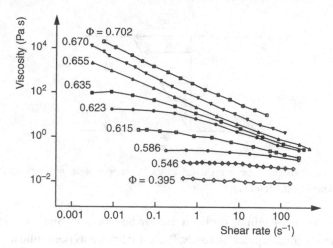

Fig. 1.3 Constant shear viscosity for suspensions of silica hard spheres with a diameter of 0.49 μm, for different values of the volume fraction Φ (Jones *et al.*, 1991).

called *plastic* deformation.[11] *Plasticity* is specific to solids and in particular to crystals, which exhibit long-range positional order in three dimensions. We will see in Chapters 3 and 4 that plastic deformation is intimately connected to the presence of topological defects within the crystal (these defects can be point-like, such as vacancies or interstitials, linear, such as dislocations, or surface-like, such as grain boundaries).

In the nineteenth century it was also realised, owing to the development of devices for measuring the viscosity (called *rheometers*), that many fluids depart from the ideal Newtonian behaviour, in that their viscosity depends on the applied shear rate $\dot{\gamma}$. In most cases, the fluids are *shear-thinning*, i.e. their viscosity decreases as $\dot{\gamma}$ increases. As examples of shear-thinning viscous fluids, one can cite solid particle suspensions as long as they are not too concentrated, and for moderate shear rates. Such an example is given in Fig. 1.3 (see also Section 7.5.1).

In other (much less common) cases, the viscosity increases with the shear rate. The fluid is then termed *shear-thickening*. Concentrated suspensions of solid particles can exhibit this type of behaviour, in a specific range of shear rate. They can even start by being shear-thinning, then shear-thickening, and then again shear-thinning, with increasing shear rate.

Liquid crystals, and in particular nematics (described in Section 1.7), can also exhibit shear-thinning or shear-thickening behaviour, depending on the way the molecules are anchored at the surface of the walls delimiting the sample (see Chapter 7).

Finally, viscoelastic liquids, consisting of a wide range of materials (polymer melts, emulsions, etc.), almost always behave under continuous shear (and in the stationary regime) as viscous non-Newtonian fluids, usually shear-thinning (see for instance Fig. 1.15).

We will now try to give precise definitions for the concepts of plastic solids and non-Newtonian viscous fluids. To this end, let us analyze the action of shear on these materials and trace some typical graphs.

---

[11] Plastic comes from the Greek adjective '$\pi\lambda\alpha\sigma\tau\iota\kappa\acute{o}\sigma$' (pronounced 'plastikos') meaning 'capable of being molded'.

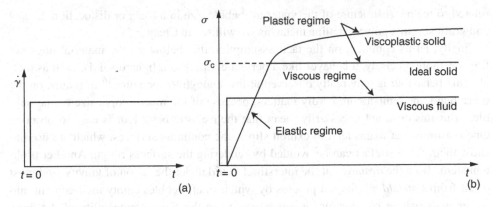

Fig. 1.4 Response of a viscous fluid, or of a solid, to the sudden onset of shear. (a) Shear rate as a function of time; (b) stress as a function of time.

1.2.1 Response to a sudden onset of shear (transient regime)

Consider a material, initially at rest, which is subjected to a constant shear rate $\dot\gamma$ starting at the moment $t = 0$, and let us plot the evolution of the stress as a function of time (Fig. 1.4).[12]

In the ideal case of a *Newtonian viscous fluid*, the stress 'immediately' reaches its stationary regime value,[13] here proportional to the shear rate ($\sigma = \eta\dot\gamma$).

For a *non-Newtonian viscous fluid*, the response to the sudden onset of shear is once again assumed to be instantaneous; namely, the stress shoots from 0 to a plateau value $\sigma(\dot\gamma)$; however, this value is no longer proportional to the shear rate.

In the case of *solids*, on the other hand, the stress increases linearly with time (an elastic regime during which $\sigma = G\dot\gamma t$ according to Hooke's law), and then saturates upon reaching a *critical stress* σ_c. At this point, the material enters the *plastic regime*.[14] The value σ_c defines the *elastic limit* of the material or, equivalently, its *critical stress*. Two situations can occur, depending on whether the stress immediately reaches saturation at σ_c, defining the *ideal plastic solid* (fairly well represented by foams and concentrated emulsions[15]) or, on the contrary, it keeps on increasing (a feature known as *work hardening*) until it saturates at a new plateau value, which depends on $\dot\gamma$: this defines the *viscoplastic solids*, comprising most ordinary solids, such as metals or minerals (Fig. 1.4).

Note that the stress σ_c is independent of $\dot\gamma$, but it does depend strongly on the material under consideration, varying from a few Pa for a foam, to several MPa or even GPa for a metal at room temperature. These extreme values show that the parameter is strongly

---

[12] Nowadays, rheometers can impose either the shear rate or the stress, while measuring the conjugate parameter. In the experiment we describe, the shear rate is fixed.

[13] This assumption will be discussed in more detail in Section 1.5.

[14] Assuming the material does not break, which means that its temperature is above the brittle–ductile transition temperature T_{BD} (see Section 5.4.1).

[15] Concentrated emulsions are foams wherein the gas (air, usually) is replaced by a liquid. The physics of foams is described in the excellent book by D. Weaire and M. A. Fortes, *The Physics of Foams*, Oxford: Oxford University Press, 1999.

related to the microstructure of the material (bubble size in a foam or dislocation density and grain size in a polycrystalline metal, as we will see in Chapter 5).

Finally, let us comment on the tacit assumption that below σ_c, the material does not flow or, more precisely, it behaves like a perfect elastic solid. In hard solids, such as metals, this behaviour is only really observed at low enough temperature.[16] In foams, on the other hand, the temperature hardly matters, because of the macroscopic size of the bubbles. But this does not necessarily mean that their elastic behaviour is easy to observe. One reason is that foams tend to exhibit slip at the confining surfaces, which are used to shear them. This artefact can be avoided by rendering the surfaces rough. Another problem stems from the *drainage* of the interstitial liquid under the action of gravity, and most of all from *Ostwald ripening*, a process by which small bubbles empty their content into larger ones (where gas pressure is lower) owing to the finite permeability of the films separating them (Gandolfo *et al.*, 1997). Under these conditions, foam structure evolves constantly with time, a process referred to as *aging*. Thus, its 'ideal' elastic behaviour can only be observed over relatively short time intervals, during which surface slippage and aging can be neglected.[17] The same problems are encountered in the case of concentrated emulsions, except that Ostwald ripening is much less pronounced than in ordinary foams. For this reason, concentrated emulsions are closer to the ideal plastic solid than foams.

Putting aside for now these experimental problems, let us again focus on describing the behaviour of stress in the stationary regime.

1.2.2 Behaviour under continuous shear (permanent regime)

In Fig. 1.5, we plot the stress measured in the stationary regime $\sigma(t \to \infty)$ as a function of $\dot{\gamma}$. Figure 1.6 shows the corresponding viscosity $\eta(t \to \infty)$ (by definition, it is given by $\sigma(t \to \infty)/\dot{\gamma}$). These graphs demonstrate that a clear distinction is to be made between fluids and solids above the 'plastic threshold'.

For a Newtonian viscous fluid, the stress increases linearly with $\dot{\gamma}$, the viscosity being constant by definition.

In a shear-thinning fluid the stress increases more slowly than $\dot{\gamma}$, such that the viscosity decreases with $\dot{\gamma}$. In this case, the viscosity often exhibits two plateau values, at low and high shear rates, denoted by η_0 and η_∞, with a 'power law' crossover behaviour. Several phenomenological equations were put forward to explain this behaviour, among them that of Cross (1965), of the form:

$$\eta = \eta_\infty + \frac{\eta_0 - \eta_\infty}{1 + (k\dot{\gamma})^p}. \tag{1.3}$$

---

[16] Typically, below a temperature of the order of $0.5T_m$, where T_m is the melting point of the body. At higher temperature, and in particular on approaching T_m, solids behave as extremely viscous Newtonian liquids. The flow is then due to the diffusion of vacancies across or at the surface of the grains in the polycrystal, as will be shown in Sections 5.3.4.1 and 5.3.4.2.

[17] The same difficulties appear under continuous shear, as soon as one tries to study the ideal plastic behaviour of foams.

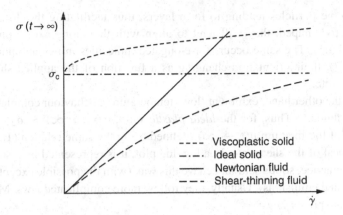

Fig. 1.5 The stress measured in the stationary regime as a function of the shear rate for a plastic solid (ideal or viscoplastic), a Newtonian viscous fluid and a shear-thinning viscous fluid.

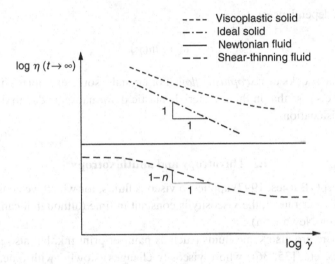

Fig. 1.6 The viscosity measured in the stationary regime as a function of the shear rate (in a log–log plot). Each graph corresponds to a different type of material.

In this expression, the constant k (with time dimensions) and the exponent p are two adjustable coefficients. Note that, over the shear rate range where $\eta_0 \gg \eta \gg \eta_\infty$, this equation reduces to:

$$\eta = \eta_0 (k\dot{\gamma})^{-p} = m\dot{\gamma}^{n-1}, \tag{1.4}$$

with $p = 1 - n$ and $m = \eta_0 k^{n-1}$. This is the *Ostwald–de Waehle law*. The exponent n is smaller than 1 for shear-thinning fluids. The coefficient m, termed 'consistence', is measured in peculiar units: $Pa\,s^n$.

We emphasise that the concept of shear-thinning is closely related to that of *structural change* affecting the sheared material. Thus, in a hard sphere suspension, this effect can be related to the particles tending to form layers, thus facilitating the flow. If the solid particles are rod-shaped, they will tend to align with their long axes along the velocity (Section 7.5.1). The same occurs for elongated molecules in nematic phases (defined in Section 1.7), their orientation changing as a function of the applied shear rate (see Section 8.1.7), etc.

Solids, on the other hand, exhibit at 'low' temperature a behaviour completely different from that of fluids.[18] Thus, for the *ideal plastic solid*, $\sigma(t \to \infty) = \sigma_c$, meaning that the viscosity of the medium $\eta(t \to \infty)$ (defined using the same relation (1.2)) decreases as the reciprocal of the shear rate. In a log–log plot, it is represented by a straight line of slope -1. In practice, solids seldom behave this way (with the possible exception of foams and concentrated emulsions). Usually, they follow more complicated laws. Most common among them are power laws:

$$\sigma - \sigma_c \propto \dot{\gamma}^{1/n} \quad \text{with} \quad n > 1, \tag{1.5}$$

or logarithmic dependences:

$$\sigma - \sigma_c \propto \ln(\dot{\gamma}). \tag{1.6}$$

In this case, one speaks of *viscoplastic flow*. For crystals, some examples will be given in Section 5.3. We stress that in these materials plastic deformation is due to defects, and in particular to dislocations.

1.3 Thixotropy and antithixotropy

This new concept (Barnes, 1997) applies to viscous fluids, for which we considered so far that, at a fixed shear rate $\dot{\gamma}$, the viscosity is constant in time (although it can vary with $\dot{\gamma}$, if the fluid is non-Newtonian).

However, there are also some fluids (such as paint or print ink, but also mud, cement, mortar, mastics, etc. [35, 36]) whose viscosity changes 'slowly' with time, under shear (at constant $\dot{\gamma}$), before reaching a constant value. This behaviour indicates that *the structure of the fluid evolves under shear*, before reaching a stationary regime. For this reason, it is not surprising that this behaviour is also manifest in non-Newtonian fluids, whose shear-thinning or shear-thickening properties are also related to a structural change under shear. One then speaks of *thixotropy* or *antithixotropy* when the fluid is shear-thinning or shear-thickening, respectively.

In the following, we consider the more common case of a shear-thinning fluid. In Fig. 1.7, we show its response to two successive jumps in shear rate. The fluid, initially at rest and in thermodynamic equilibrium, is set in motion at the instant t_1, with a shear rate $\dot{\gamma}_1$. The measured stress is zero initially, 'jumps' to the value σ_1^i, then decreases slowly, tending towards a constant value σ_1^∞. The corresponding viscosity shows the

---

[18] We emphasise that at high temperature ($T > 0.5\,T_m$), polycrystalline solids behave more like yield stress fluids (see Figure 1.10, Section 5.3.4.5 and Footnote 14).

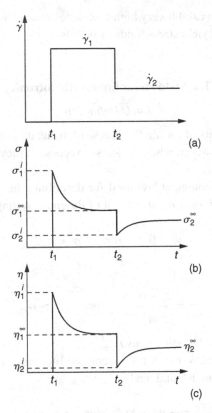

Fig. 1.7 Response of a thixotropic fluid to two successive jumps in shear rate. (a) Shear rate as a function of time; (b) stress as a function of time; (c) viscosity as a function of time.

same behaviour, tending towards an asymptotic value η_1^∞. This decrease in viscosity is due to a progressive destructuring of the material which stabilises over long times. At the instant t_2, the shear rate is reduced to $\dot\gamma_2$. The measured stress 'drops', and then increases up to a new plateau value σ_2^∞ corresponding to a viscosity η_2^∞. In this case, the viscosity increases during the transient regime since the material, strongly destructured in the beginning, becomes partially restructured at long times, $\dot\gamma_2$ being less than $\dot\gamma_1$. If the viscosities measured in the stationary regime only depend on the applied shear rate (being independent of the *history* of the sample), the fluid is termed *thixotropic*. In this case, the system returns to its initial equilibrium state once the shear is suppressed. Otherwise, one speaks of *partial thixotropy* (very common for the materials cited above).

The same concepts apply to the shear-thickening case. One then speaks of *antithixotropy* (or *rheopecty*) and of *partial antithixotropy* (or *partial rheopecty*).

It is noteworthy that the presence of thixotropy implies the need to take into account the *history of the flow*. Thus, tube flow of a thixotropic fluid is complicated by the fact that its viscosity changes along the tube.

Thixotropic fluids often exhibit very high viscosities at low shear rates, prompting us to define the new concept of yield stress fluids or liquids.

1.4 Yield stress liquids (thixotropic)

1.4.1 Definition

The low-shear-rate viscosity of certain fluids is so high that they *appear* not to flow below a certain *yield stress* σ_y. However, when the stress exceeds σ_y, they behave as viscous fluids (almost always thixotropic).

The simplest phenomenological law used for describing the rheological behaviour of these materials *in the stationary regime* is that of *Bingham*, stating that:

$$\dot{\gamma} = 0 \quad \text{for} \quad \sigma < \sigma_y, \tag{1.7}$$

while

$$\dot{\gamma} = \frac{\sigma - \sigma_y}{\eta_p} \quad \text{for} \quad \sigma > \sigma_y. \tag{1.8}$$

The coefficient η_p is called '*plastic viscosity*'.[19]

Two other models can be found in the specialised literature. That of *Herschel–Bulkley* generalises the Bingham model, and reads:

$$\sigma = \sigma_y + k\dot{\gamma}^n \quad \text{for} \quad \sigma > \sigma_y, \tag{1.9}$$

with $0 < n < 1$,[20] while that of *Casson* states that:

$$\sqrt{\sigma} = \sqrt{\sigma_y} + \sqrt{\eta_p \dot{\gamma}} \quad \text{for} \quad \sigma > \sigma_y. \tag{1.10}$$

Note that, in both cases, $\dot{\gamma} = 0$ for $\sigma < \sigma_y$.

One can also show that the Casson model does not 'contain' the Bingham model, by squaring Eq. (1.10):

$$\sigma = \sigma_y + \eta_p \dot{\gamma} + 2\sqrt{\sigma_y \eta_p \dot{\gamma}}. \tag{1.11}$$

These equations have been used, with various degrees of success, to describe the rheological properties of a wide range of materials, such as paint, mud, toothpaste, melted chocolate, mayonnaise or different kinds of cream. All these materials belong to the family of suspensions or concentrated emulsions, or to that of flocculated suspensions, where the colloidal particles gather in flocs, which are large percolation aggregates.

---

[19] Note that this kind of law does not describe the thixotropic character of these fluids, a phenomenon only appearing in the transient regime.

[20] This law is similar to the one commonly employed for describing the viscoplastic behaviour of solids above their elastic limit (see Eq. (1.5)).

1.4.2 On the difficulty of defining a yield stress

One should keep in mind that the concept of yield stress is rather *subjective*, as it often depends on the equipment used for the measurements or, which amounts almost to the same, on the application intended for the material. Indeed, imagine a fluid with rheological behaviour similar to that represented in Fig. 1.8. If the rheometer employed cannot impose shear rates below a value $\dot{\gamma}_{y2}$, the experimentalist will conclude, by extrapolating the results to zero shear rate, that within the precision of the instrument the fluid is characterised by a yield stress σ_{y2}. Consider now that another experimentalist, using a more sensitive device, which can reach a shear rate $\dot{\gamma}_{y1}$ ten times lower than $\dot{\gamma}_{y2}$, measures the rheological behaviour of the fluid. In this case, extrapolation of the experimental data to zero shear rate will lead to the conclusion that the yield stress of the fluid is σ_{y1}, smaller than σ_{y2}.

This example illustrates the difficulty of objectively defining the yield stress and raises the point of the appropriate choice of a rheometer for measuring this quantity. Obviously, this choice is related to practical considerations on the fluid, as we will show in the following concrete example.

Suppose we would like to apply on a vertical wall a layer of paint with thickness h. Under the action of gravity, the paint experiences a maximal shear stress:

$$\sigma = \rho g h, \tag{1.12}$$

where ρ is the density and g the acceleration of gravity. In practice, one requires that the deformation of the layer of paint during its drying time t_d be minimal ($\gamma < 1\%$). This criterion sets a critical shear rate $\dot{\gamma}_d = 10^{-2} t_d^{-1}$, below which the flow of the paint is negligible. The paint is thus appropriate if the yield stress σ_y, obtained by extrapolating to zero the experimental dependence $\sigma(\dot{\gamma})$ over the relevant range of the shear rate $\dot{\gamma} > \dot{\gamma}_d$, is larger than the stress due to gravity. This condition determines the choice of the rheometer, which must be able to measure the viscosity over the required range of shear rate, with a lower bound given here by $\dot{\gamma}_d$.

We should however nuance this discussion since, very often, the yield stress is quite sharp and can be defined clearly, up to a factor of 2 or 3. To demonstrate this, let us consider the typical example of an aqueous solution of bentonite (a clay made of particles consisting of an alumina layer intercalated between two layers of silica), for which the

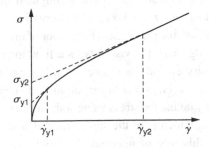

Fig. 1.8 Illustration of the difficulty of defining a yield stress.

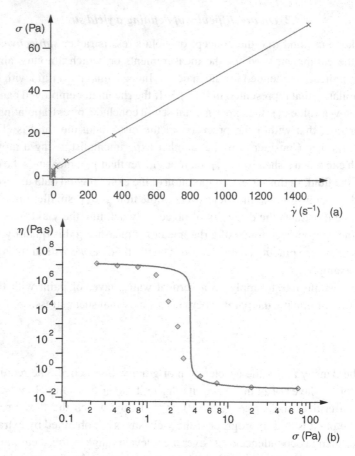

Fig. 1.9 (a) Stress as a function of shear rate for an aqueous solution of bentonite at 10 wt%; (b) the same experimental data in a different representation (here, the viscosity as a function of stress in log–log scale). The solid lines correspond to the best fit of the experimental data with the Cross law, defined by the relation (1.3) ($p = 1$, $\eta_\infty = 0.046\,\mathrm{Pa\,s}$, $\eta_0 = 1.25\,10^7\,\mathrm{Pa\,s}$ and $k = 3.6\,10^7\,\mathrm{s}$). One can see that the fit is excellent, except over the very narrow stress range where the viscosity drops (Barnes, 1999).

stress–strain curve is plotted, in linear scale, in Fig. 1.9a. It can be seen from the data that, to a good approximation, this mixture behaves as a Bingham fluid with plastic viscosity $\eta_p \approx 0.046\,\mathrm{Pa\,s}$ and a yield stress $\sigma_y = 3.5\,\mathrm{Pa}$. An alternative representation of the same experimental data, which shows the rheological behaviour of the solution much better, consists in plotting, on a log–log scale, the viscosity as a function of stress (Fig. 1.9b). One can now see that the viscosity exhibits two plateaus, at low and at high shear rate, with a very steep drop of more than seven orders of magnitude between 1 and 5 Pa. One way of defining the yield stress is by taking the stress value at the inflection point of the curve. This procedure gives $\sigma_y = 2\,\mathrm{Pa}$, which is of the same order of magnitude as in the Bingham representation. Clearly, in this type of material, the yield stress is well defined, up to a factor of 2 or 3, irrespective of the chosen definition.

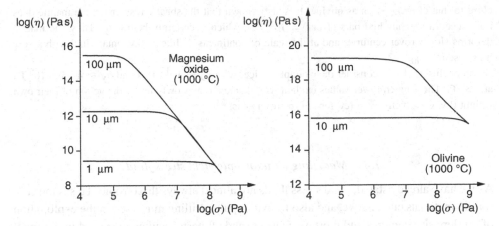

Fig. 1.10 Viscosity as a function of stress for two polycrystalline solids (magnesium oxide and olivine). The grain size is indicated for each curve (Frost and Ashby, 1982).

More importantly, these results show very clearly that a yield stress fluid is nothing other than a fluid that exhibits strong shear-thinning over a very narrow range of stress. In the following, we will keep to this definition. Thus, the experimental curve of the viscosity in Fig. 1.9 can be well described on either side of σ_y by the Cross formula (1.3), with $p = 1$. Indeed, one can show that, for $p = 1$, this formula reverts to a Bingham law at high stress, since:

$$\sigma = \frac{\eta_0}{k} + \eta_\infty \dot{\gamma} \tag{1.13}$$

for $\sigma > \sigma_y \approx \eta_0 / k$, the regime where $\eta \ll \eta_0$ and $k\dot{\gamma} \gg 1$.

Of course, not all yield stress fluids have such simple behaviour. Many of them are in fact non-Newtonian above the yield stress σ_y, and behave more closely to the Herschel–Bulkley or Casson laws.

On the other hand, the yield stress σ_y varies strongly from one system to another. Its value is generally between 10 and 100 Pa in foodstuffs such as mayonnaise or ketchup (where the viscosity η_0 is in the range 10^6 to 10^7 Pa s), but it can be much higher for concentrated suspensions of solid particles. This is the case, for instance, with pottery clay, where the yield stress σ_y is of the order of 10^5 Pa, and the viscosity η_0 close to 10^{11} Pa s.

Ultimately, one could even extend the concept of yield stress fluid to encompass polycrystalline solids which behave, at high temperature (close to their melting point) and low stress ($\sigma < \sigma_y$), as Newtonian fluids, and become strongly shear thinning above the yield stress σ_y. In this case, the values of the viscosity and the yield stress are much higher than in the previously discussed materials, as shown by the two graphs in Fig. 1.10. For instance, a polycrystal of olivine (a rock) with a grain size of 10 μm, at 1000 °C, has $\eta_0 \approx 10^{16}$ Pa s and $\sigma_y \approx 3 \times 10^8$ Pa.

Another interesting example is that of Earth's mantle, for which the geologists find much higher values: $\eta_0 \approx 10^{21}$ Pa s.[21] On the other hand, the value of σ_y is not well known, but it should be

---

[21] This value can be determined by the study of post-glacial rebounds, as we shall see in Section 3.8.6.

close to that of rocks, such as olivine. It is also known that the shear stress inside the mantle does not exceed a few hundred bars (1 bar $= 10^5$ Pa), which is certainly below σ_y. Thus, the mantle deforms slowly (over centuries and at the scale of continents[22]) like a Newtonian fluid with a very high viscosity η_0.

In conclusion, let us consider the example of ice, with $\eta_0 = 10^{13}$ Pa s and $\sigma_y \approx 4\text{--}5 \times 10^5$ Pa at -8 °C. These much lower values explain why glaciers flow slowly under the action of their own weight (by a few metres to a few tens of metres a year).[23]

1.4.3 Modelling a thixotropic yield stress fluid

As we have already stated, yield stress fluids are almost always thixotropic. This is the case for the materials cited above, and also for paper pulp, drilling mud used in the exploitation of oil deposits, cements and mortars, mastics, and all pastes and creams used in the food, cosmetics and pharmaceutical industries.

We have also seen that thixotropy (taken as perfect, for simplicity) manifests *transient behaviour*, becoming apparent after each change in shear rate. More precisely, thixotropy is associated with a *destructuring* of the fluid under shear (more pronounced as the shear rate increases) followed by a *restructuring* and a *return to equilibrium* once the shear stops.

In the simplest models, the structure of the medium is characterised by a single parameter $\lambda_c(t)$, equal to 1 when the fluid is at rest, and vanishing as the fluid becomes completely destructured, at high shear rates.

Let us assume that the fluid obeys a Herschel–Bulkley law in the stationary regime and at high shear rates. This law can be generalised to account for the thixotropic behaviour of the medium in the transient regime, as follows:

$$\sigma = \sigma_1 + \lambda_c \sigma_2 + k\dot{\gamma}^n, \tag{1.14}$$

where σ_2 is a structural stress that can depend on the shear rate. Under these conditions, $\sigma_1 + \sigma_2(\dot{\gamma} = 0)$ defines the yield stress of the material at rest, while σ_1 defines its yield stress when completely destructured (the one we measure after having vigorously sheared the sample).

As to parameter λ_c, it is given by the solution of the following differential equation:

$$\frac{d\lambda_c}{dt} = a(1 - \lambda_c) - b\lambda_c\dot{\gamma}, \tag{1.15}$$

where a and b are two material parameters describing the creation or destruction of the structure.

---

[22] The drift velocity of a continental plate (about 100 km thick) is a few cm/year. Surprisingly, its kinetic energy is no greater than that of a car moving at 20 km/h.

[23] Note that a thin water layer can form under certain glaciers, easing their flow. This phenomenon occurs in Antarctica. It is due to the fact (unusual for solids) that the melting temperature of ice decreases as the pressure increases (it is about -1.5 °C below a 2000 m thick glacier.)

To understand more clearly the implications of the model, let us solve Eq. (1.15) assuming that the fluid is initially at rest: $\lambda_c(t = 0) = 1$. The solution is given by:

$$\lambda_c = \frac{a + b\dot{\gamma}\exp[-t(a + b\dot{\gamma})]}{a + b\dot{\gamma}}. \tag{1.16}$$

This equation and Eq. (1.14) show that the stress measured immediately after applying the shear $\dot{\gamma}$, is:

$$\sigma^i(\dot{\gamma}) = \sigma_1 + \sigma_2 + k\dot{\gamma}^n. \tag{1.17}$$

It then decreases exponentially, with a decay time

$$\tau(\dot{\gamma}) = \frac{1}{a + b\dot{\gamma}}, \tag{1.18}$$

until it reaches its stationary value:

$$\sigma^\infty(\dot{\gamma}) = \sigma_1 + \frac{a\sigma_2}{a + b\dot{\gamma}} + k\dot{\gamma}^n. \tag{1.19}$$

Obviously, one can solve this problem for all initial values of the structure parameter λ_c, and the experimental data (Fig. 1.7) are fairly well described.

In the following, we will discuss another essential concept in rheology, that of '*instantaneity*'. Indeed, until now we have always assumed that the stress within the fluid varies 'instantaneously' upon changing the shear rate (see, for instance Fig. 1.4 or Fig. 1.7). This assumption deserves some serious scrutiny because the stress can never be a discontinuous function of time, even if the rheometer were capable of a sudden jump in speed. Consequently, each fluid has a typical time τ over which the stress adjusts to its new value after each jump (taken as sudden, for now) in shear rate. In fact, each rheometer also has its own response time τ_{rheo}, over which it can change its speed. In practice, when using any rheometer, a fluid will be considered purely viscous (whether Newtonian or not, thixotropic or not) if $\tau < \tau_{rheo}$. Once again, everything is relative.

In the next paragraph, we will see that τ is related to the *elasticity of the fluid*. As this elasticity becomes apparent ($\tau > \tau_{rheo}$), we will say that the fluid exhibits *viscoelastic behaviour* (and by that we mean: over the time scale of the rheometer employed).

1.5 Viscoelastic fluids

As Maxwell noted for the first time in his famous 1867 paper *On the dynamical theory of gases*, all fluids are *viscoelastic* (i.e. they possess at the same time elastic and viscous properties) over short enough time scales. Clearly, this scale is not the same for water and for the silicone-based 'Silly Putty'.[24] Let a water drop fall on the floor: it spreads rapidly

---

[24] This paste, which has no practical application other than as a toy, was discovered in 1945 by the American engineer James Wright while searching for rubber substitutes. Commercialised worldwide since 1950, one estimates the total amount sold at more than 5000 tonnes.

(in a fraction of a second) without bouncing back, if the surface of the floor is wetting (otherwise, the drop can contract again and bounce, but this effect is related to the surface tension of water (de Gennes *et al.*, 2002), not to its elasticity.) On the other hand, a ball of 'Silly Putty' will bounce several times like a ping-pong ball due to its elasticity, then end up by spreading very slowly, over the following days.

This comparison clearly shows that, in this experiment, water behaves as a Newtonian fluid, while 'Silly Putty' reacts as an elastic solid over the duration of the collision and as a very viscous fluid over a day. The behaviour of 'Silly Putty' is termed *viscoelastic*.

The 'Silly Putty' experiment also shows that a viscoelastic fluid always ends up irreversibly deformed under the action of a force, weak as it may be (here, the force of gravity), provided it is applied for long enough. From this point of view, viscoelastic fluids differ from solids, which deform reversibly under the prolonged (in principle, even 'infinite') action of a stress σ, as long as it remains below the elastic limit σ_c.[25]

Let us now see how one can define the viscoelastic relaxation time. We begin with the simplest model, that of Maxwell.

1.5.1 The Maxwell model: qualitative description

In the simplest Maxwell model, one describes the fluid by a *single* characteristic time τ, called the *viscoelastic relaxation time*.

We should point out right away that very few materials obey this model strictly. An example (classical by now) is that of 'living polymers', solutions of giant micelles appearing in lyotropic systems (which are mixtures of water and surfactant molecules), where the time τ is very well defined (see Section 7.4). One could include in the same category polycrystalline solids at high temperature, already classified as yield stress fluids.

There are however many more viscoelastic fluids, exhibiting a more or less wide distribution of relaxation times: the most common are polymer melts (such as 'Silly Putty') or polymer solutions in a solvent, surfactant solutions (formed by micelles or vesicles), as well as emulsions and solid particle suspensions (as long as they are not too concentrated). In this section, we suppose that one can define for each material a *terminal relaxation time* (the longest in its distribution), which we will take as the typical relaxation time τ in order to describe it (but this is only a rough approximation).

To gain some insight into the behaviour of a viscoelastic material, let us analyse its response to stress under the sudden action of a shear rate $\dot{\gamma}$ (Fig. 1.11a). Experiment shows that one must distinguish between the *transient regime*, when the stress increases, and the *stationary regime*, during which the stress saturates.[26] Let us describe each of them.

---

[25] We make here the implicit assumption that the temperature is low enough that vacancy diffusion, which can be responsible for some plastic deformation, is negligible. In other words, the time over which diffusion operates can be considered 'infinite' in comparison with the duration of the experiment. See Footnotes 14 and 16 as well as Sections 4.1.2.4 and 5.2.4 to 5.2.9 for a detailed discussion of the origin of the elastic limit σ_c.

[26] This behaviour is also observed in ideal and viscoplastic solids above their elastic limit (see Fig. 1.4).

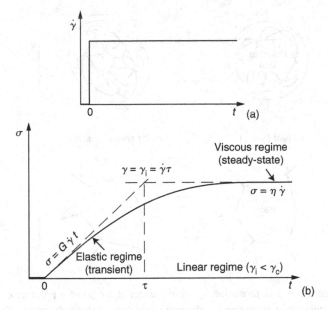

Fig. 1.11 Stress as a function of time for a viscoelastic body under the sudden action of a shear $\dot{\gamma}$. The response is elastic at short times ($t \ll \tau$) and viscous at long times ($t \gg \tau$). (a) Shear rate as a function of time; (b) stress as a function of time in the *linear regime*. After a time τ (defining the end of the elastic regime), the intrinsic deformation of the objects is more or less equal to the macroscopic deformation of the sample ($\gamma = \gamma_i = \dot{\gamma}\tau$). In the nonlinear regime, the stress can vary with time in a much more complex manner, not shown here.

1.5.1.1 The transient regime

Since the material reacts elastically over short times ($t < \tau$), the stress begins by increasing linearly with time or, equivalently, with the imposed deformation. In these conditions:

$$\sigma = G\gamma = G\dot{\gamma}t, \tag{1.20}$$

where G is the *shear modulus*.

On the other hand, the stress saturates at longer times ($t > \tau$), the hallmark of viscous behaviour, with a *viscosity η*, such that:

$$\sigma = \eta\dot{\gamma}. \tag{1.21}$$

It follows, from Fig. 1.11b and the two equations above, that the *characteristic time of viscoelastic relaxation* is simply:

$$\tau = \frac{\eta}{G}. \tag{1.22}$$

This is nothing more than the *ratio between the viscosity of the fluid and the elastic modulus*, a formula already given by Maxwell in 1867.

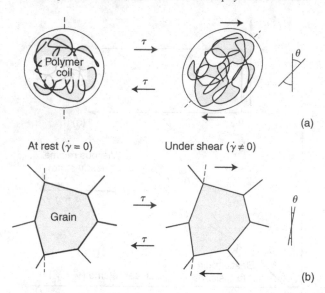

Fig. 1.12 Average intrinsic deformation $\gamma_i = \tan\theta$ under shear (a) of a polymer coil in a polymer melt of low molecular mass (unentangled chains); (b) of a grain in a polycrystal. This deformation, measured in the reference system of the centre of mass of the object, builds up or vanishes over the time τ as the shear is suddenly applied or removed.

Note that τ is the typical time during which the material preserves the *memory* of its initial shape after the flow starts. Beyond this time, the material forgets its initial shape.

One should keep in mind that this description is only strictly valid in the *linear regime* (a concept to be debated at length in Chapters 6 and 7). This requires that the shear rate be so low that the *characteristic objects in the fluid* (assumed deformable, such as micelles in a surfactant solution, polymer chains in a melt, droplets in an emulsion, or grains in a high-temperature polycrystal) under shear undergo an *elastic deformation proportional to the applied shear rate* (Fig. 1.12). In this case, the elastic modulus G is constant. The *structure of the fluid must also remain more or less unchanged under shear*, such that its viscosity η does not vary much. If the two conditions are fulfilled, one deals with the *linear regime*. Otherwise, the material enters the *nonlinear regime*. The objects are then strongly deformed and the structure of the material changes. One can then observe that the elastic modulus $G(\dot{\gamma})$ increases with increasing shear rate (with the exception of solids, where it remains constant), while the viscosity $\eta(\dot{\gamma})$ *almost* always decreases (shear-thinning behaviour).

It follows that the duration of the transient, always defined by the ratio $\eta(\dot{\gamma})/G(\dot{\gamma})$ is generally shorter in the nonlinear regime than in the linear one. It can also exhibit complex behaviour, with one or more oscillations that dampen with time (the 'overshoot' phenomenon sometimes encountered in polymers [28] or liquid crystals [27], but we will not discuss it in this book).

These two regimes (linear and nonlinear) having been identified, is there a general rule for finding the critical shear rate $\dot{\gamma}_c$ separating them?

The answer is NO! Everything depends on the material under consideration.

In the following we will write

$$\dot{\gamma}_c = \gamma_c \frac{1}{\tau}, \tag{1.23}$$

where γ_c stands for the intrinsic deformation of the objects under permanent shear, at the boundary between the linear and the nonlinear regimes. Thus, the intrinsic deformation of the objects $\gamma_i \approx \dot{\gamma}\tau$ stays below γ_c in the linear regime and exceeds γ_c in the nonlinear regime.

Keep in mind that the value of γ_c varies a lot with the nature of the material and with temperature. Thus, γ_c is of unit order in polymer melts at high temperature, but decreases to a small percentage close to their glass transition temperature. In polycrystalline solids close to their melting point, γ_c is always very small, never exceeding 10^{-3}.

Before giving concrete examples of viscoelastic fluids, let us describe their behaviour in the stationary (or viscous) regime.

1.5.1.2 The stationary regime

In Fig. 1.13, we trace the stress and the viscosity measured in the stationary regime as a function of shear rate.

As we have already pointed out, one must make a clear distinction between the linear and nonlinear regimes.

In the first ($\dot{\gamma} < \dot{\gamma}_c = \gamma_c/\tau$), the stress increases linearly with $\dot{\gamma}$, a distinctive sign of Newtonian behaviour, characterised by a constant viscosity η.

On the other hand, in the nonlinear regime ($\dot{\gamma} > \dot{\gamma}_c$) the stress usually increases more slowly than $\dot{\gamma}$ (shear-thinning behaviour). It often happens that the decrease in viscosity

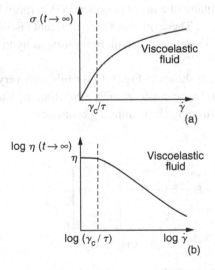

Fig. 1.13 Stress (a) and viscosity (b) in the stationary regime as a function of shear rate in a viscoelastic fluid.

follows a power law of the Ostwald–de Waehle type, with an exponent n between 0 and 1, over a wide range of shear rate (see Eq. (1.4) and Fig. 1.15), before once again saturating at very high shear rates.

Let us now give a few examples of Maxwell fluids, specifying for each of them the order of magnitude of the viscoelastic relaxation time.

1.5.2 Examples of viscoelastic Maxwell fluids

As we have already pointed out, all fluids are viscoelastic. But over what time scales τ? Let us work out some estimates, going from the shortest towards the longest times.

1.5.2.1 Simple liquids

The best example is that of water, with a viscosity $\eta = 10^{-3}$ Pa s at room temperature. Here, the elementary object is the water molecule. One should then take for the elastic modulus G, the value measured for ice (for amorphous ice, to be precise), which is of the order of 10^9 Pa. With these values, one has

$$\tau(\text{water}) \approx 10^{-12} \text{ s}. \tag{1.24}$$

This extremely short time shows that water will behave as a Newtonian fluid in usual experiments.

This conclusion holds for most simple liquids composed of small molecules (alcohols, alkanes, etc.)

1.5.2.2 Silicone oils

Silicone oils consist essentially of a linear polymer, PDMS (polydimethylsiloxane), whose formula is given in Fig. 1.14. These oils have very different viscosities, depending on their polymerisation index n. They are very useful, being used as hydraulic, damping or heating fluids, and as lubricants.

Some viscosity curves are shown in Fig. 1.15 for oils with very different molar masses. They share the same general appearance, namely a Newtonian plateau followed by a power-law decrease, the sign of strongly shear-thinning behaviour.

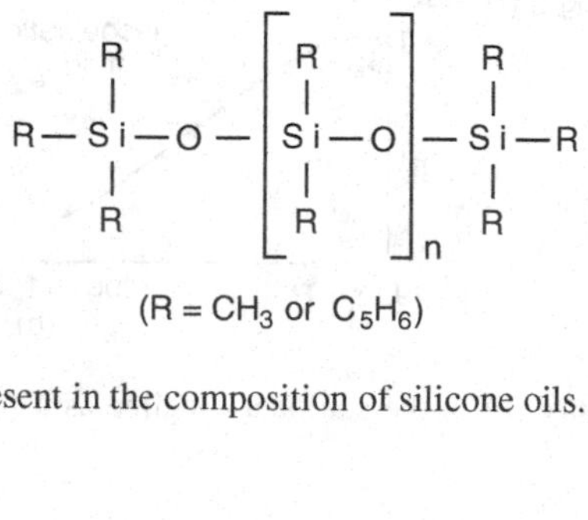

$$
R-\underset{\underset{R}{|}}{\overset{\overset{R}{|}}{Si}}-O-\left[\underset{\underset{R}{|}}{\overset{\overset{R}{|}}{Si}}-O\right]_n-\underset{\underset{R}{|}}{\overset{\overset{R}{|}}{Si}}-R
$$

$$(R = CH_3 \text{ or } C_5H_6)$$

Fig. 1.14 Linear polymers present in the composition of silicone oils. The main ingredient is PDMS ($R = CH_3$).

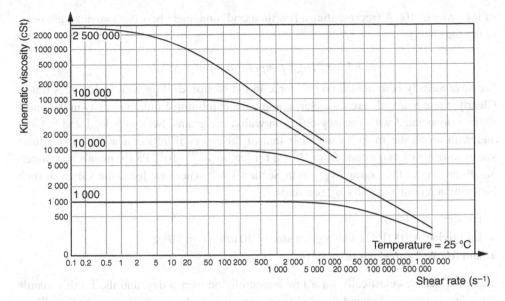

Fig. 1.15 Viscosity of silicone oils Rhodorsil 47V... (produced by Rhodia) at 25 °C. The number, replaced here by the ellipsis, indicated alongside each curve gives the linear regime viscosity of the oil in cSt (data provided by Rhodia). The cSt (centistoke) is a kinematic viscosity unit (η/ρ): 1 cSt $= 10^{-6}$ m^2/s.

Independent measurements of the elastic modulus on the rubber plateau[27] show that, for these oils (and, more generally, for polymer melts or concentrated solutions), the crossover from the linear to the nonlinear regime occurs for a value of γ_c close to 1 (far from the glass transition temperature[28]), i.e. for a shear rate close to $1/\tau$. It is then possible to estimate $1/\tau$ for each oil using its viscosity curve, and then to determine its elastic modulus G, the viscosity η on the Newtonian plateau being known. Thus, for the oil 47V 2 500 000, $\eta \approx 2500$ Pa s and $\tau \approx 0.5$ s, yielding $G \approx 5 \times 10^3$ Pa. It so happens that one obtains *roughly* the same value of the elastic modulus G for oils 47V 100 000, 10 000 and 1000, by analysing the data in Fig. 1.15. This value seems then to be universal (for this type of oil), unlike the viscosity η, which increases strongly with the molar mass M of the polymer ($\eta \propto M^{3.3}$). We will see in Chapter 7 that this feature occurs due to the chains being strongly entangled. In this case, the elastic modulus is given by the formula (Section 7.1.7.5):

$$G \sim \mathcal{N}_e \, k_B \, T, \qquad (1.25)$$

where $\mathcal{N}_e$ is the *number of entanglements* per unit volume (rubber elasticity). This number is independent of the length of the chains once they are very long. In the case of silicone oils, this formula yields $\mathcal{N}_e \approx 10^{24}$ m^{-3}, or an average distance between entanglements

---

[27] This concept will be defined more rigorously in Section 7.1.3.2.
[28] Below this temperature, the polymer forms a glass and behaves mechanically as a 'hard' solid.

of the order of 100 Å (representing a few thousand monomers between two entanglements along the chain).

1.5.2.3 'High-temperature' polycrystals

We have already commented on this topic in the section dealing with yield stress fluids. Clearly, these materials are viscoelastic in nature, exhibiting a finite viscosity in the linear regime (associated with vacancy diffusion within the grains, Section 5.3.4.1 and 5.3.4.2) and with an elastic modulus which is that of their crystal lattice. Their shear modulus varies in order of magnitude from 10^9 Pa for ice, to $10^{10}-10^{11}$ Pa for metals and minerals. Returning to the examples given in Section 1.4.2, where we listed the values of their Newtonian regime viscosity η_0, one finds:

- for ice: $\tau \approx 10^4$ s;
- for olivine at $1000\,°C$ (with a grain size of $10\,\mu m$): $\tau \approx 10^5$ s;
- and for the Earth's mantle: $\tau \approx 10^{10}$ s.

Thus, ice behaves elastically for a few hours, olivine over a day, and the Earth's mantle for a few centuries. Beyond these characteristic times, the materials start to flow like a Newtonian viscous liquid, as long as they remain within the linear regime, without exceeding the critical stress σ_y above which they enter the regime of nonlinear flow.[29]

For these materials, one can calculate the value of the critical deformation γ_c at the crossover between the linear and the nonlinear regimes, as the critical stress σ_y is known (see Fig. 1.10) as well as the value of the elastic modulus G: $\gamma_c = \sigma_y/G$. One always obtains values in the range 10^{-4} to 10^{-3}, much lower than in polymer melts.

In Chapter 4 we will explain why this value is so small. In particular, we will see that the dislocations, which are linear defects of the crystal lattice, start to propagate for $\sigma > \sigma_y$, or, equivalently, when the elastic strain of the lattice exceeds γ_c. At low stress, on the other hand, for $\sigma < \sigma_y$, only molecular diffusion across (or at the surface of) grains comes into play. These mechanisms explain the extremely high values of the viscosity η_0 and its dependence on grain size (Sections 5.3.4.1 and 5.3.4.2).

There are also other viscoelastic fluids, such as dilute polymer solutions, or dilute emulsions, that are not well described by a simple Maxwell model. One must then use the Jeffrey model, which we will briefly discuss now (Section 6.2.1.2).

1.5.3 The Jeffrey model

General definition   This model applies to systems composed of a solvent (which can be considered as purely viscous) containing elastic 'particles', i.e. easily deformed under shear. Macromolecules in dilute solution (discussed in Section 7.5.3), or dilute emulsions

---

[29] Note that in the case of polycrystalline solids we employed two different notations for the critical stress. At low temperature, we noted it by σ_c since, below this stress value, one can consider that the solid does not flow, its behaviour being purely elastic (the viscosity is so high it cannot be measured). At high temperature, we note it by σ_y, since below this stress the solid exhibits a viscosity, certainly very high, but measurable and experimentally relevant, such that it can be classified among viscoelastic yield stress liquids.

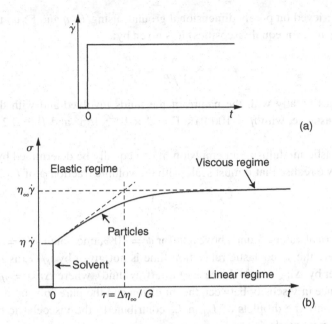

Fig. 1.16 Rheological behaviour of a Jeffrey fluid after the sudden application of a shear $\dot{\gamma}$ at the instant $t = 0$. Here, we set $\Delta\eta_\infty = \eta_\infty - \eta$.

composed of two immiscible fluids (see Section 7.5.2), belong to this new kind of viscoelastic fluids.

Obviously, these materials do not behave as simple Maxwell fluids, their response being that of a viscous fluid (the solvent), plus a viscoelastic component due to the 'particles' trapped within the solvent. One speaks of a *Jeffrey fluid*, to be defined properly in Section 6.2.1.2.

The transient behaviour of a Jeffrey fluid after the sudden application of shear is given in Fig. 1.16. It consists of an 'instantaneous' response, due to the solvent, followed by a viscoelastic response due to the deformable particles. As for the Maxwell model, one can define a viscoelastic relaxation time τ, which varies with the solvent and the particles, and which is the typical time for the onset of the stationary regime.

As to the stationary regime behaviour, it is qualitatively similar to that of a Maxwell fluid, the material exhibiting a constant viscosity (Newtonian behaviour) generally followed by a decrease (shear-thinning behaviour) as the shear rate increases.

The example of dilute emulsions   Let us consider a dilute emulsion of a liquid L_d (where subscript 'd' stands for the dispersed phase) in a liquid L_s (the 'solvent'), the two being immiscible. To simplify, let us assume that the two liquids are purely viscous, Newtonian and with the same viscosity η. Let R be the radius of the droplets formed by L_d in L_s and Γ the surface tension between the two fluids. In this case, the viscoelastic relaxation time is the time needed by a droplet of L_d, initially deformed, to recover its spherical shape. This time can be retrieved on purely dimensional grounds using R, η and Γ: up to a numerical constant (close to 2 for equal viscosities) it is given by:

$$\tau \approx \frac{\eta}{\Gamma/R} = \frac{\eta R}{\Gamma}. \tag{1.26}$$

This time varies greatly with the nature of the fluids involved and with the size of the droplets. For instance, with $\eta = 100\,\text{Pa s}$, $\Gamma = 2 \times 10^{-2}\,\text{J/m}^2$ and $R = 0.2\,\text{mm}$, one gets $\tau \approx 1$ s.

As to the elastic modulus of an emulsion, it can equally be determined by dimensional analysis once we realise that it must scale with the volume fraction ϕ of L_d in L_s:

$$G \approx \phi \frac{\Gamma}{R}. \tag{1.27}$$

With the numerical values found above and for $\phi = 10\%$, one obtains $G = 10$ Pa.

Note that here the viscoelastic relaxation time is not given by η/G (as for a Maxwell fluid) but rather by $\Delta\eta_\infty/G$ (in the case of a Jeffrey fluid), where $\Delta\eta_\infty = \eta_\infty - \eta$ stands for the difference in viscosity between the emulsion and the pure solvent in the stationary regime (since only the droplets of L_d in L_s contribute to the viscoelastic behaviour). In particular, this fact entails that

$$\Delta\eta_\infty \approx G\tau \approx \eta\phi \tag{1.28}$$

up to a numerical constant. In Chapter 7 we shall see that these predictions are accurate (to first order in ϕ) (Section 7.5.2.2).

Let us now see how one can determine the energy dissipated in a fluid under shear.

1.5.4 Energy dissipation

In the permanent deformation regime, the mechanical energy injected into the fluid is entirely dissipated as heat. The sample can thus heat up noticeably, especially if it is very viscous and if the heat conductivity of the confining surfaces is low, a phenomenon to be considered in Section 3.8.7.

Let $\sigma(t \to \infty)$ be the stress measured in the stationary regime for a shear rate $\dot{\gamma}$. The *dissipated power* $\Phi(\dot{\gamma})$ per unit volume in the fluid is equal to the work per unit time expended by the applied stress, namely:

$$\Phi(\dot{\gamma}) = \sigma(t \to \infty)\dot{\gamma}. \tag{1.29}$$

With $\eta(\dot{\gamma})$ the viscosity measured in the stationary regime, this formula becomes:

$$\Phi(\dot{\gamma}) = \eta(\dot{\gamma})\,\dot{\gamma}^2. \tag{1.30}$$

It is even more interesting to estimate the *total dissipated energy* $\bar{\Phi}(\gamma)$ as the sample undergoes a deformation of finite amplitude γ. In this case:

$$\bar{\Phi}(\gamma) = \sigma(t \to \infty)\gamma = \eta(\dot{\gamma})\,\dot{\gamma}\gamma. \tag{1.31}$$

This formula shows that, as a general rule, and in agreement with intuition, *the dissipated energy vanishes for a quasi-static deformation* (i.e. as $\dot{\gamma} \to 0$). This behaviour shows that the fluid remains, in the *quasi-static limit*, in *thermodynamic equilibrium*, which is only possible at finite temperature, in the presence of *thermal fluctuations*.

At this point, it is well worthwhile comparing the behaviour of a fluid to that of an ideal plastic solid. For the latter, $\sigma(t \to \infty) \to \sigma_c = $ Cst under continuous shear so that, using Eq. (1.31):

$$\bar{\Phi}(\gamma) = \sigma_c \gamma. \tag{1.32}$$

This formula shows that, unlike the case of a fluid (see Eq. (1.31)), *the dissipated energy remains finite for a quasi-static deformation*, as $\dot{\gamma} \to 0$. We see here that, for an ideal solid, the relevant parameter is no longer the strain rate, but rather the strain itself, the signature of *athermal behaviour*. Two examples will help us understand this concept.

In a low-temperature solid (where the self-diffusion of crystal atoms is completely negligible), this behaviour can be associated with dislocation glide in the internal stress field σ_i that they create over large scales (Section 5.3.3.2): in this case, the thermal fluctuations cannot allow the dislocations to cross the energy barriers associated with σ_i, as long as $\sigma < \sigma_c = \sigma_i$ (assuming that the internal stress field is the only obstacle to the propagation of dislocations, which is seldom the case). On the other hand, the dislocations cross these barriers when $\sigma = \sigma_c$: under these conditions, however, temperature no longer plays any role – the behaviour is indeed athermal.

In foams or concentrated emulsions, whose behaviour is close to that of ideal plastic solids, plastic deformation is mainly due to local T1 type cell rearrangements (Fig. 1.17). These elementary events are equally athermal, occurring at the macroscopic scale: ultimately, they play the same role as a dislocation crossing an internal stress barrier in a solid under the action of the external stress.

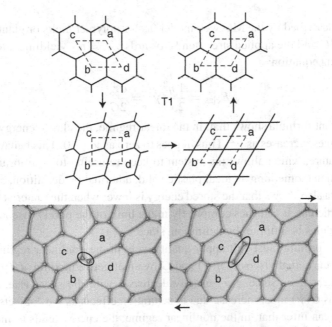

Fig. 1.17 T1 process in a two-dimensional foam (photos by G. Debrégeas).

To conclude this digression, let us point out that athermal behaviour is also observed in areas such as solid friction (Coulomb law) or partial wetting of a substrate by a liquid in the presence of hysteresis (when the advancing angle is different from the receding angle, a sign that the contact lines cannot cross the obstacles they encounter by thermal activation).

Another fundamental parameter is the elastic energy stored by the fluid during flow.

1.5.5 Stored elastic energy

As a viscoelastic fluid, initially at rest, undergoes the sudden action of a shear $\dot{\gamma}$, it stores elastic energy during the transient leading to its permanent flow regime. This effect is obviously stronger in the nonlinear regime, where the objects are strongly stretched, than in the linear regime, where their deformations are considered small. To simplify, we will only consider linear regime flows. In the following, we will perform the calculation for the Maxwell and Jeffrey viscoelastic fluids. The case of the ideal plastic solid will be considered separately, for comparison.

In the case of a Maxwell fluid, the elastic energy is only stored during the transient regime. In the linear regime, the transient lasts the same time τ, irrespective of the shear rate $\dot{\gamma}$, since the viscosity η and the modulus G are constant. To calculate the elastic energy stored E_{elas}, note that the deformation of the 'particles' in the fluid is typically given by $\gamma = \dot{\gamma}\tau$ after the time τ during which they deform elastically. Beyond this time, their deformation remains constant. Consequently, one has:

$$E_{\mathrm{elas}} \approx \frac{1}{2}G\gamma^2 = \frac{1}{2}G\dot{\gamma}^2\tau^2. \tag{1.33}$$

If the fluid is described by the Maxwell model (as for polymer melts or giant micelle solutions), $\tau = \eta/G$ and the applied stress tends towards $\sigma_\infty = \eta\dot{\gamma}$ yielding, after substitution in the preceding equation:

$$E_{\mathrm{elas}} = \frac{1}{2}\frac{\eta^2}{G}\dot{\gamma}^2 = \frac{\sigma_\infty^2}{2G}. \tag{1.34}$$

This important formula shows that, in the linear regime, the elastic energy stored by the fluid at long times increases as $\dot{\gamma}^2$. Thus, it goes to zero as $\dot{\gamma} \to 0$. This behaviour is related to the temperature, which allows the system to keep returning to equilibrium under very slow deformation (something an ideal plastic solid, athermal by definition, cannot do).

This formula also shows that the stored energy is lower when the material is stiffer (G is higher). In particular, it vanishes completely in the limit of the purely viscous fluid, where the elastic modulus is 'infinite' by definition, since $\tau = 0$.

Finally, this formula must still apply qualitatively in the nonlinear regime, provided η is replaced by the actual viscosity $\eta(\dot{\gamma})$ and G by an effective modulus $G(\dot{\gamma})$, also dependent on the shear rate. Since $\eta(\dot{\gamma})$ usually decreases with the shear rate, and since the elastic modulus $G(\dot{\gamma})$ must increase due to saturation effects as the 'objects' are strongly stretched, one can infer that, in the nonlinear regime, the energy tends to increase with $\dot{\gamma}$ more slowly than predicted by formula (1.34).

Let us now discuss the case of a Jeffrey fluid (which applies, for instance, to dilute emulsions). The viscoelastic relaxation time is then given by the expression $\tau = \Delta\eta_\infty/G$ where $\Delta\eta_\infty$ stands for the viscosity increase in the stationary regime with respect to that of the solvent η ($\Delta\eta_\infty = \eta_\infty - \eta$, see Fig. 1.16b). Under these conditions, Eq. (1.33), which is still valid, becomes:

$$E_{elas} \approx \frac{1}{2}\frac{\Delta\eta_\infty^2}{G}\dot{\gamma}^2 = \frac{\Delta\sigma_\infty^2}{2G}, \tag{1.35}$$

with $\Delta\sigma_\infty = \Delta\eta_\infty\dot{\gamma}$ the stress increase in the stationary regime due to the 'particles'.

It ensues that, in a dilute emulsion, the elastic energy varies linearly with the particle volume fraction ϕ, since from Eqs. (1.27), (1.28) and (1.35), one has:

$$E_{elas} \approx \frac{1}{2}\eta^2\phi\frac{R}{\Gamma}\dot{\gamma}^2 = \eta^2\frac{\phi}{\Delta P}\dot{\gamma}^2, \tag{1.36}$$

where we denoted by $\Delta P = 2\Gamma/R$ the capillary pressure within the droplets. As $\sigma_\infty \approx \eta(1+\phi)\dot{\gamma}$, one finds that, to first order in ϕ, the elastic energy is finally given by:

$$E_{elas} \approx \frac{\sigma_\infty^2}{\Delta P}\phi. \tag{1.37}$$

As expected, this equation shows that the smaller the droplets, the larger ΔP is, and thus the emulsion is less viscoelastic and it stores less elastic energy.

In conclusion, elastic energy is stored in a viscoelastic fluid (Maxwell or Jeffrey) under shear. This energy is only accumulated during the transient regime preceding the stationary regime. Finally, it increases in the linear regime as $\dot{\gamma}^2$ and tends to saturate in the nonlinear regime (due to the finite extensibility of the objects).

As in the paragraph above, it is worthwhile redoing this calculation for an ideal plastic solid. Let G be its elastic modulus. As the deformation varies from 0 to γ_c, the stress varies from 0 to σ_c, with $\sigma_c = G\gamma_c$. Under these conditions, the elastic energy stored during the transient regime (and then kept constant during the stationary regime) is given by:

$$E_{elas} = \frac{1}{2}G\gamma_c^2 = \frac{\sigma_c^2}{2G}. \tag{1.38}$$

We can see that this parameter is independent of $\dot{\gamma}$ and, consequently, does not go to zero in the quasi-static regime when $\dot{\gamma} \to 0$. Once again, this feature stems from the athermal character of an ideal plastic solid, its behaviour being very different from that of a viscoelastic fluid, where the intrinsic deformation of the objects and the corresponding elastic energy vanish as $\dot{\gamma} \to 0$.

In particular, this is the case for foams, where T1 type processes (represented in Fig. 1.17) are not thermally activated. Since it is impossible to relax the stress while its value is below σ_c, the foams store elastic energy until the moment they start to flow plastically, irrespective of the value of $\dot{\gamma}$, unlike ordinary viscoelastic fluids.

It so happens that the buildup of elastic energy in a viscoelastic fluid can result in spectacular effects, such as those we will describe in the following.

1.5.6 Elastic stress, normal forces and the Weissenberg effect

In practice, the elastic energy stored in the material leads to the appearance of *elastic stress* along the three directions defined by the velocity $\vec{v}$, the velocity gradient $\vec{\nabla}v$ and the direction perpendicular to them, given by the vector product $\vec{v} \times \vec{\nabla}v$.

Consider the simple shear experiment shown in Fig. 1.18. Experiment shows that most viscoelastic fluids exert a force (denoted by N_1 per unit surface) tending to push apart the plates delimiting the sample: this is the *Weissenberg effect* (Weissenberg, 1947).

We will state here, without going into further detail, that the appearance of a Weissenberg normal force is related, from the microscopic point of view, to the coupling of two effects:

- a stretching of the elastic particles under the action of local shear;
- once deformed, their tendency to align along the flow direction.

This comment helps us understand why the elastic effects and the normal force increase as the objects are more easily deformed, since in this case they store the most elastic energy.

This effect is depicted schematically in Fig. 1.19. In this case, the object is composed of a spring connecting two beads. This representation of deformable objects forms the basis of the '*dumbbell model*' widely employed in polymer rheology. We will also see that

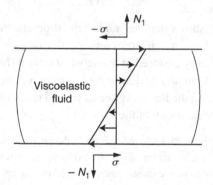

Fig. 1.18 Weissenberg effect in a droplet of viscoelastic fluid under shear. The fluid exerts on each plate a viscous drag force opposing the velocity and a normal force tending to push them apart.

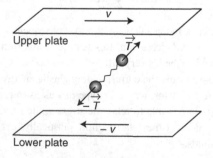

Fig. 1.19 Dumbbell model providing an explanation for the appearance of a normal force. This normal force appears due to the molecules simultaneously stretching and aligning along the flow.

the *Rouse model*, to be described in Section 7.1.6, is a generalisation to N beads of this simplified model.

It is noteworthy that the normal force must be independent of the direction along which the sample is sheared, and hence of the sign of $\dot{\gamma}$ (as one can easily see by rotating the sample through 180° about the normal to the plates). This means that the normal force N_1 must be an even function of $\dot{\gamma}$. If this force can be expanded in a power series of $\dot{\gamma}$ (this is the case for most viscoelastic materials, but not for all of them, and in particular not for liquid crystals, as we will see in Section 8.1.7), then the first non-zero term goes as $\dot{\gamma}^2$. In general, the force N_1 is thus a *nonlinear* function of the shear rate. This is why the normal forces remain moderate relative to the shear stress (proportional to $\dot{\gamma}$) in the linear regime ($\dot{\gamma} < \dot{\gamma}_c$). On the other hand, they can become dominant in the nonlinear regime ($\dot{\gamma} > \dot{\gamma}_c$), where they can reach values 100 times larger than σ in the extreme case of 'very elastic' materials.

1.5.7 Some macroscopic manifestations of the elastic stress

Anisotropic normal stresses are responsible for spectacular effects, such as that shown in the photograph of Fig. 1.20.

Here, a cylindrical rod is immersed into a viscoelastic fluid. When the rod is set in rotation the liquid starts to climb along its surface (no matter which way the cylinder turns), instead of moving outwards, as a Newtonian fluid would under the action of the inertial force. It can be shown that this phenomenon is directly related to the *Weissenberg effect* described in Fig. 1.18, and for which we presented above a qualitative physical interpretation.[30]

Another effect, extremely important in practice, is shown in Fig. 1.21. In this experiment, the viscoelastic fluid starts to swell at the exit of a tube, reaching a diameter slightly larger than that of the nozzle.[31] Strictly speaking, this swelling effect is not specific to viscoelastic materials, since it is also present in Newtonian liquids: of maximum amplitude for $Re \rightarrow 0$, namely about 13% [28, Volume 1], it decreases when the shear rate increases and leads to contraction for large Re.[32] In viscoelastic fluids, the situation is entirely different, as the effect becomes more pronounced with increasing flow rate and can lead to a diameter increase by a factor of 2 or 3.

Clearly, this swelling originates in the relaxation of the elastic stress built up in the fluid as it flows along the tube. In other words, it is due to the molecules that were stretched in the tube trying to recover their initial spherical form.[33]

---

[30] This experiment will be analysed in a simplified geometry in Section 6.3.3.1.

[31] In the polymer industry, this phenomenon is at the origin of 'die swelling'.

[32] *Re* is the Reynolds number of the flow. It is defined in Section 3.3.

[33] Note that we made here the implicit assumption of the swelling being independent of the flow rate. This is indeed the case for a long enough tube. However, at constant flow rate, the swelling is more pronounced for a shorter tube, since the elongational deformations it suffers on entering the tube do not have enough time to relax. To put it differently, the fluid 'remembers' the deformations undergone on passing the constriction. This effect can increase the swelling by a factor of 2.

Fig. 1.20 Direct manifestation of the Weissenberg effect observed for a solution of polyisobutylene in polybutene [24].

Another phenomenon, of similar origin, is that of the tubeless siphon (Fig. 1.22). It was described for the first time by Tacitus (born around AD 55) who was concerned with asphalt recovery from the Black Sea. It consists in sucking a liquid into a container using a siphon. If the experiment is performed with a Newtonian liquid, the siphon stops as soon as the tube leaves the liquid; on the contrary, for a viscoelastic fluid, the end of the tube can be lifted well above the free surface of the liquid without stopping the flow. This effect is made possible by the presence within the fluid of normal stresses, elastic in origin, which compensate the weight of the fluid column. The same thing happens when one draws vertically between two fingers a filament of saliva (which is also a viscoelastic fluid). For Newtonian fluids, where these elastic forces are absent, the column collapses immediately under its own weight as soon as the tube leaves the liquid.

Obviously, this experiment has an enormous practical significance, since it allows manufacturing of polymer fibres by the procedure described schematically in Fig. 1.23.

We will discuss in Chapter 6 other experimental illustrations of the elastic stress, such as the raising of the free surface of a viscoelastic fluid flowing down a tilted channel under the action of gravity (Tanner's experiment, see Section 6.3.3.2).

Fig. 1.21 Swelling of a jet of viscoelastic fluid on exiting a tube. Here, the fluid is a 1% polyacry-lamide solution in a 50/50 water/glycerol mixture [24].

1.6 Viscoelastic solids

As already mentioned in the introduction, there are 'soft' solids with elastic moduli much smaller than those of the usual 'hard' solids (metals, minerals or glasses). These materials also exhibit viscous dissipative phenomena at low frequency, hence the term viscoelastic. Before describing their rheological behaviour, let us give an example, in order to fix the ideas.

The most common example is *natural rubber*, which is an *elastomer* (Section 7.2). It is obtained by reacting sulphur with latex, which is the sap of hevea, a tree native to the Amazon forest. Latex is a polymer liquid containing molecules formed by long and entangled flexible chains. By boiling it with sulphur, a process discovered by Goodyear in 1830, the chains reticulate, i.e. they form chemical bonds between them at some points along their length and form a dark material with amorphous structure; this is natural rubber, still widely used to this day, most notably in car tyres. Rubber is therefore a *reticulated polymer*.

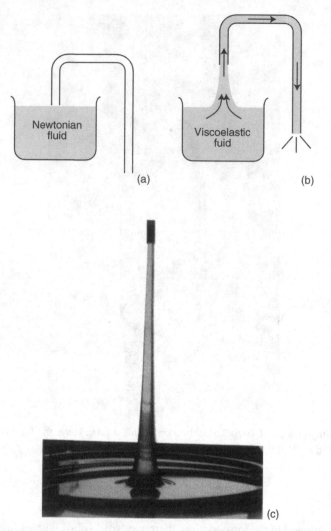

Fig. 1.22 'Tubeless siphon'. (a) In a Newtonian fluid the siphon stops as soon as the tube leaves the liquid; (b) on the contrary, a viscoelastic liquid keeps on flowing through the siphon; (c) experiment performed with a hydrocarbon polymer of high molecular mass (Peng and Landel, 1976).

The reticulation points are formed by covalent chemical bonds that can be considered as permanent (Fig. 1.24).[34]

As everyone knows, rubber is elastic: furthermore, it can sustain considerable deformation while still remaining in the linear regime, with no irreversible plastic deformation. This exceptional elasticity is *entropic* in nature, originating in the change of configuration entropy of the polymer strands connecting the different reticulation points (Section 7.2).

---

[34] Note that atmospheric oxygen has on latex the same effect as sulphur: nevertheless, it is much more reactive than sulphur and tends to cut the chains, thus deteriorating the rubber.

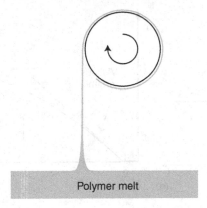

Fig. 1.23 A container filled with a polymer melt is placed under a rotating drum. By dipping a spatula in the liquid, a thread is drawn up and attached to the surface of the drum. The fluid being viscoelastic, the thread does not break and winds around the drum. Note that, in order for the thread to be usable, it must cool fast enough that it vitrifies before touching the cylinder.

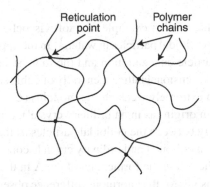

Fig. 1.24 Reticulated network.

For this reason, rubber is much 'softer' ($G \sim 10^6$ Pa) than ordinary solids ($G \sim 10^{11}$ Pa) where the elasticity is due to variations in the bond energy between atoms or molecules.

On the other hand, the response of rubber to the sudden application of stress is not instantaneous, as we assumed in the case of ordinary solids, but rather progressive, as shown in Fig. 1.25: this material is a *viscoelastic isotropic solid*, behaving as a viscous fluid at short times ($t < \tau$) and as an elastic solid over long times $t > \tau$.[35] This is where we notice the difference with respect to viscoelastic isotropic liquids, which are viscous over long times and elastic at short times.

---

[35] We note that chemical gels formed by polymers dissolved in a solvent (where the chains are also reticulated) behave in an identical way.

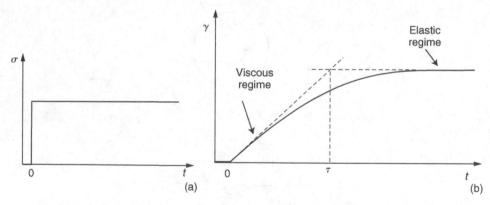

Fig. 1.25 Behaviour of a viscoelastic solid under the sudden action of stress. (a) Stress as a function of time; (b) strain as a function of time.

We will see in Section 6.2.4 that the *Kelvin–Voigt model* provides a simple description of this new type of behaviour in the linear deformation regime.[36] The time τ is typical of the material under investigation: it is still given by the ratio between the viscosity and the elastic modulus.

Colloidal crystals provide another example of a soft viscoelastic solid. These materials are obtained by dispersing colloidal particles in water.[37] In the appropriate conditions (most notably concerning the particle concentration and the ionic strength of the solution), the particles stack on a three-dimensional lattice, generally of cubic structure. A colloidal crystal is then formed, with an extremely weak elastic modulus ($G \sim 1-10^2$ Pa). In this case, the elasticity is enthalpic in origin (as in an ordinary crystal), originating in the variations of electrostatic bond energy between the colloidal particles as the crystal is deformed. An essential difference between colloidal and ordinary crystals concerns the distance between particles, of the order of the μm in the former and of the Å in the latter. We will show that this scale change is responsible for the enormous difference observed between their elastic moduli (10 orders of magnitude, typically). Colloidal crystals also have anisotropic elastic properties due to their crystalline structure (in contrast with gels and elastomers, which are elastically isotropic owing to their amorphous structure). They also exhibit strongly dissipative phenomena, since the particles are immersed in viscous liquid (unlike ordinary solids). All these properties, as well as a microscopic model allowing us to describe their rheological behaviour, will be presented in detail in Section 4.3. We should also emphasise that colloidal crystals are very fragile and can easily melt under shear. This phenomenon will be briefly discussed in Section 5.5.

Finally, we point out that waves are strongly damped in viscoelastic solids. This property is particularly important in the case of elastomers; these materials, aside from their ability

---

[36] Note that elastomers and chemical gels have an elastic limit above which the aggregate formed by the polymer chains starts to break. The material then enters a rupture regime, where it sustains irreversible damage. These phenomena will not be described in the present work.

[37] In general, one uses monodisperse spherical particles such as latex or silica beads, with a diameter ranging from a few tens of nm to a few μm (Section 4.3.1).

to sustain considerable deformation, are very good dampers for low-frequency vibrations. This property is used in numerous applications (pneumatics, for instance).

1.7 Liquid crystals

We cannot conclude this chapter without evoking liquid crystals, which will be discussed in detail in Chapter 8 [39–44].

As already stated, these materials exhibit, over a certain temperature range, *mesomorphic* phases, intermediate between solids and the isotropic liquid. The existence of these phases is a consequence of the very particular architecture of the molecules composing the liquid crystal. Best known are *calamitic* molecules: shaped as elongated rods, they possess a rigid central core (most often containing several benzene rings) upon which are grafted one or more flexible chains (Fig. 1.26a). Generally, the chains have a low chemical affinity for the core, leading to *frustration at the molecular scale*. *Nematic* and *smectic* phases, which are the most common, are a direct result of this frustration, which they relieve partially. Their structures are represented in Fig. 1.26.

In the *nematic phase* (Fig. 1.26b), the molecules tend to align parallel to each other, forming an *oriented liquid*. One speaks here of a liquid, as the centres of mass of the molecules are randomly distributed. Furthermore, each molecule can turn freely about its great axis and has the same probability of pointing 'upwards' or 'downwards'. For these reasons, each molecule can be represented by an elongated revolution ellipsoid, as in Fig. 1.26. The *orientational order* of the molecules is characterised by a unit vector the

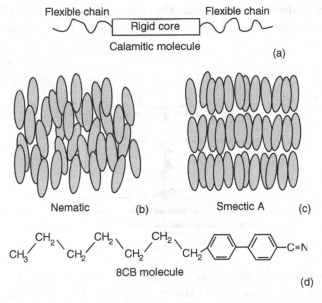

Fig. 1.26 (a) Calamitic molecule; (b) nematic phase; (c) smectic A phase; (d) example of a calamitic molecule yieding both a nematic and a smectic phase.

director $\vec{n}$, pointing along the mean direction of molecular alignment (hence, $\vec{n}$ and $-\vec{n}$ are equivalent). This is a *long-range* order, meaning that it propagates through the medium to infinity. It also means that, if a molecule is rotated at some point, all the other molecules will turn on the average by the same angle, given enough time, if no constraint is imposed on the medium. The point symmetry group of the nematic phase is $D_{\infty h}$ (Sivardière, 1995). It ensues that the nematic phase is optically birefringent, with an optical axis parallel to $\vec{n}$. The nematic phase shares this property with crystals. For this reason it is classified among *liquid crystalline phases*.

The existence of a *long-distance orientational order* in the nematic phase also has rheological consequences.

The most straightforward and spectacular one concerns the ability of a nematic to sustain *elastic torques*, as proven by the experiment described in Fig. 1.27 (Section 8.1.1). Here, the nematic is contained between two parallel plates, on which the molecules are strongly anchored. This experiment shows that, in order to maintain the twist of the director field, one must constantly exert on the two plates opposite sign torques $\vec{C}$ and $-\vec{C}$. Since the system is in equilibrium, it means that the nematic phase also exerts on the plates elastic torques that oppose the external ones. The nematic phase can thus be described as a '*torque liquid*'.

On the other hand, the nematic phase cannot sustain a constant shear force without flowing, as solids can. This behaviour is related to the absence of long-range positional order.

From this point of view, an ordinary nematic phase (composed of small molecules) behaves as a Newtonian viscous fluid, but differs from an isotropic liquid in that its viscosity under shear is directly dependent on the director orientation with respect to the velocity and the velocity gradient. This structural effect (because related to the symmetries of the phase) results in an asymmetric viscous stress tensor, leading to surprising effects, such as

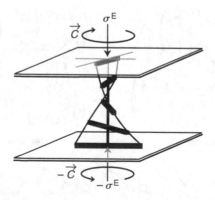

Fig. 1.27 The twisted nematic exerts a surface torque on each plate, in spite of its fluidity and the absence of long-range positional order. It also exerts a (very weak) normal force on each plate, tending to push them apart. To maintain the equilibrium, one must then apply the surface torques $\vec{C}$ and $-\vec{C}$ and the surface normal forces σ^E and $-\sigma^E$ (elastic Ericksen stress) on the two plates.

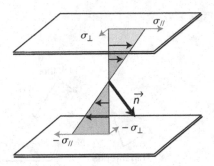

Fig. 1.28 Transverse forces in a nematic. When the director makes an angle different from 0 or $\pi/2$ with the shear plane, the nematic exerts on the two plates forces perpendicular to the velocity, which tend to push them laterally. Aside from the viscous stress $\sigma_\parallel$ and $-\sigma_\parallel$ one must therefore exert on the two plates the viscous stress $\sigma_\perp$ and $-\sigma_\perp$ perpendicular to the velocity.

the presence of *transverse forces* under shear as the director $\vec{n}$ makes an angle different from 0 or $\pi/2$ with the shear plane $(\vec{v}, \vec{\nabla}v)$ (Fig. 1.28) (Section 8.1.7.5).

Nematics can also sustain an *elastic stress*, also called the *Ericksen stress*, which is directly related to the torque elasticity of the phase (see Section 8.1.6.3). This stress (which is not related to the existence of positional order of the molecules, but rather to their orientational order) is weak, but it plays a determinant role in the dynamics of the nematic phase. It is noteworthy that this elastic stress also exists in the static regime, as shown by the example in Fig. 1.27. In this twisted geometry, it is obvious that the twist energy of the phase decreases if the two plates are pushed apart: the result is a repulsive force between the plates. This force per unit surface is termed *Ericksen stress* (Section 8.1.6.3). We emphasise that this force has strictly nothing in common with the normal stress in viscoelastic fluids that we described above in Section 1.5.6, and which are dynamic in nature, as we already pointed out.

We note, in this respect, that the normal stress related to a possible deformation of the molecules under flow is completely negligible in nematic phases of small molecules, but can become important in polymeric liquid crystals in the nonlinear regime [45]. Nevertheless, we will not approach this topic in the present work.

Finally, we will see that nematic phases exhibit defects, termed *disclinations*, which can play an important role during flow. These defects look like flexible 'threads' when observed in polarised microscopy (Fig. 1.29). It is precisely this observation that gave nematics their name, '$\nu\eta\mu\alpha$', read nema, meaning thread in Greek (Section 8.1.9).

There are also other phases that possess, aside from the nematic-type orientational order, a *long-range positional order along one or two space directions*. For this reason, their rheology is at the crossroads between nematodynamics and solid plasticity, an even more convincing justification for their denomination as '*liquid crystals*'.

The most common, and also the simplest in this category, is the *smectic A phase*,[38] where the molecules are arranged in fluid layers stacked on top of each other (Fig. 1.26b);

[38] Smectic comes from the Greek word '$\sigma\mu\eta\kappa\tau\iota\kappa\delta\sigma$' (read smectikos), which means 'endowed with cleaning properties'. This very astute choice of G. Friedel was motivated by the fact that soaps very often exhibit smectic phases.

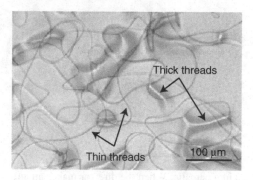

Fig. 1.29 Thread texture in a nematic (photograph taken in natural light).

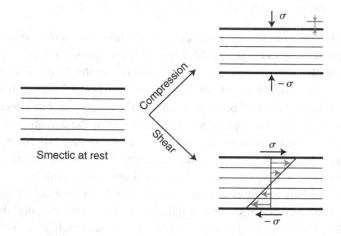

Fig. 1.30 A smectic phase undergoes elastic deformation under the action of stress normal to the layers; however, it flows viscously when sheared parallel to the layers.

on the average, the molecules are perpendicular to the layers and they can turn about their axis, pointing either upwards or downwards. Thus, the smectic A phase has the same point symmetry group as the nematic phase ($D_{\infty h}$). Consequently, it is also optically uniaxial, with an optical axis normal to the layers.

The main difference between the nematic phase and the smectic A phase is the presence in this latter of *positional order* along the layer normal. In practice (i.e. over the size of the sample) this order is *long-range*, such that smectics behave as solids under compression or dilation of the layers. However, it flows as a liquid under shear parallel to the layers (Fig. 1.30). We will see in Chapter 8 that this dual behaviour bestows on the smectic phase exceptional lubricant properties.

We will also show that the smectic phase is susceptible to mechanical instabilities, for instance the layer undulation instability (Fig. 1.31), the result of subtle competition between bending elasticity (as for nematics) and solid-type elasticity (Section 8.2.3).

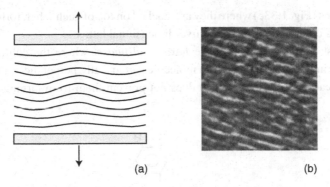

(a)   (b)

Fig. 1.31 Layer undulation instability under dilation normal to the layers. (a) Schematic representation; (b) observation in optical microscopy between crossed polarisers making an angle of $45°$ with the direction of the undulations (photograph by R. Ribotta).

Fig. 1.32 Focal conic texture in a smectic A (photograph taken between crossed polarisers).

Finally, we will see that the smectic phase possesses nematic-like defects, such as disclinations, which gather and form *focal domains* of macroscopic size (see Fig. 1.32), or solid-like defects, such as *edge or screw dislocations*, which break the layered structure at the microscopic scale (Section 8.2.9). As in the case of solids, these defects are important for explaining the rheological behaviour of the phase. Examples of flow behaviour, exemplifying either the predominance of viscous effects, or that of plastic effects, where the defects play a direct role, will be provided in Section 8.2.8.

The other family of liquid crystals combining nematic and positional order is that of *columnar phases*.

These are generally obtained using disc-shaped or *discoidal* molecules, like the truxene molecule shown in Fig. 1.33a. These molecules can form nematic phases (Fig. 1.33b), or columnar phases (Fig. 1.33c) where they are stacked on top of each other, forming columns, parallel to each other and ordered on a two-dimensional lattice.

The simplest example is that of the *hexagonal phase*, where the column lattice has hexagonal symmetry. In this phase, the molecules are usually perpendicular to the column axis, but they can also be tilted, which does not prevent them from turning freely around

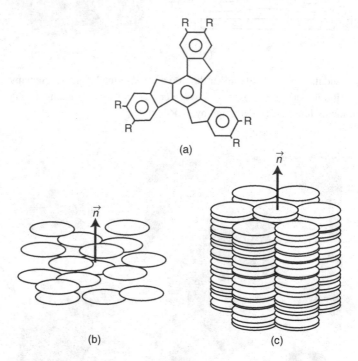

Fig. 1.33 (a) Truxene molecule; (b) 'discotic' nematic phase; (c) columnar hexagonal phase.

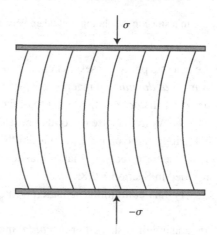

Fig. 1.34 Buckling instability of the columns under compression.

Fig. 1.35 Developable domain texture of a columnar phase (photo taken between crossed polarisers).

the column axis. The hexagonal phase is therefore optically uniaxial, with its optical axis parallel to the director $\vec{n}$, itself parallel to the axis of the columns.

Columnar phases, like smectic phases, exhibit peculiar viscoelastic properties, still in need of a satisfactory explanation. One of them, to be described in Section 8.3.2, is the presence of a buckling instability of the columns under compression parallel to the column axis (Fig. 1.34). This instability, which is similar to the layer undulation instability in smectics, is also reminiscent of the Euler instability of beams in solid mechanics (Section 4.1.10).

This phase also exhibits defects, of great rheological significance, which can be either dislocations, disclinations or π defects, which are isolated column ends (Section 8.3.2.2). As in smectics, disclinations form textures, called *developable domains*, clearly visible in polarised microscopy. An example of such texture is shown in Fig. 1.35.

As to π defects, they seem to be at the origin of surprising viscoelastic properties, still in need of an explanation, to be described in Section 8.3.2.

References

Barnes, H. A., *J. Non-Newtonian Fluid Mech.*, **70**, 1 (1997).

Barnes, H. A., *J. Non-Newtonian Fluid Mech.*, **81**, 133 (1999).

Cross, M. M., *J. Colloid Sci.*, **20**, 417 (1965).

Doraiswamy, D., The origins of rheology: a short historical excursion, *Rheol. Bull.*, **71**, Issue 1 (2002).

Frost, H. G. and M. F. Ashby, *Deformation-Mechanism Maps*, Oxford: Pergamon Press (1982).

Gandolfo, F. G. and H. L. Rosano, *J. Coll. Int. Sci.*, **194**, 31 (1997).

de Gennes, P. G., F. Brochard and D. Quéré, *Gouttes, bulles, perles et ondes*, Paris: Éditions Belin, Collection Échelles (2002).

Jones, D. A. R., B. Leary and D. V. Boger, *J. Colloid Interface Sci.*, **147**, 487 (1991).
Peng, S. T. J. and R. F. Landel, *J. Appl. Phys.*, **47**, 4255 (1976).
Sivardière, J., *La symétrie en mathématiques, physique et chimie*, Grenoble: Presses Universitaires de Grenoble (1995).
Weber, W., *Ann. Phys. Chem.*, **34**, 247 (1835).
Weissenberg, K., *Nature*, **159**, 310 (1947).

2

Overview of the mechanics of continuous media

In the previous chapter, we described the main classes of materials one can encounter and we classified them according to their rheological behaviour in several categories: elastic, plastic or viscoelastic solids, Newtonian, non-Newtonian, viscoelastic or yield stress liquids, and liquid crystals.

We also presented some of their rheological constitutive equations, which required the use of fundamental concepts such as *strain, stress* and *elastic torque*. However, they were not rigorously defined. This will be our first concern in this chapter, after having given a reminder of the conditions under which a material can be treated as a *continuous medium*.

We will then discuss the *fundamental dynamical equations* which can be used to calculate the flow of any material, provided that its rheological constitutive law is known.

Finally, we will show how to obtain formally *the constitutive equations* of a material (*rheological constitutive law* and the *Fourier law for heat diffusion*) from the general formula of *irreversible entropy production*.

The chapter is structured as follows.

We will start with the *basics of continuum mechanics*. We will then show the general way of constructing the *stress and surface torque tensors* using the '*tetrahedron method*' (Section 2.1).

We will continue by reviewing the three dynamical equations that can be used to calculate the flow of materials. We will start by proving the *law of mass conservation* and the *law of momentum conservation* using *Newton's second law* (Section 2.2).

We will then obtain the *equilibrium equation for bulk torques* from the *angular momentum theorem*. It will be shown that this third equation can predict the symmetry of the *stress tensor*, which can be symmetric or not, depending on the type of material under consideration (Section 2.3).

Finally, we will state (without proof) the general expression of *irreversible entropy production*. We will also show how this quantity can be used to construct the constitutive laws of a material, taking into account its symmetry properties (Section 2.4).

2.1 Stress tensor and surface torque tensor

A first point to be made is that the *concept of continuous medium presupposes an invariance: the independence of the resulting equations with respect to the size of the matter element used for obtaining them.* Thus, by its very nature, a continuous model ignores the detailed and discontinuous descriptions of microscopic physics, and can only be used to describe a material over *scales 'much larger' than that of its microstructure.* Consequently, the model can only be valid if the sample is much larger than the typical size of its microstructure. This condition is not always fulfilled. Some silica gels, formed by entangled fibres, themselves made of even smaller fibres, etc. provide an example, since they contain voids of all sizes. These materials are not amenable to a continuous description. One can find numerous other examples, such as sponges, mountain talus or biological tissues.

We will not deal with such materials in the following, although they are (obviously) very interesting.

To start, let us now return to the concept of internal stress and define the surface forces and torques exerted by one material element upon an adjacent element.

2.1.1 Surface forces and torques

Imagine a generic surface Σ within the material, dividing it into two sub-systems I and II which act across it upon each other.

The main assumption of the mechanics of continuous media is that each part exerts upon its neighbour at point M, a *force* and a *surface torque* which depend on M and on the orientation of the surface element at that point (Fig. 2.1). This hypothesis, sometimes termed *'Cauchy's postulate'*, is meaningful at a *'macroscopic' scale*, since the *forces between molecules* (or between the typical objects composing the material) *are very short ranged*, of the order of a few elementary distances (the elementary distance being the largest distance relevant for the microstructure).

More specifically, let $\vec{\nu}$ be the unit vector normal to Σ at point M and pointing from ① towards ②. As per our assumption, part ② exerts upon part ① at point M the force per

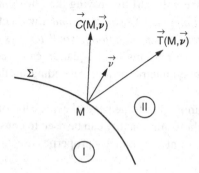

Fig. 2.1 Torsor of the surface forces acting at a point M on a surface element with the unit normal vector $\vec{\nu}$.

unit surface $\vec{T}(M, \vec{v})$ and the torque per unit surface $\vec{C}(M, \vec{v})$. Obviously, from *Newton's third law of action and reaction*, part ① exerts upon part ⑪ the force $-\vec{T}(M, \vec{v})$ and the torque $-\vec{C}(M, \vec{v})$.

Let us point out straightaway that these surface torques are only relevant for *fluid media with directors*, such as the nematic or smectic liquid crystals already cited in the first chapter [39–44]. In this case, the torques originate in the natural tendency of the elongated molecules to align parallel to one another.

Although these torques can also appear in crystalline solids,[1] they are negligible here, being always dominated by stress elasticity. Finally, on symmetry grounds, they vanish in isotropic fluids (viscoelastic or not).

This description of the internal stress in terms of surface forces and torques is not very convenient, as the vectors $\vec{T}(M, \vec{v})$ and $\vec{C}(M, \vec{v})$ depend at the same time on the point M and on the orientation $\vec{v}$ of the surface element considered. For this reason, let us introduce the *stress tensor* and *surface torque tensor*, depending only on point M. As a historical reminder, the use of tensors in elasticity to describe internal stress is due to Cauchy (first half of the nineteenth century).

2.1.2 Construction of the stress tensor

We will start by proving the existence, at each point M, of a second rank tensor $\underline{\sigma}$, named the *stress tensor*, such that:

$$\vec{T}(M, \vec{v}) = \sigma(M)\vec{v}. \tag{2.1}$$

To this end, let us construct the tetrahedron MABC of apex M and basis ABC perpendicular to the $\vec{v}$ vector (Fig. 2.2). Points A, B and C are placed on the reference axes x_1, x_2 and x_3, respectively. Let h be the height of the tetrahedron and A the area of triangle ABC. The area A_i of the face perpendicular to axis x_i is:

$$A_i = A \cos(\vec{x}_i, \vec{v}) = A v_i. \tag{2.2}$$

Applying *Newton's second law* to the tetrahedron yields directly:

$$\frac{1}{3}hA\vec{F} + A\vec{T}(M, \vec{v}) + Av_j\vec{T}_j(M) = 0. \tag{2.3}$$

In this equation,[2] $\vec{F}$ is the resultant of the bulk forces (which can include inertial forces in the dynamic regime) and $\vec{T}_j(M)$ is the surface force acting on the face of the tetrahedron perpendicular to the x_j axis.

Let h go to 0. In this limit, the bulk forces in Eq. (2.3) vanish, and it reduces to:

$$\vec{T}(M, \vec{v}) = -\vec{T}_j(M)v_j \tag{2.4}$$

---

[1] This is, in particular, the case of crystalline smectics of type B, G, E ... (for a detailed description of their structures see [43], volume 1, p. 25), since they are composed of elongated molecules tending to align parallel to one another.

[2] We use here, and throughout the rest of the book, Einstein's convention of summing over repeated indices.

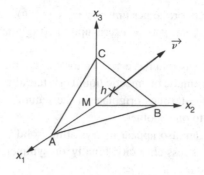

Fig. 2.2 Tetrahedron constructed about point M.

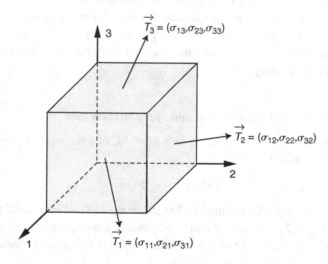

Fig. 2.3 Physical meaning of the stress σ_{ij}.

or, in index form:

$$T_i(M, \vec{v}) = -T_{ji}(M)v_j. \tag{2.5}$$

One usually writes

$$\sigma_{ij}(M) = -T_{ji}(M). \tag{2.6}$$

The tensor $\underline{\sigma}$, depending only on the point M under consideration, is the *stress tensor*. This construction shows that σ_{ij} is (up to a sign change) the ith component of the surface force acting on the face of the tetrahedron perpendicular to the x_j axis.

Thus, at each point M one can define a stress tensor $\underline{\sigma}$ containing all the information on the internal stress at that point. Conventionally, $\underline{\sigma}\vec{v}$ is the force per unit surface exerted by the medium towards which points $\vec{v}$ upon the adjacent medium. The physical meaning of σ_{ij} is depicted in Fig. 2.3.

In general, the stress tensor we have just constructed contains elastic and dissipative (viscous) terms denoted by $\underline{\sigma}^e$ and $\underline{\sigma}^v$, respectively. In liquid crystals and viscoelastic fluids it is customary to separate the hydrostatic pressure term $-P\underline{I}$ (coupled to the density) from the other elastic terms $\underline{\sigma}^E$.[3] Finally, one writes:

$$\underline{\sigma} = \underline{\sigma}^e + \underline{\sigma}^v = -P\underline{I} + \underline{\sigma}^E + \underline{\sigma}^v \tag{2.7}$$

with $\underline{I}$ the unit matrix.

2.1.3 Construction of the surface torque tensor

One can also prove, applying this time the angular momentum theorem to the tetrahedron (see Section 2.3), the existence of a second rank surface torque tensor $\underline{C}$, such that:

$$\vec{C}(M, \vec{v}) = \underline{C}(M)\vec{v}. \tag{2.8}$$

We will see that this tensor plays a fundamental role in nematic liquid crystals, which are director fluids and can sustain elastic torques without flowing.

The detailed expressions of the stress and surface torque tensors as a function of strain or strain rate represent the laws of behaviour (or constitutive laws) of the material. They will be detailed in the next chapters, keeping in mind that they depend on the material under consideration and its symmetries.

2.2 Mass and momentum conservation laws

As we said in the introduction, these two laws are fundamental, since they can be used to calculate the flow of any material, provided its rheological constitutive law is known.[4]

2.2.1 Mass conservation law

Let ρ be the density of the medium and $\vec{v}$ the velocity of the fluid in the reference system of the laboratory. Writing down the matter flux entering and leaving a *fixed volume V* delimited by a surface S, one obtains immediately:

$$\frac{d}{dt} \int_V \rho \, dV + \int_S \rho \vec{v} \cdot \vec{v} \, dS = 0, \tag{2.9}$$

---

[3] In liquid crystals, $\underline{\sigma}^E$ bears the name Ericksen stress.

[4] We emphasise, however, that these two laws are not always sufficient (in particular, in the case of liquid crystals) to calculate the flow. Indeed, in order to complete the equation system, they must be complemented by the equation of torque balance, to be demonstrated in the next section.

where $\vec{v}$ is the unit vector perpendicular to S and pointing outwards. Since the volume V is fixed, one can simply derivate under the integral; the surface integral can also be transformed into a volume integral, yielding finally:

$$\int_V \left(\frac{\partial \rho}{\partial t} + \mathrm{div}(\rho \vec{v}) \right) \mathrm{d}V = 0. \tag{2.10}$$

This relation holds for any volume V, so that:

$$\frac{\partial \rho}{\partial t} + \mathrm{div}(\rho \vec{v}) = 0. \tag{2.11}$$

This equation (written here in Eulerian form[5]) is the *mass conservation law*.

In Lagrangian form[6] it is written as follows:

$$\frac{\mathrm{D}\rho}{\mathrm{D}t} + \rho \, \mathrm{div} \, \vec{v} = 0, \tag{2.12}$$

where $\mathrm{D}/\mathrm{D}t$ represents the total (or Lagrangian) derivative with respect to time:

$$\frac{\mathrm{D}}{\mathrm{D}t} = \frac{\partial}{\partial t} + v_i \frac{\partial}{\partial x_i}. \tag{2.13}$$

Here, the derivation bears upon a *set of material particles followed along their motion*. For this reason, one also speaks of the *material derivative*.

Note that, in the special case of an *incompressible fluid*,[7] where the density of each material element remains constant during motion, one has by definition $\mathrm{D}\rho/\mathrm{D}t = 0$ or $\rho = \mathrm{Cst}$. In these conditions, Eq. (2.12) reduces to the simplified form $\mathrm{div} \, \vec{v} = 0$.

2.2.2 Momentum conservation law

To obtain this equation, let us consider a *mobile volume* V_m with its external surface S_m, composed of a collection of molecules, and follow it as it moves. Applying Newton's second law to this collection (in a *Galilean reference frame*[8]), yields:

$$\frac{\mathrm{d}}{\mathrm{d}t} \int_{V_\mathrm{m}} \rho \vec{v} \, \mathrm{d}V = \int_{S_\mathrm{m}} \underline{\underline{\sigma}} \vec{v} \, \mathrm{d}S + \int_{V_\mathrm{m}} \vec{f} \, \mathrm{d}V, \tag{2.14}$$

---

[5] In the Eulerian description, the motion of the fluid is described by its velocity field [1].

[6] In the Lagrangian description, one uses the trajectories of particles to describe fluid motion [1] (see also the appendix to this chapter).

[7] We will see in Chapter 3 that this condition is fulfilled if the Mach number of the flow (ratio between the typical velocity of the flow and the speed of sound) is much less than 1. It amounts to considering that the speed of sound in the medium is infinite. We will often make this assumption in the following.

[8] It is noteworthy that this law only holds in a Galilean reference frame, unlike the mass conservation law or the constitutive laws, which are valid in any reference frame (by the so-called 'objectivity' principle).

with $\vec{\nu}$ the unit vector normal to S_m and pointing outwards, and $\vec{f}$ the resultant of the external bulk forces (e.g. gravity). Transforming the surface integral to a bulk integral and applying the 'theorem of derivation under the integral' (see Appendix 2.A), results in:

$$\int_{V_m} \rho \frac{D\vec{\nu}}{Dt} dV = \int_{V_m} \mathrm{div}\,\underline{\sigma}\, dV + \int_{V_m} \vec{f}\, dV, \qquad (2.15)$$

where $\mathrm{div}\underline{\sigma}$ is a vector with components $(\mathrm{div}\underline{\sigma})_i = \sigma_{ij,j}$.[9] Since this relation holds for any volume V_m, it follows that:

$$\rho \frac{D\vec{\nu}}{Dt} = \mathrm{div}\,\underline{\sigma} + \vec{f}. \qquad (2.16)$$

This is one of the fundamental equations of dynamics, called *Cauchy's equation*. We recall that here:

$$\frac{D\vec{\nu}}{Dt} = \frac{\partial \vec{\nu}}{\partial t} + (\vec{\nu} \cdot \vec{\nabla})\vec{\nu}. \qquad (2.17)$$

Combining Eq. (2.16) and the mass conservation law (2.11) yields:

$$\frac{\partial \rho\vec{\nu}}{\partial t} + \mathrm{div}\,\underline{\Pi} = \vec{f}, \qquad (2.18)$$

where $\Pi_{ij} = \rho\nu_i\nu_j - \sigma_{ij}$ is the *momentum flux density tensor*. In this form, one recognises the *conservation law of the momentum* $\rho\vec{\nu}$ (the bulk force $\vec{f}$ playing the role of a source term).

2.3 Torque balance equation and symmetry of the stress tensor

The goal of this subsection is twofold: we aim to establish the *torque balance equation*, essential for describing the motion of director fluids (liquid crystals, essentially) and to show that, as a consequence, *the stress tensor is symmetric in the absence of torques* (this is the case for isotropic fluids and solids, where the torques, if they are present, can always be neglected[10]), and *asymmetric otherwise* (liquid crystals).

To determine this new equation of motion, let us apply the *angular momentum theorem* to a generic volume element V_m within the material, of surface S_m [43, volume 1]. $\vec{M}_O$ is the angular momentum at point O (origin of the *Galilean reference frame* in which one studies the flow or deformation[11]), given by:

$$\vec{M}_O = \int_{V_m} \vec{r} \times \rho\vec{\nu}\, dV, \qquad (2.19)$$

---

[9] The subscript comma indicates a partial derivative with respect to the x_j coordinate. By definition, $f_{,j} = \partial f/\partial x_j$ and $f_{,ij} = \partial^2 f/\partial x_i \partial x_j$.

[10] To our knowledge.

[11] Same caveat as in Footnote 8.

with $\vec{r}$ the vector radius, one has:

$$\frac{d\vec{M}_O}{dt} = \int_{S_m} \vec{r} \times \underline{\sigma}\vec{v} \, dS + \int_{S_m} \underline{C}\vec{v} \, dS + \int_{V_m} \vec{r} \times \vec{f} \, dV + \int_{V_m} \vec{\Gamma}^M \, dV, \qquad (2.20)$$

where $\vec{\Gamma}^M$ is an external bulk torque (it can be the result of applying an electric or magnetic field in the case of a director medium).

Using the 'theorem of derivation under the integral' (see Appendix 2.A), one obtains:

$$\frac{d\vec{M}_O}{dt} = \int_{V_m} \vec{r} \times \rho \frac{D\vec{v}}{Dt} \, dV. \qquad (2.21)$$

On the other hand, it is easily shown that

$$\int_{S_m} (\vec{r} \times \underline{\sigma}\vec{v} + \underline{C}\vec{v}) \, dS = \int_{V_m} (\vec{r} \times \text{div}\,\underline{\sigma} + \vec{\Gamma}) \, dV, \qquad (2.22)$$

where $\vec{\Gamma}$ is a bulk torque with components

$$\Gamma_i = -e_{ijk}\sigma_{jk} + C_{ij,j}. \qquad (2.23)$$

In principle, this torque contains an elastic and a viscous part:

$$\vec{\Gamma} = \vec{\Gamma}^E + \vec{\Gamma}^v \qquad (2.24)$$

to be specified depending on the material under consideration (see Chapter 8). We recall that, by definition, e_{ijk} is equal to 1 (or -1) when (i, j, k) form a circular permutation of $(1, 2, 3)$ (or $(2, 1, 3)$) and 0 otherwise (i.e. when at least two indices are equal).

Keeping in mind that $\text{div}\,\underline{\sigma} + \vec{f} = \rho\,D\vec{v}/Dt$, one obtains, after replacing in Eq. (2.22), and then in Eq. (2.20), and taking into account Eq. (2.21):

$$\int_{V_m} \left(\vec{\Gamma} + \vec{\Gamma}^M\right) dV = 0. \qquad (2.25)$$

Since this equation holds for any integration volume, it follows that:

$$\vec{\Gamma} + \vec{\Gamma}^M = 0 \qquad (2.26)$$

or, equivalently

$$\vec{\Gamma}^E + \vec{\Gamma}^v + \vec{\Gamma}^M = 0. \qquad (2.27)$$

In a medium possessing a director $\vec{n}$, this equation expresses *the balance of bulk torques* acting on $\vec{n}$. It will be discussed in detail in Chapter 8, dealing with nematic and smectic liquid crystals. At this point, let us merely note that in these materials, where the stress tensor $\underline{\sigma}$ is *not* symmetric, its antisymmetric part, denoted $\underline{\sigma}^a$, can balance (in the absence of external fields) the bulk torques associated with director disorientations. Indeed, according to Eqs. (2.23) and (2.26) when $\vec{\Gamma}^M = 0$:

$$e_{ijk}\sigma_{jk}^a + C_{ij,j} = 0. \tag{2.28}$$

In *solids* and *isotropic fluids*, on the other hand, the torque balance equation (2.26) simplifies considerably, as $\vec{\Gamma}^M = 0$ and $\underline{C} = 0$ by symmetry. It can then be written explicitly, using Eq. (2.23):

$$e_{ijk}\sigma_{jk} = 0, \tag{2.29a}$$

proving that the stress tensor is *symmetric* in these materials:

$$\sigma_{ij} = \sigma_{ji}. \tag{2.29b}$$

In conclusion, the stress tensor can only be symmetric in the absence of torques.

This condition is fulfilled exactly (by symmetry) in isotropic materials (irrespective of their nature, fluid or solid).

One can also take the stress tensor as symmetric in anisotropic solids, where elastic torques are negligible.[12]

However, the stress tensor is not symmetric in nematic or smectic liquid crystals, where elastic torques are very important (see Chapter 8).

2.4 Irreversible entropy production and constitutive laws

Irreversible entropy production is a fundamental quantity,[13] making possible the general construction of the constitutive laws for a given material, accounting for its symmetry properties.[14] Its general expression is:

$$\rho T \overset{\circ}{s} = \rho T \frac{Ds}{Dt} + T \operatorname{div}\left(\frac{\vec{q}}{T}\right), \tag{2.30}$$

where s is the entropy of the fluid per unit mass, and $\vec{q}$ is the heat flux vector. We make here a *local equilibrium assumption*, considering that s is the equilibrium entropy of the fluid at pressure P and density ρ.

Let us show that this quantity must always be *positive*. Consider a mobile volume V_m delimited by the surface S_m. According to the *second principle of thermodynamics*, one has necessarily:

$$dS \geq \frac{\delta Q}{T}, \tag{2.31}$$

---

[12] In three-dimensional solids, the director field cannot be distorted without distorting the crystal lattice itself. The elastic energy associated with lattice distortions being generally much larger than that due to distortions of the director field, torque elasticity can be neglected with respect to stress elasticity, which amounts to admitting that the stress tensor is symmetric.

[13] We emphasise straightaway that this quantity is different from viscous dissipation when the temperature is not uniform (see Eq. (2.36)).

[14] Concrete examples will be given in Chapters 3 and 8.

with δQ the heat exchanged with the environment by conduction (reversibly). In our case, this inequality is written explicitly as:

$$\frac{d}{dt}\int_{V_m} s\rho \, dV \geq \int_{S_m} \frac{-\vec{q}\cdot\vec{v}}{T}\, dS, \tag{2.32}$$

which yields, after 'derivating under the integral' and transforming the surface integral into a volume integral:

$$\int_{V_m}\left[\rho\frac{Ds}{Dt} + \mathrm{div}\left(\frac{\vec{q}}{T}\right)\right]dV \geq 0. \tag{2.33}$$

It follows that the quantity $\rho T \overset{\circ}{s}$ defined in Eq. (2.30) must always be positive:

$$\rho T \overset{\circ}{s} = \rho T \frac{Ds}{Dt} + T\mathrm{div}\left(\frac{\vec{q}}{T}\right) > 0. \tag{2.34}$$

This quantity can also be calculated in the general case (including solids and director fluids), where it is given by:

$$\rho T \overset{\circ}{s} = \sigma_{ij}^v \frac{\partial v_i}{\partial x_j} - \vec{\Gamma}^v \cdot \left(\vec{n}\times\frac{D\vec{n}}{Dt}\right) + \left(\frac{D\vec{u}}{Dt} - \vec{v}\right)\cdot\mathrm{div}\left(\underline{\sigma}^E\right) - \frac{\vec{q}}{T}\cdot\overrightarrow{\mathrm{grad}}\, T. \tag{2.35}$$

This expression can be rewritten in the following (more physical) form:

$$\rho T \overset{\circ}{s} = \sigma_{ij}^{vs} A_{ij} - \vec{\Gamma}^v \cdot \left(\vec{n}\times\vec{N}\right) + \left(\frac{D\vec{u}}{Dt} - \vec{v}\right)\cdot\mathrm{div}\left(\underline{\sigma}^E\right) - \frac{\vec{q}}{T}\cdot\overrightarrow{\mathrm{grad}}\, T, \tag{2.36}$$

where we introduced the symmetric part of the viscous stress tensor σ_{ij}^{vs} as well as two new quantities:

$$A_{ij} = \frac{1}{2}\left(\frac{\partial v_i}{\partial x_j} + \frac{\partial v_j}{\partial x_i}\right) \tag{2.37}$$

and

$$\vec{N} = \frac{D\vec{n}}{Dt} - \vec{\Omega}\times\vec{n}. \tag{2.38}$$

The first one defines the *symmetric strain rate tensor*; the second one corresponds to the *local (relative) rotation rate of the director* where $\vec{\Omega} = (1/2)\overrightarrow{\mathrm{curl}}\,\vec{v}$ is the local rotation rate of the fluid.

We will show that these two quantities vanish in the case of 'solid' rotation of the material (see Section 3.2.7 and Section 8.1.6). It follows that the dissipation is not changed by a solid rotation of the medium, in agreement with the requirements of thermodynamics.

Let us conclude by discussing the different terms in the general expression of dissipation. Four of them can be identified:

- the first one, given by $\sigma_{ij}^{vs} A_{ij}$, is well known: it corresponds to the dissipation due to viscous stress;
- the second one $\vec{\Gamma}^{v} \cdot (\vec{n} \times \vec{N})$ is important in liquid crystals, representing the dissipation due to viscous torques;
- the third one might seem rather strange, since it brings into play the difference $D\vec{u}/Dt - \vec{v}$ between the displacement derivative and the hydrodynamic velocity of the molecules. In the case of a fluid without long-range positional order, this difference vanishes because $D\vec{u}/Dt = \vec{v}$ and this term disappears. However, the situation is more subtle for crystals and ordered fluids, such as smectic or columnar liquid crystals. Indeed, the displacement $\vec{u}$ is, by definition, that of lattice points in a crystal and that of layers or columns in a smectic or columnar liquid crystal; thus, it does not necessarily coincide with the displacement of atoms or molecules, which can migrate across the lattice. In solids, this flow is contrary to the vacancy flux induced by a stress gradient within the material (see Section 5.3.4 on Nabarro–Herring creep). In smectic or columnar liquid crystals, the same phenomenon exists, leading to *permeation flow* (see Section 8.2.6). Note that this type of flow is generated by the bulk elastic force $\mathrm{div}(\underline{\sigma}^{E})$, proportional to the elastic stress gradients within the medium;
- the fourth term is the usual one corresponding to the heat flux.

Note that in *isotropic fluids* (whether Newtonian, non-Newtonian or viscoelastic) the torques vanish by symmetry, such that the viscous stress tensor is symmetric. In this case, the dissipation is simply:

$$\rho T \overset{\circ}{s} = \sigma_{ij}^{v} A_{ij} - \frac{\vec{q}}{T} \cdot \overrightarrow{\mathrm{grad}}\, T. \tag{2.39}$$

In conclusion, let us show how to construct formally the constitutive laws of a material in the framework of the *thermodynamics of irreversible processes*. To this end, one must define 'forces' $(\sigma_{ij}^{vs}, \vec{\Gamma}^{v}, \ldots)$ and 'fluxes' $(A_{ij}, \vec{n} \times \vec{N}, \ldots)$ and write linear relations between these quantities. The next step consists of simplifying these relations by taking into account, in particular, the symmetries of the medium. This procedure results in a reduction of the necessary transport coefficients (viscosities, thermal conductivities, etc.). Concrete examples will be provided in Chapters 3 and 8.

Appendix 2.A Theorem of 'derivation under the integral'

Let $f(\vec{r}, t)$ be a generic function (which can be a scalar, a vector or a tensor) and V_m a moving volume composed of particles to be followed along their motion. We will prove the following identity:

$$\frac{d}{dt} \int_{V_m} f\rho\, dV = \int_{V_m} \frac{Df}{Dt} \rho\, dV, \tag{2A.1}$$

where D/Dt represents the total (*material*) derivative with respect to time.

The simplest way of proving this theorem is by describing fluid motion in the Lagrangian formalism, i.e. by giving the trajectory of each particle in the form:

$$\vec{r} = \vec{r}(\vec{a}, t),$$
(2A.2)

where $\vec{a}$ stands for the position of the particle at a given moment, e.g. at $t = 0$. Then $\vec{a} = \vec{r}(\vec{a}, 0)$.

This equation defines a *change of coordinates from the $\vec{a}$ space to the $\vec{r}$ space*:

$$dx_i = \frac{\partial x_i}{\partial a_j} da_j,$$
(2A.3)

a volume element $d^3\vec{a}$ around $\vec{a}$ being transformed into a volume element $d^3\vec{r} = |J| d^3\vec{a}$ where J is the Jacobian (i.e. the determinant of the transformation matrix):

$$J = \det\left(\frac{\partial x_i}{\partial a_j}\right).$$
(2A.4)

In the Lagrangian description, mass conservation reads:

$$\frac{D}{Dt}\left[\rho(\vec{r}, t) d^3\vec{r}\right] = \frac{D}{Dt}\rho(\vec{r}, t)|J| d^3\vec{a} = 0,$$
(2A.5)

yielding:

$$\frac{D}{Dt}\rho(\vec{r}, t)|J| = 0.$$
(2A.6)

The theorem is proven by first transforming the integral over the moving (time-dependent) volume into an integral over the fixed volume V_a taken in the reference space:

$$\int_{V_m} f(\vec{r}, t)\rho \, dV = \int_{V_a} f\left[\vec{r}(\vec{a}, t), t\right]\rho|J| dV_a,$$
(2A.7)

then, by taking the time derivative of this new integral using the mass conservation law (2A.6) written in its Lagrangian form. The calculation is then straightforward, since the integration volume V_a does not change with time:

$$\frac{d}{dt}\int_{V_m} f\rho \, dV = \frac{d}{dt}\int_{V_a} f\rho|J| dV_a = \int_{V_a} \frac{D}{Dt}(f\rho|J|) dV_a = \int_{V_m} \frac{Df}{Dt}\rho \, dV.$$
(2A.8)

QED

More generally, one can prove that:

$$\frac{d}{dt}\int_{V_m} f(\vec{r}, t) \, dV = \int_{V_m}\left(\frac{\partial f}{\partial t} + \frac{\partial(f v_j)}{\partial x_j}\right) dV.$$
(2A.9)

This is Reynolds' theorem, which does indeed yield Eq. (2A.1) by replacing f with ρf and using the mass conservation equation (2.12).

3

Hydrodynamics of simple liquids

In this chapter we will consider isotropic Newtonian fluids, which are the simplest from the rheological point of view. Our goal is not to present a complete lecture course on hydrodynamics, but rather to review the basic notions and some essential results. In particular, we will completely ignore certain topics, such as turbulence, magnetohydrodynamics or superfluid flow. Also, simplified proofs will be given for some theorems, to avoid too complex mathematical developments. On all these issues we refer the reader to specialised works, numerous and often of excellent quality, of which we provide a (non-exhaustive) list at the end of the book [2–8].

The chapter is structured as follows.

It starts with a detailed calculation of irreversible entropy production. From this expression, we show explicitly how to construct the constitutive laws describing the behaviour of the fluid. The first one is the *rheological law*, linking the viscous stress tensor and the deformation rate tensor; the second one is *Fourier's law*, coupling the heat flux to the temperature gradient. We show that an isotropic fluid possesses *two* viscosities, the most important being the *shear viscosity*, already discussed in Chapter 1. As a supplement, we describe two microscopic theories, conceptually very different, which yield the shear viscosity of gases (*kinetic theory*) and of liquids (*Eyring theory*) (Section 3.1).

We then recall the derivation of the equations of hydrodynamics, explicitly taking into account the constitutive laws, and we determine the *speed of sound*. We then deduce the expression of the *Mach number* and discuss the incompressibility assumption (Section 3.2).

The chapter continues with an important reminder on the *Reynolds number* and the concept of *similar flows* around identically shaped bodies (Section 3.3). In particular, we give an approximate solution to the problem of a *vorticity line source* immersed in a flow that is uniform at infinity (Section 3.4). This will provide a new interpretation of the Reynolds number and show that a clear distinction must be made between flows with high and low Reynolds numbers. More precisely, this example will show that, at high Reynolds numbers, the inertial effects dominate the viscous effects outside the *wake* of the body and of a thin surface layer, called '*boundary layer*'. The fluid can then be considered as *perfect*, and hence of zero viscosity outside these two regions, i.e. almost everywhere.

Thus justified, the study of flow in perfect fluids is presented in the next part (Section 3.5). After having established the two fundamental theorems, that of *Bernoulli* and the *Kelvin circulation theorem*, we describe some *plane potential flows*, and then the *vortices*. In particular, we describe the interaction between the latter before tackling the more complex problem of their stability as they are immersed in an external flow. We will see that alongside this study we can discuss the problem of *wing lift*, which is of fundamental importance in aerodynamics.

We then present (Section 3.6) *Prandtl's theory of the viscous boundary layer* appearing at the surface of bodies for high Reynolds numbers and we discuss the conditions under which it can *separate* from the surface (a crucial phenomenon for explaining the lift loss, or stall, of wings).

Then follows a part dealing with *flow at very low Reynolds numbers* (Section 3.7). In this case, the viscous effects dominate the inertial effects everywhere, so that the equations of motion can be linearised. A certain number of remarkable properties ensue (and are briefly recalled), most important among them being the *reversibility* of the flow as the velocity at the boundaries of the fluid changes direction. These results are then applied to some classical topics of low Reynolds number flow, such as *lubrication, microorganism propulsion* and the development of the *Saffman–Taylor instability* in the *Hele–Shaw cell* (a problem directly related to that of viscous digitation in porous media).

Finally, we describe some *rheometers* currently used for measuring the viscosity of liquids (rotating, capillary and piezoelectric rheometers and the surface force apparatus) (Section 3.8). The issue of fluids *sliding* at solid surfaces is also discussed.

Two appendices close the chapter. The first one gives a succinct treatment of some aspects of discontinuity surfaces in perfect fluids (Kelvin–Helmholtz instability and shock waves). In the second one, we present the equations of hydrodynamics expressed in cylindrical and spherical coordinates.

3.1 Rheological behaviour law and the Fourier law

As already stated in the previous chapter, one must calculate the irreversible entropy production in order to find the behaviour laws of a material. In this paragraph, we show explicitly how this is done in the particular case of an isotropic fluid. One must first perform a detailed energy balance, and then an entropy balance.

3.1.1 Energy and entropy balance

Let us start by deriving the energy equation for a moving fluid.

3.1.1.1 Energy equation

Let $e(\vec{r}, t)$ be the internal energy of the fluid per mass unit, $\vec{f}$ the external bulk forces and $\vec{q}$ the heat flux vector. Applying the *first principle of thermodynamics* to a moving volume V_m of surface S_m yields immediately:

$$\frac{d}{dt} \int_{V_m} \left(\frac{1}{2} v^2 + e \right) \rho \, dV = \int_{S_m} \underline{\sigma} \vec{v} \cdot \vec{v} \, dS + \int_{V_m} \vec{f} \cdot \vec{v} \, dV + \int_{S_m} (-\vec{q} \cdot \vec{v}) \, dS, \quad (3.1)$$

where on the left-hand side one can recognise the time derivative of the total energy (internal energy + kinetic energy) of the fluid contained in the volume V_m, and on the right-hand side the power of the external forces acting on the said volume as well as the heat exchanged with the environment by conduction across the mobile surface S_m.[1] By taking the derivative under the integral and transforming the surface integrals into volume integrals, knowing that the relation holds for any moving volume V_m, we obtain:

$$\rho \frac{D}{Dt}\left(\frac{1}{2}v^2 + e\right) = \frac{\partial}{\partial x_j}(\sigma_{ij} v_i) + \vec{f} \cdot \vec{v} - \operatorname{div}\vec{q}. \tag{3.2}$$

On the other hand, the scalar product of the equation of momentum conservation with $\vec{v}$ yields:

$$\rho \frac{D\vec{v}}{Dt} \cdot \vec{v} = v_i \frac{\partial \sigma_{ij}}{\partial x_j} + \vec{f} \cdot \vec{v}. \tag{3.3}$$

The *energy equation* is obtained through term by term subtraction of these two equations:

$$\rho \frac{De}{Dt} = \sigma_{ij} \frac{\partial v_i}{\partial x_j} - \operatorname{div}\vec{q}. \tag{3.4}$$

This equation can be rewritten in the following (Eulerian) form:

$$\frac{\partial}{\partial t}\left[\rho\left(\frac{1}{2}v^2 + e\right)\right] + \frac{\partial}{\partial x_j}Q_j = 0, \tag{3.5}$$

where $Q_j = \rho v_j(\frac{1}{2}v^2 + e) - v_i\sigma_{ij} + q_j$ is the *energy flux vector*. This equation expresses *energy conservation*.

The problem of energy conservation across a discontinuity surface will be discussed in Appendix 3.A.

3.1.1.2 Entropy equation

The equation of the entropy per unit mass s can be obtained by writing that in a quasi-static reversible transformation:

$$T\frac{Ds}{Dt} = \frac{De}{Dt} + P\frac{D(1/\rho)}{Dt} = \frac{De}{Dt} + \frac{P}{\rho}\operatorname{div}\vec{v}. \tag{3.6}$$

Finally, replacing De/Dt by its expression (3.4) yields:

$$\rho T \frac{Ds}{Dt} = \sigma_{ij}^v \frac{\partial v_i}{\partial x_j} - \operatorname{div}\vec{q}. \tag{3.7}$$

This *entropy equation* yields the *irreversible entropy production* (defined in Section 2.4):

$$\rho T\overset{\circ}{s} = \rho T\frac{Ds}{Dt} + T\operatorname{div}\left(\frac{\vec{q}}{T}\right) = \sigma_{ij}^v \frac{\partial v_i}{\partial x_j} - \frac{\vec{q}}{T} \cdot \overrightarrow{\operatorname{grad}}\,T = \sigma_{ij}^v A_{ij} - \frac{\vec{q}}{T} \cdot \overrightarrow{\operatorname{grad}}\,T. \tag{3.8}$$

---

[1] Note that we neglect here any possible heating by microwaves.

The last equality is a direct consequence of the symmetry of the stress tensor in isotropic liquids (see Section 2.3).

Thus expressed, $\rho T \overset{\circ}{s}$ allows us to obtain the general expression of the material laws for an isotropic liquid. Considering as '*forces*' the quantities σ_{ij}^v and $\overrightarrow{\text{grad}}\, T$ and as their '*conjugated fluxes*' the quantities A_{ij} and $\vec{q}$, and writing *linear* relations between the forces and the fluxes (according to the rules of the *thermodynamics of irreversible processes* (de Groot, 1983), one has the two following general laws:[2]

$$\vec{q} = -\chi\, \overrightarrow{\text{grad}}\, T, \tag{3.9}$$

$$\sigma_{ij}^v = \lambda_{ijkl} A_{kl}. \tag{3.10}$$

The first one is the *Fourier law*, connecting the heat flux and the temperature gradient. The thermal conductivity χ is always positive, as heat necessarily flows from the hot towards the cold (one must have $\rho T \overset{\circ}{s} > 0$, in particular for $\vec{v} = 0$).

The second equation relates the viscous stress to the deformation rate tensor (which is the symmetrical part of the distortion rate tensor $\partial v_i / \partial x_j$, see Eq. (2.37)). This dependence is easily understood by realising that the viscous stress must vanish when the fluid motion is a solid rotation with a velocity $\vec{v} = \vec{\Omega} \times \vec{r}$ (with constant $\vec{\Omega}$). The fourth-rank tensor λ_{ijkl} is the *tensor of viscosity coefficients*. It depends on the material and must reflect its *symmetries*. This latter property can lead to considerable reduction in the number of its independent components, as we will show in the case of isotropic liquids.

3.1.2 Tensor of viscosity coefficients for an isotropic liquid

We start our search for the general expression of the λ_{ijkl} tensor by noting that it must be symmetric with respect to ij and kl interchange, since the A_{ij} and σ_{ij}^v tensors have this property:

$$\lambda_{ijkl} = \lambda_{jikl} = \lambda_{ijlk}. \tag{3.11}$$

This condition reduces the number of distinct components to 36 (instead of 81), which is still considerable.

Isotropic liquids being centrosymmetric, changing x_1 to $-x_1$ (or x_2 to $-x_2$, or x_3 to $-x_3$) must make no difference, entailing that the only non-zero components of λ_{ijkl} are those where each index makes an even number of appearances. On the other hand, index transposition (for instance, changing 1 to 2 and 2 to 1), or circular permutation (namely, changing 1 to 2, 2 to 3 and 3 to 1) should also leave the system unchanged, reducing to three the number of distinct coefficients λ_{1111}, λ_{1122} and λ_{1212} (the other non-zero components are obtained by circular permutation and index transposition). Finally, nothing

---

[2] Note that these are the simplest relations one can think of, since they only contain linear terms. This amounts to assuming that the fluid is not too far from an equilibrium state. Failing that, nonlinearities can appear. Under normal flow conditions (and even in the turbulent regime), these terms are negligible in ordinary fluids.

should be changed by a 45° rotation in the $x_1 x_2$ plane. After rotation, the old coordinates are expressed in terms of the new ones as follows:

$$x_1 = \left(x_1' - x_2'\right)/\sqrt{2}, \qquad x_2 = \left(x_1' + x_2'\right)/\sqrt{2}. \tag{3.12}$$

Taking into account the symmetries used above, $x_1 x_2 x_1 x_2$ becomes:

$$\left(x_1' - x_2'\right)\left(x_1' + x_2'\right)\left(x_1' - x_2'\right)\left(x_1' + x_2'\right)/4 = \left(2x_1'x_1'x_1'x_1' - 2x_1'x_1'x_2'x_2'\right)/4 \tag{3.13}$$

yielding an additional relation between the coefficients:

$$\lambda_{1212} = (\lambda_{1111} - \lambda_{1122})/2. \tag{3.14}$$

Usually, one sets $\lambda_{1212} = \eta$ and $\lambda_{1122} = \zeta - (2/3)\eta$, leading to the following expression for the *viscous stress tensor* in an isotropic liquid:

$$\sigma_{ij}^{v} = 2\eta\left[A_{ij} - (1/3)\delta_{ij}A_{kk}\right] + \zeta\delta_{ij}A_{kk}. \tag{3.15}$$

Substitution in the general expression of dissipation confirms that the coefficients η and ζ are necessarily positive.

The coefficient η corresponds to the *shear viscosity*. It is the most common one, and the only one that matters as long as the fluid can be considered incompressible (this is the case at low Mach number, when the flow velocity is much lower than that of sound, as already discussed in Section 2.2.1).

The second coefficient ζ defines the *bulk viscosity*. It is the only coefficient involved during uniform compression or dilation of the fluid, as in this case $\sigma_{ij}^{v} = \zeta\delta_{ij}A_{kk}$. This viscosity becomes relevant at high Mach numbers, when the flow can no longer be considered as incompressible. Its consequence is that the fluid does not return to equilibrium immediately after sudden compression or dilation, due to energy dissipation. In particular, this viscosity contributes to sound damping.

If the isotropic fluid can be assumed *incompressible*, only the shear viscosity η counts, and the irreversible entropy production reduces to the simple form:

$$\rho T \overset{\circ}{s} = \frac{\eta}{2}\left(\frac{\partial v_i}{\partial x_j} + \frac{\partial v_j}{\partial x_j}\right)^2 + \frac{\chi}{T}\left(\overrightarrow{\text{grad }T}\right)^2. \tag{3.16}$$

The first term in this expression corresponds to the energy dissipated by viscous friction within the fluid (per unit volume). We will call it *viscous dissipation* and denote it by Φ in the following. The second term is associated with heat transfer by conduction.

3.1.3 Macroscopic interpretation of the shear viscosity

We intend to show that the shear viscosity is associated with the transport of momentum.

To this end, let us consider the motion of a fluid in contact with a plate that is suddenly set in motion (Fig. 3.1). We assume that the temperature is constant (isothermal system).

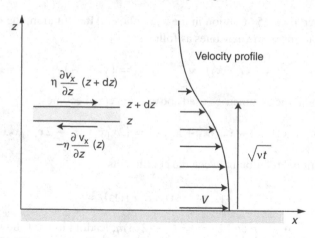

Fig. 3.1 Velocity profile observed after the plate is suddenly set in motion at the velocity V and bulk forces acting on a liquid layer of thickness dz.

The fluid occupies the half-space $z > 0$ and the surface of the plate is in the plane $z = 0$. At time $t = 0$, the plate is suddenly set in motion with the velocity $V \neq 0$ parallel to the x axis. This action exerts a shear stress on the fluid, which progressively starts to flow, as shown in Fig. 3.1.

To calculate the velocity profile $v_x(z, t)$ in the fluid at later times, let us first search for the differential equation of motion. To this end, it is enough to write the balance of forces (per unit surface) acting on the liquid layer delimited by z and $z + dz$ and to apply Newton's law to this slice, yielding:

$$\frac{\partial(\rho \, dz \, v_x)}{\partial t} = \eta \frac{\partial v_x}{\partial z}(z + dz) - \eta \frac{\partial v_x}{\partial z}(z). \tag{3.17}$$

If the density of the fluid ρ is constant, this equation becomes:

$$\frac{\partial \rho \, v_x}{\partial t} = v \frac{\partial^2 \rho \, v_x}{\partial z^2}. \tag{3.18}$$

where $v = \eta/\rho$. We recognise here a *diffusion equation for the momentum* ρv_x. The v coefficient, bearing the name *kinematic viscosity*, can thus be identified as the diffusion coefficient of momentum along the normal to the plate. It is measured in $[L]^2[T]^{-1}$ units.

Using this equation, the problem stated above can be solved, provided one knows the boundary conditions. We take a no-slip condition at the solid surface (an assumption to be discussed in Section 3.8.5):

$$v_x(z = 0, \, t > 0) = V. \tag{3.19}$$

The velocity far away in the fluid is zero:

$$v_x(z = \infty, \, t > 0) = 0. \tag{3.20}$$

Solving Eq. (3.18) with these boundary conditions yields:

$$v_x = V \left[1 - \mathrm{erf}\left(\frac{z}{2\sqrt{vt}} \right) \right],$$

(3.21)

where $\mathrm{erf}(u) = \frac{2}{\sqrt{\pi}} \int_0^u \exp(-t^2)\, dt$ is the error function (varying from 0 to 1 as u goes from 0 to $+\infty$). The velocity profile observed at a given time is drawn in Fig. 3.1. Note that the profiles obtained at different times are *similar*, i.e. they can be obtained from one another by a scale change along z. In other words, the 'scaled' velocity profile v_x/V has the same shape at all times when plotted as a function of the scaled coordinate $z/(vt)^{1/2}$.

3.1.4 Microscopic calculation of the shear viscosity

In the experiment above, the fluid transmits x momentum along the z direction perpendicular to the velocity. This flux is directed opposite to the velocity gradient.

It follows that the viscous force $F_{1 \to 2}$ exerted by a fluid layer (1) on the next one (2) equals the total momentum flux across the surface P separating them. Note that the flux is counted as positive when it occurs from (1) towards (2), and as negative in the opposite direction.

We will use the above result and the kinetic theory of gases to calculate the viscosity of a perfect gas.

3.1.4.1 Viscosity of a perfect gas

Here we adopt a molecular point of view. Assume a pure gas composed of spherical molecules with diameter d and mass m (hard spheres) and let n be their number density.

Let $\vec{v}(z)$ be the macroscopic velocity (Fig. 3.2). By definition, this velocity is an ensemble velocity, representing the average of the molecular velocities around a point situated at height z. We assume it to be very small compared to the thermal velocity, such that, *in the*

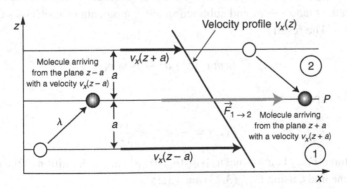

Fig. 3.2 Molecular transport of x momentum from the planes at heights $z - a$ and $z + a$ towards the plane at height z.

reference frame moving at the velocity $\vec{v}$, the gas can be considered to be in thermal equilibrium (hypothesis 1). Under these conditions, the results of the kinetic theory of gases are valid in this reference frame. Here are some of them, enumerated without demonstration:

(1) The molecules have a mean (absolute) velocity:

$$\overline{v} = \sqrt{\frac{8k_{B}T}{\pi m}},$$

(3.22)

where k_B is the Boltzmann constant. Note that this velocity is different from the mean square velocity v_m given by the relation $1/2mv_{m}^{2} = 3/2k_{B}T$.

(2) The frequency of molecular bombardment on one side of a generic surface exposed to the gas is:

$$f_b = \frac{1}{4}n\overline{v}.$$

(3.23)

(3) The mean distance covered by a molecule between successive collisions, also termed the *mean free path*, λ is:

$$\lambda = \frac{1}{\sqrt{2}\pi d^{2}n}.$$

(3.24)

(4) On average, the molecules reaching a given plane have had their latest collision in a plane a distance a away, such that

$$a = \frac{2}{3}\lambda.$$

(3.25)

Using these results, and *assuming that the molecules 'thermalise' instantly with their environment at each collision* (hypothesis 2, implying that it acquires the same average properties (translation and rotation velocities, vibration states, etc.) as its neighbours), we can calculate the viscous force exerted by layer 1, situated below the z plane, on layer 2 above it.

This force $F_{1\rightarrow2}$ is obtained by summing over the (average) x momenta of the molecules arriving from the plane $z - a$, and subtracting the x momenta of molecules coming from the plane $z + a$. The result is:

$$F_{1\rightarrow2} = f_b\, m\, v_x(z - a) - f_b\, m\, v_x(z + a)$$

(3.26)

or, after Taylor expansion:

$$F_{1\rightarrow2} = -2a\, f_b\, m\frac{\partial v_x}{\partial z}.$$

(3.27)

The viscous force $F_{1\rightarrow2}$ being equal to $-\eta(\partial v_x/\partial z)$ as per the definition of the macroscopic viscosity η, one finds, using Eqs. (3.23) and (3.25):

$$\eta = \frac{1}{3}n\, m\, \overline{v}\, \lambda = \frac{m\overline{v}}{3\sigma},$$

(3.28)

where we introduced the effective collision cross-section of the molecules $\sigma = 1/(n\lambda)$, equal to $\pi\sqrt{2}d^2$ according to Eq. (3.24). This equation was obtained by Maxwell in 1860. Combining it with Eq. (3.22) yields:

$$\eta = \frac{2}{3\pi^{3/2}} \frac{\sqrt{mk_B T}}{d^2}. \tag{3.29}$$

The formula above shows that the kinematic viscosity of a perfect gas is independent of its density, and hence of its pressure. This prediction is well verified experimentally, up to pressure values of the order of 10 atm.

At this point, it is instructive to present some orders of magnitude for the parameters used above. Thus, for oxygen at standard temperature and pressure, $d \approx 0.36$ nm, $\sigma \approx 0.57$ nm^2, $n \approx 2.5 \times 10^{25}$ m^{-3} (corresponding to an average intermolecular distance of 3.4 nm), $\lambda \approx 70$ nm, $m = 5.3 \times 10^{-26}$ kg and $\bar{v} \approx 450$ m/s, yielding $\eta = 1.4 \times 10^{-5}$ Pa s, in relatively good agreement with the experimentally measured value. This viscosity is typically 100 times lower than that of water. Note also that the mean free path is about 20 times larger than the average distance between molecules, itself 10 times larger than the molecular diameter.

The ensuing prediction of the temperature dependence is however less satisfactory, the viscosity increasing in general faster than $T^{1/2}$. One way of obtaining a more accurate prediction for the viscosity is to replace the hard-sphere interaction potential used in the perfect gas model by a more realistic one (of the Lennard-Jones kind, for instance). This type of potential, exhibiting a short-range attraction, leads to an increase of the collision cross-section σ. Obviously, this effect becomes more pronounced as the molecules move more slowly, i.e. as the temperature decreases. As a result, σ decreases with increasing temperature, leading to a viscosity increase that is steeper than $T^{1/2}$.

This type of behaviour is not observed in liquids, where the viscosity almost always decreases with increasing temperature, according to a thermal activation law. As a result, the kinetic theory cannot explain the viscosity of liquids, in particular because it ignores collisions involving more than two particles (the mean free path is assumed much larger than the particle size, which is completely false in liquids), as well as the direct interactions between molecules close to the plane P (Fig. 3.2). A change of theory is in order.

3.1.4.2 The case of simple liquids

As stated above, the kinetic theory is no longer applicable in the case of liquids, which are more similar to solids than gases at the molecular scale. The theory of liquids being much less developed than that of gases, most models for their viscosity are largely empirical. Historically, metallurgist E. N. da C. Andrade was the first to notice that the viscosity of liquids varies as $A \exp(b/T)$ (with constant A and b) over wide temperature ranges (Andrade, 1930). However, it was the theoretical chemist H. Eyring who grounded this formula by proposing the first realistic model for the viscosity of liquids, based on statistical mechanics. We will now present this model, which provides a prediction for the order of magnitude of the viscosity (Eyring, 1936) (see also Glasstone *et al.* (1941) and [2, 25]).

In a pure liquid, the molecules are in constant motion. A molecule can thus diffuse over large distances, given sufficient time. Its short-time motion, on the other hand, is restrained to vibrations within a '*cage*' formed by its nearest neighbours. Imagine now that a 'hole' forms immediately near the cage. The molecule trapped in its cage will be able to 'jump' into the hole, enabling it to move through the liquid. This diffusion process (also present in solids, see Section 5.1.1.2) is thermally activated and occurs, for each molecule, with a frequency k given by the theory of absolute reaction rates (Eyring, 1935):

$$k = \frac{k_B T}{h} e^{-\Delta g_0 / k_B T},$$ (3.30)

where h is the Planck constant, k_B the Boltzmann constant and Δg_0 an energy barrier (including the formation energy of a hole, and thus the probability of finding a hole next to the molecule). Clearly, at equilibrium, jumps have an equal probability of occurring in any space direction.

Suppose the liquid flows in the x direction with the velocity gradient $\partial v_x / \partial z$. Under these conditions, the molecules will no longer jump leftwards and rightwards along the x axis with the same probability. To quantify this effect, we place ourselves in the reference frame of a layer (A, for instance) and calculate the frequencies of 'rightwards' k^+ and 'leftwards' k^- jumps of a molecule contained in an adjacent layer, B (Fig. 3.3). To this end, we need to determine the new energy barriers corresponding to each type of jump. They can be obtained by subtracting from the equilibrium energy barrier Δg_0 the work of the viscous shear stress σ_{xz}^v as the molecule moves from the bottom of the well to the top of the barrier. For a jump to the right we obtain, denoting by a the distance between nearest-neighbour molecules:

$$\Delta g^+ = \Delta g_0 - \sigma_{xz}^v \frac{a^3}{2},$$ (3.31)

yielding the jump frequency:

$$k^+ = \frac{k_B T}{h} e^{-\Delta g^+ / k_B T}.$$ (3.32)

For a jump to the left, the same reasoning gives:

$$\Delta g^- = \Delta g_0 + \sigma_{xz}^v \frac{a^3}{2},$$ (3.33)

whence

$$k^- = \frac{k_B T}{h} e^{-\Delta g^- / k_B T}.$$ (3.34)

In conclusion, the molecules in layer B move at an average velocity:

$$v = a(k^+ - k^-)$$ (3.35)

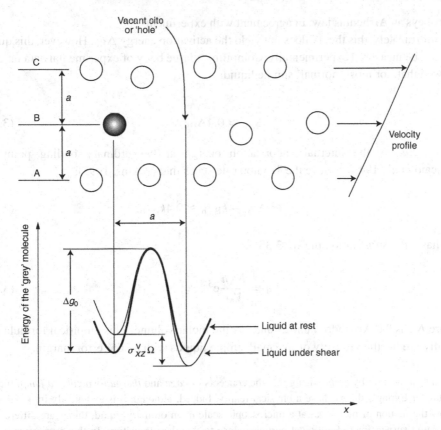

Fig. 3.3 Illustration of a thermally activated jump process in a liquid. The molecule must cross a kind of bottleneck before it can reach the vacant site. Under shear, the energy barrier that the particle must cross becomes asymmetric.

with respect to the molecules in layer A. At the scale of the interlayer distance the velocity profile is linear, such that:

$$v = a \frac{\partial v_x}{\partial y} \tag{3.36}$$

and we deduce:

$$\frac{\partial v_x}{\partial y} = k^+ - k^- = 2 \frac{k_B T}{h} e^{-\Delta g_0 / k_B T} \sinh\left(\frac{\sigma \Omega}{2 k_B T}\right), \tag{3.37}$$

with $\Omega = a^3$ the molecular volume. If the shear stress is not too high, the hyperbolic sine can be linearised, resulting in Newtonian behaviour with a viscosity:

$$\eta = \frac{h}{\Omega} e^{\Delta g_0 / k_B T} \tag{3.38}$$

that obeys an Arrhenius law, in agreement with experiments.[3]

Unfortunately, this theory does not yield the activation energy Δg_0. However, this quantity can be measured experimentally. Compiling a large body of experimental data on Δg_0 showed that, for most 'normal' simple liquids:

$$\Delta g_0 \approx 0.4 \Delta u_0, \tag{3.39}$$

where Δu_0 is the internal vaporisation energy at the ordinary boiling point T_b (Kincaid *et al.*, 1941).[4] Since the Trouton rule states that (Atkins, 1995):

$$\Delta u_0 \approx \Delta h_0 - k_B T_b \approx 9.4 k_B T_b, \tag{3.40}$$

one has, after substitution in Eq. (3.38):

$$\eta = \frac{N_A h}{V_M} e^{3.8 T_b / T}, \tag{3.41}$$

where N_A is the Avogadro number and V_M the molar volume. This empirical formula can usually predict the viscosity of 'normal' simple liquids with a 30% error margin.

We end this section by emphasising that the concepts of *cage* and *thermally activated jump*, fundamental for Eyring's theory, have a physical reality. Indeed, although it is technically impossible to follow the motion of molecules at a microscopic scale in an ordinary liquid, things are different in disordered suspensions of colloidal particles close to the glass transition. In this case, after marking the particles with fluorescent molecules, one can follow their motion by confocal microscopy (Weeks and Weitz, 2002).

In particular, one can follow the trajectory of a single particle, as shown in Fig. 3.4. This observation reveals that each particle spends most of its time within the cage formed by its neighbours, and it only moves significantly when it jumps to an adjacent cage. This experiment confirms the basis of Eyring's theory for the viscosity of simple liquids, and is undoubtedly the explanation for its success, in spite of its relative simplicity.

3.2 The equations of hydrodynamics

In this section, we use the general results proved in the previous chapter.

---

[3] One can also encounter 'anomalous' liquids, where Δg_0 decreases with increasing temperature. This occurs when the molecules in the liquid possess OH or NH functions, e.g. water, H_2O, or ethylamine, $C_2H_5NH_2$. In water, for instance, Δg_0 goes from 5060 cal/mol at 0 °C to 2800 cal/mol at 100 °C. This behaviour is due to the fact that the number of hydrogen bonds that must be broken when water flows decreases with increasing temperature. The presence of hydrogen bonds also explains why water, H_2O, or $C_2H_5NH_2$ are much more viscous than their 'normal' counterparts H_2S or ethane, C_2H_6, forming only dipolar and van der Waals bonds.

[4] In liquid metals, the volume of an atom is much larger than that of the corresponding ion (namely, of the atom stripped of its valence electrons). For this reason, the ratio between Δg_0 and Δu_0 is much smaller than the value given in relation (3.39). For more details see Ewell and Eyring (1937). We will therefore classify liquid metals among 'anomalous' liquids.

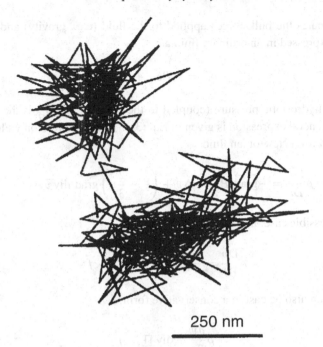

250 nm

Fig. 3.4 Trajectory of a latex bead in a disordered suspension with concentration $\phi = 0.52$ recorded over 120 min. The jump dynamics is very slow, since ϕ is close to the glass transition concentration $\phi_g = 0.58$ (image from Weeks and Weitz, 2002).

3.2.1 Mass conservation equation

We recall that it is written:

$$\frac{\partial \rho}{\partial t} + \text{div}(\rho \vec{v}) = 0 \tag{3.42}$$

or, equivalently:

$$\frac{D\rho}{Dt} + \rho \,\text{div}\, \vec{v} = 0. \tag{3.43}$$

In the limiting case of *incompressible liquids*, $\text{div}\,\vec{v} = 0$.

The problem of mass conservation across a discontinuity surface will be discussed in Appendix 3.A.

3.2.2 Momentum conservation equation (Navier–Stokes)

With complete generality, this equation can be written (see Section 2.2.2):

$$\rho \frac{D\vec{v}}{Dt} = \text{div}\,\underline{\sigma} + \vec{f}, \tag{3.44}$$

where $\vec{f}$ designates the bulk forces applied to the fluid (e.g., gravity) and $\underline{\sigma}$ is the total stress tensor, expressed in an ordinary fluid as:

$$\sigma_{ij} = -P\delta_{ij} + \sigma_{ij}^{\text{v}}. \tag{3.45}$$

Here, P is the hydrostatic pressure (coupled to the density) and σ_{ij}^{v} is the viscous stress tensor, whose general expression is given in Eq. (3.15). This expression yields the Navier–Stokes equation for a Newtonian fluid:

$$\rho \frac{\mathrm{D}\vec{v}}{\mathrm{D}t} = -\overrightarrow{\mathrm{grad}}\, P + \eta\,\Delta\vec{v} + \left(\xi - \frac{2}{3}\eta\right)\overrightarrow{\mathrm{grad}}\,\mathrm{div}\,\vec{v} + \vec{f}. \tag{3.46}$$

In the incompressible case, this equation reduces to:

$$\rho \frac{\mathrm{D}\vec{v}}{\mathrm{D}t} = -\overrightarrow{\mathrm{grad}}\, P + \eta\,\Delta\vec{v} + \vec{f}. \tag{3.47}$$

This equation can also be cast in a conservative form:

$$\rho \frac{\partial\vec{v}}{\partial t} + \mathrm{div}\,\underline{\Pi} = \vec{f} \tag{3.48}$$

with $\Pi_{ij} = \rho v_i v_j - \sigma_{ij}$ the *momentum flux density tensor*, symmetric in the case of a Newtonian fluid.

The reader can also check that, for any *fixed* volume V with surface S within the fluid, one must have (*Euler's momentum theorem*):

$$\left[\frac{\partial\rho\,\vec{v}}{\partial t}\right]_V + \left[P\vec{v} + \rho\vec{v}\,(\vec{v}\cdot\vec{v})\right]_S = \left[\vec{f}\right]_V. \tag{3.49}$$

Here $\left[\vec{a}\right]_V$ stands for the torsor engendered by the bulk density $\vec{a}$ in V, i.e. having as reduction elements:[5]

$$\vec{R} = \int_V \vec{a}\,\mathrm{d}V \quad\text{and}\quad \vec{M}_{\mathrm{O}} = \int_V \left(\overrightarrow{OM}\times\vec{a}\right)\mathrm{d}V. \tag{3.50}$$

The same definition applies for $\left[\vec{a}\right]_S$ and the surface density $\vec{a}$. Note that this theorem only applies if the stress tensor σ_{ij} is symmetric.

The problem of momentum conservation across a discontinuity surface will be discussed in Appendix 3.A.

3.2.3 Boundary conditions

Boundary conditions depend on the nature of the surface in contact with the fluid.

---

[5] We recall that a torsor is a moment vector field $\vec{M}_{\mathrm{P}}$ which satisfies, at any point P, the relation $\vec{M}_{\mathrm{P}} = \vec{M}_{\mathrm{O}} + \vec{R}\times\overrightarrow{OP}$.

For a *solid impermeable surface*, one generally writes that the velocity of the fluid equals that of the solid:

$$\vec{v} = \vec{v}^S \tag{3.51}$$

at each point of the surface. Experimentally, this condition is very well verified for simple fluids. However, it does not apply for certain polymer melts of high molecular mass, which can exhibit significant glide at the surface (see Section 3.8.5), up to several μm.

Another important case is that of two *immiscible fluids* in contact. One usually writes (the same caveat applies concerning polymers) that the two fluids have the same velocity at each point of the interface:

$$\vec{v}^{(1)} = \vec{v}^{(2)}. \tag{3.52}$$

To this boundary condition one must add the equilibrium equation for the stress forces acting on either side of the interface. This equation results from applying Newton's law to a pillbox of unit surface, centred on the interface and of vanishing width h (Fig. 3.5). The mass of the box being zero, the inertial force cancels, yielding immediately:

$$\underline{\sigma}^{(2)}\vec{v} = \underline{\sigma}^{(1)}\vec{v} \tag{3.53}$$

with $\vec{v}$ the unit vector normal to the interface, pointing, say, from (1) towards (2).

This equation is actually incomplete, insofar as it ignores the capillary force, related to the surface tension between the two fluids. This force, acting on the surface, is directed along $\vec{v}$ and its magnitude (per unit surface) is $-\gamma C \vec{v}$, where $C = 1/R_1 + 1/R_2$ is the total curvature of the interface (Fig. 3.6). With this additional force, Eq. (3.53) becomes:

$$\underline{\sigma}^{(2)}\vec{v} - \underline{\sigma}^{(1)}\vec{v} - \gamma C \vec{v} = 0. \tag{3.54}$$

Note that each radius of curvature, R_1 or R_2, is counted algebraically (negative if the corresponding centre of curvature is in the medium to which points $\vec{v}$, and positive otherwise).

We will also need in the following to express the impermeability condition on a deformable surface of equation $f(\vec{r}, t) = 0$. This surface may be, for example, a membrane or a liquid–gas interface. This condition is obtained by taking the material derivative of f

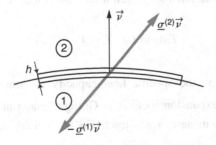

Fig. 3.5 Equilibrium of stresses at the interface between two immiscible fluids (neglecting the capillary force).

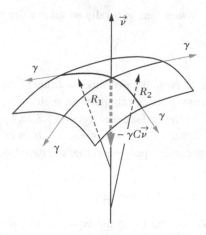

Fig. 3.6 Capillary force acting on the surface between two immiscible liquids. In this drawing, R_1 and R_2 are positive.

and writing that $(\mathrm{d}\vec{r}/\mathrm{d}t) \cdot \vec{n} = \vec{v} \cdot \vec{n}$ with $\vec{n} = \overrightarrow{\mathrm{grad}}\,f$ the vector normal to the surface and $\vec{v}$ the velocity of the liquid. This gives the kinematic equation:

$$v_i \frac{\partial f}{\partial x_i} + \frac{\partial f}{\partial t} = 0. \tag{3.55}$$

3.2.4 Heat equation

The equations obtained so far only give access to the velocity and stress fields within the fluid. However, they do not yield its local temperature, which is not uniform in general, due to viscous dissipation. In this paragraph, we will establish a new equation for calculating the temperature field.

The starting point is the entropy equation (3.7). This is not very useful by itself, since the entropy is not an experimentally accessible quantity, as opposed to the temperature. To obtain the temperature (or heat) equation, let us apply the second Clapeyron relation to a set of particles with unit mass that we follow along their motion. It yields:

$$T\,\mathrm{d}s = C_\mathrm{p}\,\mathrm{d}T - \beta \frac{T}{\rho}\mathrm{d}P, \tag{3.56}$$

where $C_\mathrm{p} = T\left(\frac{\partial s}{\partial T}\right)_P$ is the specific heat capacity at constant pressure and $\beta = \rho\left(\frac{\partial (1/\rho)}{\partial T}\right)_P$ the thermal expansion coefficient. Going to material derivatives in Eq. (3.56) and replacing $T\mathrm{D}s/\mathrm{D}t$ by its new expression in Eq. (3.7) yields the temperature equation:

$$\frac{\mathrm{D}T}{\mathrm{D}t} - \frac{\beta T}{\rho C_\mathrm{p}}\frac{\mathrm{D}P}{\mathrm{D}t} = \frac{\Phi}{\rho C_\mathrm{p}} + \Lambda\Delta T, \tag{3.57}$$

where Φ is the dissipated power, of expression:

$$\Phi = \sigma_{ij}^v \frac{\partial v_i}{\partial x_j} = \frac{\eta}{2} \left(\frac{\partial v_i}{\partial x_j} + \frac{\partial v_j}{\partial x_i} \right)^2 + \left(\zeta - \frac{2\eta}{3} \right) (\mathrm{div}\, \vec{v})^2 \qquad (3.58)$$

and

$$\Lambda = \frac{\chi}{\rho C_\mathrm{p}}, \qquad (3.59)$$

the heat diffusion coefficient.

Note that Eq. (3.57) reduces to an ordinary diffusion equation when *the fluid is at rest* ($\Phi = 0$ and $P = \mathrm{Cst}$):

$$\frac{\partial T}{\partial t} = \Lambda\, \Delta T. \qquad (3.60)$$

Finally, remember that the heat equation must be supplemented by boundary conditions in order to determine the temperature field. In practice, the temperature or the heat flux are imposed at the boundaries of the fluid.

Note that Eq. (3.57) contains a negligible pressure term if the fluid can be considered incompressible. This is certainly the case for convection experiments performed in the lab using simple liquids. On the other hand, this term becomes very important in geophysical flows, such as convection in the Earth's mantle or core. Consider the example of the mantle.[6] In its convective region, the pressure varies from 7 GPa to 130 GPa and the density increases from 3.4 to about 5.6 (see Table 4.2, page 222). Assuming that each fluid element of the mantle rises fast enough that it does not exchange heat with its environment and that viscous dissipation can be neglected, one can see from Eq. (3.57) that an 'adiabatic temperature gradient' sets in, given by:

$$\overrightarrow{\mathrm{grad}}\, T = \frac{\beta T}{\rho C_\mathrm{p}} \overrightarrow{\mathrm{grad}}\, P. \qquad (3.61)$$

This equation is nothing more than the *adiabatic expansion law*, the same phenomenon as when a gas cools down as it flows out of a pressurised container. Equation (3.61) predicts a typical temperature variation from 1700 to 2500 °C within the asthenospheric mantle. Since it is about 2800 km thick, this amounts to an average adiabatic gradient of 0.3 °C/km.[7] The adiabatic gradient also plays a role in atmospheric convection, its order of magnitude being 10 °C/km (in dry atmosphere).

We should add that in the classical Rayleigh–Bénard problem, dealing with the convective instability of a fluid heated from below and contained between two horizontal surfaces, the instability

---

[6] The ductile convective mantle is situated under the *lithosphere*. The latter constitutes the top rigid part of the mantle, situated just below the crust. Its estimated thickness is between 100 and 200 km. The temperature at the bottom of the lithosphere – which behaves in a first approximation as a thermal boundary layer in mantle convection – is about 1700 °C. Another thermal boundary layer exists between the bottom of the mantle and the liquid core. About 200 km thick, it is called the 'D'' layer'. Within it, the temperature varies from 2500 to 4000 °C, typically. It is noteworthy that the temperature gradient in these layers is 30 to 50 times larger than in the rest of the mantle. We should also mention that in the geologist's parlance *asthenosphere* designates the upper region of the convective mantle.

[7] One can prove that this estimate is accurate at high Rayleigh numbers ($Ra \gg Ra_\mathrm{c}$) or, equivalently, at a high Nusselt number Nu [4]. We recall that this latter dimensionless quantity is given by the ratio between the convective heat flux and the conductive heat flux. In mantle convection, a good estimate is $Ra \approx 10^7$ and $Nu \approx 15$ (knowing that $Nu \approx (Ra/Ra_\mathrm{c})^{1/3}$) (Turcotte and Schubert, 2002).

criterion is the same for compressible and for incompressible fluids, provided the adiabatic gradient is subtracted from the applied temperature gradient.[8] Thus, the relevant parameter for convection is the *Rayleigh number*, given by:

$$Ra = \frac{\beta G g d^4}{\nu \Lambda},$$

(3.62)

where G is the temperature gradient minus the adiabatic gradient, g the gravitational acceleration[9] and d the thickness of the fluid layer [4]. Physically, this number quantifies the competition between a destabilising driving force, the buoyancy (proportional to βg), and two stabilising damping terms, the viscosity ν and the thermal diffusion Λ. More specifically, Ra is the ratio between the heat diffusion time over the distance d: $t_{\text{diff}} = d^2/\Lambda$ and the advection time for a fluid element of unit volume that, after having expanded close to the lower surface, migrates to the upper surface under the action of the buoyancy: $t_{\text{adv}} = d/V$. The velocity V is obtained by balancing in the Navier–Stokes equation the buoyancy $\rho g \beta \Delta T$ and the viscous damping force $\eta V/d^2$, resulting (up to a prefactor) in $V = g\beta \Delta T d^2/\nu$, and then in $t_{\text{adv}} = \nu/(g \beta G d^2)$.

Note, however, that convection in the Earth's mantle is different from the usual case insofar as the heat comes not only from an external source (here, the slowly cooling core of the Earth), but mostly from the bulk disintegration of radioactive elements such as uranium, thorium or potassium. It is therefore more realistic to treat the problem of mantle convection as for a fluid layer heated from within, cooled from above and sitting atop an isolating surface. To evacuate heat, a temperature difference ΔT across the layer is needed, such that $\rho Q d = \chi \Delta T/d$, where Q is the amount of heat produced in the bulk per unit mass and time.[10] This temperature difference, calculated in the diffusive regime, plays the same destabilising role as in classical convection. In this case, the relevant Rayleigh number is obtained by replacing G with its value $\rho Q d/\chi$, yielding:

$$Ra = \frac{\beta g \rho Q d^5}{\nu \Lambda \chi}.$$

(3.63)

It can be shown that, in all cases, the fluid starts convecting as Ra exceeds a critical threshold Ra_c of the order of 2000.[11]

3.2.5 Speed of sound and the Mach number

All equations describing fluid motion being known, we are now able to calculate the speed of sound in an isotropic fluid. Knowledge of this velocity is important for justifying *ex post facto* the fluid incompressibility assumption usually employed in rheophysics (and in particular throughout this book[12]).

By definition, *a sound wave is a small-amplitude vibrating motion in a compressible fluid*. Under these conditions, one can in principle neglect the nonlinear advection terms in dynamical equations. Let us detail this important point. Let a be the amplitude of motion,

---

[8] This result is rigorously true only in the case of constant adiabatic gradient.

[9] Note that g is almost constant ($\sim 10\,\text{m}^2/\text{s}$) in Earth's mantle, since the density increases towards the centre of the Earth. More precisely, one estimates that g goes through a maximum at the interface between the mantle and the core, with a value of about $10.5\,\text{m}^2/\text{s}$. Then, the gravitational acceleration g decreases almost linearly and vanishes at the centre of the Earth.

[10] In the mantle, $Q \approx 10^{-11}$ W/kg (Turcotte and Schubert, 2002).

[11] The exact value of Ra_c depends on the boundary conditions [2, 4] and the heating mode (Turcotte and Schubert, 2002).

[12] Except for Appendix 3.A, where we will speak of shock waves in compressible fluid, and for Sections 6.2.3, 8.2 and 8.3.

ω the angular velocity of the wave and λ its wavelength. As an order of magnitude, we obtain:

$$\frac{\partial \vec{v}}{\partial t} \approx \omega^2 a, \tag{3.64}$$

$$\left(\vec{v} \cdot \vec{\nabla}\right) \vec{v} \approx \frac{\omega^2 a^2}{\lambda}. \tag{3.65}$$

Thus, one can neglect the second term compared to the first if $a \ll \lambda$, or if the velocities involved in the oscillating motion are small compared to the speed of sound: $v \approx \omega a / 2\pi \ll c = \omega\lambda / 2\pi$.

Assuming these conditions are fulfilled, let P_0 and ρ_0 be the equilibrium pressure and density of the fluid, respectively. In the presence of a wave, these quantities vary, becoming:

$$P = P_0 + P' \quad \text{with} \quad P' \ll P_0, \tag{3.66}$$

$$\rho = \rho_0 + \rho' \quad \text{with} \quad \rho' \ll \rho_0. \tag{3.67}$$

Under these conditions, the mass conservation and Navier–Stokes equations are written:

$$\frac{\partial \rho}{\partial t} + \text{div}(\rho \vec{v}) = 0, \tag{3.68}$$

$$\frac{\partial \vec{v}}{\partial t} = -\frac{1}{\rho} \overrightarrow{\text{grad}} \, P, \tag{3.69}$$

where we neglected the viscosities (which control wave damping). Linearising these equations around the equilibrium state (fluid at rest) yields the following equations for the perturbations:

$$\frac{\partial \rho'}{\partial t} + \rho_0 \, \text{div} \, \vec{v} = 0, \tag{3.70}$$

$$\frac{\partial \vec{v}}{\partial t} = -\frac{1}{\rho_0} \overrightarrow{\text{grad}} \, P'. \tag{3.71}$$

In the frequency range typical for sound waves, the motion can be considered *isentropic*, i.e. with $s(P, \rho) = \text{Cst}$, or $P = P(\rho)$. Expanding this equation around the equilibrium state yields:

$$P\left(\rho_0 + \rho'\right) = P_0 + \left(\frac{\partial P}{\partial \rho}\right)_s \rho', \tag{3.72}$$

resulting in:

$$P' = \left(\frac{\partial P}{\partial \rho}\right)_s (\rho = \rho_0) \, \rho'. \tag{3.73}$$

Taking the time derivative of this equation and using the continuity equation gives:

$$\frac{\partial P'}{\partial t} + \rho_0 \left(\frac{\partial P}{\partial \rho}\right)_s \text{div} \, \vec{v} = 0. \tag{3.74}$$

Derivating once again, and using this time the Euler equation (3.71), yields a propagation equation for the pressure perturbation:

$$\frac{\partial^2 P'}{\partial t^2} - \left(\frac{\partial P}{\partial \rho}\right)_s \Delta P' = 0. \tag{3.75}$$

It is easily shown that ρ satisfies the same equation.

This equation shows that sound propagates at a speed:

$$c = \sqrt{\left(\frac{\partial P}{\partial \rho}\right)_s}, \tag{3.76}$$

where the quantity under the square root is the *adiabatic compressibility* of the fluid. This formula can also be written in the form:

$$c = \sqrt{\gamma \left(\frac{\partial P}{\partial \rho}\right)_T}, \tag{3.77}$$

involving this time the *isothermal compressibility* and the ratio of the specific heat constants at constant pressure and volume ($\gamma = C_p/C_v$). In a *perfect gas*, with equation of state $P/\rho = RT/M$, where R is the perfect gas constant and M the molar mass, the speed of sound is:[13]

$$c = \sqrt{\gamma \frac{RT}{M}}. \tag{3.78}$$

It is also noteworthy that the sound wave is longitudinally polarised ($\vec{v} \| \vec{k}$), as one can see immediately using the Euler equation (3.71), with:

$$v = c \frac{\rho'}{\rho_0}. \tag{3.79}$$

One can also show that the local temperature field is given by the expression [3]:

$$T' = \frac{c\beta T_0}{C_p} v, \tag{3.80}$$

where β is the thermal dilation coefficient:

$$\beta = -\frac{1}{\rho}\left(\frac{\partial \rho}{\partial T}\right)_P. \tag{3.81}$$

To conclude this paragraph, let us consider a generic flow with a typical velocity V around a body of size L. One can consider the fluid as incompressible for this flow, if the velocity V is small compared to the speed of sound. This amounts to saying that the information

---

[13] Note that c in a gas is of the same order of magnitude as the mean velocity of the molecules given by Eq. (3.22).

propagates almost instantaneously in the fluid, the advection time L/V characteristic of the flow being very long compared to the sound propagation time L/c.

Thus, a fluid can be considered as *incompressible* if the Mach number of the flow under consideration is much smaller than 1:

$$Ma = \frac{V}{c} \ll 1. \tag{3.82}$$

We will assume this condition to be fulfilled throughout the rest of this chapter, except in Appendix 3.A, where we will give a few indications on shock waves and supersonic flow.

3.2.6 Dissipation theorem: drag force and torque on a moving solid

In this subsection, we will analyse the motion of a solid body in an incompressible ($M \ll 1$) viscous fluid, quiescent at infinity. We neglect the bulk forces $\vec{f}$.

We intend to prove that the drag force $\vec{F}^{\mathrm{d}}$ and the drag torque $\vec{\Gamma}^{\mathrm{d}}_G$ exerted by the fluid on the moving solid fulfil the equation:

$$\frac{\mathrm{d}E}{\mathrm{d}t} = -\vec{F}^{\mathrm{d}} \cdot \vec{v}_G - \vec{\Gamma}^{\mathrm{d}}_G \cdot \vec{\Omega} - \Phi_{\mathrm{tot}} \tag{3.83}$$

with $\vec{v}_G$ the velocity of the centre of gravity G of the solid, $\vec{\Omega}$ its rotation vector, $E = \int_{\mathrm{fluid}} \frac{1}{2}\rho\vec{v}^2 \,\mathrm{d}V$ the total kinetic energy of the fluid and $\Phi_{\mathrm{tot}} = \int_{\mathrm{fluid}} \Phi \,\mathrm{d}V$ the total power dissipated in the fluid (Fig. 3.7).

To prove this result, let us calculate $\mathrm{d}E/\mathrm{d}t$. We find, by successive use of the theorem of derivation under the integral and the Navier–Stokes equation:

$$\frac{\mathrm{d}E}{\mathrm{d}t} = \int_{\mathrm{fluid}} \rho v_i \frac{\mathrm{D}v_i}{\mathrm{D}t} \,\mathrm{d}V = \int_{\mathrm{fluid}} v_i \left(-\frac{\partial P}{\partial x_i} + \frac{\partial}{\partial x_j}\sigma^{\mathrm{v}}_{ij} \right) \mathrm{d}V. \tag{3.84}$$

Integrating by parts and letting S be the (moving) surface of the solid, one has:

$$\frac{\mathrm{d}E}{\mathrm{d}t} = \int_S \left(-Pv_iv_i + v_i\sigma^{\mathrm{v}}_{ij}v_j \right) \mathrm{d}S - \int_{\mathrm{fluid}} \left(-P\frac{\partial v_i}{\partial x_i} + \sigma^{\mathrm{v}}_{ij}\frac{\partial v_i}{\partial x_j} \right) \mathrm{d}V \tag{3.85}$$

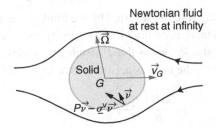

Fig. 3.7 Moving solid in a Newtonian fluid, at rest at infinity.

with $\vec{v}$ the unit vector normal to the surface of the solid and pointing inwards. The fluid is considered incompressible, div $\vec{v} = 0$ and the equation above simplifies to yield, in vector form:

$$\frac{dE}{dt} = \int_S \left(-P\vec{v} + \underline{\sigma}^v \vec{v}\right) \vec{v} \, dS - \Phi_{tot}. \tag{3.86}$$

It suffices now to write that, at each point of the solid surface, the velocity of the fluid equals that of the solid (making use of the no-slip condition):

$$\vec{v} = \vec{v}^S = \vec{v}_G + \vec{\Omega} \times \overrightarrow{GM}, \tag{3.87}$$

recalling that the resulting force and torque at point G of the forces exerted by the fluid on the solid are given by:

$$\vec{F} = \int_S \left(P\vec{v} - \underline{\sigma}^v \vec{v}\right) dS \tag{3.88}$$

and

$$\vec{\Gamma}_G = \int_S \overrightarrow{GM} \times \left(P\vec{v} - \underline{\sigma}^v \vec{v}\right) dS. \tag{3.89}$$

Replacing $\vec{v}$ by its expression (3.87) in Eq. (3.86) results in:

$$\frac{dE}{dt} = -\vec{F} \cdot \vec{v}_G - \vec{\Gamma}_G \cdot \vec{\Omega} - \Phi_{tot}. \tag{3.90}$$

This formula yields back Eq. (3.83) if we denote by $\vec{F}^d$ and $\vec{\Gamma}_G^d$ the $\vec{F}$ component along $\vec{v}_G$ and the $\vec{\Gamma}_G$ component along $\vec{\Omega}$, respectively.

Note that the drag force and torque contain an inertial contribution (related to the variation in kinetic energy dE/dt) and a purely dissipative part (associated to Φ_{tot}).

A particularly interesting case is that of a solid undergoing pure translation ($\vec{v}_G \neq 0$ and $\vec{\Omega} = 0$). If $\vec{v}_G = \overrightarrow{Cst}$ and if the Reynolds number (to be defined below, in Section 3.3) of the flow is not too high, the flow remains stationary in the reference frame of the solid.[14] Under these conditions, $dE/dt = 0$ and the dissipation theorem becomes:

$$\Phi_{tot} = -\vec{F}^d \cdot \vec{v}_G. \tag{3.91}$$

This formula yields the drag force $\vec{F}^d$ exerted by the fluid on the solid, if the dissipation is known. It also shows that there is no drag if the viscosity vanishes (in the limiting case of the perfect fluid), a result also known as the *d'Alembert paradox*.

This paradox disappears if the body is accelerated ($d\vec{v}_G/dt \neq 0$), since in this case:

$$\frac{dE}{dt} = -\vec{F}^d \cdot \vec{v}_G \neq 0. \tag{3.92}$$

---

[14] This result will be demonstrated in Section 3.3.3.

Similar results apply for a solid body in pure rotation ($\vec{v}_G - 0$ and $\vec{\Omega} \neq 0$). In particular, one still has $dE/dt = 0$ if $\vec{\Omega} = \overrightarrow{Cst}$ and if the flow is stationary in the rotating reference frame comoving with the solid, resulting in:

$$\Phi_{tot} = -\vec{\Gamma}_G^d \cdot \vec{\Omega}. \tag{3.93}$$

This equation yields the drag torque, provided the dissipation is known. We point out that in a perfect fluid this torque disappears, since the fluid glides on the solid surface without friction.

Last important comment: we never used the fact that G was the centre of gravity of the solid. Therefore, all formulas demonstrated in this paragraph are still valid if G is replaced by a generic point within the solid.

3.2.7 *Vorticity equation*

Let us first recall the kinematic significance of vorticity.

Let O be a generic point within the fluid. In the vicinity of this point, the velocity can be expanded in a power series:

$$\vec{v} = \vec{v}_O + \left(\frac{\partial v_i}{\partial x_j}\right)(\vec{v} - \vec{v}_O), \tag{3.94}$$

where $\partial v_i/\partial x_j$ are the strain rates at point O. One can always write:

$$\frac{\partial v_i}{\partial x_j} = A_{ij} + \Omega_{ij} \tag{3.95}$$

using the expression of the symmetric strain rate tensor:

$$A_{ij} = \frac{1}{2}\left(\frac{\partial v_i}{\partial x_j} + \frac{\partial v_j}{\partial x_i}\right) \tag{3.96}$$

and of the antisymmetric rotation rate tensor:

$$\Omega_{ij} = \frac{1}{2}\left(\frac{\partial v_i}{\partial x_j} - \frac{\partial v_j}{\partial x_i}\right). \tag{3.97}$$

It is easily checked that

$$\underline{\Omega}\,(\vec{r} - \vec{r}_O) = \vec{\Omega} \times (\vec{r} - \vec{r}_O) \tag{3.98}$$

with $\vec{\Omega}$ a vector with components $\Omega_1 = -\Omega_{23}$, $\Omega_2 = +\Omega_{13}$ and $\Omega_3 = -\Omega_{12}$. It ensues that $\vec{\Omega} = \vec{\omega}/2$, where $\vec{\omega} = \overrightarrow{rot}\,\vec{v}$ is, by definition, the *vorticity*.

Clearly, the vorticity $\vec{\omega}$ gives (up to a factor of 2) the *local rotation rate* $\vec{\Omega}$:

$$\vec{\omega} = 2\vec{\Omega}. \tag{3.99}$$

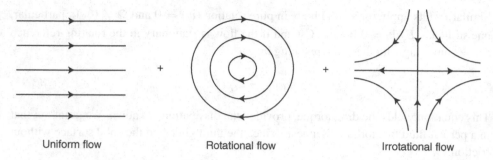

Uniform flow   Rotational flow   Irrotational flow

Fig. 3.8 The flow around a point O in the fluid can be seen as the superposition of three types of flow: constant velocity flow with $\vec{v}_O$; rotational flow; and irrotational (or elongation) flow.

Figure 3.8 details the decomposition of flow around a generic point. The tangents to the $\vec{\omega}$ vector are the vorticity lines. They can form closed loops or can end at the interface between two liquids, connecting with it at a certain angle. However, they are necessarily tangent to the surface of a solid at rest, if the no-slip condition applies.

After this reminder, let us look for the vorticity equation. To find it, take the $\overrightarrow{\text{curl}}$ of the Navier–Stokes equation (3.46), assuming constant density. Since the non-linear convection terms are given by:

$$\left(\vec{v} \cdot \vec{\nabla}\right) \vec{v} = \frac{1}{2}\overrightarrow{\text{grad}}\ \vec{v}^2 - \vec{v} \times \overrightarrow{\text{curl}}\ \vec{v}, \tag{3.100}$$

one has:

$$\frac{\partial \vec{\omega}}{\partial t} - \overrightarrow{\text{curl}}\ (\vec{v} \times \vec{\omega}) = \nu\, \Delta\vec{\omega} + \frac{1}{\rho}\overrightarrow{\text{curl}}\ \vec{f} \tag{3.101}$$

or, equivalently:

$$\frac{D\vec{\omega}}{Dt} - \left(\vec{\omega} \cdot \vec{\nabla}\right) \vec{v} = \nu\Delta\vec{\omega} + \frac{1}{\rho}\overrightarrow{\text{curl}}\ \vec{f}. \tag{3.102}$$

The second term on the left-hand side (which can also be written as $\omega(\partial v_z/\partial z)\vec{e}_z + \omega(\partial v_\perp/\partial z)\vec{e}_\perp$ by choosing $\vec{z}\|\vec{\omega}$) is important, as it describes the *stretching* and *tilting* of vortices due to the local velocity gradients $\partial v_z/\partial z$ and $\partial v_\perp/\partial z$ [2]. Equation (3.102) simplifies when the flow is *two-dimensional* in the (x, y) plane, since in this case $\vec{v} \perp \vec{\omega}$ (the only non-zero component of $\vec{\omega}$ is therefore ω_z) and $\partial/\partial z = 0$. Hence:

$$\frac{D\vec{\omega}}{Dt} = \nu\,\Delta\vec{\omega} + \frac{1}{\rho}\overrightarrow{\text{curl}}\ \vec{f}. \tag{3.103}$$

One recognises here a diffusion equation, where the kinematic viscosity $\nu = \eta/\rho$ plays the role of *vorticity diffusion coefficient*. Note also that the non-conservative forces $\vec{f}$ (i.e. those not deriving from a potential V by way of $\vec{f} = -\overrightarrow{\text{grad}}\ V$) generate vorticity: for instance, this is the case for the Coriolis force (which causes cyclones) or for electromagnetic forces (responsible for solar flares).

In the following, we will show that the vorticity is of fundamental importance in hydro-dynamics, its distribution determining many of the properties of the flow. In particular, we will see that it is very important to know where it is concentrated, and in which regions we can neglect it. This will lead us to defining the Reynolds number.

3.3 Reynolds number

3.3.1 Definition: the concept of similar flows

To fix the ideas, let us consider the set of flows around bodies of *similar shape*, namely flows related to one another by a mere scale change (Fig. 3.9).

Furthermore, we assume that $\vec{f} = 0$ and that the fluid is incompressible. It follows that the fluid is characterised by its density and the sole shear viscosity η (the bulk viscosity ζ defined by the general material law (3.15) does not come into play).

As to the body, taken as immobile in the laboratory reference frame, it is characterised by its size L.

Finally, the flow is characterised by its viscosity at infinity, V.

It is easily checked that the only dimensionless number one can build from these parameters is the Reynolds number, defined by:

$$Re = \frac{\rho V L}{\eta} = \frac{V L}{\nu},\qquad (3.104)$$

where $\nu - \eta/\rho$ is the kinematic viscosity

Hence, with L as the length unit, ρV^2 as the unit for pressure and $\rho V^2 L^2$ as the unit of force, the hydrodynamic velocity field $\vec{v}$, the pressure field P and the force F exerted by the fluid on the body are necessarily of the form:

$$\frac{\vec{v}}{V} = f\left(\frac{\vec{r}}{L}, Re\right),\qquad (3.105)$$

$$\frac{P}{\rho V^2} = g\left(\frac{\vec{r}}{L}, Re\right),\qquad (3.106)$$

$$F = \rho V^2 L\, h(Re),\qquad (3.107)$$

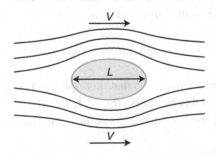

Fig. 3.9 Flow around bodies of similar shape.

where the functions f, g and h must be determined by solving the Navier–Stokes equation.

Equation (3.105) shows, in particular, that if two flows have the same Reynolds number, their dimensionless velocity fields $\vec{v}/V$ are identical functions of $\vec{r}/L$. Such flows are termed *similar*, i.e. they can be obtained from one another by a simple scale change of the lengths and velocities involved.

The laws (3.105)–(3.107) are called *Reynolds' similarity laws*. They are of fundamental importance in hydrodynamics, allowing the study of flow around large objects using smaller mock-ups (scaled models) of similar shape, provided the Reynolds numbers are conserved.

3.3.2 Physical meaning of the Reynolds number

To understand the meaning, let us rewrite the Navier–Stokes equation in the stationary case:

$$\rho \left(\vec{v} \cdot \vec{\nabla} \right) \vec{v} = -\overrightarrow{\text{grad}}\, P + \eta\, \Delta \vec{v}. \tag{3.108}$$

This equation expresses the balance between the inertial force and the viscous force, with their respective orders of magnitude:

$$\rho \left(\vec{v} \cdot \vec{\nabla} \right) \vec{v} \approx \rho \frac{V^2}{L}, \tag{3.109}$$

$$\eta \Delta \vec{v} \approx \eta \frac{V}{L^2}. \tag{3.110}$$

The ratio of these two forces yields exactly the Reynolds number:

$$Re = \frac{\text{inertial force}}{\text{viscous force}}. \tag{3.111}$$

Thus, the Reynolds number quantifies the relative importance of the inertial and viscous effects.

More precisely, at low Reynolds numbers ($Re \ll 1$) the viscous effects dominate the inertial effects and can be felt over distances much larger than the size of the body. The flow is then well described by the *linear equation*:

$$\overrightarrow{\text{grad}}\, P = \eta \Delta \vec{v}. \tag{3.112}$$

In particular, we will see in Section 3.7.5 that this equation is relevant for lubrication flow.

Conversely, at high Reynolds numbers ($Re \gg 1$) the viscous terms become negligible and are only felt close to the boundaries. This time, the flow obeys *Euler's equation of perfect fluids*:

$$\rho \left(\vec{v} \cdot \vec{\nabla} \right) \vec{v} = -\overrightarrow{\text{grad}}\, P. \tag{3.113}$$

This equation is nonlinear, rendering the calculations uniquely complicated, since there is no general method for solving it.

In the next paragraph, we show that the Reynolds number is also the relevant parameter for determining flow stability.

3.3.3 Flow stability and critical Reynolds number

We start by describing von Karman's wire experiment. This is a basic experiment in hydrodynamics, described in many books. A cylindrical wire, with diameter D, is placed in a flow of velocity V. The axis of the wire is perpendicular to the velocity.

Experiment shows (Fig. 3.10) that different flow regimes develop behind the wire as the Reynolds number increases. More specifically, one observes that [2, 4]:

(1) at low velocity ($Re < 1$), the flow is laminar and perfectly symmetric (Fig. 3.10a);
(2) at intermediate velocity ($1 < Re < 47$), two fixed counter-rotating vortices develop downstream of the cylinder (recirculation flow, Fig. 3.10b);
(3) at high velocity ($Re > 47$), the flow is no longer stationary, meaning that the velocity at a given point depends explicitly on the time; in this case, eddies of alternating sign are

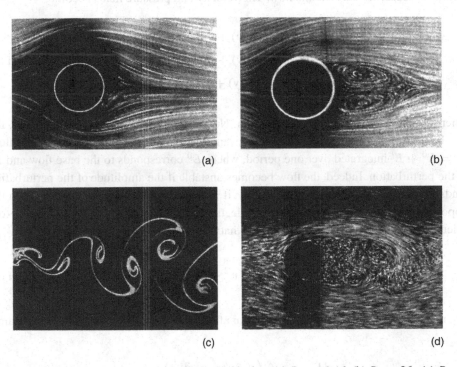

(a) (b)

(c) (d)

Fig. 3.10 Different flow regimes behind a cylindrical wire: (a) $Re = 0.16$; (b) $Re = 26$; (c) $Re = 200$; (d) $Re = 10^4$. Flow visualisation is achieved by doping the fluid with fine elongated particles with anisotropic reflective properties ([2, 8]; image (d) by Werlé, ONERA).

periodically emitted downstream (Fig. 3.10c). These eddies (or vortices) form a double row called a 'Bénard–von Karman vortex street'. At very high Reynolds numbers, this row loses its stability and becomes turbulent (Fig. 3.10d).

This experiment, which can be redone after changing the size of the obstacle, shows that the flow invariably loses its stationary character above a perfectly defined critical Reynolds number, here $R_c \approx 47$. Why? This is what we will try to explain now.

To simplify the demonstration, we assume that the flow around the cylinder is invariant under translation along the z axis, parallel to the axis of the cylinder. We are interested in the Reynolds number at which the initial stationary flow:

$$
\begin{aligned}
v_x &= v_x^0(x, y), \\
v_y &= v_y^0(x, y), \\
P &= P^0(x, y)
\end{aligned}
\tag{3.114}
$$

becomes unstable.

Suppose that a periodical perturbation along the x direction (parallel to the velocity at infinity) is superimposed on this flow. The velocity and pressure fields become:

$$
\begin{aligned}
v_x &= v_x^0(x, y) + v_x'(x, y, t), \\
v_y &= v_y^0(x, y) + v_y'(x, y, t), \\
P &= P^0(x, y) + P'(x, y, t),
\end{aligned}
\tag{3.115}
$$

where $\vec{v}'$ and P' describe the perturbation. Note that, at the surface of the cylinder, the no-slip condition imposes $\vec{v}' = 0$. The relevant quantity is the kinetic energy of the fluid $E = E^0 + E'$ integrated over one period, where E^0 corresponds to the base flow and E' to the perturbation. Indeed, the flow becomes unstable if the amplitude of the perturbation tends to increase with time or, equivalently, if E' increases with time: $dE'/dt > 0$. In the opposite case ($dE'/dt < 0$) the flow is stable, the perturbation tending to damp out. Before calculating dE'/dt, one should first notice that:

$$
E' = \frac{1}{2}\rho \int \left(v_x'^2 + v_y'^2 \right) dx\, dy,
\tag{3.116}
$$

where the terms linear in v_x' and v_y' disappear after integration over the period. One obtains:

$$
\frac{dE'}{dt} = \rho \int \left(v_x' \frac{\partial v_x'}{\partial t} + v_y' \frac{\partial v_y'}{\partial t} \right) dx\, dy
\tag{3.117}
$$

and then, by linearising the Navier–Stokes equation and only retaining the perturbation terms:

$$\frac{\partial v'_x}{\partial t} + v^0_x \frac{\partial v'_x}{\partial x} + v'_x \frac{\partial v^0_x}{\partial x} + v^0_y \frac{\partial v'_x}{\partial y} + v'_y \frac{\partial v^0_x}{\partial y} = -\frac{1}{\rho}\frac{\partial P'}{\partial x} + \nu \Delta v'_x, \tag{3.118a}$$

$$\frac{\partial v'_y}{\partial t} + v^0_x \frac{\partial v'_y}{\partial x} + v'_x \frac{\partial v^0_y}{\partial x} + v^0_y \frac{\partial v'_y}{\partial y} + v'_y \frac{\partial v^0_y}{\partial y} = -\frac{1}{\rho}\frac{\partial P'}{\partial y} + \nu \Delta v'_x. \tag{3.118b}$$

Finally, replacing $\partial v'_x/\partial t$ and $\partial v'_y/\partial t$ by their expressions in Eq. (3.117), and then using the incompressibility condition:

$$\frac{\partial v'_x}{\partial x} + \frac{\partial v'_y}{\partial y} = 0 \tag{3.119}$$

and a few more manipulations yield:

$$\frac{dE'}{dt} = -\rho \int v'_x v'_y \left(\frac{\partial v^0_x}{\partial y} + \frac{\partial v^0_y}{\partial x} \right) dx\, dy - \eta \int \omega'^2 \, dx\, dy, \tag{3.120}$$

where ω' is the vorticity:

$$\omega' = \frac{\partial v'_y}{\partial x} - \frac{\partial v'_x}{\partial y}. \tag{3.121}$$

The $-\eta \int \omega'^2 \, dx\, dy$ term is always negative, representing a decrease in kinetic energy by viscous dissipation: consequently, it has a stabilising effect on the flow. However, the other term, $-\rho \int v'_x v'_y \left(\frac{\partial v^0_x}{\partial y} + \frac{\partial v^0_y}{\partial x} \right) dx\, dy$ can be positive if the quantities $-\rho v'_x v'_y$ and $\frac{\partial v^0_x}{\partial y} + \frac{\partial v^0_y}{\partial x}$ have the same sign over a dominant part of the period. This term stands for the *transfer of kinetic energy* between the initial flow and the perturbation. The quantity $-\rho v'_x v'_y$ has stress dimensions: it is called the *Reynolds stress*. It plays a significant role in the theory of turbulence.

In conclusion, the flow is unstable if:

$$\rho \int v'_x v'_y \left(\frac{\partial v^0_x}{\partial y} + \frac{\partial v^0_y}{\partial x} \right) dx\, dy > -\eta \int \omega'^2 \, dx\, dy. \tag{3.122}$$

In terms of the dimensionless quantities:

$$x^* = \frac{x}{D}, \qquad v'^*_x = \frac{v'_x}{V}, \qquad \text{etc.,} \tag{3.123}$$

where D is the diameter of the cylinder and V the velocity of the fluid at infinity, and recalling that $Re = \rho V D / \eta$, the instability condition (3.122) can be rewritten formally as:

$$Re > \text{Min}_K \frac{\int \omega'^{*2} \, dx^* \, dy^*}{-\int v'^*_x v'^*_y \left(\frac{\partial v^{0*}_x}{\partial y^*} + \frac{\partial v^{0*}_y}{\partial x^*} \right) dx^* \, dy^*} = Re_c, \tag{3.124}$$

where K designates the set of all kinematically admissible velocity fields (i.e. fulfilling the conditions $v_x^{0*} = v_y^{0*} = 0$ at the surface of the cylinder, but not necessarily the Navier–Stokes equation).

This example shows clearly that the Reynolds number is the relevant parameter for characterising the loss of system stability.

3.4 Boundary layer and wake: the example of the vorticity line source

In this paragraph, we consider the simplified problem of a vorticity line source in a uniform flow with velocity U (Fig. 3.11).

In practice, this line can be a thin wire of radius R placed across the flow. Indeed, we will show in Section 3.6, dedicated to the study of the boundary layer, that the vorticity is continuously generated at the surface of the wire.

Our goal is to show that the vorticity is concentrated within a finite-size layer around the wire and in its wake, extending downstream.

To this end, let us write the vorticity equation. If the wire is thin enough, it will not perturb the flow too much, and we can write, in a first approximation:

$$\frac{\partial \omega}{\partial x} = \frac{v}{U} \Delta \omega. \tag{3.125}$$

To solve this equation, we search for a solution of the form:

$$\omega = e^{kx} f(x, y), \tag{3.126}$$

where we set:

$$k = L_v^{-1} = \frac{U}{2v}. \tag{3.127}$$

The distance L_v is called the *viscous length*. Replacing ω by its expression (3.126) in the vorticity equation (3.125) yields:

$$\Delta f = k^2 f. \tag{3.128}$$

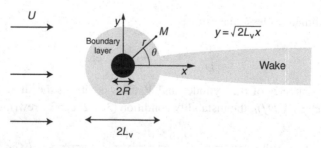

Fig. 3.11 Vorticity line source in a uniform flow with velocity U. The vorticity is concentrated within a disc of radius $L_v = 2v/U$ around the line and in the parabolic wake extending downstream.

In polar coordinates, the solution to this equation is:

$$f = f_0 K_0(kr), \tag{3.129}$$

where f_0 is a constant, and K_0 the modified Bessel function of order 0. It can be shown that the asymptotic behaviour of this function is:

$$K_0(kr) \approx \sqrt{\frac{\pi}{2}} \frac{e^{-kr}}{\sqrt{kr}} \quad \text{for} \quad kr \gg 1, \tag{3.130}$$

so, keeping in mind that $x = r \cos\theta$, with θ the polar angle, the vorticity field is given by:

$$\omega \approx \omega_0 \frac{1}{\sqrt{kr}} e^{-kr(1-\cos\theta)} \quad \text{for} \quad kr \gg 1. \tag{3.131}$$

This formula shows that the vorticity is mainly concentrated within a disc of radius L_v (as long as $\theta \neq 0$) and in the *wake* of equation $y < (2L_v x)^{1/2}$ (centred on the $\theta = 0$ plane), where it is advected by the flow.

Note that the Reynolds number of the flow is here given by:

$$Re = \frac{R}{L_v}. \tag{3.132}$$

Consequently, $R \ll L_v$ at low Reynolds number ($Re \ll 1$), implying that the body is immersed in a vorticity 'sea'. The viscous effects dominate the flow.

On the other hand, $R \gg L_v$ at high Reynolds number ($Re \gg 1$). Under these conditions, the vorticity is concentrated within a thin layer around the body (termed the boundary layer) and in its wake. These two regions being very 'thin' at high Reynolds number, the flow is similar to that of a perfect fluid almost everywhere. Note however that neither the viscous effects nor the inertial ones can be neglected within the boundary layer and the wake, where they are of the same order of magnitude, by definition.

3.5 High Reynolds number flow: the perfect fluid approximation

We have just seen that, at high Reynolds numbers, the flow around a body tends to resemble that of a perfect fluid, with zero viscosity. In this limit, and for an incompressible fluid (requiring that the velocities involved be small compared to the speed of sound), the Navier–Stokes equation reduces to the following form:

$$\rho \frac{D\vec{v}}{Dt} = -\overrightarrow{\text{grad}}\, P + \vec{f}. \tag{3.133}$$

This is *Euler's equation of perfect fluids*.

3.5.1 Bernoulli's theorem

We assume that the bulk forces derive from a potential V:

$$\rho^{-1}\vec{f} = -\overrightarrow{\text{grad}}\ V \tag{3.134}$$

and that the flow is stationary:

$$\frac{\partial \vec{v}}{\partial t} = 0. \tag{3.135}$$

Under these conditions, Euler's equation (3.133) becomes:

$$\frac{1}{2}\overrightarrow{\text{grad}}\ v^2 - \vec{v} \times \overrightarrow{\text{curl}}\ \vec{v} = -\frac{1}{\rho}\overrightarrow{\text{grad}}\ P - \overrightarrow{\text{grad}}\ V. \tag{3.136}$$

Taking the scalar product with $\vec{t}$, the unit vector tangent to the streamlines ($\vec{t} = \vec{v}/|\vec{v}|$) yields:

$$\frac{\partial}{\partial\vec{t}}\left(\frac{1}{2}v^2 + \frac{P}{\rho} + V\right) = 0, \tag{3.137}$$

where $\partial/\partial\vec{t} = \vec{t}\cdot\overrightarrow{\text{grad}}$ designates the derivative along the direction $\vec{t}$. This equation proves that the *head* H, given by:

$$H = \frac{1}{2}v^2 + \frac{P}{\rho} + V, \tag{3.138}$$

is constant along a streamline. We restate our assumptions, namely that the fluid is incompressible, that the flow is stationary and that the bulk forces derive from a potential.

This theorem, fundamental in hydrodynamics, expresses nothing other than energy invariance under a reference frame change. It yields, for instance, the lift of an airplane wing, as we will show in Section 3.5.3.

Bernoulli's theorem leads to situations that might seem paradoxical. We will discuss three examples.

The first one involves two plastic discs and a source of compressed air. The air is injected into a tube attached to the centre of one of the discs. Experiment shows that, if this disc is brought close enough to the second disc, the latter can be lifted, although air is blowing from the tube (Fig. 3.12a). In the second example, a ping-pong ball remains suspended at the bottom of a reversed funnel when air is blown in from the top (Fig. 3.12b). The last example involves a ping-pong ball levitating in a tilted air flow, pointing upwards (Fig. 3.12c). This experiment proves the presence of a position (under the flow axis) where the air exerts on the ball drag and lift forces that exactly balance its weight. The ball also turns on itself, due to the strong shear. This phenomenon is known by the name of the *Coanda effect*.[15] These experiments are only successful if the air flow is strong enough (but not strong enough to lead to the development of instabilities).

---

[15] Henri Coanda (1886–1972) was a Romanian engineer who pioneered the development of the reaction engine (its principle had been discovered in 1905 by the Frenchman René Lorin). After studies in Belgium and France, he presented the first model of a reaction aircraft in 1910 in Paris, at the second International Aeronautic Salon. His plane crashed into a wall on the first test. After this incident, he designed a revolutionary nozzle, based on a principle still in use today. His research led him to discovering the effect bearing his name.

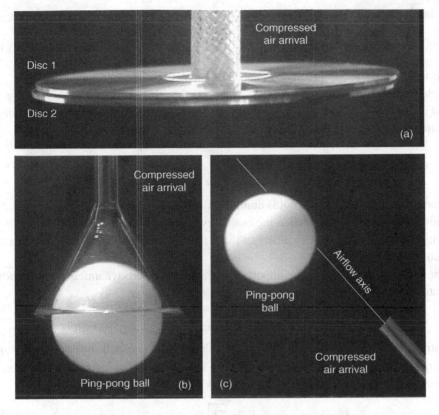

Fig. 3.12 Three simple experiments illustrating Bernoulli's theorem.

3.5.2 Kelvin's circulation theorem

This is another important theorem of hydrodynamics. Once again, we assume that the fluid is incompressible and that the bulk forces derive from a potential V (Eq. (3.134)). Let C_m be a *mobile* contour within the fluid (i.e. composed of fluid particles that we follow along their motion) and let Γ be the circulation of the velocity along this contour:

$$\Gamma = \int_{C_m} \vec{v} \cdot d\vec{l}. \tag{3.139}$$

We intend to prove that the circulation Γ is conserved on this circuit.

To this end, let us calculate dE/dt. Using Euler's equation and the fact that C_m is a mobile contour, we find:

$$\frac{d\Gamma}{dt} = \int_{C_m} \frac{D\vec{v}}{dt} d\vec{l} + \int_{C_m} \vec{v} \frac{Dd\vec{l}}{Dt}$$

$$= \int_{C_m} \left(-\frac{1}{\rho} \overrightarrow{\mathrm{grad}}\ \vec{P} - \overrightarrow{\mathrm{grad}}\ V \right) d\vec{l} + \int_{C_m} \vec{v} \cdot d\vec{v} \tag{3.140}$$

yielding immediately:

$$\frac{d\Gamma}{dt} = 0, \tag{3.141}$$

since the velocity $\vec{v}$, the pressure P and the potential V are continuous functions. This completes the proof of the theorem.

An obvious corollary is that, in a perfect fluid, if the flow is irrotational at some time, it remains irrotational at all subsequent times.

3.5.3 Irrotational (or potential) flow

We have seen that, at high Reynolds numbers, flow around a body is described by Euler's equation for perfect fluids outside the boundary layer and the wake. On the other hand, Kelvin's theorem states that, in a perfect fluid, the irrotational character of a flow is conserved in time.

These two features explain the importance attached by hydrodynamicists to the study of irrotational flow in perfect fluids.

3.5.3.1 General points

We still assume that the fluid is incompressible and that the bulk forces derive from a potential V. If the flow is irrotational ($\overrightarrow{\text{curl}}\,\vec{v} = 0$), there exists a function φ, called the *velocity potential*, such that:

$$\vec{v} = \overrightarrow{\text{grad}}\,\varphi, \tag{3.142}$$

hence the term *potential flow*.

With these conditions, Euler's equation becomes:

$$\frac{\partial \vec{v}}{\partial t} + \frac{1}{2}\overrightarrow{\text{grad}}\,\vec{v}^2 = -\overrightarrow{\text{grad}}\,\frac{P}{\rho} - \overrightarrow{\text{grad}}\,V, \tag{3.143}$$

which yields, after replacing $\vec{v}$ by its expression (3.142) and integrating once:

$$\frac{\partial \varphi}{\partial t} + \frac{\vec{v}^2}{2} + \frac{P}{\rho} + V = f(t), \tag{3.144}$$

where $f(t)$ is an arbitrary function of the time. Since the potential φ is only defined up to a time-dependent function, one can always arrange things such that $f(t) = 0$. Equation (3.144) represents a first integral of the potential motion equations. Note that this equation reduces to Bernoulli's theorem in the case of stationary flow.

3.5.3.2 Plane potential flow

The assumptions are the same as in the above paragraph, and we furthermore suppose that the flow is restricted to the (x, y) plane.

The fluid being incompressible, div $\vec{v} = 0$, so there exists a function ψ, termed the *stream function*, such that:

$$\vec{v} = -\vec{k} \times \overrightarrow{\text{grad}}\ \psi, \tag{3.145}$$

where $\vec{k}$ is the unit vector normal to the (x, y) plane. Furthermore, since $\vec{v} = \overrightarrow{\text{grad}}\ \varphi$, one can write:

$$v_x = \frac{\partial \varphi}{\partial x} = \frac{\partial \psi}{\partial y} \quad \text{and} \quad v_y = \frac{\partial \varphi}{\partial y} = -\frac{\partial \psi}{\partial x}. \tag{3.146}$$

Mathematically speaking, the relations between the derivatives of the functions φ and ψ are nothing other than the well-known Cauchy–Riemann conditions, stating that the combination

$$f = \varphi + i\psi \tag{3.147}$$

is a *holomorphic* function of the complex variable $z = x + iy$. This function is called the *complex potential*. Its derivative:

$$\frac{df}{dz} = \frac{\partial \varphi}{\partial x} + i\frac{\partial \psi}{\partial x} = v_x - iv_y = ve^{-i\alpha} \tag{3.148}$$

is the *complex velocity*, completely defining the velocity vector (of amplitude v and making an angle α with the x axis).

If the flow is stationary, the complex potential is a function of the unique variable z. The streamlines, coinciding with the trajectories of fluid particles, are given by the equation $\psi = \text{Cst}$. It follows that along a wall (necessarily fixed, if the flow is stationary), the ψ function, i.e. the imaginary part of $f(z)$, must take a constant value. This condition states that the velocity must be tangent to the surface of the body in a perfect fluid (impermeability condition).

Let us now give a few examples of potential flow.

The simplest one is that of a uniform flow with a velocity v_0 parallel to the x axis. In this case:

$$f(z) = v_0 z. \tag{3.149}$$

For a radial flow originating at the $x = y = 0$ axis:

$$f(z) = \frac{D}{2\pi} \ln z, \tag{3.150}$$

where D is the flow rate per unit length (we have a source for $D > 0$, and a sink for $D < 0$). We recall that, by definition, $\ln(z) = \ln(r) + i\theta$, with $z = re^{i\theta}$. This type of flow is important, because by summing the complex potentials of a sink and a source placed at a certain distance one can simulate the flow around an airfoil.

The complex potential:

$$f(z) = v_0 \left(z + \frac{R^2}{z} \right) \tag{3.151}$$

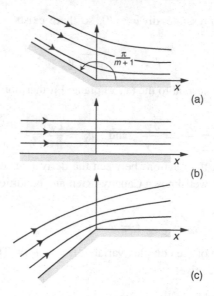

Fig. 3.13 Flow around a wedge, given by the complex potential $f(z) = Cz^m$. (a) $m > 0$; (b) $m = 0$; (c) $m < 0$.

is also interesting, describing a flow that is uniform at infinity and deviated by the presence of a cylinder with radius R.

Another useful potential, to be discussed in Section 3.6.4, is the following:

$$f(z) = Cz^{m+1}, \qquad (3.152)$$

where $-1/2 \le m < \infty$. It describes the flow around the wedge formed by the two planes $\theta = 0$ and $\theta = \pi/(m+1)$.

The different possible cases are depicted in Fig. 3.13. It is easily checked that on the x axis (in $\theta = 0$), the velocity v_x follows a power law:

$$v_x = (m+1)Cx^m. \qquad (3.153)$$

Our last example will be that of a vortex, described by the complex potential:

$$f(z) = -\frac{i\Gamma}{2\pi} \ln z. \qquad (3.154)$$

Here, the streamlines are circles centred on the $x = y = 0$ axis. The velocity is orthoradial, with components:

$$v_\theta = \frac{\Gamma}{2\pi r}. \qquad (3.155)$$

This flow is irrotational outside the axis. However, the circulation of the velocity along a closed contour encircling the vortex axis equals Γ. Hence, all the vorticity is concentrated on the vortex axis, where it is, in fact, infinite (and so is the velocity).

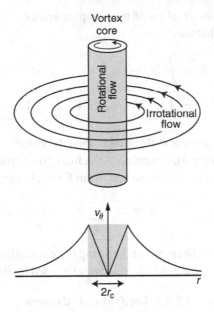

Fig. 3.14 Rankine vortex.

In the Rankine model (Fig. 3.14), this singular line is replaced by a vorticity tube with radius r_c, called the *vortex core*, within which the fluid turns as a solid, at a velocity:

$$v_\theta = \frac{\Gamma}{2\pi r_c^2} r.$$ (3.156)

In this model, the r derivative of the velocity $v_\theta(r)$ has a discontinuity at the core surface, at $r = r_c$. Therefore, this model can only work in a fluid with zero viscosity, where the tangential stress $\sigma_{r\theta}$ vanishes everywhere.

Finally, note that even though the velocity field is continuous outside the vortex axis, this does not hold for the velocity potential, which is, from Eq. (3.154):

$$\varphi = \frac{\Gamma\theta}{2\pi}.$$ (3.157)

This formula shows that in the $\theta = 0$ plane the potential exhibits a discontinuity equal to the circulation Γ. We will see in the next chapter that a vortex is mathematically equivalent to a dislocation in a solid, the velocity potential playing the role of the displacement u, and the circulation Γ the role of the Burgers vector b.

To conclude this section, we give the expression for the velocity circulation over a closed contour C. From the residue theorem one can write:

$$\int_C \frac{df}{dz}\, dz = 2\pi i \sum_k A_k,$$ (3.158)

where the A_k values are the residues of the complex velocity corresponding to the poles encircled by C. On the other hand:

$$\int_C \frac{df}{dz}\, dz = \int_C (v_x - i\, v_y)\,(dx + i\, dy)$$

$$= \int_C v_x\, dx + v_y\, dy + i \int_C v_x\, dy - v_y\, dx. \tag{3.159}$$

The real part of this expression is nothing other than the velocity circulation over the contour C. As to the imaginary part, it represents the fluid flow across the contour C. If there is no source within C this flow is zero so that, from Eq. (3.158):

$$\Gamma = 2\pi i \sum_k A_k. \tag{3.160}$$

In the next subsection, we will show that knowing the circulation over the contour C one can find the lift force (also called the *Magnus force*) exerted by the flow on this contour.

3.5.3.3 The Joukovski theorem

To fix the ideas, consider that the contour C belongs to an airplane wing placed in a flow that is uniform at infinity, with a velocity $\vec{v}_\infty$ parallel to the x axis. The flow is considered invariant along z.

Using Bernoulli's theorem, we will show that the perfect fluid exerts on the wing a *lift force* perpendicular to $\vec{v}_\infty$, of algebraic value (with respect to the y axis):

$$F = -\rho v_\infty \Gamma, \tag{3.161}$$

where Γ is the velocity circulation over the positively oriented contour (the Joukovski theorem).

A simplified demonstration of this result can be given assuming that the flow is laminar and that the fluid streamlines are everywhere 'adherent' to the surface of the wing. In technical terms, this means that the viscous boundary layer formed on the upper surface of the wing separates exactly at the trailing edge (the so-called *Chaplygin–Kutta condition*) (Fig. 3.15).

Let v_u be the velocity on the upper surface of the wing and v_l the velocity on the lower surface. From Bernoulli's equation, we have:

$$P_u + \frac{1}{2}\rho v_u^2 = P_l + \frac{1}{2}\rho v_l^2 = \text{Cst}, \tag{3.162}$$

where P_u and P_l designate the pressure values (generally different) on the upper and lower surfaces, respectively. This pressure difference creates a lift force, perpendicular to the velocity, and of amplitude:

$$F = \int_0^a (P_l - P_u)\, dx, \tag{3.163}$$

where a is the width of the wing.

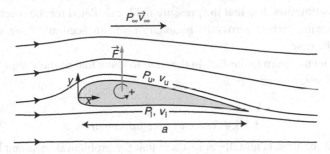

Fig. 3.15 Flow around an airplane wing. One assumes that the fluid streamlines are 'adherent' to the surface of the wing over its entire width.

Applying Eq. (3.162) yields (per unit length along z):

$$F = \int_0^a \frac{1}{2}\rho(v_u - v_l)(v_u + v_l)\,dx \approx \rho v_\infty \int_0^a (v_u - v_l)\,dx. \qquad (3.164)$$

This formula directly gives the Joukovski equation (3.161), knowing that the velocity circulation over the (positively oriented) contour is, by definition:

$$\Gamma = \int_0^a v_l\,dx + \int_a^0 v_u\,dx = \int_0^a (v_l - v_u)\,dx. \qquad (3.165)$$

As a complement, we present a useful formula (demonstrated in [3]) for calculating explicitly the velocity circulation around a thin wing:

$$\Gamma = v_\infty \int_0^a [\xi_l(x) + \xi_u(x)]\sqrt{\frac{x}{a-x}}\,dx. \qquad (3.166)$$

Note that here the $y = \xi_l(x)$ and $y = \xi_u(x)$ functions are the profiles of the lower and upper wing surface in the (x, y) reference frame attached to the wing (Fig. 3.15).

Using this formula and the Joukovski equation (3.161) one can calculate, for instance, the lift of an airplane wing, given by the equation $\xi_l(x) = \xi_u(x) = \alpha(a - x)$ and making an angle α with the velocity of the fluid at infinity:

$$F = \pi \alpha a \rho v_\infty^2. \qquad (3.167)$$

This force is proportional to the square of the velocity at infinity, to the wing width a and to the attack angle α (as long as it is small enough).

In conclusion, let us point out that the lift force can be rigorously proven to be perpendicular to the velocity at infinity, using the dissipation theorem (Section 3.2.6). Indeed, we know that in a perfect fluid $\Phi = 0$, and that in the stationary regime $dE/dt = 0$. As per this theorem, it follows that $\vec{F} \cdot \vec{v}_\infty = 0$.

It is also noteworthy that, in order to prove the Joukovski theorem, aside from the Chaplygin–Kutta condition on the separation of the boundary layer, we implicitly made an additional assumption. It is that the pressure field, calculated for the potential flow, acts directly on the wing surface across the boundary layer. In Section 3.6 we will show that this is indeed the case.

Before that, let us return to the Kelvin theorem to show that as the wing (or the fluid) is set in motion, vortices must appear.

3.5.3.4 Theory of the initial vortex

We assume that the fluid is initially at rest and that the problem is invariant by translation along the wingspan (2D flow). To show that a vortex is emitted as the fluid is set in motion, let us calculate the velocity circulation Γ_{total} over a very large (tending to infinity) contour encircling the wing.

This quantity, starting from a zero value in the fluid at rest, must remain null during time, as per the Kelvin theorem. It follows that if for any reason (which can only be related to viscous effects in the boundary layer formed at the surface of the wing) the velocity circulation Γ around the contour of the wing becomes non-zero, a vortex of circulation $\Gamma_{\text{v}} = -\Gamma$ must have appeared during the transient regime, since the total circulation remains zero (Fig. 3.16):

$$\Gamma_{\text{total}} = \Gamma_{\text{v}} + \Gamma = 0. \tag{3.168}$$

This initial vortex theory is obviously very simplified, since it ignores completely the finite wingspan. Experiment shows that, in practice, vortices attached to the wing profile appear downstream in the permanent regime. These trailing vortices are perpendicular to the wing profile and tend to stretch with time.

To show the existence of these vortices qualitatively, start by assuming that the different sections of the wing create an identical lift force. The wingspan being finite, the onset of the permanent regime no longer leads to the formation of two infinitely long vortices of opposite sign, but rather to the appearance of a unique vortex loop, closing upon itself 'at infinity' (in the permanent regime). Moreover, the velocity circulation around this loop must be conserved, according to the Kelvin theorem. Physically, the formation of the two trailing vortices attached to the wing tips is due to the pressure difference in the fluid between the lower (high pressure) and upper (low pressure) surfaces at the tip of the wing (Fig. 3.17a).

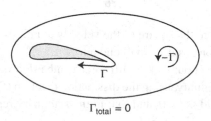

$$\Gamma_{\text{total}} = 0$$

Fig. 3.16 Two-dimensional theory of the initial vortex. During the transient regime leading to the stationary regime, a vortex with circulation $-\Gamma$ is emitted in the wake of the wing.

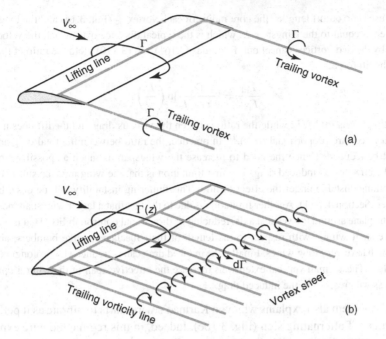

Fig. 3.17 'Lifting line model' for a finite-span wing. (a) Case of a uniform-section wing and the two trailing vortices; (b) case of a wing with variable section along z and the corresponding vortex sheet.

In practice, the different sections of the wing are not identical, resulting in a lift force that varies along the wingspan (in the z direction). The effect is a change in circulation along the wing:

$$d\Gamma = \frac{\partial \Gamma}{\partial z} dz, \tag{3.169}$$

to which are associated vorticity lines with a circulation $d\Gamma$, trailing downstream (Fig. 3.17b). This hydrodynamic model of the wing, consisting of an attached vortex with a variable circulation along the wingspan (the so-called 'lifting line') and a semi-infinite sheet of vorticity lines is called the *lifting line model*.

Note that this 3D model breaks the d'Alembert paradox. Indeed, the vorticity is continuously injected into the fluid, since the length of the trailing vortices constantly increases. Thus, the wing experiences a finite drag force, since it must provide (in the stationary regime) an amount of work equal to the increase in the kinetic energy of the fluid. This force is called the *induced drag*. The result above is however not in contradiction with the general dissipation theorem because the appearance of trailing vortices and of a non-zero lift force is in fact due to the viscous forces acting in the boundary layer enveloping the wing (see Section 3.6).

In conclusion, let us give an order of magnitude for the induced drag F_{id}. In the example of Fig. 3.17a, this force is simply 2τ, where τ is the line tension of each wing-tip vortex. The line tension is defined below, in Eq. (3.178). A simple calculation yields:

$$F_{id} = \frac{\rho \Gamma^2}{2\pi} \ln\left(\frac{L}{r_c}\right), \tag{3.170}$$

where we used two cutoff lengths, the core radius of each vortex r_c (Fig. 3.14) and the long-distance cutoff, taken as equal to the wingspan L, which is the typical distance over which the velocity fields generated by the two vortices cancel out. Formulas (3.161) and (3.170) yield the ratio of the induced drag and the lift force:

$$\frac{F_{id}}{F_{lift}} = \frac{\Gamma}{2\pi v_\infty L} \ln\left(\frac{L}{r_c}\right). \tag{3.171}$$

This ratio goes as $\ln(L)/L$, while the ratio between the viscous drag and the lift does not depend on L (both forces are proportional to L).[16] In practice, the ratio between the total drag and the lift force must be reduced. Hence the need to increase the wingspan as much as possible, in order to decrease the effect of its induced drag. The first limitation is that the wing must be sturdy enough to avoid becoming unstable under the effect of flow. This fluttering instability will be described in the next chapter (Section 4.1.11). Another limitation is due to the fact that a larger wingspan increases the inertia of the plane around the vertical axis, reducing its directional controllability. That is why fighter planes have short wings (with respect to the length of their fuselage) unlike bombers, airliners or gliders, which have very long wings. Finally, the induced drag can be reduced by prolonging the wing with winglets. These small vertical extensions increase the effective span of the wing, improving its effectiveness with respect to the induced drag.

Kelvin's theorem also explains why von Karman's wire starts to vibrate as it periodically emits vortices of alternating sign (Fig. 3.10c). Indeed, in this regime the wire experiences a Magnus (or lift) force, perpendicular to the velocity at infinity, which changes sign with each newly emitted vortex. The wire can then resonate and start to 'sing' at the frequency of vortex emission.

In the next subsection, we will analyse the behaviour of vortices in an external flow and the interaction between them.

3.5.3.5 Dynamics of a collection of straight and parallel vortices

Experiment shows that, in the first approximation, a vortex immersed in a uniform flow is carried along at the velocity of the current. It also shows that two vortices with the same circulation immersed in a fluid at rest at infinity move along a circle with opposite velocities, while two vortices with contrary circulations ($\Gamma_1 = -\Gamma_2$) undergo translation, with the same velocity, in a direction perpendicular to the line connecting them.

These phenomena are particularly easy to explain in the case of irrotational flow in a perfect fluid, where it is enough to add together the complex potentials f_i of the vortices and of the uniform flow to describe their coupled dynamics. Indeed, formula (3.148) shows immediately that the velocity field associated to the complex potential $f = \Sigma f_i$ is nothing more than a simple superposition of all velocity fields corresponding to the complex potentials f_i. Consequently, the resulting flow is also irrotational and fulfils the incompressibility condition.

Therefore, the ith vortex will move with a velocity $\vec{v}_i$ equal to the sum of all velocities engendered at its position by the uniform flow and by all the other vortices j, with $j \neq i$.

Thus, we predict that a vortex immersed in a uniform flow with velocity $\vec{v}$ will simply be carried along at the velocity $\vec{v}$.

---

[16] The calculation of the viscous drag will be given in Section 3.6.

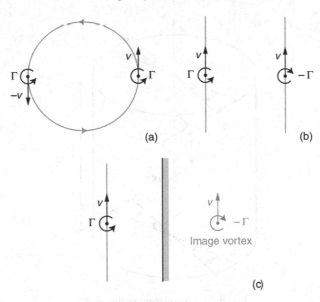

Fig. 3.18 Trajectories described in a perfect fluid at rest at infinity by two vortices with identical circulations (a) or opposite circulations (b) and by a vortex in the presence of a wall (c). The trajectories are shown in grey and the vortices are perpendicular to the page.

Two vortices with identical circulation ($\Gamma = \Gamma_1 = \Gamma_2$) in a fluid at rest at infinity will turn on a circle of constant diameter d, at the angular velocity (Fig. 3.18a):

$$\Omega = \frac{\Gamma}{\pi d^2}. \tag{3.172}$$

On the other hand, they will describe straight and parallel trajectories, with a velocity:

$$v = \frac{\Gamma}{\pi d} \tag{3.173}$$

if they have opposite circulations ($\Gamma = \Gamma_1 = -\Gamma_2$), where d is the separation distance (Fig. 3.18b).

One can also predict that a vortex placed at a distance d from a solid wall will move parallel to the wall at a velocity:

$$v = \frac{\Gamma}{4\pi d}. \tag{3.174}$$

To prove this result, it is enough to add an image vortex with the opposite circulation, symmetrically across the wall (Fig. 3.18c): this way, the velocity at the wall is tangent to it, as required by the impermeability condition.

In the next subsection, we will analyse the stability of a vortex immersed in an outer flow. This problem is formally equivalent to the stability of a dislocation in an external stress

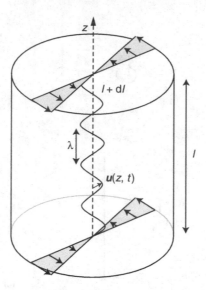

Fig. 3.19 Meandering vortex line.

field, to be addressed in Chapter 8, dealing with liquid crystals (we will discuss the example of the helical instability of a screw dislocation in a smectic phase under compression normal to the layers).

3.5.3.6 Stability of a vortex in an outer flow

We start by analysing the stability of a vortex, with circulation Γ, aligned parallel to the z axis, in the absence of external flow. Suppose that, under the action of a small perturbation, the core of the vortex moves by (Fig. 3.19):

$$\vec{u}(z, t) = \vec{u}_0 \sin(qz) \exp(i\omega t). \tag{3.175}$$

To find the differential equation describing the evolution of the line, and hence that of $\vec{u}$, let us look for the (effective) forces acting on the line.

The first one, well known in the theory of vibrating strings, is directly related to the line tension τ of the vortex. Directed along the z axis, it is proportional to the local curvature of the line and reads, in the small displacement limit:

$$\vec{F}_\tau = \tau \frac{\partial^2 \vec{u}}{\partial z^2}. \tag{3.176}$$

This force opposes stretching of the vortex: it is always stabilising. Note that, by definition, the line tension is:

$$\tau = \frac{dE}{dl}, \tag{3.177}$$

where dE is the increase in the kinetic energy of the line as it stretches by dl. The calculation yields:

$$\tau = \frac{\rho \Gamma^2}{4\pi} \ln\left(\frac{2\pi}{qr_c}\right), \tag{3.178}$$

where we neglected the core energy, proportional to the core radius r_c. We emphasise that this calculation follows closely that of the line tension of dislocations in solids, given in Section 5.1.2.6.

The other force acting on the line is the *Magnus force*. It can be destabilising, as we will see later. One obtains it by applying the Joukovski theorem to the vortex core. In our case, it states that:

$$\vec{F}_m = \rho \vec{\Gamma} \times \dot{\vec{u}}, \tag{3.179}$$

where $\dot{\vec{u}}$ is the velocity of the line with respect to the fluid at infinity.

Finally, the equation describing the equilibrium of (very real) forces acting on the vortex core is given by:

$$\rho r_c^2 \ddot{\vec{u}} = \rho \vec{\Gamma} \times \dot{\vec{u}} + \tau \frac{\partial^2 \vec{u}}{\partial z^2}, \tag{3.180}$$

where the first term represents the inertia force.

To find the dispersion relation $\omega(q)$, it is enough to replace $\vec{u}$ by its expression (3.175) in the equation above. Neglecting the inertia force (which is always possible for a sufficiently small r_c) yields:

$$\tau q^2 u_{0x} + i\omega \rho \Gamma u_{0y} = 0, \tag{3.181a}$$

$$i\omega \rho \Gamma u_{0x} - \tau q^2 u_{0y} = 0. \tag{3.181b}$$

This system of homogeneous equations has a non-trivial solution only if its determinant is zero. This condition gives the dispersion relation:

$$\omega = \pm \frac{\tau q^2}{\rho \Gamma} = \pm \frac{\Gamma}{4\pi} q^2 \ln\left(\frac{2\pi}{qr_c}\right). \tag{3.182}$$

Since ω is always real, the line is *stable*, because the amplitude of the perturbation does not increase with time.

The situation can change if the line is *immersed in an external flow*.

As a simple illustration, imagine a stationary flow in the (x, y) plane. Let $\vec{v}$ be the corresponding hydrodynamic velocity field. Close to the z axis, this flow can be written in the following general form:

$$v_x = v_{0x} + Sx + \Omega y, \tag{3.183a}$$

$$v_y = v_{0y} - \Omega x - Sy, \tag{3.183b}$$

where S is the local elongation rate and Ω the local rotation rate.

Since the vortex is carried along by the mean flow, at a velocity (v_{0x}, v_{0y}), the stability calculation must be performed in this reference frame (which is also that of the vortex). In this case it amounts to setting $v_{0x} = v_{0y} = 0$, so we can preserve the same notations. The vortex then feels a Magnus force:

$$\vec{F}_{\mathrm{m}} = \rho\vec{\Gamma} \times \left(\dot{\vec{u}} - \vec{v}\right),$$ (3.184)

where $\vec{v}$ is the velocity in $x = u_x$ and $y = u_y$ given by Eq. (3.183) (where one must set $v_{0x} = v_{0y} = 0$). Rewriting the equation of force equilibrium to account for the external flow yields:

$$\rho\vec{\Gamma} \times \left(\dot{\vec{u}} - \vec{v}\right) + \tau\frac{\partial^2\vec{u}}{\partial z^2} = 0.$$ (3.185)

Consequently, the displacement $\vec{u}$ must obey the new equations:

$$\left(\tau q^2 + \rho\Gamma\Omega\right) u_{0x} + (i\omega\rho\Gamma + \rho\Gamma S) u_{0y} = 0,$$ (3.186a)

$$(i\omega\rho\Gamma - \rho\Gamma S) u_{0x} - \left(\tau q^2 + \rho\Gamma\Omega\right) u_{0y} = 0,$$ (3.186b)

resulting in the following general dispersion relation:

$$\omega^2 = \omega_0^2 + 2\Omega\omega_0 + \Omega^2 - S^2,$$ (3.187)

where $\omega_0 = \tau q^2/\rho\Gamma$ is the typical pulsation in the absence of the external flow (see Eq. (3.182)).

The expression above shows that this time the line can be linearly unstable, since ω^2 can become negative.

This occurs, for instance, under *simple shear* (corresponding to $\Omega = S \neq 0$), when the vortex circulation Γ and the local rotation rate Ω of the external flow are of opposite signs (counter-rotating flows). This can be seen immediately by rewriting the dispersion equation (3.187) in the form:

$$\omega^2 = aQ^2 + bQ,$$ (3.188)

where we set $Q = q^2$, $a = \tau^2/(\rho^2\Gamma^2)$ and $b = 2\Omega\tau/(\rho\Gamma)$. This relation shows that a band of unstable wavelengths exists for $b < 0$, i.e. if Γ and Ω have opposite signs. In this case, the most unstable wave vector of the vortex is given by:

$$q_{\mathrm{Max}} = \sqrt{\frac{\rho}{\tau}|\Omega\Gamma|}$$ (3.189)

with a maximum growth rate set only by the external flow, since:

$$i\omega_{\mathrm{Max}} = |\Omega|.$$ (3.190)

3.6 Boundary layer theory

We have already referred several times to the fundamental role of the boundary layer. In this section, we will try to highlight its essential properties [2, 4, 6].

3.6.1 The Prandtl equations

We start by assuming stationary flow in an incompressible fluid. To fix the ideas, we also assume that the body is a well-profiled airplane wing, of width a, placed in a uniform flow with velocity V at infinity (Fig. 3.20).

If the Reynolds number is high ($Re = \rho V a / \eta \gg 1$), we know that the flow can be considered as potential almost everywhere, except in the wake and within a thin layer at the surface of the body.

In this latter region, known as the *boundary layer*, the viscous terms cannot be neglected, since it is precisely due to them that the tangential velocity of the fluid goes to zero at the surface of the body. For this reason, and according to the Navier–Stokes equation, we will have in the boundary layer:

$$v_x \frac{\partial v_x}{\partial x} + v_y \frac{\partial v_x}{\partial y} = -\frac{1}{\rho}\frac{\partial P}{\partial x} + v\left(\frac{\partial^2 v_x}{\partial x^2} + \frac{\partial^2 v_x}{\partial y^2}\right), \tag{3.191a}$$

$$v_x \frac{\partial v_y}{\partial x} + v_y \frac{\partial v_y}{\partial y} = -\frac{1}{\rho}\frac{\partial P}{\partial y} + v\left(\frac{\partial^2 v_y}{\partial x^2} + \frac{\partial^2 v_y}{\partial y^2}\right). \tag{3.191b}$$

To these two equations one must add the incompressibility condition:

$$\frac{\partial v_x}{\partial x} + \frac{\partial v_y}{\partial y} = 0. \tag{3.192}$$

To simplify these equations, we start by noticing that, as long as the fluid streamlines do not separate from the surface of the body, the v_y component of the velocity is always very small compared to v_x:

$$v_y \ll v_x. \tag{3.193}$$

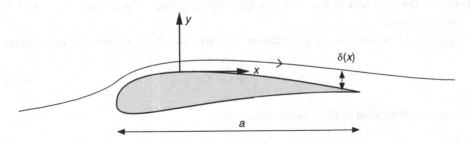

Fig. 3.20 Boundary layer close to a well-profiled body.

Under these conditions, we can neglect the v_y component in Eq. (3.191b) and write simply:

$$\frac{\partial P}{\partial y} = 0. \tag{3.194}$$

This relation is fundamental, showing that the pressure in the boundary layer is the same as in the main flow:

$$P = P(x), \tag{3.195}$$

where $P(x)$ is obtained by solving the problem of the potential flow in the perfect fluid around the body.

Let us now see how to simplify the first Navier–Stokes equation. We denote by δ the typical thickness of the boundary layer. The incompressibility condition (3.192) indicates that, as an order of magnitude

$$\frac{v_x}{a} \approx \frac{v_y}{\delta}, \tag{3.196}$$

where $\delta = \delta(x = a)$. It follows that the inertial advection term and the viscous term in the first Navier–Stokes equation are, respectively:

$$v_x \frac{\partial v_x}{\partial x} + v_y \frac{\partial v_x}{\partial y} \approx \frac{V^2}{a} + \frac{V}{\delta}\frac{\delta}{a}V \approx \frac{V^2}{a}, \tag{3.197}$$

$$\nu \frac{\partial^2 v_x}{\partial y^2} \approx \nu \frac{V}{\delta^2}. \tag{3.198}$$

In the boundary layer, these two terms must be of the same order of magnitude. Besides, this condition stems from the mathematics of the problem, because the differential equation for v_x must be second-order if the boundary conditions at the surface of the body and of the boundary layer are to be satisfied. Indeed, one must have $v_x = 0$ in $y = 0$ and $v_x = v(x, y)$ in $y = \delta$, where v is the velocity calculated for the main flow by solving the problem of potential flow in a perfect fluid. Hence, writing that the inertial and viscous terms are comparable yields:

$$\delta \approx \frac{a}{\sqrt{Re}}. \tag{3.199}$$

Therefore, the thickness of the boundary layer varies as the reciprocal square root of the Reynolds number.

To conclude this section, we note that, in the main flow, the Bernoulli equation yields:

$$P + \frac{1}{2}\rho v^2 = \text{Cst.} \tag{3.200}$$

Taking the x derivative of this relation gives:

$$\frac{dP}{dx} = -\rho v \frac{dv}{dx} \tag{3.201}$$

so that the *Prandtl equations for the boundary layer* can be written as:

$$\frac{\partial P}{\partial y} = 0,$$ (3.202)

$$v_x \frac{\partial v_x}{\partial x} + v_y \frac{\partial v_x}{\partial y} = v \frac{dv}{dx} + \nu \frac{\partial^2 v_x}{\partial y^2},$$ (3.203)

$$\frac{\partial v_x}{\partial x} + \frac{\partial v_y}{\partial y} = 0.$$ (3.204)

Unfortunately, there is no systematic method for solving these nonlinear equations.

3.6.2 Qualitative study of separation

We have already noted the separation problem several times. Indeed, it is when the boundary layer separates from the upper surface of the wing that the plane stalls.

This occurs when the incidence angle of the wing with respect to the velocity at infinity exceeds a critical value, or *stall angle*. Under these conditions, the lift vanishes and the streamlines around the wing look as in Fig. 3.21. This diagram shows that the boundary layer separates ahead of the trailing edge of the wing, so that the Chaplygin–Kutta condition is no longer fulfilled.

The separation phenomenon is obviously fundamental in aerodynamics, since it allows an airplane or a bird to land. On the other hand, it must be avoided during takeoff!

To find out in which flow region the boundary layer can separate, let us investigate the velocity profile in the vicinity of the separation point S. Its representation in Fig. 3.22 shows that, downstream from the separation point, the profile always exhibits an inflection point, I. This means that, in order for an inflection point to appear in the profile, the curvature of the velocity profile must necessarily be positive at the wall, knowing that it becomes negative again far from the wall. This simple geometric argument shows that a necessary condition for separation is that:

$$\frac{\partial^2 v_x}{\partial y^2} > 0 \quad \text{at the wall with } y = 0.$$ (3.205)

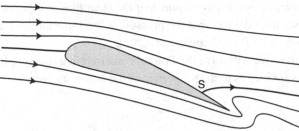

Fig. 3.21 Shape of the streamlines as the wing stalls. The boundary layer separates on the upper surface of the wing, ahead of the trailing edge, at a point S (separation point). Under these conditions, wing lift vanishes.

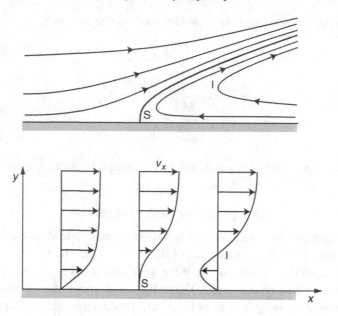

Fig. 3.22 Shape of the streamlines and of the velocity profile $v_x(y)$ close to the separation point S.

This condition can simply be rewritten as a condition on the pressure in the main flow. Indeed, the second Prandtl equation (3.203) yields:

$$\eta \frac{\partial^2 v_x}{\partial y^2} = \frac{dP}{dx} \quad \text{at the wall with } y = 0. \qquad (3.206)$$

A *necessary condition for separation* (but not sufficient) is thus that

$$\frac{dP}{dx} > 0 \quad \text{or} \quad \frac{dv}{dx} < 0 \quad \text{in the main flow.} \qquad (3.207)$$

The second inequality results from the direct application of Bernoulli's theorem.
Thus, the boundary layer can only separate if the main flow is sufficiently decelerated.
Note that here we used the fact that the pressure is the same on the wall as in the main flow, according to the first Prandtl equation (3.202).
In the next sections, we will solve the Prandtl equations in a few simple cases. This will give us a more precise idea as to the properties of the boundary layer.

3.6.3 Blasius profile

We consider the problem of the boundary layer formed at the surface of a thin semi-infinite plate, immersed in a flow with velocity at infinity v_∞ parallel to the plate (Fig. 3.23).

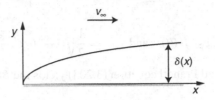

Fig. 3.23 Blasius boundary layer.

As a first step, let us look for the Reynolds number of the flow. The plate being considered infinite in the x direction, the only characteristic distance at point x will be the distance x itself. Consequently, the local Reynolds number is:

$$Re = \frac{v_\infty x}{\nu},\tag{3.208}$$

where ν is the kinematic viscosity.

This relation shows that the Reynolds number increases as x. The boundary layer must therefore become unstable and turbulent far enough from the leading edge.

To analyse the velocity profile in the laminar region, one must solve the Prandtl equations (3.202)–(3.204). To this end, we introduce the dimensionless variable:

$$\theta = \frac{y}{\delta},\tag{3.209}$$

where δ is the thickness of the boundary layer at a distance x from the leading edge, obtained by writing the equality of the viscous and inertial stresses (see Eq. (3.199)):

$$\delta = \frac{x}{\sqrt{Re}} = \sqrt{\frac{\nu x}{V}}.\tag{3.210}$$

We assume that the velocity profiles at two different points are similar and can be deduced from each other using δ as the length scale along y. More precisely, we are looking for a solution of the form:

$$\frac{v_x}{v_\infty} = g(\theta).\tag{3.211}$$

Since the flow is two-dimensional and the fluid is incompressible, there exists a stream function ψ such that:

$$v_x = \frac{\partial \psi}{\partial y},\tag{3.212}$$

$$v_y = -\frac{\partial \psi}{\partial x}.\tag{3.213}$$

Let $f(\theta)$ be a primitive of $g(\theta)$ ($f' = g$). Integrating Eq. (3.212) yields:

$$\psi = v_\infty \delta \, f(\theta)\tag{3.214}$$

resulting in:

$$v_x = v_\infty f'(\theta) \quad \text{and} \quad v_y = \frac{1}{2}\sqrt{\frac{v v_\infty}{x}}(\theta f' - f).$$ (3.215)

Substituting into the second Prandtl equation (3.203) yields the *Blasius equation*:

$$ff'' + 2f''' = 0.$$ (3.216)

This equation being third order, three boundary conditions are needed to solve it. Writing that $v_x = v_y = 0$ on the plate with $y = 0$ and that $v_x = v_\infty$ at $y = \infty$ yields:

$$f(0) = f'(0) = 0$$ (3.217)

and

$$f'(\infty) = 1.$$ (3.218)

The solution $f'(\theta)$ to this problem represents the velocity profile. It can be calculated numerically. We plot it in Fig. 3.24.

We can also calculate the drag force R exerted by the fluid on the plate. It is obtained by integrating the shear stress on both surfaces of the plate, taken as having a total width a:

$$R = 2 \int_0^a \eta \left(\frac{\partial v_x}{\partial y}\right)_{y=0} dx.$$ (3.219)

Using Eq. (3.215) the calculation gives:

$$R = 4 f''(0) v_\infty \sqrt{\eta \rho a v_\infty} \approx 1.33 \sqrt{\eta \rho a} v_\infty^{3/2}.$$ (3.220)

This formula shows that, in the laminar regime, the drag force increases as $v_\infty^{3/2}$.

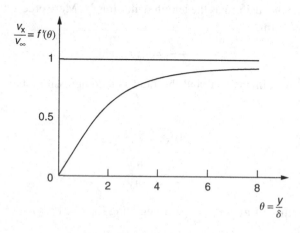

Fig. 3.24 Blasius function.

This dependence is different from that obtained at low Reynolds number, where the drag is proportional to v_∞ (see Section 3.7.7).

We conclude this section with a few general remarks, emphasising first of all that the thickness of the boundary layer at point x varies as $(vt)^{1/2}$, where t is the advection time of the vorticity over a distance x: $t = x/v_\infty$. This result should not come as a surprise, since the thickness of the boundary layer is fixed by the competition between vorticity advection and diffusion (see Section 3.4).

It follows, in particular, that the thickness of the boundary layer close to a planar disc turning at the angular velocity Ω is constant and given by:

$$\delta \approx \sqrt{\frac{v}{\Omega}}, \tag{3.221}$$

the characteristic time here being $1/\Omega$.

In the same way, the thickness of the boundary layer will remain finite close to a plate vibrating at a frequency ω:

$$\delta \approx \sqrt{\frac{v}{\omega}}. \tag{3.222}$$

Finally, the thickness of the Blasius boundary layer can be limited by aspirating the fluid through the wall. If v_a is the aspiration velocity, the thickness of the boundary layer saturates at a constant value, independent of the velocity v_∞ in the main flow:

$$\delta_a \approx \frac{v}{v_a}. \tag{3.223}$$

This value is reached beyond a typical distance $\delta_a(v_\infty/v_a)$ from the leading edge of the plate. Aspiration of the boundary layer helps stabilise it and prevents it from becoming turbulent, thus decreasing the drag. It also delays the separation when the flow decelerates. This technique was used to increase the lift of the turbosail invented by Malavard.

3.6.4 Falkner–Skan equation

One can solve in the same manner the problem of the boundary layer for a flow within a dihedral angle or around a wedge (see Fig. 3.13).

In this case, the main flow is described by a power law $v = Ax^m$ (see Section 3.5.3.2). Defining the variable

$$\theta = \frac{y}{\delta} \tag{3.224}$$

and knowing that

$$\delta = \sqrt{vt} \quad \text{with} \quad t = \frac{x}{Ax^m} \tag{3.225}$$

one obtains, with

$$\frac{v_x}{Ax^m} = f'(\theta), \tag{3.226}$$

the following *Falkner–Skan equation*:

$$f''' + \frac{1}{2}ff'' + m(1 - f'^2) = 0. \tag{3.227}$$

The boundary conditions are the same as for the Blasius problem (which corresponds to the particular case $m = 0$):

$$f(0) = f'(0) = 0 \quad \text{and} \quad f'(\infty) = 1. \tag{3.228}$$

This equation can be solved numerically. The solutions are depicted schematically in Fig. 3.25. We can see that the velocity profile has the same shape as for the Blasius problem as long as $m > m_c \approx -0.091$. Below this critical value, the velocity changes sign at the wall, showing that the boundary layer separates from the solid surface at the tip of the wedge (Fig. 3.26). This result is compatible with the general conclusions drawn in Section 3.6.2, since the flow is decelerated for $m < 0$ ($dv/dx = mAx^{m-1} < 0$).

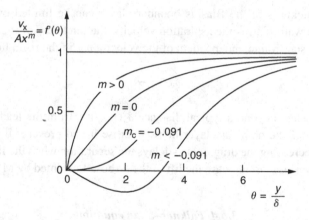

Fig. 3.25 Shape of the velocity profiles within the boundary layer for a main flow of the type $v_x = Ax^m$, for different values of the exponent (Falkner–Skan functions).

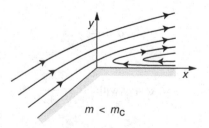

Fig. 3.26 Flow diagram for $m < m_c$. The boundary layer separates at the tip of the wedge.

3.6.5 *Integral equation of the boundary layer*

We are still concerned with the flow around a semi-infinite plane immersed in the main flow $v(x)$ (with $v(x) = \mathrm{Cst}$ in the Blasius problem and $v(x) = Ax^m$ in the Falkner–Skan case).

Von Karman introduced the two 'integral lengths' δ_1 and δ_2 defined as [6]:

$$\delta_1 v = \int_0^\delta (v - v_x)\,dy, \tag{3.229}$$

$$\delta_2 v^2 = \int_0^\delta v_x(v - v_x)\,dy. \tag{3.230}$$

The first one, δ_1, is called the *displacement thickness* and the second one, δ_2, *momentum thickness*.

We will show that the shear stress at the wall:

$$\sigma_0 = \eta \left(\frac{\partial v_x}{\partial y} \right)_{y=0} \tag{3.231}$$

is connected to these two lengths by the following equation:

$$\frac{\sigma_0}{\rho} = \frac{d}{dx}\left(v^2 \delta_2\right) + \delta_1 v \frac{dv}{dx}. \tag{3.232}$$

This is the *von Karman integral equation*.

To prove it, let us integrate the Prandtl equation (3.203) over the thickness of the boundary layer. Using the fact that $\partial v_x / \partial y = 0$ in $y = \delta$, one has:

$$\int_0^\delta \left(v_x \frac{\partial v_x}{\partial x} + v_y \frac{\partial v_x}{\partial y} - v \frac{\partial v}{\partial x} \right) dy = -\frac{\sigma_0}{\rho}. \tag{3.233}$$

But, from the continuity equation:

$$v_y = -\int_0^y \frac{\partial v_x}{\partial x}\,dy, \tag{3.234}$$

yielding, after substitution in Eq. (3.233):

$$\int_0^\delta \left[v_x \frac{\partial v_x}{\partial x} - \frac{\partial v_x}{\partial y} \left(\int_0^y \frac{\partial v_x}{\partial x}\,dy \right) - v \frac{\partial v}{\partial x} \right] dy = -\frac{\sigma_0}{\rho}. \tag{3.235}$$

Integrating by parts the second term on the left-hand side finally leads to:

$$\int_0^\delta \left(2v_x \frac{\partial v_x}{\partial x} - v \frac{\partial v_x}{\partial x} - v \frac{\partial v}{\partial x} \right) dy = -\frac{\sigma_0}{\rho}. \tag{3.236}$$

or, similarly:

$$\int_0^\delta \frac{\partial}{\partial x} [v_x(v - v_x)] \, dy + \frac{dv}{dx} \int_0^\delta (v - v_x) \, dy = \frac{\sigma_0}{\rho}. \tag{3.237}$$

This is nothing other than the von Karman equation (3.232).

This equation turns out to be very interesting, since it provides a fairly precise estimate of the shear stress at the wall σ_0, without solving the problem numerically.

To show this explicitly, let us do the calculation in the case of the Blasius problem. Here, $v = v_\infty$, and the von Karman equation reduces to

$$\frac{\sigma_0}{\rho} = v^2 \frac{d\delta_2}{dx}. \tag{3.238}$$

Since δ_2 is an integral quantity, it is only moderately sensitive to the choice of the velocity profile. The simplest option is to consider a linear profile:

$$v_x = \frac{v_\infty y}{\delta}. \tag{3.239}$$

In this case, one has:

$$\sigma_0 = \eta \frac{v_\infty}{\delta} \tag{3.240}$$

and

$$\delta_2 = \frac{\delta}{6}, \tag{3.241}$$

which yields, after substitution in the von Karman equation (3.238):

$$\frac{v}{v_\infty \delta} = \frac{1}{6} \frac{d\delta}{dx}. \tag{3.242}$$

The integration results in:

$$\delta = 2\sqrt{3} \sqrt{\frac{vx}{v_\infty}}, \tag{3.243}$$

which finally gives access to the stress at the wall:

$$\sigma_0 = \frac{\rho}{2\sqrt{3}} \sqrt{\frac{v}{x}} v_\infty^{3/2} \tag{3.244}$$

and to the drag force on a plate of width a:

$$R = 2 \int_0^a \sigma_0 \, dx = \frac{2}{\sqrt{3}} \sqrt{\eta \rho a} v_\infty^{3/2} = 1.15 \sqrt{\eta \rho a} v_\infty^{3/2}. \tag{3.245}$$

As expected, this formula is very close to the exact result, except for the numerical prefactor (1.15 instead of 1.33, see Eq. (3.220)).

3.6.6 Thermal boundary layer

We will not treat this problem in detail, but only present a few results that we find interesting; their demonstration is left as an exercise to the reader.

We are concerned here with heat transfer between the fluid and the plate in the Blasius problem, when they are not at the same temperature. To fix the ideas, let T_1 be the temperature of the fluid in the main flow and T_0 the temperature of the plate. Assuming these two temperatures to be constant, what will be the temperature profile in the fluid in the stationary regime?

From the heat equation (3.57) one can show that a *thermal boundary layer* of thickness δ', and with the same $x^{1/2}$ profile as the vorticity boundary layer, develops close to the plate (Fig. 3.27). This result can be obtained by writing that the two terms corresponding to temperature advection and diffusion are of the same order of magnitude. This line of reasoning implies that the energy dissipated by viscous friction is negligible compared to these two terms, which is legitimate if the absolute value of the temperature difference $\Delta T = |T_1 - T_0|$ is large enough:

$$C_p \, \Delta T \gg v_\infty^2. \tag{3.246}$$

One can then show that the thickness of the thermal boundary layer varies as:

$$\delta' \approx \frac{\delta}{Pr^{1/3}} \tag{3.247}$$

when $Pr \gg 1$, and as:

$$\delta' \approx \frac{\delta}{Pr^{1/2}}. \tag{3.248}$$

for $Pr \ll 1$, where Pr is the *Prandtl number* of the fluid, defined by:

$$Pr = \frac{v}{\Lambda}. \tag{3.249}$$

In this definition, v is the kinematic viscosity of the fluid and Λ its thermal diffusion coefficient (given by Eq. (3.59)). This number is simply the ratio of the typical diffusion

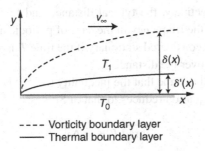

Fig. 3.27 Temperature and vorticity boundary layers in a fluid of high Prandtl number.

Table 3.1. *Some values of the Prandtl number. From [3]*

Material	Pr
Air	0.73
Water	6.75
Alcohol	16.6
Glycerine	7250
Mercury	0.044
Earth's mantle	$\sim 10^{21}$

times l^2/Λ and l^2/ν of the temperature and vorticity fluctuations, respectively, over the same distance l. It varies strongly with the fluid, as shown in Table 3.1.

We can see that, in materials such as glycerine (where the vorticity diffuses much faster than heat since $Pr \gg 1$), the thermal boundary layer is much thinner (by about a factor of 20, from Eq. (3.247)) than the vorticity boundary layer. The opposite is true for mercury, where the heat diffuses much faster than the vorticity ($Pr \ll 1$): in this material, the thermal boundary layer is typically five times thicker than the vorticity boundary layer, according to Eq. (3.248). In air, on the other hand, the two boundary layers have roughly the same thickness, since $Pr \approx 1$.

3.7 Flow at low Reynolds number

At low Reynolds number, the nonlinear component of the inertial force $\rho(\vec{v} \cdot \vec{\nabla})\vec{v}$ is, by definition, negligible compared to the viscous force $\eta \Delta \vec{v}$. Experiment shows that, for these conditions, the typical flow velocity is always very small compared to the speed of sound, so that the fluid can be considered incompressible.

However, a low Reynolds number does not imply a stationary flow. For this additional assumption to hold, the inertial term $\rho \, \partial \vec{v}/\partial t$ must be negligible compared to the viscous term $\eta \Delta \vec{v}$. This occurs for:

$$T \gg \frac{L^2}{\nu}, \tag{3.250}$$

where L and T are, respectively, the typical distance and time for the experiment (for instance, the reciprocal of the frequency in the case of periodic motion).

Thus, the motion can be considered stationary if the time T is very long compared to the time of vorticity diffusion over the distance L.

In the following, we will assume that the inequality (3.250) is fulfilled. For these conditions, the Navier–Stokes equation reduces to that of Stokes:

$$\overrightarrow{\text{grad}}\, P = \eta \, \Delta \vec{v} \tag{3.251}$$

neglecting the bulk forces $\vec{f}$.

Taking the div and $\overrightarrow{\text{curl}}$ of this equation, one can easily see that $\Delta P = 0$ and $\Delta \vec{\omega} = 0$, where $\vec{\omega}$ is the vorticity. These two equations are alternative forms of the Stokes equation. Below, we recall some of the essential properties of flows obeying the Stokes equation.

3.7.1 Unicity and additivity

The Stokes equation being linear and second order, there is a unique flow solution that obeys the equation and the boundary conditions imposed on the surfaces (see Section 3.2.3).

This result is not at all trivial, since the general Navier–Stokes has a multitude of solutions as soon as the nonlinearities become important (namely, at high Reynolds number).

Another consequence of the linearity of the Stokes equation is that the velocity fields corresponding to different flows around the same object can be superposed linearly, as long as the velocities on the walls of the object are combined in the same way.

3.7.2 Reversibility

As the Stokes equation is linear, its general solution is of the form:

$$v_i(\vec{r}) = \int_S G_{ij}(\vec{r}, \vec{r}') V_j(\vec{r}') \, dS(\vec{r}'), \qquad (3.252)$$

where S represents the boundary surface, $\vec{V}$ the velocity at this border and G a Green function. This expression shows that, by changing $\vec{V}$ into $-\vec{V}$ on the border, the velocity of the fluid is reversed everywhere, since $\vec{v}$ changes to $-\vec{v}$ from Eq. (3.252).

3.7.3 Minimal dissipation

To find the flow giving an extremum of the total dissipation:

$$\Phi_{\text{tot}} = \int_V \frac{\eta}{2} \left(\frac{\partial v_i}{\partial x_j} + \frac{\partial v_j}{\partial x_i} \right)^2 dV, \qquad (3.253)$$

and that obeys the incompressibility condition div $\vec{v} = 0$, we must look for the extrema of the functional

$$\overline{\Phi}_{\text{tot}} = \int_V \left[\frac{\eta}{2} \left(\frac{\partial v_i}{\partial x_j} + \frac{\partial v_j}{\partial x_i} \right)^2 - \lambda \, \text{div} \, \vec{v} \right] dV, \qquad (3.254)$$

where λ is the Lagrange multiplier associated with the incompressibility condition. The corresponding Lagrange equation immediately yields the Stokes equation, provided that we set $P = \lambda/2$.

Consequently, the velocity field satisfying the Stokes equation is, among all kinematically allowed velocity fields (i.e. those obeying the boundary conditions and the incompressibility condition), the one resulting in an extremum of the dissipation Φ_{tot}. One can show that this extremum is a minimum (see, for instance, [2]).

3.7.4 The reciprocity theorem

Consider two Stokes flows $v_i^{(1)}, \sigma_{ij}^{(1)}$ and $v_i^{(2)}, \sigma_{ij}^{(2)}$ around a body with surface S. One can show that:

$$\int_S \sigma_{ij}^{(1)} v_j^{(2)} \mathrm{d}S = \int_S \sigma_{ij}^{(2)} v_j^{(1)} \mathrm{d}S \tag{3.255}$$

with σ_{ij} the total stress tensor.

It follows that, if $\vec{F}$ and $\vec{\Gamma}$ are, respectively, the resultant force and torque at any point O of the forces acting on a body that moves at velocity $\vec{U}$ and rotates at angular velocity $\vec{\Omega}$, one has, for two realisations 1 and 2 of the flow:

$$\vec{U}^{(1)} \cdot \vec{F}^{(2)} + \vec{\Omega}^{(1)} \cdot \vec{\Gamma}^{(2)} = \vec{U}^{(2)} \cdot \vec{F}^{(1)} + \vec{\Omega}^{(2)} \cdot \vec{\Gamma}^{(1)}. \tag{3.256}$$

3.7.5 Lubrication theory

In this section, we consider the flow of a fluid confined in a very narrow gap between two solid surfaces (lubrication geometry). The lower (immobile) surface coincides with the $y = 0$ plane; the upper surface, which is mobile, is described by the equation $y = h(x, t)$. The fluid fills the space $0 < x < a$ between the two surfaces. We assume that $a \gg h$ and that the problem is invariant under translation along z (Fig. 3.28).

Let us start by writing the boundary conditions. If the fluid does not slip over the surfaces, then:

$$\vec{v} = 0 \quad \text{at} \quad y = 0 \tag{3.257}$$

and

$$\vec{v} = (U, V) \quad \text{at} \quad y = h(x, t) \tag{3.258}$$

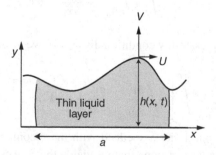

Fig. 3.28 Fluid confined between two adjoining surfaces.

with

$$V = \frac{\partial h}{\partial t} + \frac{\partial h}{\partial x} U. \tag{3.259}$$

As a second step, let us simplify the Navier–Stokes equations in order to establish the Reynolds equation for the evolution of the pressure field.

We notice first of all that $v_y \ll v_x$, since from the incompressibility equation:

$$\frac{v_x}{a} \approx \frac{v_y}{h}. \tag{3.260}$$

We can therefore neglect the v_y component in the second Navier–Stokes equation (along y) and write simply:

$$\frac{\partial P}{\partial y} = 0. \tag{3.261}$$

In the first Navier–Stokes equation (along x), we must consider the order of magnitude of the different terms. Using Eq. (3.260) we find, in order:

$$\rho \frac{\partial v_x}{\partial t} \approx \rho \frac{U^2}{a}, \tag{3.262}$$

$$\rho \left(v_x \frac{\partial v_x}{\partial x} + v_y \frac{\partial v_x}{\partial y} \right) \approx \rho \frac{U^2}{a}, \tag{3.263}$$

$$\eta \Delta \vec{v} \approx \eta \frac{U}{h^2}. \tag{3.264}$$

Based upon the general definition of the Reynolds number (3.111), we find:

$$Re = \frac{\rho U h^2}{\eta a} \tag{3.265}$$

and we see that the characteristic length scale of the problem is here the non-trivial combination h^2/a. For $h \ll a$, $Re \ll 1$ so that the two inertial terms can be neglected in the first Navier–Stokes equation, which becomes simply:

$$\frac{\partial P}{\partial x} = \eta \frac{\partial^2 v_x}{\partial y^2}. \tag{3.266}$$

Solving Eqs. (3.261) and (3.266) immediately results in:

$$v_x = \frac{1}{2\eta} \frac{\partial P}{\partial x} y(y - h) + U \frac{y}{h} \tag{3.267}$$

taking into account the boundary conditions $v_x(y = 0) = 0$ and $v_x(y = h) = U$. The general solution is thus a superposition of a Couette flow (linear term) and a Poiseuille flow (quadratic term). This equation also yields the flow rate:

$$Q = \int_0^h v_x \, dy = \frac{1}{12\eta} \frac{\partial P}{\partial x} h^3 + U \frac{h}{2}. \tag{3.268}$$

Using the incompressibility equation and the boundary conditions, the first equation gives:

$$\frac{\partial Q}{\partial x} = U\frac{\partial h}{\partial x} + \int_0^h \frac{\partial v_x}{\partial x}dy = U\frac{\partial h}{\partial x} - V \tag{3.269}$$

and, using Eq. (3.259):

$$\frac{\partial Q}{\partial x} + \frac{\partial h}{\partial t} = 0. \tag{3.270}$$

This is the flow rate equation. Replacing Q by its expression (3.268) yields the *Reynolds equation* for the pressure:

$$\frac{\partial}{\partial x}\left(h^3\frac{\partial P}{\partial x}\right) = 6\eta\left(h\frac{\partial U}{\partial x} + U\frac{\partial h}{\partial x} + 2\frac{\partial h}{\partial t}\right). \tag{3.271}$$

Let us now apply this formula to the lubrication problem. The basic geometry is depicted in Fig. 3.29. The mobile upper surface is planar and makes an angle α with the lower one. It moves along the x axis at a constant velocity U. Under these conditions:

$$V = 0, \quad \frac{\partial h}{\partial x} = \alpha \quad \text{and} \quad \frac{\partial h}{\partial t} = -\alpha U. \tag{3.272}$$

Substituting into the Reynolds equation, and assuming that $h \gg \alpha a$, yields:

$$\frac{\partial^2 P}{\partial x^2} = -\frac{6\eta\alpha U}{h^3}. \tag{3.273}$$

The integration is straightforward once we notice that $dx = dh/\alpha$ and gives, writing that the pressure within the fluid equals the atmospheric pressure P_a at the free edges:

$$P - P_a = \frac{3\eta U}{\alpha}\frac{(h - h_1)(h_2 - h)}{hh_1h_2}, \tag{3.274}$$

where h_1 and h_2 stand for the thickness of the liquid layer at $x = 0$ and $x = a$, respectively.

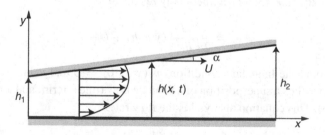

Fig. 3.29 Lubrication geometry and shape of the velocity profile within the fluid layer.

Using this equation we can calculate the normal and tangent forces exerted by the liquid on the mobile surface:

$$F_N = \int_{h_1}^{h_2} (P - P_a) \frac{dh}{\alpha} = \frac{3\eta U}{\alpha^2} \left[\frac{1}{2} \frac{h_2^2 - h_1^2}{h_1 h_2} - \ln\left(\frac{h_2}{h_1}\right) \right], \tag{3.275}$$

$$F_T = -\eta \int_{h_1}^{h_2} \left(\frac{\partial v_x}{\partial y}\right)_{y=h} \frac{dh}{\alpha} = -\eta \frac{U}{\alpha} \left[\left(1 + \frac{3}{2\alpha}\right) \ln\left(\frac{h_2}{h_1}\right) + \frac{3}{4\alpha} \frac{h_2^2 - h_1^2}{h_1 h_2} \right]. \tag{3.276}$$

These formulas simplify as the angle α tends to 0 becoming, to lowest order in α:

$$F_N = \frac{\eta U}{2} \alpha \left(\frac{a}{h}\right)^3 + 0(\alpha^3), \tag{3.277}$$

$$F_T = -\eta U \frac{a}{h} + 0(\alpha^2). \tag{3.278}$$

These forces are in the ratio

$$\frac{F_N}{F_T} \approx \alpha \left(\frac{a}{h}\right)^2. \tag{3.279}$$

This factor can be much larger than 1 if the liquid layer is very thin compared to its lateral extension ($a \gg h$). Under these conditions, the liquid exerts on the mobile surface a normal force, also known as the lubrication force, much larger than the tangent friction force. This is the principle of lubrication.

In the chapter on liquid crystals we will see that lamellar phases, due to their layered structure, are much better lubricants than isotropic fluids.

We will also see in Sections 3.8.3 and 3.8.4 two other applications of the Reynolds equation to the cases of the piezoelectric rheometer and the surface force apparatus.

3.7.6 Spreading of a liquid on a rotating plate (spin-coating)

Let us place a drop of a viscous liquid atop a planar disc turning quickly around its revolution axis (Fig. 3.30). Experiment shows that, under the action of the centrifugal force, the drop will spread very fast and form a film of homogeneous thickness over the disc surface.

To prove this let us write the equations of motion in cylindrical coordinates (r, θ, z) (see Appendix 3.B). Let ω be the angular velocity of the plate. In the asymptotic regime, $v_\theta = \omega r$. We assume that the Reynolds number is small (a hypothesis we will check a posteriori).

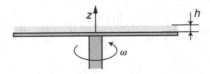

Fig. 3.30 Liquid film on a rotating plate.

Preserving the centrifugal inertia term, which is the driving force for the spreading, the equations of motion can be written explicitly as:

$$\frac{\partial P}{\partial r} = \eta \left(\frac{\partial^2 v_r}{\partial r^2} + \frac{1}{r} \frac{\partial v_r}{\partial r} - \frac{v_r}{r^2} + \frac{\partial^2 v_r}{\partial z^2} \right) + \rho \omega^2 r, \tag{3.280}$$

$$\frac{\partial P}{\partial z} = \eta \left(\frac{\partial^2 v_z}{\partial r^2} + \frac{1}{r} \frac{\partial v_z}{\partial r} + \frac{\partial^2 v_z}{\partial z^2} \right), \tag{3.281}$$

to which one must add the incompressibility equation:

$$\frac{\partial v_r}{\partial r} + \frac{v_r}{r} + \frac{\partial v_z}{\partial z} = 0. \tag{3.282}$$

We can search for a solution of the form:

$$v_r = -Cz^2 r + Drz, \tag{3.283}$$

$$v_z = 2C\frac{z^3}{3} - Dz^2, \tag{3.284}$$

where C and D are constants. This velocity field obeys the incompressibility equation and the no-slip condition on the disc at $z = 0$.

Replacing v_z by its expression in the equation of motion (3.281) and knowing that at the surface of the film (at $z = h$):

$$P - 2\eta \frac{\partial v_z}{\partial z} = P_a \tag{3.285}$$

where P_a is the atmospheric pressure, one finds the pressure in the film:

$$P = 2\eta C(z^2 + h^2) - 2\eta D(z + h) + P_a. \tag{3.286}$$

Now, replacing v_r by its expression in the first equation of motion (3.280), one finds the constant C:

$$C = \frac{\rho \omega^2}{2\eta}. \tag{3.287}$$

To calculate the constant D, let us write that the shear stress vanishes at the free surface:

$$\sigma_{zr}^v = \eta \left(\frac{\partial v_r}{\partial z} + \frac{\partial v_z}{\partial r} \right) = 0 \quad \text{(at } z = h\text{)} \tag{3.288}$$

yielding

$$D = 2Ch. \tag{3.289}$$

Hence, this calculation shows that the film has a uniform thickness h, evolving with time according to the following equation:

$$\frac{dh}{dt} = v_z(z = h) = -\frac{4}{3}Ch^3. \tag{3.290}$$

Integrating this equation is straightforward, resulting in:

$$\frac{1}{h_0^2} - \frac{1}{h^2} = -\frac{4}{3}\frac{\rho\omega^2}{\eta}t. \tag{3.291}$$

Obviously, this calculation is only valid for very low Reynolds numbers, which amounts to:

$$Re = \frac{\rho Ch^4}{\eta} = \left(\frac{\rho\omega h^2}{\eta}\right)^2 \ll 1. \tag{3.292}$$

This condition is very quickly fulfilled, provided the film is thin.

To fix the ideas, take the example of a fluid with a viscosity of 10 Pa s and an initial thickness of 100 μm on top of a plate turning at the considerable velocity of 10 000 rpm. Equation (3.291) shows that its thickness is typically 1 μm after 1 min, 0.33 μm after 10 min and 0.14 μm after 1 hour. Note that the film thickness very quickly becomes independent of the initial thickness (a condition fulfilled as soon as $t \gg \eta/\rho\omega^2 h_0^2$, yielding here $t \gg 10^{-2}$ s).

3.7.7 Stokes formula

We have shown in Section 3.6.3, dealing with the Blasius boundary layer, that the drag force on a plate moving at the velocity V in a fluid at rest at infinity is proportional to $V^{3/2}$ for $Re \gg 1$. How does this law change at low Reynolds number?

To answer this question, let us first point out that the f function in Reynolds' similarity law (3.105) is independent of the Reynolds number in the $Re \ll 1$ limit. Indeed, let us multiply by λ the velocity V of the body while keeping its size L fixed. The Reynolds number of the corresponding flow is multiplied by λ. But we know from Section 3.7.2 that multiplying V by λ results in a similar magnification of the hydrodynamic velocity field $\vec{v}$ (this result follows from the linearity of the Stokes equation). Consequently:

$$f\left(\frac{\vec{r}}{L}, \lambda Re\right) = f\left(\frac{\vec{r}}{L}, Re\right) \quad \forall\lambda \tag{3.293}$$

proving that f is independent of Re. Reynolds' similarity law becomes:

$$\frac{\vec{v}}{V} = f\left(\frac{\vec{r}}{L}\right) \quad \text{for} \quad Re \ll 1. \tag{3.294}$$

To find out how the drag force $\vec{F}^{\mathrm{d}}$ acting on the body varies, one just needs to remember that, in the stationary regime, the dissipation theorem (3.83) states that:

$$\vec{F}^{\mathrm{d}} \cdot \vec{V} = \Phi_{\mathrm{tot}}. \tag{3.295}$$

This force, opposing the velocity of the body, is given (in absolute value) by:

$$F^{\mathrm{d}} = \frac{\Phi_{\mathrm{tot}}}{V} = \frac{\eta}{2V} \int_{\mathrm{fluid}} \left(\frac{\partial v_i}{\partial x_j} + \frac{\partial v_j}{\partial x_i} \right)^2 dV, \tag{3.296}$$

which yields, after using the law (3.294) and rendering the integral dimensionless:

$$F^{\mathrm{d}} = A\eta L V, \tag{3.297}$$

where A is a constant depending only on the shape of the body and its orientation with respect to the flow.

The same line of reasoning can be followed for a rotating body. One proves in exactly the same way that the fluid exerts on the body a drag torque $\vec{\Gamma}^{\mathrm{d}}$, opposing the rotation vector $\vec{\Omega}$, with a magnitude:

$$\Gamma^{\mathrm{d}} = B\eta L^3 \Omega, \tag{3.298}$$

where B is a constant depending only on the shape of the body and on its orientation with respect to the rotation vector.

One can show that, for a sphere of radius R moving at a velocity V (*Stokes formula*):

$$F^{\mathrm{d}} = 6\pi \eta R V \tag{3.299}$$

yielding $A = 6\pi$ with our notation (and taking $L = R$). This formula holds up to Reynolds numbers close to one. We recall that the velocity and pressure fields around the sphere are given by the following formulas:

$$\vec{v} = \frac{3}{4} R \frac{\vec{V} + \vec{e}_r (\vec{V} \cdot \vec{e}_r)}{r} + \frac{R^3}{4} \frac{\vec{V} - 3\vec{e}_r (\vec{V} \cdot \vec{e}_r)}{r^3}, \tag{3.300}$$

$$P = \mathrm{Cst} + \frac{3}{2} \eta \frac{\vec{V} \cdot \vec{e}_r}{r^2} R, \tag{3.301}$$

where $\vec{e}_r = \vec{r}/r$ is the radial unit vector, taking the coordinate origin at the centre of the sphere. Equation (3.300) shows that the velocity decreases slowly with the distance, as $1/r$. The *hydrodynamic interactions* between two moving spheres are thus *long ranged*. This result is very important in the study of suspensions.

One can also show that, for a sphere rotating at the angular velocity Ω:

$$\Gamma^{\mathrm{d}} = 8\pi \eta R^3 \Omega \tag{3.302}$$

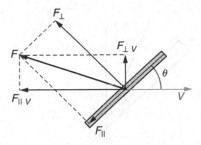

Fig. 3.31 Force acting on a cylinder (with circular section) moving at a velocity V.

yielding $B = 8\pi$ (again, with $L = R$). As to the velocity field, it is given by the expression:

$$\vec{v} = \frac{R^3}{r^3}\vec{\Omega} \times \vec{r}. \qquad (3.303)$$

This time, the velocity field drops as $1/r^2$, faster than for a translation.

Another useful formula, for which we will not give the demonstration, gives the resistance force on an elongated cylinder with radius R and length L (taken as the characteristic length). This force depends on whether the velocity of the cylinder is parallel ($F_\parallel$) or perpendicular ($F_\perp$) to its revolution axis. A complete calculation yields [2]:

$$F_\parallel = \Lambda_\parallel \eta L V_\parallel = \frac{4\pi \eta L V_\parallel}{\ln(L/R) - 0.72} \qquad (3.304)$$

and

$$F_\perp = A_\perp \eta L V_\perp = \frac{8\pi \eta L V_\perp}{\ln(L/R) + 0.5}. \qquad (3.305)$$

Note that $A_\perp \approx 2A_\parallel$; this relation becomes exact in the $L/R \to \infty$ limit.

These two formulas also give the force exerted by the fluid on the cylinder as its revolution axis makes a generic angle θ with its velocity $\vec{V}$. Indeed, it is then sufficient to decompose the velocity into its parallel and perpendicular components with respect to the cylinder axis ($V_\parallel = V\cos\theta$ and $V_\perp = V\sin\theta$), to calculate the associated drag forces $F_\parallel$ and $F_\perp$ using the formulas above, and then to sum these forces (this is allowed at low Reynolds number, the Stokes equation being linear). This calculation shows that the force exerted by the fluid on the cylinder is contained in the symmetry plane defined by the cylinder axis and the velocity. This force has a component antiparallel to the velocity, of magnitude:

$$F_{\parallel V} = \left(A_\parallel \cos^2\theta + A_\perp \sin^2\theta\right)\eta L V \qquad (3.306)$$

and a component perpendicular to the velocity, given by:

$$F_{\perp V} = |A_\parallel - A_\perp|(\sin\theta\cos\theta)\eta L V. \qquad (3.307)$$

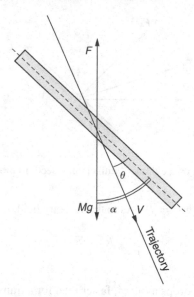

Fig. 3.32 Oblique fall of an elongated rod with circular section in a viscous fluid.

This latter component, vanishing by symmetry for $\theta = 0$ or $\pi/2$, corresponds to a 'viscous' lift force (since, unlike the Magnus force, it is not inertial in origin).

A very instructive exercise is to study the fall of a rod in a viscous fluid, when its long axis initially makes an angle α with the vertical (Fig. 3.32). Experiment shows that, if the rod exhibits revolution symmetry around its long axis, it does not fall vertically, but sideways, preserving its initial angle with the vertical. This is a direct result of symmetry, the fluid exerting no torque on the rod (for a demonstration, see for instance [2]). It is then easily checked, by balancing the viscous force F and the weight, that the velocity of the rod makes a constant angle $\alpha - \theta$ with the vertical, given by the relation:

$$\tan(\alpha - \theta) = \frac{\tan \alpha}{2 + \tan^2 \alpha}. \tag{3.308}$$

We note that the maximum of the angle of fall $\alpha - \theta$ is obtained for $\tan \alpha = 2^{1/2}(\alpha = 57.7°)$, namely $19.5°$.

3.7.8 Microorganism propulsion

We will now try to understand how microorganisms move (Purcell, 1977). Their movement is obviously important, our existence depending on it. Where would we be if spermatozoa weren't able to swim in the seminal liquid?

To better understand the subtle points of propulsion at very low Reynolds numbers, let us start by observing a fish swimming in a river. It can be schematised as a rigid body with an oscillating tail. Clearly, it can move using this periodic motion.

We will show that this is only possible because the Reynolds number associated with the flow around the fish is always much higher than 1. Indeed, even for a millimetric fry, one has typically:

$$Re \approx 10 \qquad (3.309)$$

with $\rho \approx 1000$ kg/m^3, $\eta \approx 10^{-3}$ Pa s, $L \approx 1$ mm and $U \approx 1$ cm/s (which is fairly slow). For a big pike, this number would be much larger (10^6), showing that for all fish the lift (or Magnus) forces associated with the inertial effects are largely dominant over the viscous forces in their propulsion mechanisms.

Let us now imagine our fish being immersed in a very viscous liquid, so that its Reynolds number falls below 1. In this case (provided it survives) it would be incapable of swimming over long distances. A graphical explanation is presented in Fig. 3.33. We can see that the fish advances with respect to the fluid (which is at rest at infinity) as its tail beats upwards in schema (a), but moves back at exactly the same speed as its tail beats downwards in schema (b). This follows directly from the reversibility of the Stokes equation at low Reynolds number (see Section 3.7.2). The fish would then start to oscillate in place, so that it would need to come up with another propulsion mode in order to advance.

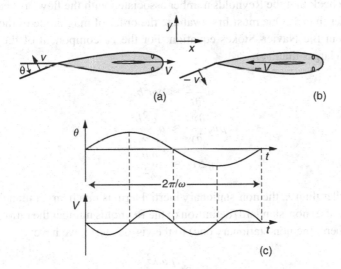

(a) (b)

(c)

Fig. 3.33 A fish immersed in a very viscous liquid would start to oscillate in place as it moves its tail. The two schemas (a) and (b) depict the fish in two identical positions. In (a), its tail beats upwards; the fish moves to the right (with respect to the fluid, considered at rest at infinity). In (b), its tail beats downwards; this time it moves to the left, as per the theorem on the reversibility of motion. In (c) we have plotted the rotation angle of the tail and the horizontal component of the fish's velocity as a function of time. Note that the fish oscillates at double the frequency of its tail's motion.

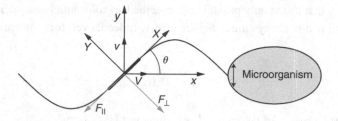

Fig. 3.34 Wave propagating along the flagellum of a microorganism and the forces acting on a flag-ellum element.

This is exactly what microorganisms did by getting equipped with a flagellum.[17] In this case, the motion is no longer reversible since a wave propagates along the flagellum (Taylor, 1951, 1958; [2]).

Before estimating the velocity of a flagellum, let us present a few orders of magnitude. In practice, a flagellum is a very thin cylindrical tube, a few tens of nanometres in diameter and a few micrometres in length. In the reference frame of the microorganism, the flagellum can thus be represented by a curve of equation (Fig. 3.34):

$$y = b \sin(kx + \omega t) \tag{3.310}$$

with $\lambda \approx 2\pi/k \approx 1$ μm and $\omega \approx 100$ Hz, typically.

Let us first check that the Reynolds number associated with the flow around a flagellum is much smaller than 1. One must first evaluate the order of magnitude of the inertial and viscous terms in the Navier–Stokes equation. For the v_y component of the velocity (in practice, much larger than v_x):

$$\rho \frac{\partial v_y}{\partial t} \approx \rho \omega^2 b, \tag{3.311}$$

$$\rho v_y \frac{\partial v_y}{\partial y} \approx \rho \frac{\omega^2 b^2}{\lambda}, \tag{3.312}$$

$$\eta \Delta v_y \approx \eta \frac{\omega b}{\lambda^2}. \tag{3.313}$$

Since b is smaller than λ, the non-stationary inertial term is much larger than the advection term. Taking as the (non-standard) definition of the Reynolds number the ratio of the largest inertial term (here, the non-stationary one) to the viscous term, we have:

$$Re \approx \frac{\rho \omega \lambda^2}{\eta}. \tag{3.314}$$

The values given above for λ and ω yield $Re \approx 10^{-4}$ taking water as the reference fluid ($\eta = 10^{-3}$ Pa s and $\rho = 1000$ kg/m³).

---

[17] Propulsion modes other than the tail or the flagellum also exist, for instance, the ciliary mode. Used by small mollusca in water, it is particularly effective at intermediate Re numbers, neither too high nor too low ($10^{-2} < Re < 10$). For more details, see (Childress, 1981).

We will now attempt a qualitative derivation of the flagellum's velocity. To that end, we only need to write down the normal and tangential forces acting on a small flagellum element. Let us do the calculation in the laboratory frame of reference, where the fluid is at rest at infinity. Let V be the translation velocity of the flagellum along x, θ the angle made by the flagellum element with the x axis and v its velocity along the y direction. The fluid exerts on the element a drag force (per unit length) given by the formulas (3.304) and (3.305). With $v_\parallel$ and $v_\perp$ the velocity of the flagellum element parallel and perpendicular to its long axis respectively, we find, assuming that θ remains small:

$$F_\parallel = -Kv_\parallel = -K(V + v\theta), \qquad (3.315)$$

$$F_\perp = -2Kv_\perp = -2K(v - V\theta), \qquad (3.316)$$

where K is a constant (in the limit of an infinitely slender flagellum).

If the inertia of the flagellum is negligible, the x projection of the forces acting upon each of its elements must vanish. It follows that:

$$F_\parallel - F_\perp\theta = 0, \qquad (3.317)$$

which, using relations (3.315) and (3.316), yields:

$$-K(V + v\theta) + 2Kv\theta = 0, \qquad (3.318)$$

resulting in:

$$V = v\theta. \qquad (3.319)$$

Since we know that:

$$v = b\omega\cos(kx + \omega t) \qquad (3.320)$$

and

$$\theta = bk\cos(kx + \omega t), \qquad (3.321)$$

we conclude that the flagellum moves with a finite average velocity:

$$\overline{V} \approx b^2 k\omega/2. \qquad (3.322)$$

The velocity is positive, meaning that the microorganism moves against the wave propagating along its flagellum (a wave propagating in the opposite direction, i.e. $y = b\sin(\omega t - kx)$, would give $V \approx -b^2 k\omega/2$).

As to the actual velocity, we find typically $\overline{V} \approx 5$ μm/s with $b \approx 0.1$ μm and the values given above for k and ω. This is indeed the order of magnitude observed in the kingdom of protozoa.

Formula (3.319) remains applicable to the case of the fish and helps characterise its motion. Indeed, the equation of its tail is that of a straight line:

$$y = \theta_0 x \sin\omega t \quad \text{with} \quad -L < x < 0, \qquad (3.323)$$

where L is the length of the tail, yielding:

$$v = \theta_0 \omega x \cos \omega t \tag{3.324}$$

and

$$\theta = \theta_0 \sin \omega t. \tag{3.325}$$

From Eq. (3.319) we deduce that the fish oscillates in place at double the frequency of its tail's beat, since:

$$V \approx -\theta_0^2 \omega L \sin(2\omega t). \tag{3.326}$$

This law is compatible with schema (c) of Fig. 3.33 and predicts that the fish is overall stationary:

$$\overline{V} = 0. \tag{3.327}$$

3.7.9 Poiseuille flow and porous media

Let us first consider the flow of an incompressible viscous fluid, of viscosity η, in a cylindrical tube with radius R and length L as a constant pressure difference $\Delta P = P_0 - P_L$ is imposed between the two ends of the tube. If the tube is long enough (we will see later what this means), the flow can be seen as invariant along the z direction, parallel to the tube axis. If the flow remains laminar, the only non-zero component of the velocity is v_z. Thus, in cylindrical coordinates (r, θ, z), the Navier–Stokes equation reduces to:

$$-\frac{\partial P}{\partial z} + \eta \frac{1}{r} \frac{d}{dr} \left(r \frac{dv_z}{dr} \right) = 0. \tag{3.328}$$

This equation is easily integrated, because the pressure gradient is constant $\left(-\frac{\partial P}{\partial z} = \frac{\Delta P}{L} \right)$ and gives, assuming a no-slip condition at the wall of the tube:

$$v_z = \frac{(P_0 - P_L)R^2}{4\eta L} \left[1 - \left(\frac{r}{R} \right)^2 \right]. \tag{3.329}$$

The velocity profile is parabolic, meaning that the viscous shear stress $\sigma_{zr}^v = \eta \frac{\partial v_z}{\partial r}$ has its maximum on the surface of the tube and decreases linearly to 0 in the centre of the tube (Fig. 3.35).

One can also determine the flow rate Q and the average velocity $\langle v_z \rangle$ of the fluid in the tube (*Poiseuille law*):

$$Q = \pi R^2 \langle v_z \rangle = \int_0^R v_z 2\pi r \, dr = \frac{\pi(P_0 - P_L)R^4}{8\eta L}. \tag{3.330}$$

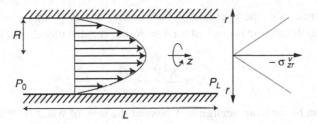

Fig. 3.35 Poiseuille flow in a cylindrical tube. Velocity profile and (absolute value of the) shear stress.

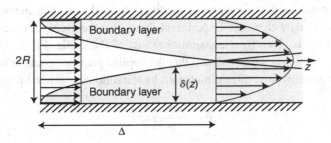

Fig. 3.36 'Plug flow' at the inlet of a cylindrical tube. The parabolic velocity profile is reached as the boundary layers coalesce.

This very important formula shows that the average velocity of the fluid is proportional to the pressure gradient (up to a sign change) and inversely proportional to the viscosity of the fluid:

$$\langle v_z \rangle = -\frac{K}{\eta} \frac{\partial P}{\partial z}. \tag{3.331}$$

The K coefficient, with units of L^2, is proportional to the cross-section of the tube:

$$K = \frac{R^2}{8}. \tag{3.332}$$

It quantifies its permeability.

Note that these formulas apply as long as the Reynolds number of the flow, defined by $Re = 2R\rho\langle v_z \rangle/\eta$, is below 2100. Above this value, the flow becomes generally turbulent.

These laws assume that the flow is invariant along the tube. In practice, this is not strictly true, since the velocity profile at the entry of the tube is not parabolic, but closer to a plug flow (Fig. 3.36). In these conditions, a vorticity boundary layer develops at the tube wall, its thickness δ increasing away from the inlet. According to Prandtl's theory of the boundary layer:

$$\delta \approx \sqrt{\nu \frac{z}{V}} \tag{3.333}$$

with z the distance from the inlet and V the velocity at this point ($V = Q/\pi R^2$). The asymptotic parabolic regime is reached for $\delta \approx R$, i.e. at a distance Δ away from the inlet:

$$\Delta \approx \frac{R^2 V}{\nu} = \frac{Q}{\pi \nu}. \tag{3.334}$$

This distance can be far from negligible. Consider the flow of water ($\nu \approx 10^{-6}$ m^2/s) in a hose of radius 1 cm. In this case, $\Delta \approx 1$ m for $V = 1$ cm/s ($Q \approx 1100$ litres/hour). On the other hand, for oil with a viscosity of 10 Pa s and a capillary tube with a radius of 0.1 mm, one has $\Delta \approx 0.1$ mm, with the same velocity $V = 1$ cm/s.

Let us now consider flow in a porous medium, defined as a bulk medium filled with pores interconnected by channels. Experiment shows that the average velocity of the fluid within the medium is given by a law similar to that of Poiseuille (Eq. (3.331)), provided the medium is completely saturated, and that the applied pressure gradient is not too high. This is *Darcy's law*, given in three dimensions by the following general formula:

$$\vec{v} = -\frac{K}{\eta}\overrightarrow{\text{grad}}\, P. \tag{3.335}$$

The constant K, with units of L^2, is the *permeability* of the medium. It is a scalar for an isotropic material and a tensor otherwise.

The permeability is proportional to the square of the average channel radius and to the porosity of the medium, defined by:

$$\phi = \frac{\text{pore volume}}{\text{total volume}}. \tag{3.336}$$

This result is trivial for a one-dimensional porous medium, depicted in Fig. 3.37. We are dealing here with a solid material pierced along the z direction by circular holes with radius R, separated from each other by a distance L. In this case:

$$\phi = \frac{\pi R^2}{L^2}. \tag{3.337}$$

The permeability along the z direction is found immediately, using the Poiseuille law:

$$K = \phi \frac{R^2}{8}. \tag{3.338}$$

3.7.10 Saffman–Taylor instability in a Hele–Shaw cell

An important practical problem is the retrieval of oil trapped within porous rock. The usual solution consists of digging two wells, injecting water under pressure into the first one and retrieving the oil out of the second well. However, experiment shows that using

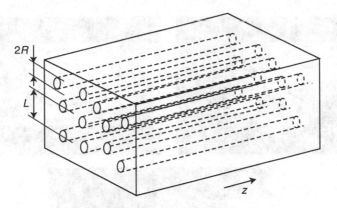

Fig. 3.37 Example of a one-dimensional porous medium.

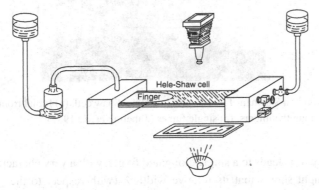

Fig. 3.38 Hele–Shaw cell used by Saffman and Taylor to study viscous fingering (Saffman and Taylor, 1958).

this technique one cannot completely extract the oil from the deposit, but only a limited proportion, which decreases as the imposed flow rate increases.

This difficulty is related to the development, in the porous medium, of an instability at the interface between the two fluids. This instability leads very quickly to the formation of 'fingers', and the oil remains trapped between them.

To explain this instability, P. G. Saffman and G. I. Taylor devised a similar experiment in a Hele and Shaw cell, which replaces the porous medium (Fig. 3.38). This cell consists of two rectangular glass plates, fitted with spacers, which form an elongated channel of a thickness b and width $L \gg b$. In this experiment, the channel is filled with oil (it is preferable to use a glass-wetting liquid to avoid meniscus problems related to wetting hysteresis), then emptied by injecting air (or water) at one end.

The behaviour of the interface between the two fluids is shown in Fig. 3.39. At the beginning, the interface is flat (we neglect its curvature along the cell thickness). As air under pressure is injected in the cell, the interface starts to move. At the same time, an undulation develops (a), and then amplifies, giving rise to fingers (b). Very quickly, the fingers hinder each other and successively stop growing, except for the largest one; in the

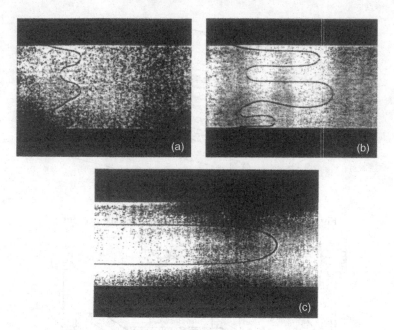

Fig. 3.39 Formation of a Saffman–Taylor finger in a Hele–Shaw cell. (a) Initial front destabilisation; (b) competition among the fingers; (c) single finger (Tabeling *et al.*, 1987).

stationary regime, this leads to a single elongated finger, with a very characteristic rounded tip (c). Experiment shows that its relative width λ (with respect to the channel width) decreases as its velocity increases, from a value close to 1 (at very low flow rate) down to an asymptotic value close to 1/2 (at high flow rate).

In the following, we will give a few details on the Hele–Shaw cell. We will then explain why the flat interface is unstable and determine the shape of the Saffman–Taylor finger in the permanent regime as a function of its relative width λ.

3.7.10.1 Some details concerning the Hele–Shaw cell

We will prove that a Hele–Shaw cell behaves as a two-dimensional porous medium, as long as the flow remains laminar between the two plates and their size is much larger than the distance between them. Indeed, under these conditions we have, denoting by z the normal to the plates:

$$v_z \ll v_x,\ v_y \quad \text{and} \quad \frac{\partial}{\partial z} \gg \frac{\partial}{\partial x},\frac{\partial}{\partial y}. \tag{3.339}$$

On the other hand:

$$Re = \frac{\rho U b^2}{\eta L} \ll 1 \tag{3.340}$$

taking for U the typical velocity at which the fingers advance. Using these inequalities simplifies the Navier–Stokes equation which becomes, in the plane of the plates (x, y):

$$0 = -\overrightarrow{\text{grad}}_{\parallel} P + \eta \frac{\partial^2 \vec{v}_{\parallel}}{\partial z^2} \tag{3.341}$$

with $P = P(x, y)$. This equation yields the mean fluid velocity $\langle \vec{v}_{\parallel} \rangle$, averaged across the thickness of the cell:

$$\langle \vec{v}_{\parallel} \rangle = -\frac{b^2}{12\eta} \overrightarrow{\text{grad}}_{\parallel} P. \tag{3.342}$$

This is a two-dimensional Darcy law. In the following, we set $\vec{v} = \langle \vec{v}_{\parallel} \rangle$ and $\overrightarrow{\text{grad}}_{\parallel} = \overrightarrow{\text{grad}}$, keeping in mind that we work in the (x, y) plane and that the velocity is *averaged across the cell thickness*.

It is noteworthy that a Hele–Shaw cell can be used to *simulate* planar potential flow of a perfect fluid, since $\text{div}\, \vec{v} = 0$ and $\overrightarrow{\text{curl}}\, \vec{v} = 0$. This result might seem paradoxical at first sight, since we are in the $Re \ll 1$ limit, a regime where viscous effects dominate. The explanation is that, since we averaged the velocity across the cell thickness, we must do the same for the vorticity. One can then verify very easily that $\langle \vec{\omega} \rangle = 0$ (since the vorticity $\vec{\omega}$ has opposite signs on the two plates), removing the paradox.

3.7.10.2 Origin of the Saffman–Taylor instability: qualitative study

Suppose a fluid of viscosity η_1 displaces a fluid of viscosity η_2.

The equations of the problem reduce to the Darcy law and to the incompressibility condition in each medium:

$$\vec{v}_{1,2} = -\frac{b^2}{12\eta} \overrightarrow{\text{grad}}\, P_{1,2}, \tag{3.343}$$

$$\text{div}\, \vec{v}_{1,2} = 0, \tag{3.344}$$

to which must be added two boundary conditions at the interface between the two fluids:

$$\vec{v}_1 \cdot \vec{n} = \vec{v}_2 \cdot \vec{n} = \vec{v}_i \cdot \vec{n}, \tag{3.345}$$

$$P_1 - P_2 = \frac{\gamma}{R}. \tag{3.346}$$

The first boundary condition expresses immiscibility of the two fluids (here, $\vec{n}$ is the unit vector normal to the interface, pointing from 1 towards 2, and $\vec{v}_i$ is the velocity of the interface). The second equation is the Laplace law. It states that a pressure difference exists between the two fluids as the interface, with a surface tension γ, is curved (R is taken as positive when the centre of curvature of the interface is in medium 1).

Before solving these equations, let us first give a qualitative explanation for the front becoming unstable as the displacing fluid (1) is less viscous than the displaced fluid (2). To simplify the problem, we assume that $\eta_1 \ll \eta_2$. Under these conditions, $\text{grad}\, P_1 \ll$

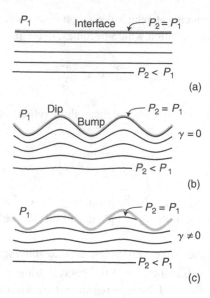

Fig. 3.40 Qualitative explanation of the Saffman–Taylor instability.

grad P_2, so that we can set $P_1 = \text{Cst}$. In liquid (2), on the other hand, the pressure decreases linearly along the cell from P_1 to P_a, the ambient pressure at the outlet of the cell.

In Fig. 3.40a, we plot the isobars (equipressure lines) within the cell, for a straight interface. Imagine now that the interface undulates. If the surface tension is zero, the interface remains an isobar. Under these conditions, the other isobars deform, as shown in Fig. 3.40b. We see immediately that the isobars crowd together ahead of a bump and fan out ahead of a depression. But we know from Darcy's law that the velocity of the fluid is proportional to the local pressure gradient. We conclude that, ahead of a bump, the velocity increases, since the amplitude of the gradient increases, while the velocity decreases ahead of a depression due to a decrease in the amplitude of the gradient. This mechanism is completely *destabilising*, since bumps accelerate, and depressions slow down.

In fact, this destabilising action of the pressure field is opposed by the surface tension, which tends to smooth the undulations. Indeed, the Laplace equation shows that, if $\gamma \neq 0$, the surface is no longer an isobar on the side of fluid (2). It follows that the isobars cut the interface (Fig. 3.40c), reducing the amplitude of the pressure gradient ahead of a bump and increasing it in a depression. This is a *stabilising* mechanism.

Consequently, the Saffman–Taylor instability is the result of direct competition between the destabilising pressure field and the stabilising surface tension.

3.7.10.3 *Linear stability analysis*

Let us now search for the dispersion relation of the planar front without making any particular assumption on the viscosities η_1 and η_2.

Consider a planar front propagating in the cell at a velocity V. The two fluids being incompressible, they flow at the same velocity:

$$\vec{\overline{v}}_1 = \vec{\overline{v}}_2 = V\vec{x}, \tag{3.347}$$

where $\vec{x}$ is the unit vector along the x direction, parallel to the channel axis. The overbar distinguishes quantities related to the unperturbed base flow. The Darcy law yields the corresponding pressure field:

$$\overline{P}_{1,2} = \overline{P}_0 - \frac{12\eta_{1,2}}{b^2} V x, \tag{3.348}$$

where P_0 is the pressure at the planar front, given by $x = \xi(y) = 0$.

To study the linear stability of this solution, we superimpose on the front a perturbation with amplitude ε and wave vector k (Fig. 3.41):

$$\xi = \varepsilon \exp(\omega t + iky) \tag{3.349}$$

and calculate its growth rate ω. If there are k values for which $\omega > 0$, we say that the front is linearly unstable at these k values. On the other hand, the front is linearly stable if $\omega < 0$ for all k.

To find the dispersion relation $\omega = \omega(k)$, let us first search for non-stationary perturbations of the velocity and pressure fields created by this undulation. We set:

$$\vec{v}_i = \vec{\overline{v}}_i + \vec{v}'_i \quad \text{and} \quad P_i = \overline{P}_i + P'_i \quad (i = 1, 2), \tag{3.350}$$

where the primed terms refer to the perturbations.

From the Darcy law, $\vec{v}'_i \propto \overrightarrow{\text{grad}}\, P'_i$ and, from the incompressibility condition, div $\vec{v}'_i = 0$. It follows that the pressure perturbations obey the equation:

$$\Delta P'_i = 0 \tag{3.351}$$

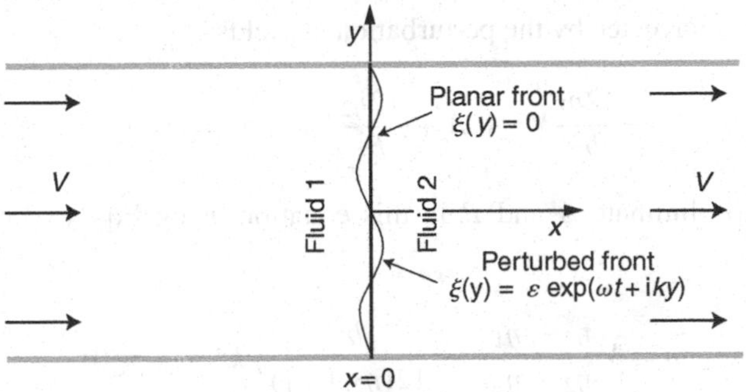

Fig. 3.41 Diagram of the experiment.

and can be written generically as:

$$P_i' = f_i(x) \exp(\omega t + iky). \tag{3.352}$$

Substituting in Eq. (3.351) yields:

$$f_i'' - k^2 f_i = 0 \tag{3.353}$$

and finally, preserving only the physically acceptable solutions:

$$P_1' = A \exp(kx) \exp(\omega t + iky) \qquad (x < 0), \tag{3.354}$$

$$P_2' = B \exp(-kx) \exp(\omega t + iky) \qquad (x > 0) \tag{3.355}$$

for $k > 0$. The Darcy law then yields the associated velocity perturbations:

$$v_{1x}' = -\frac{b^2}{12\eta_1} Ak \exp(kx) \exp(\omega t + iky), \tag{3.356}$$

$$v_{2x}' = -\frac{b^2}{12\eta_2} Bk \exp(-kx) \exp(\omega t + iky). \tag{3.357}$$

The dispersion relation is obtained by writing that the perturbations also obey the boundary conditions at the interface.

Condition (3.345) gives, to first order in ε:

$$v_{1x}' = v_{2x}' = \frac{\partial \xi}{\partial t} \tag{3.358}$$

so that:

$$-\frac{b^2}{12\eta_1} Ak = \frac{b^2}{12\eta_2} Bk = \varepsilon \omega. \tag{3.359}$$

As to the Laplace law (3.346), recalling that $1/R = -\partial^2 \xi / \partial y^2$, and keeping in mind that the base solution is advected by the perturbation, it yields:

$$-\frac{12\eta_1}{b^2} V\varepsilon + A + \frac{12\eta_2}{b^2} V\varepsilon - B = \gamma k^2 \varepsilon. \tag{3.360}$$

It suffices then to eliminate A and B in this equation using Eqs. (3.359) to obtain the dispersion relation:

$$\omega = V \frac{\eta_2 - \eta_1}{\eta_2 + \eta_1} k - \frac{b^2}{12(\eta_2 + \eta_1)} \gamma k^3 \qquad (k > 0). \tag{3.361}$$

This formula shows that:

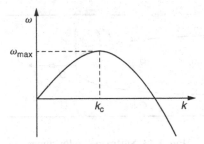

Fig. 3.42 Dispersion relation $\omega(k)$ for $\eta_1 < \eta_2$.

- if $\eta_1 > \eta_2$, then $\omega < 0$ for all k values. This means that the front is linearly stable when the displacing fluid is the most viscous. This result is in agreement with the experiment.
- if $\eta_1 < \eta_2$, there is always a wave vector range for which $\omega > 0$ (Fig. 3.42). The front is thus systematically unstable when the displacing fluid is the least viscous, a result once again confirmed by the experiments. In practice, it is the fastest growing perturbation that dominates, being the first to develop. Its wave vector is given by:

$$k_c = \frac{2}{b}\sqrt{\frac{V(\eta_2 - \eta_1)}{\gamma}} = 2\frac{\sqrt{Ca}}{b} \qquad (3.362)$$

with Ca the *capillary number*, defined by:

$$Ca = \frac{V(\eta_2 - \eta_1)}{\gamma}. \qquad (3.363)$$

This dimensionless number can be seen as the ratio of the viscous and capillary forces.

The wave vector k_c corresponds to the *most unstable wavelength of the planar front l_c*, also called the *capillary length*:

$$l_c = \pi b\sqrt{\frac{\gamma}{V(\eta_2 - \eta_1)}} = \frac{\pi b}{\sqrt{Ca}}. \qquad (3.364)$$

The capillary length l_c sets the characteristic length scale of the Saffman–Taylor problem. This formula reveals the competition between the capillary forces, tending to increase the capillary length, and the viscous forces, which tend to reduce it.

This linear theory describes the behaviour of the front as the instability appears. It does not, however, predict its subsequent evolution, which is dominated by nonlinear effects. In the next subsection, we will analyse the Saffman–Taylor finger that forms after the transient regime.

3.7.10.4 Shape of the finger as a function of its relative width

Experiment shows that, over a wide velocity range, a finger with well-defined width and shape develops in the centre of the cell (Fig. 3.39c). Let λ be its width relative to the channel width L, and U the velocity of its tip (Fig. 3.43).

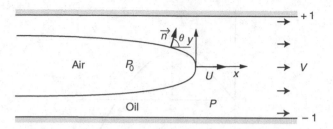

Fig. 3.43 Saffman–Taylor finger.

Far ahead of the finger, the velocity of the fluid is V. Flux conservation imposes:

$$V = \lambda U. \tag{3.365}$$

Saffman and Taylor showed that the shape of the fingers can be calculated analytically, as long as the surface tension γ is negligible.

To simplify the calculation even further, we assume that the displacing fluid is air, of negligible viscosity. Under these conditions, the pressure within the finger is constant, equal to P_0. On the other hand, the velocity $\vec{v}$ and pressure P vary in the displaced liquid (we will omit the subscript 2 in the following), obeying equations:

$$\vec{v} = \overrightarrow{\mathrm{grad}}\ \varphi \tag{3.366}$$

with

$$\varphi = -\frac{b^2}{12\eta}(P - P_0), \tag{3.367}$$

where φ is the velocity potential. With this choice, we see that:

$$\varphi = 0 \quad \text{at the interface,} \tag{3.368}$$

since we neglect capillary effects. At the other borders of the domain, we must have:

$$\varphi = Vx \quad \text{at} \quad x = +\infty, \tag{3.369}$$

$$\frac{\partial \varphi}{\partial y} = 0 \quad \text{at} \quad y = \pm 1. \tag{3.370}$$

From now on, we choose as length unit the semi-width of the channel ($x \equiv 2x/L$ and $y \equiv 2y/L$).

We now introduce the stream function ψ, such that:

$$v_x = \frac{\partial \varphi}{\partial x} = \frac{\partial \psi}{\partial y}, \quad v_y = \frac{\partial \varphi}{\partial y} = -\frac{\partial \psi}{\partial x}. \tag{3.371}$$

With this choice, the condition $\mathrm{div}\ \vec{v} = 0$ is automatically fulfilled.

The boundary conditions at infinity and at the edges of the cell require that:

$$\psi = Vy \quad \text{at} \quad x = +\infty, \tag{3.372}$$

$$\psi = \pm V \quad \text{at} \quad y = \pm 1. \tag{3.373}$$

The second condition states that the edges of the cell are stream lines ($\psi = \text{Cst}$, the value of the constant being set by the other boundary condition).

Finally, the following kinematic condition must be satisfied at the interface (it states simply that the two fluids are not miscible):

$$\vec{v} \cdot \vec{n} = \vec{v}_i \cdot \vec{n} = U \cos \theta, \tag{3.374}$$

where θ is the angle between the x axis and the normal $\vec{n}$ to the interface, pointing outwards (Fig. 3.43).

Let $\vec{t}$ be the unit vector tangent to the interface ($\vec{t} = (-n_y, n_x)$ since $\vec{n} \cdot \vec{t} = 0$). At the interface we have, using Eq. (3.371):

$$\vec{v} \cdot \vec{n} = \frac{d\varphi}{d\vec{n}} = \frac{d\psi}{d\vec{t}} = \frac{d\psi}{ds} \tag{3.375}$$

denoting by s the curvilinear abscissa. Since $\cos \theta = \frac{dy}{ds}$, we deduce that, on the interface:

$$\frac{d\psi}{dy} = U. \tag{3.376}$$

The difficulty of the problem resides now in finding the shape of the interface that satisfies all these equations.

To solve this *free boundary problem*, note first of all that the complex potential $w = \varphi + i\psi$ is a holomorphic function of the complex variable $z = x + iy$, since it obeys the Cauchy conditions (3.371).

Saffman and Taylor's trick was to use as variables φ and ψ instead of the x and y coordinates, since these functions take simple values on the boundary of the fluid domain, which becomes rectangular, as shown in Fig. 3.44.

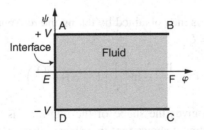

Fig. 3.44 Flow domain in the φ and ψ variables.

On the other hand, we know from the inverse Cauchy relations that

$$\frac{\partial x}{\partial \varphi} = \frac{\partial y}{\partial \psi} \quad \text{and} \quad \frac{\partial x}{\partial \psi} = -\frac{\partial y}{\partial \varphi}, \tag{3.377}$$

yielding:

$$\Delta_{\varphi,\psi} x = \Delta_{\varphi,\psi} y = 0. \tag{3.378}$$

We see that y is a harmonic function of the variables θ and ψ, taking a value of -1 on DC ($\psi = -V$), $+1$ on AB (in $\psi = V$), Ψ/U on AD ($\varphi = 0$, this segment corresponding to the interface we are trying to describe) and tending towards ψ/V for $\varphi \rightarrow +\infty$. It is enough now to develop y over the Laplacian eigenfunctions associated with the rectangle ABCD:

$$y = \frac{\psi}{V} + \sum_{n=1}^{+\infty} A_n \sin\left(\frac{n\pi\psi}{V}\right) \exp\left(\frac{-n\pi\varphi}{V}\right). \tag{3.379}$$

The coefficients A_n are obtained by writing that $y = \psi/U$ at $\varphi = 0$, giving:

$$\frac{\psi}{U} = \frac{\psi}{V} + \sum_{n=1}^{+\infty} A_n \sin\left(\frac{n\pi\psi}{V}\right). \tag{3.380}$$

Multiplying this expression by $\sin\left(\frac{n\pi\psi}{V}\right)$, then integrating over a period $2V/n$ yields:

$$A_n = \frac{2}{n\pi}(1 - \lambda) \quad \text{where} \quad \lambda = \frac{V}{U}. \tag{3.381}$$

One can also calculate x using the inverse Cauchy relations (3.377) and the fact that $\varphi = \psi = 0$ for $x = y = 0$ (namely, at the tip of the finger):

$$x = \frac{\varphi}{V} + \sum_{n=1}^{+\infty} A_n \left[1 - \cos\left(\frac{n\pi\psi}{V}\right) \exp\left(-\frac{n\pi\varphi}{V}\right)\right]. \tag{3.382}$$

From the relations (3.379) and (3.382) we obtain $z = x + iy$ as a function of $w = \varphi + i\psi$:

$$z = \frac{w}{V} + \frac{2}{\pi}(1 - \lambda) \ln\left[\frac{1}{2}\left[1 + \exp\left(-\frac{\pi w}{\lambda}\right)\right]\right]. \tag{3.383}$$

The shape of the interface is then obtained by taking the real part of the expression above and letting $\varphi = 0$ and $\psi = Uy$:

$$x = \frac{1 - \lambda}{\pi} \ln\left[\frac{1}{2}\left[1 + \cos\left(\frac{\pi y}{\lambda}\right)\right]\right]. \tag{3.384}$$

This remarkable equation gives the shape of the interface as a function of the relative width of the finger. Experiment shows that this shape, although calculated by neglecting the surface tension, is very close to the real one.

This theory is nevertheless incomplete, insofar as it fails to predict the width of the finger as a function of its velocity. This result should not come as a surprise, since by letting $\lambda = 0$ we removed from the problem the relevant length scale, namely the capillary length l_c given by Eq. (3.364). For this reason, the theory presented above predicts the existence of a *continuum of solutions*, uniquely parameterised by λ. To identify among this continuum the effectively observed solution, one must once again allow for capillary effects.

3.7.10.5 The selection problem

Experiment shows that the width of the finger is perfectly reproducible and that it decreases with increasing velocity.

An even more remarkable result, which the Saffman and Taylor calculation does not predict either, is that the relative width of the finger tends towards 1/2 at high velocity.

As already stated, these phenomena can only be explained by taking capillarity into account. We can see this by using the following dimensional argument by P. Pelcé (Pelcé, 2000). We write that the pressure variation along the finger, of the order of $(\Delta P)_{flow} \approx \lambda L (\eta / b^2) V$ (since the velocity varies from 0 to V over a distance of the order of λL) is of the same order of magnitude as the pressure variation due to the surface tension, estimated by $(\Delta P)_{surface} \approx \gamma / \lambda L$. Putting these two terms equal, and reinstating the capillary number, yields:

$$\lambda \approx \frac{1}{Ca} \left(\frac{b}{L} \right)^2 . \tag{3.385}$$

This relation shows that the finger width λ must decrease as the capillary number, and hence the finger velocity, increases. Obviously, the saturation of λ at high velocity cannot be predicted by this simple dimensional argument, whose only virtue is showing that an additional relation must exist between these parameters when surface tension is taken into account.

However, this formula shows that λ must be a universal function of a new dimensionless number, denoted by $1/B$ in the literature, and defined by $1/B = 12Ca(L/b)^2$. This dependence is quite logical, since $1/B$ is proportional to the squared ratio of the cell width to the most unstable wavelength of the planar front (or capillary length) given by formula (3.364), which is the relevant length in the Saffman–Taylor problem:

$$\frac{1}{B} = 12\pi^2 \left(\frac{L}{l_c} \right)^2 . \tag{3.386}$$

This prediction is well verified experimentally, as shown in Fig. 3.45. Note that in this plot the asymptotic value of λ is slightly less than 1/2. This difference with respect to the theoretical 1/2 value predicted by the two-dimensional model is due to the presence of wetting films on the limiting plates inside the finger (three-dimensional effects).

We will not discuss the selection problem any further, due to its mathematical complexity. The interested reader can find all the details in the excellent book by P. Pelcé, already cited above.

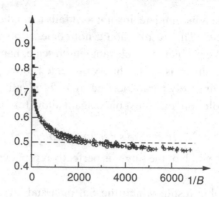

Fig. 3.45 Relative width of the finger as a function of $1/B$.

3.8 Viscosity measurements

Throughout this chapter, we have assumed that the viscosity of the medium was known. But how does one go about measuring it? This is what we will find out in this paragraph on rheometers, which are devices specially conceived for viscosity measurements.

3.8.1 Rotating rheometers

This is the most common type. Here we describe the two geometries most frequently employed.

3.8.1.1 Double-cylinder geometry

In this geometry, the fluid is contained between two coaxial cylinders with different radii, which can rotate about their axis. These cylinders can be made out of metal or even glass, if one should need to observe the sample visually during the experiment. This feature is essential when studying complex fluids, whose structure changes under shear, but is unimportant when the liquid of interest is Newtonian and isotropic.

In practice, the viscosity is obtained from the torque dependence on the angular velocity. This curve can be obtained by imposing the angular velocity and measuring the torque or, conversely, by imposing the torque and measuring the velocity.

For Newtonian fluids the two methods are equivalent, since the two quantities are related by a univocal relation. This does not always hold for complex fluids, such as gels or liquid crystals, where 'bistable' behaviour can be encountered. More specifically, this means that the same applied stress can correspond to two different angular velocities, associated with two different states of the system. Clearly, for such materials the two types of rheometers will provide different, but complementary, information.

This said, let us calculate the flow between two cylinders in the simple case of a Newtonian fluid (Fig. 3.46). To fix the ideas, we assume that the inner cylinder, with a radius R_i, turns at the angular velocity Ω, while the other cylinder, with a radius $R_e > R_i$, is fixed.

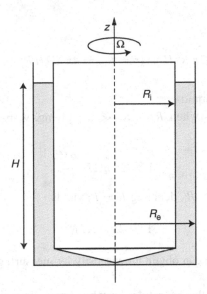

Fig. 3.46 Double-cylinder rheometer. When the gap between the two cylinders is small compared to the cylinder radius, the shear rate is almost constant. Under these conditions, the bottom of the inner cylinder can be cut in a cone shape, so that the shear rate in this region is the same as in the gap between the two cylinders. In this (Mooney) geometry, the end effect can be 'properly' accounted for.[18]

To find the torque one must apply to the inner cylinder to keep it turning, one must first solve the Navier–Stokes equation. In the permanent regime, and under the assumption of laminar flow, one has (see Appendix 3.B):

$$\frac{\partial^2 v_\theta}{\partial r^2} + \frac{1}{r}\frac{\partial v_\theta}{\partial r} - \frac{v_\theta}{r^2} = 0 \qquad (3.387)$$

with the boundary conditions:

$$v_\theta(r = R_i) = \Omega R_i \quad \text{and} \quad v_\theta(r = R_e) = 0. \qquad (3.388)$$

The solution to this equation reads:

$$v_\theta = \frac{\Omega R_i^2}{R_e^2 - R_i^2}\left(\frac{R_e^2}{r} - r\right). \qquad (3.389)$$

The shear stress is then $\sigma_{\theta r}$:

$$\sigma_{\theta r} = \eta\left(\frac{\partial v_\theta}{\partial r} - \frac{v_\theta}{r}\right) = -2\eta\Omega\frac{R_i^2}{R_e^2 - R_i^2}\frac{R_e^2}{r^2}, \qquad (3.390)$$

---

[18] Another, simpler, technique consists of using an inner cylinder with a recess instead of a cone at its tip. An air bubble is then trapped below the cylinder as it is lowered into the liquid, strongly limiting the friction in this region since the viscosity of air is generally much lower than that of the measured liquids. The end effect then becomes negligible.

yielding for the torque:

$$\Gamma = -2\pi R_i^2 H \sigma_{\theta r}(r = R_i) = 4\pi \eta \Omega H R_i^2 \frac{R_e^2}{R_e^2 - R_i^2} \tag{3.391}$$

with H the height of the liquid.

This formula simplifies when $R_e - R_i \ll R_e$ giving, with $h = R_e - R_i$ and $R = (R_e + R_i)/2$:

$$\Gamma = 2\pi \eta \Omega H \frac{R^3}{h}. \tag{3.392}$$

In the opposite limit, $R_e \gg R_i$, denoting $R = R_i$ yields:

$$\Gamma = 2\pi \eta \Omega H R^2. \tag{3.393}$$

Using these formulas, one can obtain the viscosity by measuring the torque as a function of the angular velocity.

In classical rheometers, these formulas must be corrected for end effects, inherent to this kind of setup. These corrections are made by calibrating the rheometer using liquids of well-known viscosities (such as water).

One should also check that the flow between the two cylinders remains laminar, avoiding in particular the development of eddies within the rheometer. This instability, studied by Taylor in the last century, can occur only if the inner cylinder rotates, the outer one remaining fixed, which is the case for most rheometers (in the opposite case, of a fixed inner cylinder and a rotating outer one, this instability does not occur). As the instability develops, the viscosity of the fluid seems to increase, because more energy must be injected with respect to the laminar regime, to sustain the eddies. If the gap between the cylinders h is small compared to their radius R, it can be shown that the Taylor instability appears for a high enough Taylor number ([2]; Chandrasekhar, 1981):

$$Ta = \frac{h}{R} Re^2 = \frac{\rho^2 \Omega^2 R h^3}{\eta^2} > Ta_c = 1712. \tag{3.394}$$

Note that the Reynolds number is given here by $Re = \dfrac{\rho(\Omega R)h}{\eta}$.

In commercial rheometers, this instability can develop for low-viscosity liquids, typically below 10 mPa s (keeping in mind that, for water, $\eta \approx 1$ mPa s at 20.2 °C).

We should point out here that formula (3.392) still holds in the stationary regime for viscoelastic fluids. In this case, η stands for the viscosity value $\eta(\dot\gamma)$ at the shear rate $\dot\gamma = \Omega R/h$. However, these fluids can exhibit a convective instability, formally similar to that of Taylor, but of completely different origin. Indeed, the driving force is no longer the centrifugal force, but the elastic forces. The relevant dimensionless number that characterises this *elastic instability* is the new combination $(h/R)Wi^2$, where Wi is the Weissenberg number, defined by:

$$Wi = \frac{\tau}{1/\dot\gamma}. \tag{3.395}$$

This number compares the viscoelastic relaxation time τ of the material with the time needed by a fluid element to undergo deformation of unit order (in polymers, $Wi \simeq 1$ sets the crossover from the linear regime to the nonlinear one). A linear stability calculation shows that the instability develops if (Larson *et al.*, 1990):

$$\frac{h}{R} Wi^2 > 35.4. \tag{3.396}$$

In practice, this condition is much more restrictive than condition (3.394) for viscoelastic fluids, such as polymer melts.

3.8.1.2 Cone-plate geometry

One of the shortcomings in the previous geometry was that the shear rate was not constant throughout the gap between the cylinders (unless the gap h is much smaller than R, in which case the shear rate is more or less constant, with a value close to $\Omega R/h$).

In the cone-plate geometry described in Fig. 3.47, this condition is better satisfied. In general, the cone is truncated a few micrometres from the tip to avoid touching the plate during the measurement. Nevertheless, the 'virtual' tip of the cone must be exactly on the plate, which is obtained by first bringing the two sides into mechanical contact, and then separating the cone from the plate by the distance required to satisfy this condition as closely as possible. As in the previous case, the viscosity can be measured by imposing either the torque or the angular velocity, and measuring the other quantity.

For a Newtonian fluid, one can easily find the relation between the torque and the angular velocity when the angle of the cone θ_0 is very small (Fig. 3.47). For these conditions, the velocity field within the fluid varies linearly from 0 to $\Omega r \cos\theta_0 \approx \Omega r$ from the plate towards the cone, remaining at a distance r from the tip of the cone. The sample thickness being $r\theta_0$ at that point, we deduce that the shear rate is locally constant, of (absolute) value:

$$\dot{\gamma} = \frac{\Omega}{\theta_0}. \tag{3.397}$$

The torque needed to keep the cone turning at the angular velocity Ω is therefore (see Appendix 3.B for the expression of the $\sigma_{\theta\varphi}$ stress component):

$$\Gamma = \int_0^{2\pi} \int_0^R \eta \dot{\gamma} r^2 \, d\varphi \, dr = \frac{2\pi}{3\theta_0} \eta R^3 \Omega. \tag{3.398}$$

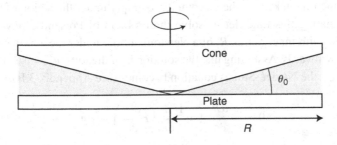

Fig. 3.47 Cone-plate rheometer.

This formula requires that the virtual tip of the cone be exactly on the plate. In the measurements, this condition must be strictly respected to obtain accurate results. As an example, a 10 μm error in the positioning of a cone with radius $R = 5$ cm and angle $\theta_0 = 1°$ results in an error of about 1% on the viscosity.

Sometimes, the plate-plate geometry is also used. In this case, the shear rate is no longer constant, varying linearly with the distance to the centre. This poses no problems as long as the fluid is Newtonian. Then, the torque is given as a function of the angular velocity by the following formula:

$$\Gamma = \frac{\pi}{2} \frac{R^4}{h} \eta \Omega, \tag{3.399}$$

where R is the common radius of the plates and h the gap between them.

Once again, we emphasise that the relation (3.398) is still valid in the stationary regime for a viscoelastic fluid. In this case, η stands for the viscosity $\eta(\dot{\gamma})$ measured at the shear rate $\dot{\gamma} = \Omega/\theta_0$. The cone-plate geometry can also exhibit an elastic oscillating instability. It develops when the following criterion is fulfilled (Olagunju, 1995):

$$\tau^2 \Omega^2 / \theta_0 > 21.2 \tag{3.400}$$

with τ the viscoelastic relaxation time. This formula shows that the angle of the cone θ_0 must be as small as possible if we want to measure the viscosity of a viscoelastic fluid in the strongly nonlinear regime ($\dot{\gamma} \gg 1/\tau$).

3.8.2 Capillary rheometer

The Poiseuille flow in a cylinder, studied in Section 3.7.9, can also be used for viscosity measurements. In this flow, the shear rate is not constant, reaching a maximum at the walls of the cylinder and vanishing in the centre, which might be problematic for a non-Newtonian fluid.

A viscometer based on this principle is shown in Fig. 3.48. Its main element is a long capillary tube, surmounted by a reservoir and immersed at its lower end into a large container filled with the liquid to be measured. The liquid is sucked in until it reaches the upper mark, and the reservoir is then opened to the ambient pressure. The viscosity is determined by measuring the time taken for the viscometer to empty under the action of gravity.

Before calculating this time, let us solve the problem of Poiseuille flow in a vertical cylindrical tube, with inner radius R, of a Newtonian fluid under the action of gravity. The z axis points downwards. Assuming that the solution is of the form $v_z = v_z(r)$, $v_\theta = v_r = 0$ and $P = P(r, z)$, the Navier–Stokes equation becomes (see Appendix 3.B):

$$0 = -\frac{\partial P}{\partial z} + \eta \frac{1}{r} \frac{d}{dr} \left(r \frac{dv_z}{dr} \right) + \rho g \tag{3.401}$$

with the boundary condition $v_z(R) = 0$.

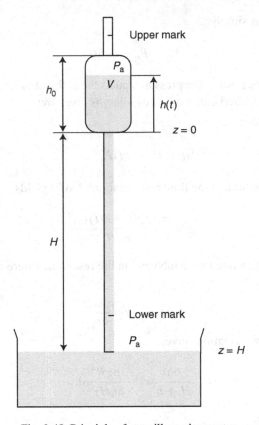

Fig. 3.48 Principle of a capillary viscometer.

Using the 'modified pressure' $\overline{P} = P - \rho g z$, the Navier–Stokes equation can be rewritten as:

$$\frac{\partial \overline{P}}{\partial z} = \eta \frac{1}{r} \frac{d}{dr} \left(r \frac{dv_z}{dr} \right). \tag{3.402}$$

This equation is solved exactly as in the absence of gravity (see Section 3.7.9), provided the pressure P is replaced by the modified pressure $\overline{P}$. The result is a 'modified flow rate law':

$$Q = \frac{\pi \left(\overline{P}_0 - \overline{P}_H \right) R^4}{8 \eta H}. \tag{3.403}$$

Let us now calculate the flow time t_f taken by the viscometer to empty. Let h_0 be the total height of the reservoir, r its radius, $h(t)$ the liquid level in the reservoir at time t, H the length of the capillary and R its radius. In practice, $r \gg R$ so that dissipation in the reservoir is negligible compared to that occurring in the capillary. Under these conditions, we can consider that the fluid in the reservoir is almost at rest, so that the pressure at the inlet of the capillary is simply:

$$P_0 = P_a + \rho g h(t). \tag{3.404}$$

At the outlet of the capillary the pressure equals the ambient pressure, so that the modified pressure at the inlet and outlet of the capillary is given by:

$$\overline{P}_0 = P_0 = P_a + \rho g h(t), \tag{3.405}$$

$$\overline{P}_H = P_a - \rho g H. \tag{3.406}$$

Using these two formulas and the flow rate equation (3.403) yields:

$$Q = \frac{\pi \rho g \left[H + h(t) \right] R^4}{8 \eta H}. \tag{3.407}$$

But the flow rate is the same in the tube and in the reservoir, where it can also be written as:

$$Q = -\pi r^2 \frac{dh}{dt}. \tag{3.408}$$

Comparing the last two equations gives:

$$\frac{dh}{H+h} = -\frac{\rho g R^4}{8 \eta H r^2} dt, \tag{3.409}$$

whence the flow time:

$$t_f = -\frac{8 \eta H r^2}{\rho g R^4} \int_{h_0}^{0} \frac{dh}{H+h} = \frac{8 \eta H r^2}{\rho g R^4} \ln \left(\frac{h_0 + H}{H} \right). \tag{3.410}$$

For $h_0 \ll H$, this formula reduces to:

$$t_f = \frac{8 \eta V}{\pi \rho g R^4}, \tag{3.411}$$

with V the volume of the reservoir.

These equations yield the viscosity of the fluid. They are only approximate and must be corrected for end effects, in particular because Poiseuille flow sets in inside the capillary only at a certain distance from the extremities. This distance was determined in Section 3.7.9. Writing that it must remain much smaller than the total length of the capillary yields the following condition for neglecting the end effects:

$$H \gg \frac{\rho^2 g R^4}{8 \eta^2}. \tag{3.412}$$

Taking, for instance, water and a capillary radius of 0.5 mm requires $H \gg 8$ cm, so a very long tube is needed. In this case, a better choice would be a capillary with radius 0.2 mm and length 10 cm. With a reservoir volume of 1 cm^3, the flow time would be of the order of $t_f \approx 7$ min, which is reasonable.

Clearly, formulas (3.411) and (3.412) are very useful for choosing the appropriate viscometer.

3.8.3 Piezoelectric rheometer

This rheometer also works on the principle of Poiseuille flow. The liquid is confined between two parallel glass plates attached to two piezoelectric ceramic elements. One of them is used as an actuator, compressing or dilating the sample, while the other serves as a sensor, measuring the lubrication force. The advantage of this setup is that of using very small deformation amplitudes (of the order of one Å to a few tens of Å using a single ceramic, and more by using a stack) in the sinusoidal mode, at frequencies typically between 0.1 Hz and 10 kHz. This type of setup is mainly used to study the linear viscoelasticity of complex fluids, but can obviously be employed for measuring the viscosity of Newtonian liquids, as we will show in the following.

Let R be the radius of the circular plates and h the gap between them (Fig. 3.49). We are interested in the rheological law $F = F(\dot{\varepsilon})$, where F is the force applied by the upper ceramic element to the top plate and $\dot{\varepsilon} = \dfrac{1}{h}\dfrac{dh}{dt}$ the deformation rate of the sample.

To solve this problem, we assume that $h \ll R$. Under these conditions, $v_z \ll v_r$ and $\partial v_r / \partial r \ll \partial v_r / \partial z$. If the frequency is not too high, one can also neglect the inertial terms compared to the viscous terms ($\rho \partial v_r / \partial t \ll \eta \partial^2 v_r / \partial z^2$). Finally, the equations of motion and mass conservation are written, in cylindrical coordinates (see Appendix 3.B):

$$0 = -\frac{\partial P}{\partial r} + \eta \frac{\partial^2 v_r}{\partial z^2} \qquad \text{(motion along } r\text{)}, \qquad (3.413)$$

$$0 = -\frac{\partial P}{\partial z} \qquad \text{(motion along } z\text{)}, \qquad (3.414)$$

$$\frac{1}{r}\frac{\partial}{\partial r}(r v_r) + \frac{\partial v_z}{\partial z} = 0 \qquad \text{(continuity)}. \qquad (3.415)$$

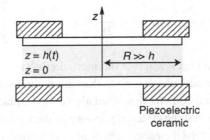

Fig. 3.49 Acoustic piezoelectric rheometer.

The boundary conditions for the velocity are:

$$v_r(z = 0) = v_z(z = 0) = 0, \quad v_r(z = h) = 0, \quad v_z(z = h) = \dot{h} \tag{3.416}$$

and for the pressure:

$$P(r = R) = P_a, \tag{3.417}$$

where P_a is the ambient pressure.

It is easily verified that the solution of these equations takes the form:

$$v_r = 3\frac{\dot{h}}{h}r\frac{z(z-h)}{h^2}, \tag{3.418}$$

$$v_z = \frac{\dot{h}}{h^3}z^2(3h - 2z), \tag{3.419}$$

$$P - P_a = -3\eta\frac{\dot{h}}{h^3}\left[1 - \left(\frac{r}{R}\right)^2\right]R^2. \tag{3.420}$$

One can deduce the force to be applied on the top plate in order to deform the sample:

$$F = -2\pi\int_0^R (P - P_a - \sigma_{zz}^v)\, r\, dr = \frac{3\pi R^4}{2h^3}\eta\dot{h} = \frac{3\pi R^4}{2h^2}\eta\dot{\varepsilon}. \tag{3.421}$$

This relation, also known as the *Stefan equation*, shows that, for a sinusoidal deformation, the response of the sample is in quadrature with the excitation (viscous response). Also note that the force varies as the reciprocal thickness cubed (at fixed h) and as the fourth power of the radius. Hence the importance of accurately measuring these two quantities for a precise determination of the viscosity. In practice, the geometrical ratio R^4/h^3 is obtained by calibrating the rheometer with a liquid of known viscosity, such as water.

In high-frequency experiments, inertial effects can become important. The equation above can be corrected for small inertial effects via a perturbative approach. More specifically, it involves calculating the inertial term $\rho\,\partial v_r/\partial t + v_r(\partial v_r/\partial r) + v_z(\partial v_r/\partial z)$ using the velocity profile given by Eqs. (3.418) and (3.419), and then going through the calculations above to find the perturbations to v_r, v_z and P. The final result for F is:

$$F = \frac{3\pi R^4}{2h^2}\eta\dot{\varepsilon}\left[1 - \frac{5\rho}{28\eta}h^2\dot{\varepsilon} + \frac{\rho}{10\eta}\frac{h^2\ddot{\varepsilon}}{\dot{\varepsilon}}\right]. \tag{3.422}$$

Using this formula one can check whether the inertial effects are negligible and, if this is not the case, take them into account as long as they are not too great.

Finally, the piezoelectric rheometer described above can be placed into a hydrostatically pressured vessel, in order to test the effect of pressure on the viscosity. These measurements are particularly important for lubricants, which are often used in extreme conditions of temperature and pressure (several kbar).

3.8.4 Surface force apparatus

This device was developed for the study of interactions (electrostatic, van der Waals, steric, etc.) between two surfaces close to contact, as well as for investigating highly confined fluids. Indeed, one may wonder down to what length scale the continuous theory still applies, hence the idea of building an apparatus for rheological measurements in highly confined media. One might think of reducing the distance between the plates in the previous setup. In practice, it is almost impossible to go below a few micrometres, due to the difficulty in achieving parallelism better than 10^{-4} rad (amounting to a thickness variation of 1 μm over a distance of 1 cm). On the other hand, the force F becomes considerable at very small gap values due to its $1/h^2$ dependence (at fixed $\dot{\varepsilon}$), leading to mechanical problems (indeed, one must be able to deform the sample, which requires that the setup be much more rigid than the sample!).

A simple solution to these problems consists in confining the sample between a plane and a sphere. The contact now being point-like as the sphere touches the plane, the parallelism constraints no longer apply. The experimental geometry is shown in Fig. 3.50. The sphere, with a radius R, is at a distance h from the plane. The fluid under investigation is placed between the two elements.

The experiment once again consists in advancing or withdrawing the sphere from the plane, either in a motion of uniform translation, or by oscillating about a fixed position. The second method is obviously the most convenient for measuring the viscosity. In practice,

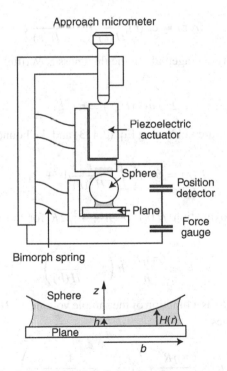

Fig. 3.50 Schematics of a surface force apparatus.

the sphere is advanced to the plane until mechanical contact (detected using a force gauge), and then withdrawn to the desired distance using a piezoelectric actuator. The same actuator can be used to impose the sinusoidal motion. Extremely sensitive force gauges (interferometric or capacitive) are then used to measure the force as a function of the excitation.

To find the rheological response function $F = F(\dot{\varepsilon})$ relating the force applied to the sphere F to the deformation rate $\dot{\varepsilon} = \dfrac{1}{h}\dfrac{dh}{dt}$, one must solve the Navier–Stokes equation in spherical-plane geometry, for the case of a Newtonian fluid. Once again, the calculation is not too complicated as long as the flow remains in the lubrication regime. In this case, the equations remain the same as for the plane-plane geometry (Eqs. (3.413)–(3.415)) with the new boundary conditions:

$$v_r(z = 0) = v_z(z = 0) = 0, \tag{3.423}$$

$$v_r(z = H(r)) = 0, \quad v_z(z = H(r)) = \dot{h} \tag{3.424}$$

and

$$P(r = b) = P_a. \tag{3.425}$$

Here, b is the sample radius and $z = H(r)$ the equation describing the sphere surface.

To solve this problem, let us write that the radial velocity profile is parabolic, given by the expression:

$$v_r(r, z) = 6u(r)\frac{z}{H(r)}\left(1 - \frac{z}{H(r)}\right), \tag{3.426}$$

where $u(r)$ is the velocity averaged along the thickness z. Writing that the fluid is incompressible yields:

$$2\pi r u(r)H(r) = \pi r^2 \dot{h}, \tag{3.427}$$

which then gives the pressure field using Eq. (3.413) and the boundary condition (3.425):

$$P(r) = 6\pi \eta \dot{h} \int_r^a \frac{r}{H^3(r)}dr + P_a. \tag{3.428}$$

Using a parabolic approximation $H(r) = h + r^2/2R$ for the equation of the sphere results in:

$$F = \frac{6\pi \eta R^2}{h}\dot{h}\left(1 - \frac{h}{H(b)}\right). \tag{3.429}$$

One can also express $H(b)$ as a function of the sample volume V: $H(b) = (h^2 + V/\pi R)^{1/2}$. The viscous force becomes:

$$F = \frac{6\pi \eta R^2}{h}\dot{h}\left(1 - \frac{1}{\sqrt{1 + V/\pi Rh^2}}\right)^2. \tag{3.430}$$

For a liquid drop with a volume of 1 mm³, a sphere radius of 2 mm and a gap of 1 μm, the second term between parentheses is less than 1%. Thus, under the usual working conditions of this setup, *the measured force does not depend on sample volume* (a major advantage, since one must no longer measure this parameter to obtain the viscosity) and is given by:

$$F = \frac{6\pi \eta R^2}{h}\dot{h} = 6\pi \eta R^2 \dot{\varepsilon}. \tag{3.431}$$

According to this relation, the viscous force diverges as h tends to 0. That is why a ball falling onto a plane in a fluid of finite viscosity theoretically needs an infinite time to reach the plane. In practice, the contact occurs between small asperities present on each surface.

Formula (3.429) also shows that the overwhelming contribution to the viscous force is made by the fluid found in the central region, within a disc of a typical radius $4\sqrt{Rh}$ (corresponding to a maximum thickness of $10h$). By advancing the sphere to the plane, one can therefore check whether the viscosity depends on the confinement. This type of measurement confirmed that the viscosity of liquids remains unchanged down to scales of a few molecular distances.

3.8.5 Wall slippage

It is worth noting that, in all our calculations, we assumed that the fluid did not slip over the surfaces. Recent experiments on simple fluids such as water or glycerol show this assumption to be valid if the fluid has a good affinity for the surface, i.e. if it wets the surface. In the opposite case (non-wetting fluid), recent theoretical work predicts that the fluid exhibits slippage at the surface (Barrat and Boquet, 1999). Consequently, the velocity profile penetrates within the solid over a certain distance l_{pen}, as shown in Fig. 3.51.

This slippage length can reach about ten times the molecular size, a value that is apparently confirmed by recent SFA experiments (Cottin–Bizonne *et al.*, 2002).

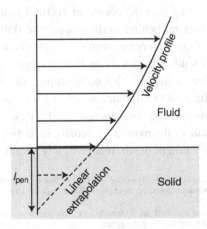

Fig. 3.51 Definition of the slippage (or penetration) length of the velocity profile.

In comparison, polymers can exhibit much larger wall slippage, since the molecules are much larger (Gay, 1999). Slippage lengths of several micrometres are frequently encountered in such systems and must be taken into account for samples less than about 100 micrometres thick.

Finally, this effect becomes dramatic in concentrated emulsions or suspensions. At low shear rates, depletion layers appear at the walls (with a lower concentration of droplets or solid particles). These layers, albeit very thin compared to the sample thickness, are much less viscous than the rest of the sample and 'concentrate' the shear. Bulk flow is therefore of the 'plug' type (little change in the velocity) with apparent slippage lengths that can easily exceed the sample thickness (for a review, see Barnes, 1995).

This discussion shows the importance of measuring in real time the velocity profile within the rheometer for finding out the real shear rate of the sample. Different techniques, based, for instance, on ultrasound waves (Manneville *et al.*, 2004), dynamic light scattering (Salmon *et al.*, 2003) or NMR spectroscopy (López–González *et al.*, 2006), can be used nowadays to determine the complete velocity profile 'in real time',[19] even in the case of very turbid materials, such as foams or emulsions (where direct observation by particle tracking (PIV) is impossible).

3.8.6 *Viscosity measurements in geophysics*

In the first chapter we mentioned that, over geological time scales, the Earth's mantle behaves as a Newtonian fluid of huge viscosity: $\eta \simeq 10^{21}$ Pa s. However, although we speak of a fluid, the mantle is almost entirely composed of crystalline rock (not molten, as in the case of magma[20]) creeping extremely slowly according to specific mechanisms, different from those of liquids (see the Nabarro–Herring, Coble or Harper–Dorn creep models described in Section 5.3.4).

The best method for estimating the average viscosity of the mantle consists in studying post-glacial rebounds. More precisely, it relies on studying the uplift of shorelines in Scandinavia, Canada or Siberia after the sudden melt of glacial caps 18 000 years ago. These caps, after having built up over the course of 100 000 years reaching a thickness of 2–3 km, melted suddenly, leaving behind shallow seas (the Baltic Sea and Hudson Bay) whose bottom still rises by about 1 cm/year, even today. This is the conclusion of very thorough geological paleoshore studies begun about 40 years ago. These seas are predicted to dry up completely over a few millennia. This delay in the viscous response of the Earth's mantle to ice removal can be used to estimate, by solving the equations of hydrodynamics, the order of magnitude of its average viscosity: $\eta_0 \simeq 10^{21}$ Pa s.

Although the average value of the mantle's viscosity is fairly well known, this is not the case for its variation as a function of depth. It is generally admitted that it increases by one

---

[19] Using acoustics, for instance, one second is typically needed to probe the flow over a thickness of one millimetre with a space resolution of 40 µm. Since this technique is making rapid progress, these specifications will have been largely exceeded by the time this book is published.

[20] The magma is molten rock containing dissolved gas. It forms locally in the mantle and rises to the surface of the Earth's crust along faults or fissures. In particular, this occurs at the location of volcanoes and of mid-oceanic ridges, where the tectonic plates diverge. Nevertheless, we emphasise that the mantle is made up almost exclusively of crystallised rocks, and only 5% of magma. Above all, one should not imagine the continents drifting on top of a layer of molten magma.

or two orders of magnitude between the top and the base of the convective mantle. These estimates are obtained via numerical models of mantle convection, striving to reproduce as closely as possible the available experimental data: geoid shape, velocity of the tectonic plates, density profile obtained by seismic tomography, etc. It should be noted that the viscosity increases with depth in the mantle in spite of the temperature rise[21] due to the increase in pressure.[22]

Similar effects are present in 'ordinary' liquids, and we will discuss them in the next subsection from the point of view of Eyring's theory.

3.8.7 Liquid viscosity as a function of temperature and pressure: consequences for rheological measurements

To close this chapter, let us briefly discuss experimental results on the variation in viscosity for 'ordinary' liquids as a function of temperature and pressure, as well as its rheological consequences.

Concerning temperature, Eyring's theory predicts an Arrhenius-type behaviour (Eq. (3.38)). On the other hand, experiment shows that the more viscous a liquid, the faster its viscosity varies with temperature.

This effect can complicate viscosity determinations for very viscous fluids, which heat up in the rheometer.

For an estimate of how much a fluid heats up inside a rheometer, let us do the calculations in the simplest case, that of Couette flow between two parallel plates positioned at $z = 0$ and $z = h$. The lower plate is fixed, and the other one moves at a speed V. For simplicity's sake, let us assume that both plates are maintained at the same temperature T_0 and that the viscosity of the fluid is independent of temperature. Obviously, in these conditions the velocity profile remains linear and the heat equation becomes, in the stationary regime:

$$\chi \frac{\partial^2 T}{\partial z^2} + \Phi = 0, \tag{3.432}$$

where Φ is the dissipation per unit volume and χ the thermal conductivity. Since

$$\Phi = \eta \left(\frac{V}{h} \right)^2, \tag{3.433}$$

we can solve Eq. (3.432) immediately, to obtain the temperature profile:

$$T = T_0 + \frac{\eta}{2\chi} \left(\frac{V}{h} \right)^2 z(h - z). \tag{3.434}$$

---

[21] Indeed, the temperature increases by about 800 °C (outside the thermal boundary layers) in the mantle due to the adiabatic gradient (see Footnote 6, Section 3.2.4).

[22] The pressure within the mantle is given in the next chapter (Table 4.2, page 222).

As expected, the heating is more intense in the centre of the sample, where it is given by:

$$\Delta T = \frac{\eta}{8\chi} h^2 \left(\frac{V}{h}\right)^2. \tag{3.435}$$

This heating can be significant. Indeed, let us consider silicone oil with a viscosity $\eta = 10^6$ Pa s and a thermal conductivity $\chi = 0.25$ W/m K. Assuming $h = 10^{-3}$ m yields $\Delta T = 0.25\dot{\gamma}^2$, where γ is the shear rate. In this case, thermal effects become important for shear rates above a few s^{-1}.

Clearly, thermal effects are very important in the industrial processing of very viscous materials. Flow prediction must then account for local heating and the associated viscosity change, rendering the calculations much more complicated.

The viscosity also depends on hydrostatic pressure. It almost always increases with pressure (except for water, where it starts by decreasing with increasing pressure below 30 °C, before behaving 'normally' at higher temperature), since increasing the pressure reduces the free volume, and hence the jump frequency of molecules from one cage to the next (Section 3.1.4.2). This effect is less spectacular than that of temperature, since liquids must be compressed very strongly in order to achieve a significant increase in viscosity.

The graph in Fig. 3.52 gives the viscosity of a mineral naphthenic oil as a function of pressure. The pressure scale might seem impressive, ranging from 0.1 MPa (1 bar or 1 atm) to 750 Mpa (or 7.5 kbar), but keep in mind that such pressure values are common in gearboxes and can even exceed 1 GPa inside a rolling mill or in the dies of a wire drawing bench.

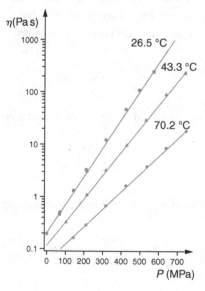

Fig. 3.52 Viscosity variation for a mineral naphthenic oil as a function of pressure (Schmidt *et al.*, 2000).

This graph also shows that the viscosity–pressure dependence is linear in a log–linear plot, meaning that the viscosity varies exponentially with the pressure. This result is easily explained in the framework of Eyring's theory by developing the free activation energy Δg of the viscosity (given by Eq. (3.38)) as a function of the pressure to give:

$$\Delta g(P) = \Delta g_0 + (P - P_a)v_{act}, \tag{3.436}$$

where P_a is the ambient pressure, $\Delta g_0 = \Delta g(P_a)$ the free activation energy at ambient pressure, and v_{act} an *activation volume*, generally expressed by:

$$v_{act} = \left(\frac{\partial \Delta g}{\partial P}\right)_T. \tag{3.437}$$

From Eqs. (3.436) and (3.38), we infer that the viscosity η at pressure P satisfies the relation:

$$\ln(\eta) = \ln(\eta_a) + \frac{v_{act}}{k_B T}(P - P_a), \tag{3.438}$$

where η_a is the viscosity measured at ambient pressure. This law is confirmed by experiments, proving that, in a first approximation, the activation volume (given directly by the slope of the experimental curve in a semi-log plot) is independent of the pressure. For the mineral oil whose viscosity dependence is given in Fig. 3.52, the activation volume is typically $v_{act} \approx 35 \text{ Å}^3$, i.e. the volume of a sphere 2 Å in radius, and corresponds roughly to the volume of a monomer.[23]

In conclusion, let us note that the flow-induced pressure variations in classical rheometers are much smaller than those given in Fig. 3.52. Hence, the corresponding viscosity variations are in general negligible and much less important than thermal effects.

Appendix 3.A Tangential discontinuities and shock waves (in perfect fluids)

The flow situations considered in this chapter were such that all quantities (such as velocity, pressure, density, etc.) varied continuously in space. Under some conditions, however, these quantities can exhibit *discontinuous* distributions. This occurs along certain surfaces known as *discontinuity surfaces*. It should be noted that, if a discontinuity surface appears within a fluid (a shock wave, for instance), *its velocity is generally different from that of the fluid: it follows that the fluid particles can very well cross this surface.*

In this appendix, we describe some of the properties of these special surfaces. For simplicity's sake, we assume that the fluids under consideration are *perfect* (but not necessarily incompressible).

---

[23] For this particular oil, the activation volume is constant over a large pressure range. This is not always the case; in general, the activation volume decreases with increasing pressure (Schmidt *et al.*, 2000).

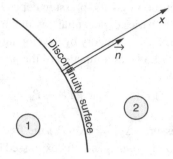

Fig. 3.A.1 Element of a discontinuity surface and the associated reference frame. Fluids (1) and (2) can be different or have the same nature (see below).

3.A.1 Boundary conditions on a discontinuity surface

Consider an element of a generic discontinuity surface and equip it with a reference frame such that the x axis points along the normal $\vec{v}$ of the element (Fig. 3.A.1).

To find the appropriate boundary conditions, let us write down the conservation of energy, mass and momentum across this surface.

3.A.1.1 Energy conservation

The energy flow across the surface must be continuous: the energy going in on one side must go out on the other side, which can be expressed by:

$$\left[\vec{Q}\cdot\vec{v}\right] = Q_{1x} - Q_{2x} = 0,\tag{3A.1}$$

where $\vec{Q}$ is the energy flux vector given by formula (3.5). For perfect fluids, this equation becomes (with $\vec{q} = 0$):

$$\left[\rho v_x\left(\frac{v^2}{2}+h\right)\right] = 0,\tag{3A.2}$$

where $h = e + P/\rho$ is the mass enthalpy of the fluid(s) involved.

3.A.1.2 Mass conservation

Mass conservation can be written as:

$$[\rho v_x] = 0.\tag{3A.3}$$

This relation states that the mass of the matter entering on one side equals that exiting on the other side.[24]

---

[24] This condition does not imply that the two fluids be the same, since chemical reactions can take place 'within' the discontinuity surface, as for a flame. We exclude this possibility in the following.

3.A.1.3 Momentum conservation

The momentum flux must also be continuous, as the forces between the fluids on either side of the discontinuity surface must be equal. Hence:

$$[\Pi \vec{v}] = 0 \tag{3A.4}$$

with $\underline{\Pi}$ the momentum flux density tensor given by Eq. (3.48). More specifically, in the case of a perfect fluid this equation becomes:

$$\left[P + \rho v_x^2\right] = 0 \tag{3A.5}$$

for the x component, and:

$$\left[\rho v_x v_y\right] = 0 \quad \text{and} \quad [\rho v_x v_z] = 0 \tag{3A.6}$$

for the y and z components.

Equations (3A.2), (3A.3), (3A.5) and (3A.6) represent a complete system of boundary conditions for the discontinuity surface. They reveal the presence of two types of discontinuities, distinguished by the presence or absence of matter flow across them. We will study them separately.

3.A.2 Tangential discontinuities

3.A.2.1 Definition

There is no matter flow across the surface. Under these conditions, it is easily shown that:

$$v_{1x} = v_{2x} = 0, \qquad P = 0. \tag{3A.7}$$

As for the tangential velocities v_y and v_z and the density (as well as all other thermodynamic quantities, except for pressure), they can exhibit arbitrary jumps.

3.A.2.2 Kelvin–Helmholtz (or tangential discontinuity) instability

We assume that the two fluids (1) and (2) are different (immiscible) and *incompressible*. It is easy to prove that the surface separating them (taken as planar at the start) is absolutely unstable if the tangential velocities of the two fluids are different. The geometry is sketched in Fig. 3.A.2.

The principle of the demonstration consists of introducing at the surface an infinitesimal perturbation, given by:

$$\xi = \xi_0 \exp\left[i(kz - \omega t)\right] \tag{3A.8}$$

with an arbitrarily small amplitude ξ_0. Without any loss of generality, one can set $v_1 = v$ and $v_2 = 0$. The dispersion relation connecting ω to k is obtained by writing the boundary conditions at the interface (kinematic conditions and stress continuity). It is given by:

$$\omega = kv \frac{\rho_1 \pm i\sqrt{\rho_1 \rho_2}}{\rho_1 + \rho_2}. \tag{3A.9}$$

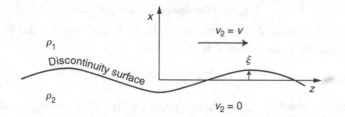

Fig. 3.A.2 Geometry of the Kelvin–Helmholtz instability.

One can see that ω is a complex quantity, and that there always exist some ω values with positive imaginary component. Thus, tangential discontinuities are always unstable with respect to infinitesimal perturbations of the surface. This instability is responsible for wave formation at the water surface when the wind blows. In this case, however, both the surface tension γ and the gravity play a stabilising role. Indeed, it can be shown that the surface is absolutely unstable provided the following condition is fulfilled:

$$v^4 > \frac{4\gamma g(\rho_2 - \rho_1)(\rho_1 + \rho_2)^2}{\rho_1^2 \rho_2^2}. \tag{3A.10}$$

In particular, this relation shows that the surface is always unstable if the heavier fluid is on top of the lighter one ($\rho_2 < \rho_1$). In the opposite case, there exists a finite velocity value below which the interface remains planar.

We should also point out that the instability does not develop immediately, if the viscosity of the fluids is taken into account: the velocity variations take place in a layer of finite thickness. The flow y is laminar at low velocity, but rapidly becomes turbulent as v increases.

3.A.3 Shock waves

Let us start by defining this concept in a perfect compressible fluid.

3.A.3.1 Definition

The flow of matter across the discontinuity surface is finite. From Eqs. (3A.3) and (3A.6) it follows that:

$$v_y = v_z = 0. \tag{3A.11}$$

The tangential velocity is thus continuous at the discontinuity surface. One can therefore write, from Eqs. (3A.2), (3A.3) and (3A.5) that

$$[\rho v_x] = 0, \tag{3A.12a}$$

$$\left[\frac{v_x^2}{2} + h\right] = 0, \tag{3A.12b}$$

$$\left[P + \rho v_x^2\right] = 0. \tag{3A.12c}$$

The discontinuities that obey these conditions are called *shock waves*.

Let us now present some of their properties.

3.A.3.2 Shock adiabatic

We choose a reference frame in which the discontinuity surface element is at rest, the tangent component of the fluid's velocity on either side of the surface being zero. In the following, we denote by v the x component of the velocity (instead of v_x) and we choose v_1 and v_2 to be positive, designating by (1) the fluid (a gas, most commonly) towards which the shock wave moves and by (2) the one it leaves behind (Fig. 3.A.3).

We now establish some general relations ensuing from the conservation laws (3A.12). Instead of the density ρ, we use the specific volume $V = 1/\rho$. Equation (3A.12a) can be rewritten as:

$$\frac{V_1}{V_2} = \frac{v_1}{v_2}. \tag{3A.13}$$

From Eqs. (3A.13) and (3A.12c), we deduce:

$$v_1^2 = V_1^2 \frac{P_2 - P_1}{V_1 - V_2}, \tag{3A.14}$$

$$v_2^2 = V_2^2 \frac{P_2 - P_1}{V_1 - V_2}, \tag{3A.15}$$

whence, by subtraction:

$$v_1 - v_2 = \sqrt{(P_2 - P_1)(V_1 - V_2)}. \tag{3A.16}$$

Note that we took here the + sign in front of the square root to enforce $v_1 - v_2 > 0$, a result to be demonstrated below from thermodynamic considerations.

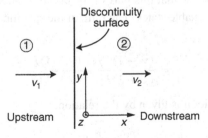

Fig. 3.A.3 Conventions adopted for describing the shock wave.

Let us also present the expression of the difference in the kinetic energy of the fluid across the discontinuity:

$$\frac{1}{2}\left(v_1^2 - v_2^2\right) = \frac{1}{2}(P_2 - P_1)(V_1 + V_2).$$
(3A.17)

Using Eq. (3A.12b), this relation can be rewritten as:

$$h_2 - h_1 = \frac{1}{2}(P_2 - P_1)(V_1 + V_2).$$
(3A.18)

This is the *Hugoniot relation*, also known as the *shock adiabatic* since it is formally equivalent to the equation relating the pressure and the specific volume of a fluid adiabatically compressed from an initial to a final state. The Hugoniot curve is given by the function:

$$P_2 = H(V_2, P_1, V_1).$$
(3A.19)

In many practically relevant situations, when the thermodynamic function $h = h(P, V)$ has a simple form, this relation can be written explicitly. In particular, this is the case for a perfect gas with constant specific heat parameters.

3.A.3.3 Shock adiabatics in a perfect gas with constant specific heat capacity

Replacing the enthalpy in Eq. (3A.18) by its expression:

$$h = C_p T = \frac{\gamma}{\gamma - 1} PV$$
(3A.20)

yields the equation of the Hugoniot curve in closed form:

$$\frac{P_2}{P_1} = \frac{(\gamma + 1)V_1 - (\gamma - 1)V_2}{(\gamma + 1)V_2 - (\gamma - 1)V_1}.$$
(3A.21)

It is a branch of a hyperbola (Fig. 3.A.4).

In principle, this curve can be extrapolated to pressure values below the initial one ($P_2 < P_1$). We will see below that this part of the curve (plotted as a dashed line) corresponds to physically unreachable states. The ratio of the specific volumes is obtained from Eq. (3A.21):

$$\frac{V_2}{V_1} = \frac{(\gamma - 1)P_2 + (\gamma + 1)P_1}{(\gamma + 1)P_2 + (\gamma - 1)P_1}.$$
(3A.22)

As for the temperature ratio, it is given by the relation:

$$\frac{T_2}{T_1} = \frac{P_2 V_2}{P_1 V_1}.$$
(3A.23)

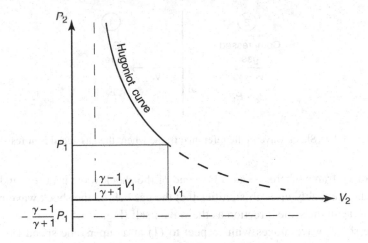

Fig. 3.A.4 Hugoniot curve in a perfect gas.

To express the velocities as a function of the pressures and the initial specific volume, one uses Eqs. (3A.14), (3A.15) and (3A.22):

$$v_1^2 = \frac{V_1}{2}[(\gamma - 1)P_1 + (\gamma + 1)P_2], \tag{3A.24}$$

$$v_2^2 = \frac{V_1}{2}\frac{[(\gamma + 1)P_1 + (\gamma - 1)P_2]^2}{(\gamma - 1)P_1 + (\gamma + 1)P_2}. \tag{3A.25}$$

It is worthwhile comparing these velocities to the speed of sound on either side of the discontinuity. Since, from Eq. (3.76)

$$c^2 = \left(\frac{\partial P}{\partial \rho}\right)_s = \frac{\gamma P}{\rho} = \gamma PV \tag{3A.26}$$

we have:

$$M_1^2 = \left(\frac{v_1}{c_1}\right)^2 = \frac{(\gamma - 1) + (\gamma + 1)(P_2/P_1)}{2\gamma}, \tag{3A.27}$$

$$M_2^2 = \left(\frac{v_2}{c_2}\right)^2 = \frac{(\gamma - 1) + (\gamma + 1)(P_1/P_2)}{2\gamma}, \tag{3A.28}$$

where M_1 and M_2 are the Mach numbers on each side of the shock wave. Below, we will prove that $P_2 > P_1$. Equations (3A.17), (3A.22), (3A.27) and (3A.28) yield:

$$V_2 < V_1, \quad \rho_2 > \rho_1, \quad v_1 > v_2, \quad M_1 > 1, \quad M_2 < 1. \tag{3A.29}$$

These relations state that, in the reference frame comoving with the discontinuity, *the flow is supersonic ahead of the shock wave and subsonic behind it.*

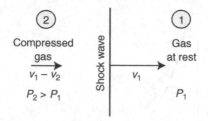

Fig. 3.A.5 Shock wave in the reference frame where the gas ahead is at rest.

In a reference frame where the gas (1) ahead of the shock wave is at rest and the wave is moving, the inequality $v_1 > v_2$ signifies that the gas behind the shock wave moves at a velocity $v_1 - v_2$ in the same direction as the wave itself (Fig. 3.A.5).

Since the shock wave moves with respect to (1) at a supersonic speed ($v_1 > c_1$), it is obvious that no perturbation originating on the shock wave can penetrate into (1). To state this differently, the presence of the shock wave does not affect the state of the gas ahead of it.

We still need to prove the strong inequality

$$P_2 > P_1, \tag{3A.30}$$

to be satisfied upon crossing the shock wave (which, in fact, is always a compression wave).

We know that the entropy of a perfect gas with constant specific heat parameters is given by (up to an additive constant):

$$s = C_v \ln(PV^\gamma). \tag{3A.31}$$

When crossing the shock wave, the entropy can only increase. The entropy variation is:

$$s_2 - s_1 = C_v \ln \left(\frac{P_2 V_2^\gamma}{P_1 V_1^\gamma} \right) = C_v \ln \left[\frac{P_2}{P_1} \left(\frac{(\gamma - 1)(P_2/P_1) + (\gamma + 1)}{(\gamma + 1)(P_2/P_1) + (\gamma - 1)} \right)^\gamma \right]. \tag{3A.32}$$

This quantity is easily shown to be positive for $P_2 > P_1$ and negative otherwise. Thus, the condition of entropy increase must necessarily lead to the inequality (3A.30) (which gives rise to the inequalities (3A.29)).

We should point out here that the presence of shock waves leads to an entropy increase for a flow that can everywhere be considered that of a perfect fluid, with no viscosity or thermal conduction. Thus, the discontinuities lead to energy dissipation for the flow of a perfect fluid. Of course, this dissipation takes place in the very thin fluid layers materialising the physical shock waves. It is nevertheless remarkable that the magnitude of this dissipation is completely determined by the sole conservation laws for mass, energy and momentum, applied on either side of these layers. This occurs precisely because the thickness of these layers is selected such as to yield the entropy increase required by these laws. Its value is generally of the order of a few micrometres.

3.A.3.4 Shock waves in practice

We have already shown that shock waves can only appear in the supersonic regime. This result is well known by ballisticians, who observed these shock waves at the end of the nineteenth century during the motion of projectiles travelling at supersonic speeds.

To understand the formation of a shock wave under these conditions, one can adopt the following reasoning: in supersonic flow, perturbations originating on the body can only propagate downstream. Consequently, a uniform supersonic flow should remain unperturbed up to the very leading edge of the body. This is impossible, since then the normal component of the gas velocity would not vanish at the surface of the body. Only the appearance of a shock wave can remove this contradiction, the gas flow between it and the front of the body becoming subsonic.

Ahead of the shock wave, the flow is uniform. Behind it, the motion is modified, and the flow goes around the body.

If the front of the body is rounded, the shock wave is always detached (Fig. 3.A.6a).

If, on the other hand, the body is pointed, the shock wave can make contact with it. In this case, the wave exhibits a taper point coinciding with the tip of the body (Fig. 3.A.6b).

Note that, for a body of a given shape, such a flow regime is only possible at high enough speeds; below a certain speed threshold the shock wave detaches, the presence of the tip notwithstanding.

Finally, we emphasise that the energy dissipated by the shock wave gives rise to a drag force acting on the body and termed *shock wave drag*. For a body of revolution with a

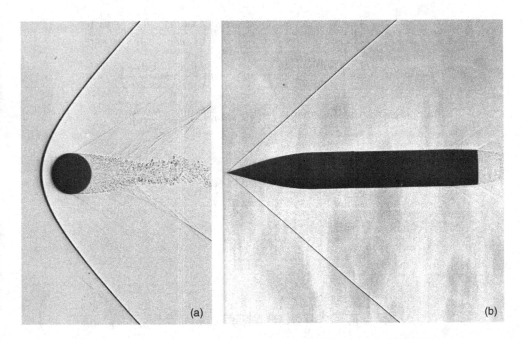

Fig. 3.A.6 (a) Detached shock wave; (b) attached shock wave (from [8]).

length L along the flow direction and a section S in the perpendicular plane, one can show that the order of magnitude of this force is given by:

$$F \approx \frac{\rho_1 v_1^2 S^2}{L^2}.$$
(3A.33)

Thus, the d'Alembert paradox disappears in the case of shock-wave flow.

Appendix 3.B Hydrodynamic equations for an incompressible fluid

3.B.1 Cylindrical coordinates (r, θ, z)

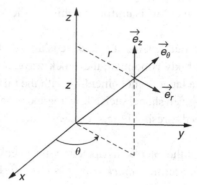

Viscous stress tensor

$$\sigma_{rr} = -P + 2\eta \frac{\partial v_r}{\partial r} \qquad \sigma_{r\theta} = \eta \left(\frac{1}{r}\frac{\partial v_r}{\partial \theta} + \frac{\partial v_\theta}{\partial r} - \frac{v_\theta}{r} \right)$$

$$\sigma_{\theta\theta} = -P + 2\eta \left(\frac{1}{r}\frac{\partial v_\theta}{\partial \theta} + \frac{v_r}{r} \right) \qquad \sigma_{\theta z} = \eta \left(\frac{\partial v_\theta}{\partial z} + \frac{1}{r}\frac{\partial v_z}{\partial \theta} \right)$$

$$\sigma_{zz} = -P + 2\eta \frac{\partial v_z}{\partial z} \qquad \sigma_{zr} = \eta \left(\frac{\partial v_z}{\partial r} + \frac{\partial v_r}{\partial z} \right)$$

Continuity equation

$$\frac{\partial v_r}{\partial r} + \frac{1}{r}\frac{\partial v_\theta}{\partial \theta} + \frac{\partial v_z}{\partial z} + \frac{v_r}{r} = 0$$

Navier–Stokes equation ($\vec{f} = 0$)

$$\frac{\partial v_r}{\partial t} + v_r \frac{\partial v_r}{\partial r} + \frac{v_\theta}{r}\frac{\partial v_r}{\partial \theta} + v_z \frac{\partial v_r}{\partial z} - \frac{v_\theta^2}{r}$$

$$= -\frac{1}{\rho}\frac{\partial P}{\partial r} + v \left(\frac{\partial^2 v_r}{\partial r^2} + \frac{1}{r^2}\frac{\partial^2 v_r}{\partial \theta^2} + \frac{\partial^2 v_r}{\partial z^2} + \frac{1}{r}\frac{\partial v_r}{\partial r} - \frac{2}{r^2}\frac{\partial v_\theta}{\partial \theta} - \frac{v_r}{r^2} \right)$$

$$\frac{\partial v_\theta}{\partial t} + v_r \frac{\partial v_\theta}{\partial r} + \frac{v_\theta}{r} \frac{\partial v_\theta}{\partial \theta} + v_z \frac{\partial v_\theta}{\partial z} + \frac{v_r v_\theta}{r}$$

$$= -\frac{1}{\rho r} \frac{\partial P}{\partial \theta} + \nu \left(\frac{\partial^2 v_\theta}{\partial r^2} + \frac{1}{r^2} \frac{\partial^2 v_\theta}{\partial \theta^2} + \frac{\partial^2 v_\theta}{\partial z^2} + \frac{1}{r} \frac{\partial v_\theta}{\partial r} + \frac{2}{r^2} \frac{\partial v_r}{\partial \theta} - \frac{v_\theta}{r^2} \right)$$

$$\frac{\partial v_z}{\partial t} + v_r \frac{\partial v_z}{\partial r} + \frac{v_\theta}{r} \frac{\partial v_z}{\partial \theta} + v_z \frac{\partial v_z}{\partial z} = -\frac{1}{\rho} \frac{\partial P}{\partial z} + \nu \left(\frac{\partial^2 v_z}{\partial r^2} + \frac{1}{r^2} \frac{\partial^2 v_z}{\partial \theta^2} + \frac{\partial^2 v_z}{\partial z^2} + \frac{1}{r} \frac{\partial v_z}{\partial r} \right)$$

3.B.2 Spherical coordinates (r, θ, φ)

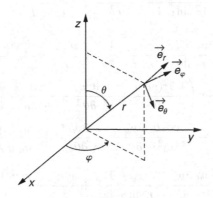

Stress tensor

$$\sigma_{rr} = -P + 2\eta \frac{\partial v_r}{\partial r}$$

$$\sigma_{\varphi\varphi} = -P + 2\eta \left(\frac{1}{r \sin\theta} \frac{\partial v_\varphi}{\partial \varphi} + \frac{v_r}{r} + \frac{v_\theta \cot\theta}{r} \right)$$

$$\sigma_{\theta\theta} = -P + 2\eta \left(\frac{1}{r} \frac{\partial v_\theta}{\partial \theta} + \frac{v_r}{r} \right)$$

$$\sigma_{r\theta} = \eta \left(\frac{1}{r} \frac{\partial v_r}{\partial \theta} + \frac{\partial v_\theta}{\partial r} - \frac{v_\theta}{r} \right)$$

$$\sigma_{\theta\varphi} = \eta \left(\frac{1}{r \sin\theta} \frac{\partial v_\theta}{\partial \varphi} + \frac{1}{r} \frac{\partial v_\varphi}{\partial \theta} - \frac{v_\varphi \cot\theta}{r} \right)$$

$$\sigma_{\varphi r} = \eta \left(\frac{\partial v_\varphi}{\partial r} + \frac{1}{r \sin\theta} \frac{\partial v_r}{\partial \varphi} - \frac{v_\varphi}{r} \right)$$

Continuity equation

$$\frac{\partial v_r}{\partial r} + \frac{1}{r}\frac{\partial v_\theta}{\partial \theta} + \frac{1}{r\sin\theta}\frac{\partial v_\varphi}{\partial \varphi} + \frac{2v_r}{r} + \frac{v_\theta \cot\theta}{r} = 0$$

Navier–Stokes equation ($\vec{f} = 0$)

$$\frac{\partial v_r}{\partial t} + v_r\frac{\partial v_r}{\partial r} + \frac{v_\theta}{r}\frac{\partial v_r}{\partial \theta} + \frac{v_\varphi}{r\sin\theta}\frac{\partial v_r}{\partial \varphi} - \frac{v_\theta^2 + v_\varphi^2}{r}$$

$$= -\frac{1}{\rho}\frac{\partial P}{\partial r} + \nu\left(\frac{1}{r}\frac{\partial^2(rv_r)}{\partial r^2} + \frac{1}{r^2}\frac{\partial^2 v_r}{\partial \theta^2} + \frac{1}{r^2\sin^2\theta}\frac{\partial^2 v_r}{\partial \varphi^2} + \frac{\cot\theta}{r^2}\frac{\partial v_r}{\partial \theta}\right.$$

$$\left. -\frac{2}{r^2}\frac{\partial v_\theta}{\partial \theta} - \frac{2}{r^2\sin\theta}\frac{\partial v_\varphi}{\partial \varphi} - \frac{2v_r}{r^2} - 2\frac{v_\theta \cot\theta}{r^2}\right)$$

$$\frac{\partial v_\theta}{\partial t} + v_r\frac{\partial v_\theta}{\partial r} + \frac{v_\theta}{r}\frac{\partial v_\theta}{\partial \theta} + \frac{v_\varphi}{r\sin\theta}\frac{\partial v_\theta}{\partial \varphi} + \frac{v_r v_\theta}{r} - \frac{v_\varphi^2 \cot\theta}{r}$$

$$= -\frac{1}{\rho r}\frac{\partial P}{\partial \theta} + \nu\left(\frac{1}{r}\frac{\partial^2(rv_\theta)}{\partial r^2} + \frac{1}{r^2}\frac{\partial^2 v_\theta}{\partial \theta^2} + \frac{1}{r^2\sin^2\theta}\frac{\partial^2 v_\theta}{\partial \varphi^2}\right.$$

$$\left. +\frac{\cot\theta}{r^2}\frac{\partial v_\theta}{\partial \theta} - \frac{2\cos\theta}{r^2\sin^2\theta}\frac{\partial v_\varphi}{\partial \varphi} + \frac{2}{r^2}\frac{\partial v_r}{\partial \theta} - \frac{v_\theta}{r^2\sin^2\theta}\right)$$

$$\frac{\partial v_\varphi}{\partial t} + v_r\frac{\partial v_\varphi}{\partial r} + \frac{v_\theta}{r}\frac{\partial v_\varphi}{\partial \theta} + \frac{v_\varphi}{r\sin\theta}\frac{\partial v_\varphi}{\partial \varphi} + \frac{v_r v_\varphi}{r} + \frac{v_\theta v_\varphi \cot\theta}{r}$$

$$= -\frac{1}{\rho r\sin\theta}\frac{\partial P}{\partial \varphi} + \nu\left(\frac{1}{r}\frac{\partial^2(rv_\varphi)}{\partial r^2} + \frac{1}{r^2}\frac{\partial^2 v_\varphi}{\partial \theta^2} + \frac{1}{r^2\sin^2\theta}\frac{\partial^2 v_\varphi}{\partial \varphi^2}\right.$$

$$\left. +\frac{\cot\theta}{r^2}\frac{\partial v_\varphi}{\partial \theta} + \frac{2}{r^2\sin\theta}\frac{\partial v_r}{\partial \varphi} + \frac{2\cos\theta}{r^2\sin^2\theta}\frac{\partial v_\theta}{\partial \varphi} - \frac{v_\varphi}{r^2\sin^2\theta}\right)$$

References

Andrade, E. N. da C., *Nature*, **125**, 309 (1930).

Atkins, P. W., *Physical Chemistry*, Oxford: Oxford University Press (1995).

Barnes, H. A., *J. Non-Newtonian Fluid Mech.*, **56**, 221 (1995).

Barrat, J.-L. and L. Boquet, *Phys. Rev. Lett.*, **82**, 4671 (1999); *Faraday Discuss.*, **112**, 119 (1999).

Chandrasekhar, S., *Hydrodynamic and Hydromagnetic Stability*, New York: Dover (1981).

Childress, S., *Mechanics of Swimming and Flying*, Cambridge Studies in Mathematical Biology, Cambridge: Cambridge University Press (1981).

Cottin–Bizonne, C., S. Jurine, J. Baudry, J. Crassous, F. Restagno and E. Charlaix, *Eur. Phys. J. E*, **9**, 47 (2002).

Ewell, R. H. and H. Eyring, *J. Chem. Phys.*, **5**, 726 (1937).

Eyring, H., *J. Chem. Phys.*, **3**, 107 (1935).

Eyring, H., *J. Chem. Phys.*, **4**, 283 (1936).

Gay, C., *Eur. Phys. J. B*, **7**, 251 (1999).

Glasstone, S., K. J. Laidler and H. Eyring, *Theory of Rate Processes*, New-York: McGraw-Hill (1941) Chapter 9.

de Groot, S. R. and P. Mazur, *Non-Equilibrium Thermodynamics*, New York: Dover (1983).

Kincaid, J. F., H. Eyring and A. E. Stearn, *Chem. Rev.*, **28**, 301 (1941).

Larson, R. G., S. J. Muller and E. S. G. Shaqfeh, *J. Fluid. Mech*, **218**, 573 (1990).

López–González, M. R., W. M. Holmes and P. T. Callaghan, *Soft Matter*, **2**, 855 (2006).

Manneville, S., L. Bécu and A. Colin, *Eur. Phys. J. Appl. Phys.*, **28**, 361 (2004).

Olagunju, D. O., Z. *Angew. Math. Phys.*, **46**, 224 (1995).

Pelcé, P., *Théorie des formes de croissance*, Les Ulis: EDP Sciences/CNRS Éditions (2000).

Purcell, E. M., *Am. J. Phys.*, **45**, 3 (1977).

Saffman, P. G. and G. I. Taylor, *Proc. Roy. Soc. A*, **245**, 312 (1958).

Salmon, J.-B, S. Manneville, A. Colin and B. Pouligny, *Eur. Phys. J. Appl. Phys.*, **22**, 143 (2003).

Schmidt, A., P. W. Gold, C. Assmann, H. Dicke and J. Loos, *Viscosity-pressure-temperature behaviour of mineral and synthetic oils, 12th International Colloquium Tribology 2000-Plus*, Stuttgart/Ostfildern, January 11–13 (2000).

Tabeling, P., G. Zocchi and A. Libchaber, *J. Fluid. Mech.*, **177**, 67 (1987).

Taylor, G. I., *Proc. Roy. Soc. A*, **209**, 447 (1951).

Taylor, G. I., *Proc. Roy. Soc. A*, **245**, 312 (1958).

Turcotte, D. L. and G. Schubert, *Geodynamics*, Cambridge: Cambridge University Press (2002).

Weeks, E. R. and D. A. Weitz, *Phys. Rev. Lett.*, **89**, 095704 (2002).

4

Elasticity of solids

In the previous chapter, we studied the behaviour of Newtonian fluids. In this chapter and the next (as well as in Section 7.2), we will consider the other extreme case, that of *solids*. These are generally *crystals*, where the atoms (or molecules) occupy the sites of a *three-dimensional Bravais lattice*. Hence, these materials exhibit *positional order* (and also orientational order, when the molecules are not spherical) *extending over a long range in the three space directions*.

There are also materials with a *disorganised structure*, similar to that of a liquid, but which can be considered *frozen* over the typical time scales of an experiment. Consequently, the mechanical behaviour of these materials will be similar to that of solids, as long as their temperature is low enough, i.e. below their *glass transition temperature*, and provided the time scale of the observation is not too long. In this family one finds the silica glasses used for window panes, metallic glasses obtained by thermally quenching a molten metal at cooling rates between 10^6 and 10^8 K/s, and certain plastics, which are polymeric materials to be considered in detail in Chapter 7.

As we have already emphasised in Chapter 1, a distinction must be made between 'hard' solids (*metals, minerals, glasses*, etc.), 'soft' solids (*colloidal crystals* and *elastomers* such as rubber, gels, etc.). Since the latter are also *viscoelastic*, exhibiting very specific properties, we will discuss them separately. More precisely, we will dedicate the final section of this chapter to colloidal crystals, which are three-dimensional ordered arrangements of sol particles dispersed in a fluid. As to elastomers, formed by reticulated polymers, their elastic properties will be treated in Section 7.2. One could also classify the *blue phases* of cholesteric liquid crystals (which are chiral nematics) among extremely soft solids. However, their case is even more peculiar, since here the long-range three-dimensional order of the elongated molecules is purely orientational; we refer the reader to volume 1 of reference [43] for the description of their mechanical properties.

Notwithstanding their wide diversity, all these materials share the presence of an *elastic limit*, below which their deformation is *reversible* (*elastic regime*) whereas above this threshold their deformations become *permanent*, and thus *irreversible*. On the other hand, the way a solid behaves beyond its elastic limit depends crucially on its structure (for instance, a crystal will behave completely unlike a glass or an elastomer) and on the temperature. In particular, one can often define a temperature corresponding to the *brittle–ductile transition* (T_{BD}) above which a solid starts by *creeping* before breaking, while below this temperature it will *break* directly;[1] in the first case, one speaks of *plasticity* and *ductile fracture*, while in the second case one refers to *brittle fracture*. Crystal plasticity and the brittle fracture of glasses (in particular) will be analysed in the next chapter.

After this quick reminder, let us outline the plan of the chapter.

Hard solids being by and large the most common and the most important for applications, we start by describing their *elastic properties in the static regime* (Section 4.1). In this part, entitled '*Elastostatics of ordinary hard solids*', we recall the essential concepts of *stress and strain tensors* and then give the complete expression of *Hooke's law*, which provides a linear connection between these two tensors. We will also see how this law simplifies in the case of isotropic, and then anisotropic materials, as a consequence of the point symmetries of the material. We will then return to the notions of *normal and tangential stress*, leading to the construction of the *Mohr circles* and to the *Tresca criterion*, setting the limits of elasticity theory. This will be followed by the solution of a few classical elasticity problems (traction and twist of a beam and bending of a plate), and finally by a detailed analysis of the *Euler instability of beams* ('buckling' under compression) and of the flutter of an airplane wing (an instability of the flexion–torsion type induced by a circulatory lift force).

The second part is concerned with the *elastodynamics of ordinary hard solids* (Section 4.2). As the name indicates, this theory deals with the dynamic problems of elasticity. As an example, we will discuss the case of *elastic waves*. We will see that their polarisation can be transverse or longitudinal and we will calculate their propagation speed. We will show that their frequency is *quantified* in finite media (meaning that it can only have a countable infinity of values) thus forming a *discrete spectrum*. The case of *surface Rayleigh waves*, which are *evanescent waves* propagating at the surface of a body will also be treated. We will see that these waves are a very particular combination of transverse and longitudinal waves. At the end, we will give a few examples of practical applications for waves. For instance, we will show that they can be used to probe bulk materials in depth and to measure their elastic constants. We will also see how the study of their propagation through the Earth allowed seismologists to establish a model for its internal structure.

In conclusion of the chapter, we will speak of *colloidal crystals*, which are *soft viscoelastic solids* (Section 4.3). After a reminder of their conditions of formation, we will explain why only shear waves can propagate within them, provided their frequency is not too high. In particular, we will describe an original experimental setup that can be used to detect them.

4.1 Elastostatics of ordinary hard solids

In this section we will give the basis of the theory of elasticity [9–11]. This theory was developed by Cauchy and Poisson in the 1820s.

---

[1] The brittle–ductile transition temperature was discovered during the Second World War, due to the breaking of landing barges in the North Sea. Ignorance of this phenomenon is also among the causes of the wreck of the *Titanic*, where the steel hull had a brittle–ductile transition temperature well above $0\,^{\circ}\mathrm{C}$.

Elasticity of solids

4.1.1 Strain tensor

4.1.1.1 Definition

Consider a point M of a solid body and denote it by its vector radius $\vec{r}$. After deformation of the body, this point moves to a new position M$'$, with a vector radius $\vec{r}'$. The difference

$$\vec{u}(\vec{r}) = \vec{r}' - \vec{r} \tag{4.1}$$

defines the *displacement vector*. It is a function of $\vec{r}$, since $\vec{r}'$ depends on $\vec{r}$.

Due to the deformation, the distance between the points of the solid body varies. Let dl be the distance between two neighbouring points, with coordinates (x_1, x_2, x_3) and $(x_1 + dx_1, x_2 + dx_2, x_3 + dx_3)$ prior to the deformation. One has:

$$dl^2 = dx_1^2 + dx_2^2 + dx_3^2. \tag{4.2}$$

After deformation, these two points will have moved to the new coordinates (x_1', x_2', x_3') and $(x_1' + dx_1', x_2' + dx_2', x_3' + dx_3')$, such that the distance between them becomes:

$$dl'^2 = dx_1'^2 + dx_2'^2 + dx_3'^2. \tag{4.3}$$

To calculate dl', let us expand Eq. (4.1) to first order in the distortion $\delta_{ij} = \partial u_i / \partial x_j$. We find:

$$dx_j' = dx_i + \frac{\partial u_i}{\partial x_j} dx_j, \tag{4.4}$$

yielding, after substitution in Eq. (4.3):

$$dl'^2 = dl^2 + 2\frac{\partial u_i}{\partial x_j} dx_i \, dx_j. \tag{4.5}$$

These formulas apply in the case of very small distortion: $\partial u_i / \partial u_j \ll 1$ (which does not imply very small displacements).

In elasticity, it is convenient to define the strain tensor $\underline{\varepsilon}$, which is the symmetric part of the distortion tensor $\underline{\delta}$:

$$\varepsilon_{ij} = \frac{1}{2}\left(\frac{\partial u_i}{\partial x_j} + \frac{\partial u_j}{\partial x_i}\right). \tag{4.6}$$

Hence, this tensor is *symmetric* by construction:

$$\varepsilon_{ij} = \varepsilon_{ji}. \tag{4.7}$$

Using this new tensor, formula (4.5) can be rewritten as:

$$dl'^2 = dl^2 + 2\varepsilon_{ij} \, dx_i \, dx_j. \tag{4.8}$$

Note that the antisymmetric part $\underline{\omega}$ of the distortion tensor:

$$\omega_{ij} = \frac{1}{2}\left(\frac{\partial u_i}{\partial x_j} - \frac{\partial u_j}{\partial x_i}\right) \tag{4.9}$$

describes a solid rotation, playing no role in elasticity.

4.1.1.2 Geometric interpretation

Let $\vec{A}$ and $\vec{B}$ be two orthogonal unit vectors ($|\vec{A}| = |\vec{B}| = 1$ and $\vec{A} \cdot \vec{B} = 0$) 'attached at the extremities' to the undeformed solid body. After deformation, these vectors change in length and are no longer perpendicular, having transformed into $\vec{A}' = \vec{A} + \delta\vec{A}$ and $\vec{B}' = \vec{B} + \delta\vec{B}$.

Denoting by $\pi/2 - \theta$ the angle made by the new vectors $\vec{A}'$ and $\vec{B}'$ (Fig. 4.1), one can easily check that:

$$\frac{|\vec{A}'| - |\vec{A}|}{|\vec{A}|} = L(\vec{A}, \vec{A}) \tag{4.10}$$

and that:

$$\theta = 2L(\vec{A}, \vec{B}), \tag{4.11}$$

where L is the bilinear form associated with the strain tensor $\underline{\varepsilon}$:

$$L(\vec{A}, \vec{B}) = \varepsilon_{ij} A_i B_j = \vec{A}^t \underline{\varepsilon} \vec{B}. \tag{4.12}$$

These formulas show immediately that ε_{ii} (no summation over the i index) represents the *unit elongation* in the x_i direction, while ε_{ij} ($i \neq j$) describes *shear* in the (x_i, x_j) plane.

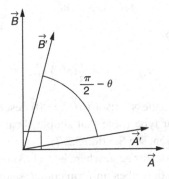

Fig. 4.1 Transformation of two orthogonal unit vectors attached to the solid.

Also note that, the strain tensor being symmetric (real), there exists a basis of eigenvectors in which it is expressed by a diagonal matrix:

$$\underline{\varepsilon} = \begin{pmatrix} \varepsilon_1 & 0 & 0 \\ 0 & \varepsilon_2 & 0 \\ 0 & 0 & \varepsilon_3 \end{pmatrix}. \tag{4.13}$$

The eigenvalues ε_i, known as *principal unit elongations*, are obtained by solving the equation:

$$\det \left(\underline{\varepsilon} - \varepsilon \, \underline{I} \right) = 0. \tag{4.14}$$

4.1.1.3 Bulk dilation

Consider now a volume element $dV = dx_1 d x_2 d x_3$ of the undeformed body. To calculate the new value $dV' = dx'_1 dx'_2 dx'_3$ after deformation, let us go to the reference frame of the principal axes of the strain tensor at the current point. Writing that $dx'_1 = dx_1(1+\varepsilon_1)$, etc. yields immediately:

$$dV' = dV \left[1 + \mathrm{tr} \left(\underline{\varepsilon} \right) \right]. \tag{4.15}$$

The quantity:

$$\frac{dV'}{dV} = 1 + \mathrm{tr} \left(\underline{\varepsilon} \right) = 1 + \mathrm{div}\, \vec{u} \tag{4.16}$$

is called *bulk dilation*.

One can also define a *relative volume increase*:

$$\frac{dV' - dV}{dV} = -\frac{\delta\rho}{\rho} = \mathrm{div}\, \vec{u} \tag{4.17}$$

with ρ the density of the body.

4.1.1.4 Spherical and deviator components

Sometimes, one sets:

$$\varepsilon_{ij} = e_{ij} + e\delta_{ij}. \tag{4.18}$$

The tensor $\underline{e}$, with components:

$$e_{ij} = \varepsilon_{ij} - \frac{1}{3}\mathrm{tr} \left(\underline{\varepsilon} \right) \delta_{ij}, \tag{4.19}$$

is termed the *deviator*. This traceless tensor $(\mathrm{tr}(\underline{e}) = 0)$ describes constant-volume deformations of elongation or shear type (note that simple shear can be decomposed into an elongation followed by a rotation, see Section 3.2.7).

The other tensor, with components $e\delta_{ij}$, where $e = (1/3)\mathrm{tr}(\underline{\varepsilon})$, corresponds to the *spherical part* of the strain tensor. It describes uniform compression or dilation deformations of the sample.

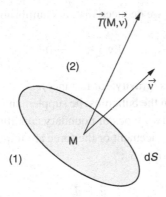

Fig. 4.2 Stress exerted by (2) onto (1) across the surface element dS within the material.

4.1.1.5 Fundamental (or elementary) invariants

One can show that any polynomial of rank n in the ε_{ij} elements which is invariant under a reference change can be expressed using the three following *fundamental (or elementary) invariants*:

$$\mathrm{tr}\left(\underline{\varepsilon}\right), \qquad \frac{1}{2}\left[\mathrm{tr}\left(\underline{\varepsilon}^2\right) - \left(\mathrm{tr}\,\underline{\varepsilon}\right)^2\right], \qquad \det\left(\underline{\varepsilon}\right). \tag{4.20}$$

Another possible choice is:

$$\mathrm{tr}\left(\underline{\varepsilon}\right), \qquad \mathrm{tr}\left(\underline{\varepsilon}^2\right), \qquad \mathrm{tr}\left(\underline{\varepsilon}^3\right). \tag{4.21}$$

4.1.2 Stress tensor

The existence of the stress tensor has already been proven in Chapter 2 using the tetrahedron construction. Hence, at this point, we will only recall its properties briefly.

4.1.2.1 Properties

Let $\underline{\sigma}$ be the stress tensor at point M, dS a surface element and $\vec{v}$ the unit vector perpendicular to dS.

We have shown that the medium pointed at by $\vec{v}$ exerts onto its complement, across the surface element dS, a stress (Fig. 4.2):

$$\vec{T}(M, \vec{v}) = \underline{\sigma}(M)\vec{v}. \tag{4.22}$$

Using the angular momentum theorem, we have also shown that the elastic stress tensor is *symmetric* in solids,[2] and hence *diagonalisable*. In the following, we denote the *principal stress values* by σ_1, σ_2 and σ_3.

---

[2] This result requires that the torques vanish, or at least be negligible. This assumption is verified in most anisotropic solids, where the elastic energy associated with the torques is several orders of magnitude lower than that associated with the stress.

Finally, from Newton's law we deduced that, at equilibrium:

$$\text{div}\,\underline{\sigma} + \vec{f} = 0 \tag{4.23}$$

with $\vec{f}$ the external bulk forces (gravity, for instance).

This equilibrium equation in the bulk must be supplemented by the boundary conditions on the surface S of the body. Two types of boundary conditions are usually distinguished, depending on whether the displacement or the force are imposed at the surface. Thus, one can have on one part of the surface S_u:

$$\vec{u} = \vec{u}_s, \tag{4.24}$$

where $\vec{u}_s$ is the imposed displacement, while on its complement S_f:

$$\underline{\sigma}\vec{v} = \vec{f}_s, \tag{4.25}$$

with $\vec{f}_s$ the imposed surface force and $\vec{v}$ the unit vector perpendicular to S_f and pointing outwards.

4.1.2.2 Normal and tangent stress: the Mohr circles

At each point M of the solid body, and for a given orientation $\vec{v}$, the stress force can be decomposed into its tangent and normal components:

$$\vec{T}(M, \vec{v}) = T_n\vec{v} + \vec{T}_t, \tag{4.26}$$

where

$$T_n = \vec{T} \cdot \vec{v} \tag{4.27}$$

is the *normal stress* and

$$\vec{T}_t = \vec{T} - T_n\vec{v} \tag{4.28}$$

the *tangent stress*.

Let us now consider a given point M of the solid and try to find all possible values of the normal and tangent stresses at this point, assuming the principal stress values σ_1, σ_2 and σ_3 are known. For each orientation $\vec{v}$ one gets, by going to the reference frame defined by the eigenvectors of the stress tensor:

$$T_n = \sigma_i v_i^2, \tag{4.29}$$

$$T^2 = \sigma_i^2 v_i^2, \tag{4.30}$$

$$v_i v_i = 1. \tag{4.31}$$

These three equations yield:

$$v_1^2 = \frac{T_t^2 + (T_n - \sigma_2)(T_n - \sigma_3)}{(\sigma_1 - \sigma_2)(\sigma_1 - \sigma_3)}, \tag{4.32}$$

$$v_2^2 = \frac{T_t^2 + (T_n - \sigma_3)(T_n - \sigma_1)}{(\sigma_2 - \sigma_3)(\sigma_2 - \sigma_1)}, \tag{4.33}$$

$$v_3^2 = \frac{T_t^2 + (T_n - \sigma_1)(T_n - \sigma_2)}{(\sigma_3 - \sigma_1)(\sigma_3 - \sigma_2)}. \tag{4.34}$$

By relabelling the principal axes such that $\sigma_3 \leq \sigma_2 \leq \sigma_1$, one can see from these equations that, in the (T_n, T_t) plane, the only allowed values for the normal and tangent stresses are contained within the lunules situated inside the large circle, defined by the equation $T_t^2 + (T_n - \sigma_3)(T_n - \sigma_1) = 0$ and outside the small circles given by the equations $T_t^2 + (T_n - \sigma_2)(T_n - \sigma_3) = 0$ and $T_t^2 + (T_n - \sigma_1)(T_n - \sigma_2) = 0$. These are the *Mohr circles* shown in Fig. 4.3.

In particular, this graph shows that the maximum tangent stress is given by:

$$T_t^{max} = \frac{\sigma_1 - \sigma_3}{2} \tag{4.35}$$

and for a normal stress:

$$T_n = \frac{\sigma_1 + \sigma_3}{2}. \tag{4.36}$$

These values correspond to the orientations $\vec{v}$ given by Eqs. (4.32)–(4.34):

$$v_1^2 = v_3^2 = \frac{1}{2} \quad \text{and} \quad v_2^2 = 0. \tag{4.37}$$

This calculation shows that the maximum in tangent stress is obtained for the directions of the bisectors of the angle $(\vec{x}_1, \vec{x}_3)$, formed by the two principal directions corresponding to the extremal eigenvalues of the stress tensor (Fig. 4.4).

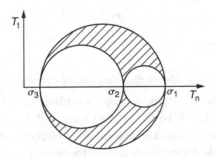

Fig. 4.3 Mohr circles. All possible values for the normal and tangent stress are contained within the cross-hatched region.

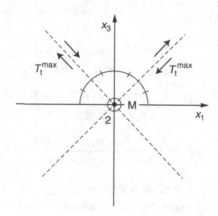

Fig. 4.4 Maximum shear stress at a given point. Directions x_1 and x_3 correspond to the extremal eigenvalues of the stress tensor at that point.

4.1.2.3 Tresca criterion

The theory of elasticity makes the assumption of small and reversible deformations. This means that, once the applied stress is removed, the body returns to its initial state. On the other hand, the body deforms *plastically* if permanent deformation ensues.

In practice, the deformation remains elastic as long as the applied stress is not too strong. Obviously, this is not a very precise criterion. To refine it, one must realise that crystalline solids (we exclude from the discussion the case of glasses, which are more likely to break, see Section 5.4) undergo plastic deformation due to the atomic planes gliding over each other. At least, this is what happens at low temperature (i.e. for $T < 0.5\, T_m$, typically, with T_m the melting temperature of the material). Hence, a more precise criterion for the elastic limit consists in requiring that, for each point of the body, the maximum tangent (or shear) stress remain below a certain threshold value σ_c, also called the *critical shear stress*, or *elastic limit*:

$$T_t^{\text{max}} < \sigma_c. \tag{4.38}$$

This is the *Tresca criterion*, which can also be expressed as:

$$(\sigma_1 - \sigma_3)^2 < 4\sigma_c^2 \tag{4.39}$$

using the result of Eq. (4.35).

4.1.2.4 Measuring the critical shear stress

This parameter can be obtained by performing a traction test on a cylindrical test bar of the investigated material. Let F be the applied force and S the cross-section of the bar (Fig. 4.5). Experiment shows that, above a minimal traction force, the test bar deforms plastically. This irreversible deformation leads to the appearance of steps on the surface of the material, revealing that atomic planes have slipped on top of each other. As shown by the photographs in Fig. 4.6, these steps are tilted with respect to the direction of traction.

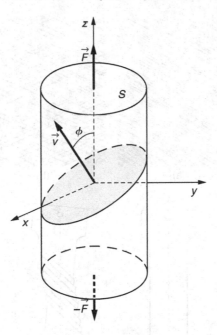

Fig. 4.5 Traction test.

To explain this result, let us determine the maximum shear stress within the test bar. It is easy to verify that the stress tensor is:

$$\underline{\sigma} = \begin{pmatrix} 0 & 0 & 0 \\ 0 & 0 & 0 \\ 0 & 0 & \sigma \end{pmatrix} \qquad (4.40)$$

with $\sigma = F/S$.

Hence, the tangent stress within the bar, for a direction $\vec{v}$ making an angle ϕ with the traction direction, is given by:

$$\vec{T}_t = \vec{T} - (\vec{T} \cdot \vec{v})\vec{v}$$
$$= \sigma \cos\phi \, \vec{z} - \sigma \cos^2\phi \, \vec{v} \qquad (4.41)$$

with an absolute value:

$$T_t^2 = \sigma^2 \cos^2\phi \sin^2\phi. \qquad (4.42)$$

Thus, the tangent stress reaches its maximum for $\phi = 45°$, at a value:

$$T_t^{\max} = \frac{\sigma}{2} = \frac{1}{2}\frac{F}{S}. \qquad (4.43)$$

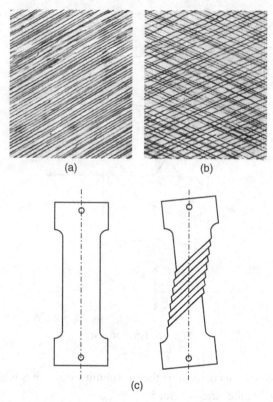

(a)   (b)

(c)

Fig. 4.6 Slip lines observed in reflection optical microscopy at the surface of an aluminium monocrystal (×150): (a) after primary slip; (b) after the successive activation of two slip systems; (c) sketch of a monocrystalline test bar under traction, before and after slip of the crystal planes. Since the grip ends remain aligned, the slip planes rotate with respect to the deformation axis ([13], Volume 5, p. 1370).

It is then enough to measure the minimum traction force F_{min} above which the bar deforms plastically in order to obtain σ_c, since:

$$\sigma_c = \frac{1}{2} \frac{F_{min}}{S}. \tag{4.44}$$

Its order of magnitude is $\sigma_c \approx 1$ kgf/mm^2 or, in SI units, $\sigma_c \approx 10^7$ Pa.

4.1.2.5 Theoretical critical shear stress

Can one predict theoretically this experimental value of the critical shear? To answer this question, let us find out the stress that must be applied for two atomic planes to slip over each other in a cooperative manner.

Let u be the displacement of a plane with respect to the next, while c and a are the sizes of the crystal lattice along and perpendicular to the direction of shear, respectively

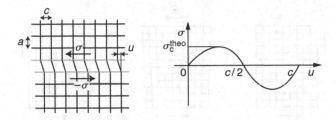

Fig. 4.7 Principle of calculating the theoretical critical shear stress.

(Fig. 4.7). As long as the displacement $u \ll a$, the shear stress remains proportional to the strain:

$$\sigma = \mu \frac{u}{a},\qquad(4.45)$$

where μ is the shear modulus, by definition.

On the other hand, crystal periodicity imposes:

$$\sigma(u + c) = \sigma(u).\qquad(4.46)$$

The simplest function $\sigma(u)$ that satisfies these two relations is:

$$\sigma = \frac{\mu c}{2\pi a} \sin\left(2\pi \frac{u}{c}\right).\qquad(4.47)$$

This function goes through a maximum, defining the *theoretical critical shear stress*:

$$\sigma_c^{theo} = \frac{\mu c}{2\pi a} \approx \frac{\mu}{2\pi} \quad (\text{if } c \approx a).\qquad(4.48)$$

In aluminium, for instance, $\mu \approx 2.7 \times 10^{10}$ Pa, yielding a theoretical value close to 5×10^9 Pa. This value is typically three orders of magnitude above the experimental one.

This profound discrepancy between theory and experiment remained unexplained for a long time. It was only at the beginning of the twentieth century, more exactly in 1929, that Y. Yamaguchi put forward the idea that the planes do not slip one over the other collectively, but rather progressively, by the propagation of linear defects, or *dislocations* (Fig. 4.8a) (Yamaguchi, 1929). In this case, the slip is facilitated, as shown by the analogy with the Mott carpet (Fig. 4.8b).

Keep in mind that, at the time, these defects were already well known by some Italian mathematicians, among them V. Volterra, who had begun their study in 1907 (Volterra, 1907). Yamaguchi's idea was then revived in the 1930s by G. I. Taylor, E. Orowan and M. Polanyi, who published separately and almost simultaneously three fundamental papers in 1934 (Taylor, 1934; Orowan, 1934; Polanyi, 1934), but it was only in 1939 that J. M. Burgers showed clearly the link between the defects of Volterra and those of Yamaguchi (Burgers, 1937). One had to wait for the 1950s and progress in electron microscopy for W. Bollmann in Switzerland (Bollmann, 1956) and P. B. Hirsch

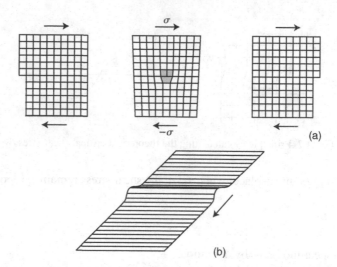

Fig. 4.8 (a) Slip of two atomic planes by glide of an edge dislocation (for more details, see Section 5.2.4); (b) Mott carpet: it is much easier to drag a carpet over the floor by making a fold and then edging it across than by pulling the whole carpet. The same principle applies to dislocations.

in England (Hirsch *et al.*, 1956) (who founded the famous Cambridge school) to provide direct experimental evidence for the existence of these defects (in stainless steel for Bollmann and in aluminium for Hirsch) and to convincingly establish their role in plasticity. Note that other, more indirect, methods can also be used to visualise these defects (see Section 5.1.2.3).

These concepts will be detailed in the following chapter, dedicated to dislocations and to plastic deformation. To conclude this paragraph let us cite an excellent historical article on the concept of dislocations by B. Jouffrey (Jouffrey, 1979).

4.1.3 Thermodynamics of the deformation

We will now prove that a universal relation connects the stress tensor to the deformation.

To this end, consider a generic volume V, delimited by a surface S, within the solid. Assume that, under the action of the applied forces, the displacement varies from $\vec{u}$ to $\vec{u} + \delta\vec{u}$. During this quasi-static elementary deformation, the bulk forces $\vec{f}$ and the internal stress forces $\vec{T}(\vec{v})$ acting on the surface S do the work:

$$\delta W = \int_V f_i \, \delta u_i \, dV + \int_S T_i(\vec{v}) \delta u_i \, dS. \tag{4.49}$$

This work can be rewritten, by using Eqs. (4.22) and (4.23) and transforming the surface integral into a volume integral:

$$\delta W = \int_V \sigma_{ij} \, du_{i,j} \, dV, \tag{4.50}$$

where we set $u_{i,j} = \partial u_i / \partial x_j$.

However, we know that for a reversible quasi-static transformation:

$$dE = \delta W + T \, dS, \tag{4.51}$$

with E the internal energy of the volume V and S its entropy. By switching to the volume density of entropy and internal energy (s and e, respectively), the two equations above yield:

$$de = T \, ds + \sigma_{ij} \, du_{i,j}. \tag{4.52}$$

This equation can be rewritten in the equivalent form:

$$df = -s \, dT + \sigma_{ij} \, du_{i,j}, \tag{4.53}$$

where $f = e - Ts$ is the free energy per unit volume.

From these two equations we can deduce the following relations:

$$\sigma_{ij} = \left(\frac{\partial e}{\partial u_{i,j}} \right)_s = \left(\frac{\partial f}{\partial u_{i,j}} \right)_T. \tag{4.54}$$

These thermodynamic relations are very general, applying even when the stress tensor is asymmetric, as for liquid crystals.

We know however that this tensor is symmetric in solids, allowing us to write:

$$\sigma_{ij} \, du_{i,j} = \frac{1}{2} \left(\sigma_{ij} \, du_{i,j} + \sigma_{ji} \, du_{j,i} \right) = \sigma_{ij} \, d\varepsilon_{ij}, \tag{4.55}$$

where we brought back the strain tensor ε_{ij}. Under these conditions, formulas (4.54) become:

$$\sigma_{ij} = \left(\frac{\partial e}{\partial \varepsilon_{ij}} \right)_s = \left(\frac{\partial f}{\partial \varepsilon_{ij}} \right)_T. \tag{4.56}$$

We emphasise that, unlike Eqs. (4.54), these formulas are not always valid.

4.1.4 Hooke's law

Now that we have found the thermodynamic relation connecting the stress tensor to the strain, let us try to use it for constructing the constitutive laws of the material. We start by expanding the deformation free energy in a power series of the strain ε_{ij}. Since the free energy must have a minimum at equilibrium, in the absence of deformation, the first non-zero term is of second order in u_{ij}. Keeping only this term results in:

$$f = \frac{1}{2} \lambda_{ijkl} \varepsilon_{ij} \varepsilon_{kl}, \tag{4.57}$$

yielding directly, through use of Eq. (4.56) and taking $\lambda_{ijkl} = \lambda_{klij}$:[3]

$$\sigma_{ij} = \lambda_{ijkl}\varepsilon_{kl}. \tag{4.58}$$

This is *Hooke's law*, stating that stress is proportional to strain (that is why one speaks of *linear elasticity*). The λ_{ijkl} parameters are the *elastic moduli*. In general there are 45 of them, but their number can be reduced to 21, as they must obey the following additional relations:

$$\lambda_{ijkl} = \lambda_{jikl} = \lambda_{ijlk}, \tag{4.59}$$

stemming from the symmetry of the stress and strain tensors.

This number can be reduced even further by taking into account the symmetry of the crystal. We will see that an isotropic material only has two distinct moduli, while there are five in a hexagonal crystal and three in a cubic one.

Before showing how to find the elastic moduli, note that the free energy can also be written in the general form:

$$f = \frac{1}{2}\sigma_{ij}\varepsilon_{ij} \tag{4.60}$$

in a Cartesian reference frame and provided the linear elasticity approximation still holds.[4]

4.1.4.1 Isotropic material

In this case, the tensor λ_{ijkl} of the elastic moduli must have the same expression (i.e. the same components) in any orthonormal reference frame. The free energy is then a quadratic invariant of the strain tensor ε_{ij}, so it can be expressed as a function of the fundamental invariants. Several choices are possible for these invariants, as stated in Section 4.1.1.5. Let us take, for instance, $\mathrm{tr}(\underline{\varepsilon})$, $\mathrm{tr}(\underline{\varepsilon}^2)$ and $\mathrm{tr}(\underline{\varepsilon}^3)$; it is clear that one can only construct two independent quantities quadratic in the strain, for instance: $\mathrm{tr}(\underline{\varepsilon})^2$ and $\mathrm{tr}(\underline{\varepsilon}^2)$. The most general form of the elastic free energy for an isotropic material is therefore a linear combination of these two quantities:

$$f = \frac{1}{2}\lambda \left[\mathrm{tr}(\underline{\varepsilon})\right]^2 + \mu\,\mathrm{tr}(\underline{\varepsilon}^2) = \frac{1}{2}\lambda\varepsilon_{ii}^2 + \mu\varepsilon_{ij}\varepsilon_{ij}. \tag{4.61}$$

The constants λ and μ are the *Lamé coefficients* of the material.

Using the thermodynamic relation (4.56), one then finds the constitutive law of the material, relating the stress tensor to the strain tensor:

$$\sigma_{ij} = \lambda\varepsilon_{kk}\,\delta_{ij} + 2\mu\varepsilon_{ij}. \tag{4.62}$$

---

[3] Something we can always do, without restricting the generality of Eq. (4.57).
[4] This formula does not apply in cylindrical or spherical coordinates.

One can also express the free energy as a function of $\text{tr}(\varepsilon)^2$ and the trace of the squared deviator $\text{tr}(e^2)$ (Section 4.1.1.4). Comparison with formula (4.61) yields:

$$f = \mu e_{ij} e_{ij} + \frac{K}{2}\varepsilon_{ii}^2, \tag{4.63}$$

where K is a new modulus, given by:

$$K = \lambda + \frac{2}{3}\mu. \tag{4.64}$$

Equation (4.62) can also be rewritten in the following form:

$$\sigma_{ij} = K\varepsilon_{kk}\,\delta_{ij} + 2\mu e_{ij}. \tag{4.65}$$

This choice allows a clear distinction to be made between the shear deformation (associated with the deviator) and the deformations involving uniform compression or dilation of the material (associated with the volume variations given by the trace of the strain tensor).

The coefficient μ is thus identified with the *shear modulus*, while K is the *compression modulus* (associated with uniform compression or dilation).

4.1.4.2 Anisotropic material

In this subsection, we will show how to find the elasticity tensor of a material, accounting for its symmetry. For simplicity's sake, let us introduce the notations [10]:

$$\begin{pmatrix} \varepsilon_{11} & \varepsilon_{12} & \varepsilon_{13} \\ \varepsilon_{21} & \varepsilon_{22} & \varepsilon_{23} \\ \varepsilon_{31} & \varepsilon_{32} & \varepsilon_{33} \end{pmatrix} = \begin{pmatrix} \varepsilon_1 & (1/2)\varepsilon_6 & (1/2)\varepsilon_5 \\ (1/2)\varepsilon_6 & \varepsilon_2 & (1/2)\varepsilon_4 \\ (1/2)\varepsilon_5 & (1/2)\varepsilon_4 & \varepsilon_3 \end{pmatrix}. \tag{4.66}$$

We are allowed to do this, the tensor ε_{ij} being symmetric by construction. Similarly, we can write:

$$\begin{pmatrix} \sigma_{11} & \sigma_{12} & \sigma_{13} \\ \sigma_{21} & \sigma_{22} & \sigma_{23} \\ \sigma_{31} & \sigma_{32} & \sigma_{33} \end{pmatrix} = \begin{pmatrix} \sigma_1 & \sigma_6 & \sigma_5 \\ \sigma_6 & \sigma_2 & \sigma_4 \\ \sigma_5 & \sigma_4 & \sigma_3 \end{pmatrix}, \tag{4.67}$$

since the stress tensor is also symmetric in solids.

With these notations, Hooke's law takes the simplified form:

$$\sigma_i = \lambda_{ij}\,\varepsilon_j, \tag{4.68}$$

where we set

$$\lambda_{ijkl} = \lambda_{mn} \quad \text{with} \quad i, j, k, l = 1, 2, 3, \quad \text{and} \quad m, n = 1, 2, \ldots, 6, \tag{4.69}$$

while the elastic energy becomes:

$$f = \frac{1}{2}\lambda_{ij}\varepsilon_i\varepsilon_j. \tag{4.70}$$

Let us now analyse the effect of symmetry. We must account for the fact that the tensor λ_{mn} is *invariant with respect to all (point) symmetries of the material*. This procedure reduces considerably the number of distinct components of the elasticity tensor λ_{mn}.

We will demonstrate the effect for the case of a crystal with cubic symmetry (very common in nature). For instance, let us look for the effect of a 4_m symmetry axis (rotation by $90°$ around the x_3 axis followed by reflection in a mirror perpendicular to this axis). As a result of this operation, the axes transform as follows:

$$1 \to 2, \quad 2 \to -1, \quad 3 \to -3. \tag{4.71}$$

Hence, in the 4-index tensor notation, the index pairs become:

$$11 \to 22, \quad 22 \to 11, \quad 33 \to 33, \quad 23 \to 13, \quad 31 \to -32, \quad 12 \to -21 \tag{4.72}$$

yielding, in the 2-index matrix notation:

$$1 \to 2, \quad 2 \to 1, \quad 3 \to 3, \quad 4 \to 5, \quad 5 \to -4, \quad 6 \to -6. \tag{4.73}$$

Consequently, the matrix λ_{ij} undergoes the following transformation:

$$
\begin{pmatrix}
11 & 12 & 13 & 14 & 15 & 16 \\
 & 22 & 23 & 24 & 25 & 26 \\
 & & 33 & 34 & 35 & 36 \\
 & & & 44 & 45 & 46 \\
 & & & & 55 & 56 \\
 & & & & & 66
\end{pmatrix}
\to
\begin{pmatrix}
22 & 21 & 23 & 25 & -24 & -26 \\
 & 11 & 13 & 15 & -14 & -16 \\
 & & 33 & 35 & -34 & -36 \\
 & & & 55 & -54 & -56 \\
 & & & & 44 & 46 \\
 & & & & & 66
\end{pmatrix}. \tag{4.74}
$$

Since these two matrices must be equal, we deduce that some coefficients are equal, while others are zero, etc.

This procedure helps reduce the number of elastic moduli; it shows that a cubic material is described by only three elastic moduli. In this particular case, the free energy is written explicitly:

$$f = \frac{1}{2}\lambda_{11}\left(\varepsilon_{11}^2 + \varepsilon_{22}^2 + \varepsilon_{33}^2\right) + \lambda_{12}\left(\varepsilon_{11}\varepsilon_{22} + \varepsilon_{11}\varepsilon_{33} + \varepsilon_{22}\varepsilon_{33}\right)$$
$$+ 2\lambda_{66}\left(\varepsilon_{12}^2 + \varepsilon_{13}^2 + \varepsilon_{23}^2\right), \tag{4.75}$$

where we took axes 1, 2 and 3 along the sides of the cube.

In the following, we will assume that the materials are isotropic (Eqs. (4.61) or (4.63)). This is true for glasses (below T_g, the glass transition temperature) or certain gels, and also for polycrystals, as long as we are only interested in deformations over scales very large compared to the size of the crystallites. One can then show that the tensor of the elastic moduli of an 'isotropic' polycrystal is given, in the first approximation, by the isotropic part of the moduli tensor for the monocrystal (Lifshitz and Rosentsveig, 1946). We emphasise that this result is not at all trivial and that it only applies if the monocrystal is not too anisotropic. This holds for many metals, but not for all materials. As counter-examples one can cite graphite, talc or the smectic B phase of certain mesomorphic materials, which are very anisotropic owing to their lamellar structure. In these materials, which are nevertheless exceptions to the rule, the elastic modulus for shear parallel to the lamellae is much weaker than all other moduli, with far-reaching consequences on their elastic and plastic properties.

4.1.4.3 Inverse formulas

We have shown that a thermodynamic relation exists between the stress tensor and the strain tensor, provided the former is symmetric (Eq. (4.56)). Can this formula be inverted?

The answer is 'yes', as long as Hooke's law applies. Indeed, the energy is then a quadratic form of the σ_{ij}, so that we can write, from Euler's theorem:

$$f = \frac{1}{2} \frac{\partial f}{\partial \sigma_{ij}} \sigma_{ij}. \tag{4.76}$$

We know however that $f = (1/2)\varepsilon_{ij}\sigma_{ij}$. Thus, by simple identification:

$$\varepsilon_{ij} = \left(\frac{\partial f}{\partial \sigma_{ij}} \right)_T. \tag{4.77}$$

For completeness, let us present the inverse formulas in isotropic elasticity. For the energy:

$$f = \frac{1}{2} \left[\frac{1+\nu}{E} \sigma_{ij}\sigma_{ij} - \frac{\nu}{E} \sigma_{ii}^2 \right], \tag{4.78}$$

yielding, by application of Eq. (4.77):

$$\varepsilon_{ij} = \frac{1+\nu}{E} \sigma_{ij} - \frac{\nu}{E} \sigma_{kk} \, \delta_{ij}. \tag{4.79}$$

In these formulas, E is the *Young modulus*, of expression:

$$E = \mu \frac{3\lambda + 2\mu}{2(\lambda + \mu)} \tag{4.80}$$

and ν is the *Poisson coefficient*:

$$\nu = \frac{\lambda}{2(\lambda + \mu)}. \tag{4.81}$$

The first has elastic modulus units, while the second is dimensionless. Below, in Section 4.1.9.1, we will give a simple interpretation of these parameters.

Also note that the Young modulus E and the compression modulus K can be expressed very simply as a function of the shear modulus μ and the Poisson coefficient ν as:

$$E = \mu(1+\nu) \quad \text{and} \quad K = \frac{2}{3}\mu \frac{1+\nu}{1-2\nu}. \tag{4.82}$$

Table 4.1

	C	Al	Si	Fe	Cu	Ag	W	Au
$\mu \times 10^{-10}$ (Pa)	56.5	2.7	7.55	8.3	4.8	3	16	2.78
ν		≈ 0.1	0.34	0.27	0.29	0.35	0.37	0.30	0.42

The value given for diamond is imprecise, because it is very difficult to measure.
From [16].

Finally, the elastic energy must always be positive, on thermodynamic stability grounds. The following inequalities ensue:

$$\lambda > 0, \quad \mu > 0, \quad E > 0, \quad K > 0 \quad \text{and} \quad -1 < \nu < \frac{1}{2}. \tag{4.83}$$

Experiment shows that ν is very often positive, as shown by Table 4.1. There are however exceptions to this rule. As examples we can cite the open-cell isotropic foams discovered by Lakes in 1987 (where ν can reach -0.7) (Lakes, 1987) and a silica phase, cristobalite, where ν is -0.5 in certain directions and -0.16 on the average (Yeganeh-Haeri *et al.*, 1992).

Some values for the shear modulus and the Poisson coefficient are given for diamond and metals in Table 4.1. One can see that the Poisson coefficient is often close to 0.3 (except for diamond), showing that in these materials, from Eq. (4.82), $E \approx 1.3\mu$ and $K \approx 2\mu$.

4.1.5 The Navier equation

Once the constitutive laws of the material are known, one can write the differential equation satisfied by the displacement vector $\vec{u}$ at equilibrium. Using Eqs (4.23) and (4.62), one finds for an isotropic material:

$$(\lambda + 2\mu)\overrightarrow{\text{grad}}\,\text{div}\,\vec{u} - \mu\,\overrightarrow{\text{curl}}\,\overrightarrow{\text{curl}}\,\vec{u} + \vec{f} = 0. \tag{4.84}$$

This is the *Navier equation*. Alternatively, it can also be expressed as:

$$(\lambda + \mu)\overrightarrow{\text{grad}}\,\text{div}\,\vec{u} + \mu\,\Delta\vec{u} + \vec{f} = 0. \tag{4.85}$$

4.1.6 How to solve an elasticity problem

This is obviously a very good question; we will first present the most general approach before giving specific examples.

The problem is finding the displacement field obeying the Navier equation and the boundary conditions at the surface S of the body, namely: $\vec{u} = \vec{u}_s$ on S_u and $\underline{\sigma}\vec{\nu} = \vec{f}_s$ on S_f, the complement of S_u.

One can prove mathematically that this problem is well posed and that it admits a unique solution (a so-called 'regular' problem). Two methods are available for solving it, and we will now summarise them.

4.1.6.1 The displacement method

This method proceeds through the following steps:

(1) try a displacement field $\vec{u}$ of a certain form, compatible with the boundary conditions on S_u (such a field is called *kinematically admissible*);
(2) calculate $\underline{\varepsilon}$;
(3) calculate $\underline{\sigma} = \lambda(\mathrm{tr}\underline{\varepsilon})\underline{I} + 2\mu\underline{\varepsilon}$;
(4) check that $\underline{\sigma}\vec{v} = \vec{f}_s$ on S_f and that $\mathrm{div}\,\underline{\sigma} + \vec{f} = 0$ in the bulk (i.e. that the Navier equation is satisfied).

If these conditions are fulfilled, we can be sure that the test field $\vec{u}$ is the correct one.

4.1.6.2 The stress method

This method is complementary to the displacement method. Its successive steps towards the solution can be summarised thus:

(1) try a stress field $\underline{\sigma}$ of a certain form such that $\underline{\sigma}\vec{v} = \vec{f}_s$ on S_f and $\mathrm{div}\,\underline{\sigma} + \vec{f} = 0$ in the bulk (a *statically admissible* field);
(2) construct the $\underline{\varepsilon}$ tensor;
(3) search for a function $\vec{u}$ such that $\varepsilon_{ij} = (u_{i,j} + u_{j,i})/2$. This is not a trivial problem. One can prove that such a function exists if and only if the stress field satisfies the *Beltrami compatibility conditions*:

$$\Delta\sigma_{ij} + \frac{1}{1+v}(\mathrm{tr}\,\underline{\sigma})_{,ij} + f_{i,j} + f_{j,i} + \frac{v}{1+v}f_{k,k}\,\delta_{ij} = 0; \qquad (4.86)$$

(4) if this is the case, calculate the displacement $\vec{u}$ and check that it obeys the boundary conditions on S_u.

Once again, the problem is solved if all these steps are successfully completed.

4.1.7 Saint-Venant's principle

As a general rule, elasticity problems (however simple) can only be solved for very specific boundary conditions, which are often impossible to implement experimentally.

Hence the importance of *Saint-Venant's principle*,[5] stating that:

If a force distribution $\vec{f}_s$ acting on one part S_f of the boundary is replaced by a second distribution $\vec{f}_s'$ also acting on S_f, these two distributions forming equal torsors[6] and the

---

[5] One should call it a theorem, since it has been demonstrated.
[6] This means that the two distributions have the same resultant and the same torque at any point O.

boundary conditions on the complement of S_f remaining unchanged, then in any region sufficiently far away from S_f, the strain and stress fields are practically unchanged.

In practice, this ensures that the strain and stress fields within a test bar, such as the one shown in Fig. 4.9 (far from the upper and lower extremities), are not sensitive to the specifics of grip end fixation, provided the total traction force is always the same.

4.1.8 General theorems

Before going on to solve some classical elasticity problems, we will recall some general theorems.

4.1.8.1 Work theorem

This is the most important. Let:

$$W = \int_S \vec{f_S} \cdot \vec{u} \, dS + \int_V \vec{f} \cdot \vec{u} \, dV. \tag{4.87}$$

This quantity represents the *work* of bulk and surface forces during the displacement $\vec{u}$. The work theorem states that, in linear elasticity:

$$W = 2 \int_V f \, dV. \tag{4.88}$$

To prove this theorem, let us calculate separately the work of the surface force and that of the bulk force. We find, using Eq. (4.22) and the symmetry of the stress tensor:

$$\int_S \vec{f_S} \cdot \vec{u} \, dS = \int_S \sigma_{ij} v_j u_i \, dS = \int_V (\sigma_{ij} u_i)_{,j} \, dV$$

$$= \int_V \sigma_{ij,j} u_i \, dV + \int_V \sigma_{ij} u_{i,j} \, dV$$

$$= \int_V \sigma_{ij,j} u_i \, dV + \int_V \sigma_{ij} \varepsilon_{ij} \, dV. \tag{4.89}$$

On the other hand, Eq. (4.23) yields:

$$\int_V \vec{f} \cdot \vec{u} \, dV = - \int_V \sigma_{ij,j} u_i \, dV. \tag{4.90}$$

It is then sufficient to add these two contributions, keeping in mind that in linear elasticity $f = \frac{1}{2}\sigma_{ij}\varepsilon_{ij}$, to prove the work theorem.[7]

---

[7] The origin of the factor 2 in Eq. (4.88) is left as an exercise to the reader.

4.1.8.2 Reciprocity theorem

Consider two different stress states, (1) and (2), of the same body. One can prove that if Hooke's law applies then:

$$\int_V \vec{f}^{(1)} \cdot \vec{u}^{(2)} \, dV + \int_S \vec{f}_s^{(1)} \cdot \vec{u}^{(2)} \, dS = \int_V \vec{f}^{(2)} \cdot \vec{u}^{(1)} \, dV + \int_S \vec{f}_s^{(2)} \cdot \vec{u}^{(1)} \, dS. \quad (4.91)$$

4.1.8.3 Potential energy theorems

We will not prove these theorems, but it is worthwhile mentioning them here, as they allow bracketing the solutions when these latter are not amenable to analytical calculations. These theorems are mostly useful for numerical simulations.

Let $\{\vec{u}'\}$ be a *kinematically admissible displacement field* and $\{\sigma''\}$ a *statically admissible stress field*. We recall that, by definition, the former satisfies the boundary conditions:

$$\vec{u}' = \vec{u}_s \quad \text{on} \quad S_u, \quad (4.92)$$

while the latter satisfies:

$$\underline{\sigma}'' \vec{v} = \vec{f}_s \quad \text{on} \quad S_f \quad (4.93)$$

and

$$\text{div} \, \underline{\sigma}'' + \vec{f} = 0 \quad \text{in the bulk.} \quad (4.94)$$

One calls *potential energy of a statically admissible stress field*, the quantity:

$$V^* \left(\underline{\sigma}'' \right) = - \int_V f \left(\underline{\sigma}'' \right) dV + \int_{S_u} \vec{u} \cdot \vec{f}_s'' \, dS. \quad (4.95)$$

In the same way, one defines the *potential energy of a kinematically admissible displacement field* as:

$$V \left(\vec{u}' \right) = \int_V f \left(\vec{u}' \right) dV - \int_V \vec{f} \cdot \vec{u}' \, dV - \int_{S_f} \vec{f}_s \cdot \vec{u}' \, dS. \quad (4.96)$$

One can prove the following inequalities, representing the *potential energy theorem*:

$$V^* \left(\underline{\sigma}'' \right) \le V \left(\vec{u} \right) = V^* \left(\underline{\sigma} \right) \le V \left(\vec{u}' \right). \quad (4.97)$$

Note that the equality in the middle corresponds to the work theorem.

4.1.9 Application to some simple problems

4.1.9.1 Simple compression of a cylindrical beam with generic section

We have already considered this problem in Section 4.1.2.4. We assume here that the beam is cylindrical. The shape of its section is unspecified, but its area S is constant.

To begin with, we will solve the problem of the beam compressed at both ends with a force F. For simplicity's sake, we assume that the stress is uniformly spread over the surfaces S_0 and S_1 of its two end sections and that no force acts on its lateral surface S_2 (Fig. 4.9a).

It is easily checked that the stress tensor:

$$\underline{\sigma} = \begin{pmatrix} 0 & 0 & 0 \\ 0 & 0 & 0 \\ 0 & 0 & F/S \end{pmatrix} \tag{4.98}$$

verifies the bulk equilibrium equation $\operatorname{div}(\underline{\sigma}) = 0$, as well as the boundary conditions at both ends and on the lateral surface of the beam.

Knowing the stress field, one can then calculate the deformation field using the inverse formulas (4.79):

$$\underline{\varepsilon} = \begin{pmatrix} -\frac{\nu}{E}\sigma & 0 & 0 \\ 0 & -\frac{\nu}{E}\sigma & 0 \\ 0 & 0 & \frac{1}{E}\sigma \end{pmatrix}, \tag{4.99}$$

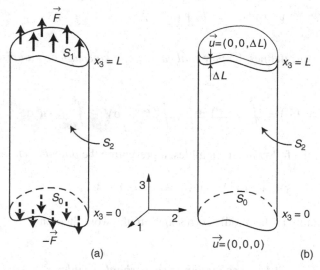

Fig. 4.9 Two ways of deforming a beam in traction or in compression: (a) imposed force; (b) imposed displacement.

where $\sigma = F/S$. The displacement is obtained by integration, up to an additive constant:

$$u_1 = -\frac{\nu}{E}\sigma x_1,$$

$$u_2 = -\frac{\nu}{E}\sigma x_2, \tag{4.100}$$

$$u_3 = \frac{\sigma}{E}x_3.$$

Let L be the length of the beam. Letting $x_3 = L$ in the last equation above yields:

$$\sigma = E\frac{\Delta L}{L}, \tag{4.101}$$

where $\Delta L/L$ is the relative elongation of the beam. This equation shows that the Young modulus is nothing other than the compression modulus of the beam.

We can also determine the relative variation of the beam cross-section $\Delta S/S$:

$$\frac{\Delta S}{S} = \varepsilon_{11} + \varepsilon_{22} = -2\frac{\nu}{E}\sigma = -2\nu\frac{\Delta L}{L}, \tag{4.102}$$

or, for a circular beam of radius R, the relative variation of its radius:

$$\frac{\Delta R}{R} = -\nu\frac{\Delta L}{L}. \tag{4.103}$$

These formulas show that the Poisson coefficient ν is directly connected to the variation of the beam cross-section (or radius). For a positive ν, which is almost always the case, the beam cross-section decreases under traction, while it increases under compression.

We end with the expression of the bulk free energy:

$$f = \frac{1}{2}E\left(\frac{\Delta L}{L}\right)^2. \tag{4.104}$$

What happens now if, instead of imposing the force $\vec{F}$, one imposes a rigid displacement at the two ends? This would happen, for instance, if the beam were welded onto cross bars that are much more rigid than itself. The boundary conditions are then $\vec{u} = (0, 0, 0)$ on S_0 and $\vec{u} = (0, 0, \Delta L)$ on S_1, again with $\sigma\vec{\nu} = 0$ on the lateral surface S_2.

It so happens that this new problem does not have a simple solution, in spite of the very simple boundary conditions.

However, we will show that it is possible to bracket the force F one must apply in order to obtain the displacement ΔL by using the potential energy theorem.

We start by noting that, using the notations of the previous paragraph, $S_u = S_0 \cup S_1$ and $S_f = S_2$. On the other hand, straightforward application of the work theorem yields:

$$V(\vec{u}) = \int_V f\,dV = \frac{1}{2}F\Delta L. \tag{4.105}$$

It is then enough to find a bracketing of $V(\vec{u})$ to bracket the applied force F.

To this end, consider first of all the following kinetically admissible displacement field:

$$\vec{u}' = \left(0, 0, \frac{\Delta L}{L} x_3\right).$$

(4.106)

Obviously, this field is a solution to the problem when L is much larger than the lateral size of the beam, as per Saint-Venant's principle.

For this field we obtain, using expression (4.61) of the elastic energy:

$$V(\vec{u}') = \int_V f(\vec{u}') \, dV = \frac{\lambda + 2\mu}{2} \frac{(\Delta L)^2}{L} S,$$

(4.107)

yielding, by use of Eq. (4.97), the following first inequality:

$$\frac{F}{S} \leq (\lambda + 2\mu) \frac{\Delta L}{L}.$$

(4.108)

With that, we have an upper bound for the applied force F.

To find the lower bound, consider the following statically admissible stress field:

$$\underline{\sigma}'' = \begin{pmatrix} 0 & 0 & 0 \\ 0 & 0 & 0 \\ 0 & 0 & \sigma \end{pmatrix}$$

(4.109)

with σ constant. This field obeys the equilibrium condition $\operatorname{div} \underline{\sigma}'' = 0$ and the boundary condition $\underline{\sigma}'' \vec{\nu} = 0$ on S_f. One calculates for this field:

$$V^*(\underline{\sigma}'') = \int_{S_u} \vec{u} \cdot \vec{f}_s'' \, dS - \int_V f(\underline{\sigma}'') \, dV = \Delta L \sigma S - \frac{1}{2} S L \frac{\sigma^2}{E}.$$

(4.110)

We can see that $V^*(\underline{\sigma}'')$ reaches its maximum $\frac{1}{2} E S \frac{(\Delta L)^2}{L}$ for $\sigma = E \frac{\Delta L}{L}$. Hence the following lower bound for F:

$$\frac{F}{S} \geq E \frac{\Delta L}{L}.$$

(4.111)

Regrouping the two inequalities (4.108) and (4.111), and expressing the Lamé coefficients λ and μ as a function of the Young modulus E and the Poisson coefficient ν, yields the following bracketing:

$$E \frac{\Delta L}{L} \leq \frac{F}{S} \leq \frac{1 - \nu}{1 - \nu - 2\nu^2} E \frac{\Delta L}{L}.$$

(4.112)

Let B be the elastic modulus of the beam, defined by the relation:

$$\frac{F}{S} = B \frac{\Delta L}{L}.$$

(4.113)

This modulus depends on exactly how the forces are applied at the boundaries and takes values between the following two bounds:

$$E \leq B \leq \frac{1-\nu}{1-\nu-2\nu^2} E. \qquad (4.114)$$

With $\nu = 0.3$, this results in $E \leq B \leq 1.35\,E$. The minimum is reached when the length of the beam is much larger than its diameter; for these conditions, the detailed distribution of forces at the extremities does not matter, owing to Saint-Venant's principle. Conversely, the maximum is reached when the beam length is much shorter than its diameter and when $u_1 = u_2 = 0$ at the ends. In this limit, the section area remains unchanged, explaining the increase in the modulus B.

4.1.9.2 Torsion of a rod with simply connected cross-section

This is another extremely classical problem in elasticity. To solve it, we will assume that $\vec{f} = 0$ along with the following boundary conditions (Fig. 4.10):

$$\vec{f_s} = \vec{0} \quad \text{on} \quad S_2, \qquad (4.115)$$

$$u_1 = u_2 = 0 \quad \text{and} \quad f_{s3} = 0 \quad \text{on} \quad S_0, \qquad (4.116)$$

$$u_1 = -\alpha L x_2, \quad u_2 = \alpha L x_1 \quad \text{and} \quad f_{s3} = 0 \quad \text{on} \quad S_1. \qquad (4.117)$$

The last two boundary conditions state that the face S_1 turned by an angle αL around the x_3 axis with respect to the face S_0. This is a *torsion* deformation (α being the rate of twist per unit length of the beam).

Although the boundary conditions are mixed at the surfaces S_0 and S_1, this is still a regular problem, admitting a unique solution for the stress and strain (up to a translation).

We will show that the torque C needed to twist the rod by an angle α is:

$$C = D\alpha, \qquad (4.118)$$

with D the *torsion modulus*, proportional to the shear modulus μ and to a geometric factor d with L^4 units:

$$D = 2\mu d \quad \text{with} \quad d = \int_{S_1} \theta \, dS. \qquad (4.119)$$

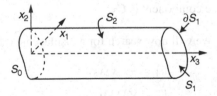

Fig. 4.10 Rod with simply connected cross-section.

In this formula, $\bar{\theta}(x_1, x_2)$ is a function (with L^2 units) termed the *torsion stress function*, satisfying the equation

$$\Delta\bar{\theta} + 2 = 0 \tag{4.120}$$

and the boundary condition:

$$\bar{\theta} = 0 \quad \text{on} \quad S_2. \tag{4.121}$$

To prove this general result, we will proceed in three steps, applying in succession the stress method and the displacement method, before reconciling the two.

Stress method   Let us first search for a stress field of the form:

$$\underline{\sigma} = \begin{pmatrix} 0 & 0 & \theta_{,2} \\ 0 & 0 & -\theta_{,1} \\ \theta_{,2} & -\theta_{,1} & 0 \end{pmatrix}. \tag{4.122}$$

The function $\theta(x_1, x_2)$ is taken to vanish on S_2. It is easily checked that $\operatorname{div}\underline{\sigma} = 0$ and that $\underline{\sigma}\vec{v} = 0$ on S_2 (since $\underline{\sigma}\vec{v} = (0, 0, \theta_{,2} v_1 - \theta_{,1} v_2) = (0, 0, d\theta/d\vec{s})$, where $\vec{\tau}$ is the tangent vector at the surface; since $\theta = 0$ on S_2, $d\theta/d\vec{\tau} = 0$ on S_2, also). One can also show that $(\underline{\sigma}\vec{v})_3 = f_{s3} = 0$ on S_1 and prove that the overall resultant force $\vec{R}$ at this extremity is zero, since:

$$R_1 = \int_{S_1} \theta_{,2} \, dS = \int_{\partial S_1} -\theta \, dx_1 = 0, \tag{4.123}$$

$$R_2 = \int_{S_1} -\theta_{,1} \, dS = \int_{\partial S_1} -\theta \, dx_2 = 0. \tag{4.124}$$

We are finally in a position to determine the required torque $\vec{C}$. Its components are:

$$C_1 = \int_{S_1} x_3\theta_{,1} \, dS = \int_{S_1} (x_3\theta)_{,1} \, dS = \int_{\partial S_1} x_3\theta \, dx_2 = 0, \tag{4.125}$$

$$C_2 = \int_{S_1} x_3\theta_{,2} \, dS = 0, \tag{4.126}$$

$$C_3 = -\int_{S_1} (x_1\theta_{,1} + x_2\theta_{,2}) \, dS = 2\int_{S_1} \theta \, dS. \tag{4.127}$$

As expected, the only finite component is C_3.

Displacement method   We will now search for a displacement field of the form:

$$u_1 = -\alpha x_2 x_3,$$
$$u_2 = \alpha x_1 x_3, \tag{4.128}$$
$$u_3 = \alpha\varphi(x_1, x_2, x_3),$$

where φ is an unknown function, to be determined. Starting from this field, one can calculate the strain tensor:

$$\underline{\varepsilon} = \begin{pmatrix} 0 & 0 & \frac{\alpha}{2}(\varphi_{,1} - x_2) \\ 0 & 0 & \frac{\alpha}{2}(\varphi_{,2} + x_1) \\ \frac{\alpha}{2}(\varphi_{,1} - x_2) & \frac{\alpha}{2}(\varphi_{,2} + x_1) & \alpha\varphi_{,3} \end{pmatrix}. \tag{4.129}$$

Comparison of the two methods   We must now confront the two methods. Using the constitutive law of the material ($\sigma_{ij} = \lambda \varepsilon_{kk} \delta_{ij} + 2\mu \varepsilon_{ij}$), one obtains using Eq. (4.129):

$$\underline{\sigma} = \begin{pmatrix} \alpha\lambda\varphi_{,3} & 0 & \alpha\mu(\varphi_{,1} - x_2) \\ 0 & \alpha\lambda\varphi_{,3} & \alpha\mu(\varphi_{,2} + x_1) \\ \alpha\mu(\varphi_{,1} - x_2) & \alpha\mu(\varphi_{,2} + x_1) & \alpha(\lambda + 2\mu)\varphi_{,3} \end{pmatrix}. \tag{4.130}$$

Comparing this expression to the result of the stress method (Eq. (4.122)) yields the following relations:

$$\varphi_{,3} = 0,$$
$$\theta_{,2} = \alpha\mu(\varphi_{,1} - x_2), \tag{4.131}$$
$$-\theta_{,1} = \alpha\mu(\varphi_{,2} + x_1).$$

First of all, we can see that φ depends on x_1 and x_2 alone, implying that $tr(\varepsilon) = 0$ and $\underline{\sigma} = 2\mu\underline{\varepsilon}$. Hence, only shear strain is involved. On the other hand, eliminating φ from the last two equations yields:

$$\Delta\theta + 2\mu\alpha = 0, \tag{4.132}$$

giving back Eq. (4.120) with $\bar{\theta} = \theta/\mu\alpha$. This concludes the proof of the theorem, knowing that the torque is given by Eq. (4.127).

Applications   To end this paragraph, let us apply this method to a few particular cases.

Among the simplest situations is that of a rod with an elliptic section given by the equation $\frac{x_1^2}{a_1^2} + \frac{x_2^2}{a_2^2} = 1$ (Fig. 4.11a). It is easily checked that the function

$$\bar{\theta} = \frac{a_1^2 a_2^2}{a_1^2 + a_2^2} \left(1 - \frac{x_1^2}{a_1^2} - \frac{x_2^2}{a_2^2} \right) \tag{4.133}$$

is the solution of the problem, yielding the torsion modulus of an elliptical section rod:

$$D = \pi\mu \frac{a_1^3 a_2^3}{a_1^2 + a_2^2}. \tag{4.134}$$

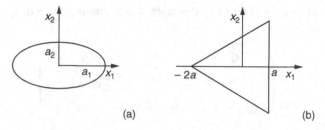

Fig. 4.11 Rod with (a) elliptical; (b) triangular section.

In the particular case of a circular rod with radius R, this formula becomes:

$$D = \frac{\pi}{2}\mu R^4. \tag{4.135}$$

Another interesting case is that of a rod whose cross-section is an equilateral triangle (Fig. 4.11b). It is easily checked that the following function:

$$\bar{\theta} = -\frac{1}{6a}(x_1 - a)(x_1 - x_2\sqrt{3} + 2a)(x_1 + x_2\sqrt{3} + 2a) \tag{4.136}$$

is the solution to the problem. One calculates then:

$$D = \frac{9\sqrt{3}}{10}\mu a^4. \tag{4.137}$$

One can use this method to treat more complicated cases, such as that of a circular rod with a triangular cutout (which also has an analytical solution). There is however no simple analytical solution for a rod with rectangular (or even square) cross-section.

4.1.9.3 *Small bending of a plate*

Another important problem in elasticity is the bending of a thin plate, i.e. one whose thickness e is negligible compared to its lateral size. We will show that, in the case of a *spherical deformation* of low amplitude compared to the thickness of the plate (in the so-called *small bending* approximation), the energy per unit surface of the plate (taken as planar at rest) is:

$$f_c = \frac{1}{2}k\left(\frac{2}{R}\right)^2, \tag{4.138}$$

where R is the radius of curvature (the total curvature being $2/R$) and k its spherical curvature modulus, given by:

$$k = \frac{1}{24}\frac{E}{1-\nu}e^3. \tag{4.139}$$

For a simple demonstration of this result, assume that the edge of the plate is acted upon by a linear torque $\vec{C}$, tangent to the edge and of constant absolute value (Fig. 4.12). To solve

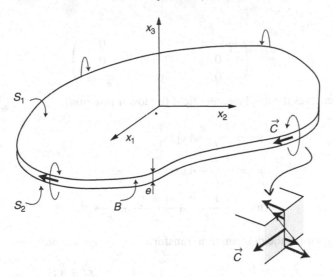

Fig. 4.12 Thin plate experiencing a constant linear torque along its edge B. Note that for the case shown here $A < 0$.

this problem, we will again try to guess the solution. For instance, let us try the following stress field:

$$\underline{\sigma} = \begin{pmatrix} Ax_3 & 0 & 0 \\ 0 & Ax_3 & 0 \\ 0 & 0 & 0 \end{pmatrix}. \qquad (4.140)$$

Clearly, with this choice $\underline{\sigma} = 0$ over the midplane of the plate, situated at $x_3 = 0$. This field satisfies the equilibrium condition in the bulk $\mathrm{div}\,\underline{\sigma} = 0$, as well as the boundary conditions $\underline{\sigma}\vec{v} = 0$ at the free surfaces S_1 and S_2 (as long as $\vec{v}$ stays close to $\vec{x}_3$, which is precisely the small bending approximation). At the edge B, however, the stress is:

$$\begin{aligned} f_{s1} &= Ax_3 v_1, \\ f_{s2} &= Ax_3 v_2, \\ f_{s3} &= 0, \end{aligned} \qquad (4.141)$$

yielding the linear torque:

$$C = A\frac{e^3}{12}. \qquad (4.142)$$

This stress field also satisfies the Beltrami compatibility equations (Eq. (4.86)), the stress components σ_{ij} being affine functions of the coordinates. This ensures the existence of a displacement field associated with this stress field. To calculate it, let us construct the strain tensor using the elasticity equation (4.79). We find:

$$\varepsilon = \begin{pmatrix} \frac{1-v}{E}Ax_3 & 0 & 0 \\ 0 & \frac{1-v}{E}Ax_3 & 0 \\ 0 & 0 & -\frac{2v}{E}Ax_3 \end{pmatrix}. \tag{4.143}$$

Integration then gives the displacement field (up to a translation):

$$u_1 = \frac{1-v}{E}Ax_1x_3,$$

$$u_2 = \frac{1-v}{E}Ax_2x_3, \tag{4.144}$$

$$u_3 = -\frac{1-v}{E}A\frac{x_1^2+x_2^2}{2} - \frac{v}{E}Ax_3^2.$$

From the last equation, the deformation transforms the midplane into:

$$\zeta(x_1, x_2) \equiv u_3(x_3 = 0) = -\frac{1-v}{E}A\frac{x_1^2+x_2^2}{2}, \tag{4.145}$$

which is a spherical cap, of total curvature:

$$C = \frac{2}{R} = \Delta\zeta = -2A\frac{1-v}{E}, \tag{4.146}$$

where $\Delta = \partial^2/\partial x_1^2 + \partial^2/\partial x_2^2$. One can then determine the energy per unit volume:

$$f = \frac{1}{2}\sigma_{ij}\varepsilon_{ij} = \frac{1-v}{E}A^2x_3^2 \tag{4.147}$$

and the energy per unit surface of the plate, by integrating over the thickness:

$$f_c = \int_{-e/2}^{+e/2} f\,dx_3 = \frac{1-v}{E}A^2\frac{e^3}{12}, \tag{4.148}$$

which can finally be written as:

$$f_c = \frac{1}{48}\frac{E}{1-v}e^3C^2. \tag{4.149}$$

This is indeed the sought-for formula in the case of *spherical deformation*.

This formula can be extended to a generic deformation of the plate, remaining in the small bending approximation. In this case, we know that the midplane of the plate is characterised at each point by two topological invariants: the total curvature $C = 1/R_1 + 1/R_2$ and the Gaussian curvature $G = 1/(R_1 R_2)$, where R_1 and R_2 are the two principal curvature radii of the surface at that point. One can then prove that the deformation energy per unit surface can be expressed in the following general form [9]:

$$f_c = \frac{1}{2}KC^2 + \frac{1}{2}\bar{K}G, \tag{4.150}$$

with

$$K = \frac{1}{12} \frac{E}{1 - v^2} e^3 \quad \text{and} \quad \bar{K} = -\frac{1}{6} \frac{E}{1 + v} e^3. \tag{4.151}$$

One can check that the case of purely spherical deformation, where $R_1 = R_2 = R$, gives $k = K + \bar{K}/4$, yielding back formula (4.139) that we have already proven. Finally, we note that, if the midplane equation is $x_3 = \zeta(x_1, x_2)$, the total and Gaussian curvatures in the small bending limit are given by the following formulas:

$$C = \frac{\partial^2 \zeta}{\partial x_1^2} + \frac{\partial^2 \zeta}{\partial x_2^2} \quad \text{and} \quad G = \frac{\partial^2 \zeta}{\partial x_1^2} \frac{\partial^2 \zeta}{\partial x_2^2} - \left(\frac{\partial^2 \zeta}{\partial x_1 \partial x_2} \right)^2. \tag{4.152}$$

To illustrate the use of these equations, let us consider the standard problem of a horizontal circular plate with radius R_0 and embedded boundary. Since the problem has cylindrical symmetry, the displacement ζ only depends on the polar coordinate r. The geometrical embedding condition imposes:

$$\zeta(R_0) = 0 \quad \text{and} \quad \frac{d\zeta}{dr}(R_0) = 0. \tag{4.153}$$

To find the equilibrium equation of the plate in ζ, we set the variation in bending elastic energy equal to the work of the external forces acting upon it, as ζ varies by $\delta\zeta$. Let $\vec{P}$ be the normal force per unit surface acting on the plate. Since the plate boundary is embedded, this is the only force that does work, so from the first principle of thermodynamics we have:

$$\delta \int_{\text{plate}} f_c \, dx_1 dx_2 = \int_{\text{plate}} P \delta\zeta \, dx_1 dx_2. \tag{4.154}$$

Before performing any calculations, one should first notice that the Gaussian curvature term in f_c is the div of a certain vector (actually determining this vector is left as an exercise to the reader[8]). Consequently, its surface integral over the plate reduces to a contour integral over the boundary, and vanishes for an embedded plate. Thus, the only relevant term is the total curvature, in $(\Delta\zeta)^2$. The energy variation is then easily determined giving, after integrating by parts twice:

$$\delta \int_{\text{plate}} f_c \, dx_1 dx_2 = \int_{\text{plate}} K\Delta^2 \zeta \, \delta\zeta \, dx_1 dx_2. \tag{4.155}$$

The two equations above yield the desired bulk equation:

$$K\Delta^2 \zeta - P = 0, \tag{4.156}$$

where Δ^2 is the bi-Laplacian. We now need to solve this equation with the boundary conditions (4.153). We will consider two cases.

---

[8] The result will be given in Chapter 8 (see Eq. (8.159)).

(1) Assume first that the only force acting on the plate is its weight. In this case, $P = \rho e g$, with ρ the density of the material, e the thickness of the plate and g the gravity. In cylindrical coordinates, $\Delta = \dfrac{1}{r}\dfrac{d}{dr}\left(r\dfrac{d}{dr}\right)$. One can check that the general solution of Eq. (4.156) is given by:

$$\zeta(r) = \frac{\rho g e}{64K}r^4 + ar^2 + b + cr^2 \ln\left(\frac{r}{R_0}\right) + d\ln\left(\frac{r}{R_0}\right). \tag{4.157}$$

Here, $d = 0$ because the displacement cannot diverge in $r = 0$. Moreover $c = 0$, although this term goes to zero in $r = 0$, because its Laplacian diverges in $r = 0$ (this would correspond to a force applied in the centre of the plate, as we will see later). Finally, the two integration constants a and b are given by the embedding conditions (4.153), so in the end:

$$\zeta(r) = \frac{\rho g e}{64K}\left(R_0^2 - r^2\right)^2. \tag{4.158}$$

(2) Suppose now that the plate is only acted upon by a force F applied at its centre. Since $P = 0$ everywhere, the general form of the solution is:

$$\zeta(r) = ar^2 + b + cr^2 \ln\left(\frac{r}{R_0}\right). \tag{4.159}$$

This time, the three constants a, b and c are given by the embedding conditions (4.153) and by the work theorem (see Eq. (4.88)) stating that $Fb = 2E_c$, where $E_c = \int_0^{R_0} \frac{1}{2}K(\Delta\zeta)^2\, 2\pi r\, dr$ is the total bending energy. Finally:

$$\zeta(r) = \frac{F}{8\pi K}\left[\frac{1}{2}\left(R_0^2 - r^2\right) + r^2 \ln\left(\frac{r}{R_0}\right)\right]. \tag{4.160}$$

We end this digression by noting that the bulk equation (4.156) applies irrespective of the boundary conditions, and only involves the total curvature constant K. However, the boundary conditions are very complicated when the edge of the plate is free or supported, i.e. simply resting on an immobile support, without being fixed to it. In this case, the Gaussian curvature constant $\bar{K}$ is explicitly involved. This delicate problem is discussed in detail in [9].

4.1.10 Buckling instability of a beam

This is a well-known instability, discovered by L. Euler [9, 11]. Indeed, let us place a beam vertically upon a table and press down on the free end.

Experiment shows that, above a certain compression threshold, the beam buckles spontaneously (Fig. 4.13). This instability is an example of a broken-symmetry bifurcation, since by 'buckling' the revolution symmetry of the problem is broken (in the case of a beam with

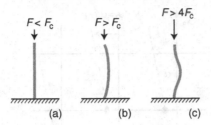

Fig. 4.13 A beam, shown below (a) and above its buckling threshold when its ends are free to turn (b), or not (c).

circular cross-section, which we will assume in the following, although this feature is not essential).

To calculate the critical buckling force, we need to minimise the elastic energy f_L of the beam under compression. Assume that it buckles in the (x_1, x_3) plane: the only finite components of the displacement are u_1 and u_3. If the length of the beam is much larger than its radius ($L \gg R$) and the displacement remains small (rigorously speaking, compared to the beam radius),[9] it only depends on the x_3 coordinate, meaning that we can write:

$$f_L = \int_0^L \left[\frac{1}{2} B \left(\frac{\partial u_3}{\partial x_3} \right)^2 + \frac{1}{2} K \left(\frac{\partial^2 u_1}{\partial x_3^2} \right)^2 \right] dx_3. \qquad (4.161)$$

This expression contains two terms:

(1) The first one is proportional to the square of the relative elongation of the beam and to the compression modulus, calculated in Section 4.1.9.1:

$$B = ES = \pi R^2 E \qquad \text{in the } L \gg R \text{ limit.} \qquad (4.162)$$

Hence, it is a pure compression term.

(2) The second term is proportional to the square of the local curvature of the beam (equal to Δu_1, in the small bending approximation). Thus, it corresponds to the bending energy of the beam. One can prove (and we will admit it in the following) that the bending modulus K is given by formula [9]:

$$K = EI = \pi \frac{R^4}{4} E, \qquad (4.163)$$

where I is the moment of inertia of the cross-section of the beam with respect to a diameter.

Clearly, Eq. (4.161) cannot explain the Euler instability because, by bending the beam, one can only increase its bending energy, with no possible energy gain.

---

[9] This is the *small bending* approximation, which is more restrictive than the small strain assumption.

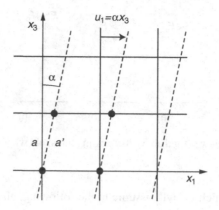

Fig. 4.14 Origin of the anharmonic correction.

To explain the buckling we must therefore search for the source of the gain in energy as the beam bends. Let us consider the sketch in Fig. 4.14, showing the crystal lattice before and after being sheared by $u_1 = \alpha x_3$. During this displacement, the distance between two neighbouring atoms goes from a to a', with:

$$\frac{a' - a}{a} = \frac{a/\cos\alpha - a}{a} \approx \frac{\alpha^2}{2} = \frac{1}{2}\left(\frac{\partial u_1}{\partial x_3}\right)^2. \tag{4.164}$$

This leads to a slight change in the expression of the strain ε_{33}, which becomes, in the general case:[10]

$$\varepsilon_{33} = \frac{\partial u_3}{\partial x_3} + \frac{1}{2}\left(\frac{\partial u_1}{\partial x_3}\right)^2 + \frac{1}{2}\left(\frac{\partial u_2}{\partial x_3}\right)^2. \tag{4.165}$$

Accounting for this *anharmonic correction* – which amounts simply to considering the elongation of the beam as it bends at a constant projected length along the x_3 axis – leads to the following (more complete) expression for the elastic energy (4.161):

$$f_L = \int_0^L \left[\frac{1}{2}B\left(\frac{\partial u_3}{\partial x_3} + \frac{1}{2}\left(\frac{\partial u_1}{\partial x_3}\right)^2\right)^2 + \frac{1}{2}K\left(\frac{\partial^2 u_1}{\partial x_3^2}\right)^2\right] dx_3. \tag{4.166}$$

One can see right away from this formula that energy can only be gained during buckling of the beam if the double product $\dfrac{\partial u_3}{\partial x_3}\left(\dfrac{\partial u_1}{\partial x_3}\right)^2$ is negative, requiring that:

$$\frac{\partial u_3}{\partial x_3} < 0. \tag{4.167}$$

As expected, the Euler instability can only occur in compression.

---

[10] In practice, the crystal lattice also turns by an angle α, which does not affect the elastic energy, obviously.

To determine its threshold, let us calculate the variation in elastic energy as the two displacement components vary by δu_1 and δu_3. Denoting the integrand in Eq. (4.166) by $f(u_{1,3}, u_{1,33}, u_{3,3})$, and writing that

$$\delta \int_0^L f(u_{1,3}, u_{1,33}, u_{3,3}) \, dx_3 = \delta W, \qquad (4.168)$$

where δW is the work of the external forces and torques applied to the beam[11] yields (neglecting gravity), after a double integration by parts, the following equilibrium equations in the bulk:

$$\frac{\partial^2}{\partial x_3^2} \frac{\partial f}{\partial u_{1,33}} - \frac{\partial}{\partial x_3} \frac{\partial f}{\partial u_{1,3}} = 0 \quad (u_1 \text{ variation}), \qquad (4.169)$$

$$-\frac{\partial}{\partial x_3} \frac{\partial f}{\partial u_{3,3}} = 0 \quad (u_3 \text{ variation}). \qquad (4.170)$$

The last equation gives the first integral:

$$\frac{\partial u_3}{\partial x_3} + \frac{1}{2} \left(\frac{\partial u_1}{\partial x_3} \right)^2 = \text{Cst}, \qquad (4.171)$$

where the integration constant is simply F/B.

As to the other minimisation equation, it can be written explicitly as:

$$K \frac{\partial^4 u_1}{\partial x_3^4} - B \frac{\partial}{\partial x_3} \left[\left(\frac{\partial u_3}{\partial x_3} + \frac{1}{2} \left(\frac{\partial u_1}{\partial x_3} \right)^2 \right) \frac{\partial u_1}{\partial x_3} \right] = 0, \qquad (4.172)$$

and yields, taking into account the first integral:

$$K \frac{\partial^4 u_1}{\partial x_3^4} - F \frac{\partial^2 u_1}{\partial x_3^2} = 0. \qquad (4.173)$$

This fourth-order linear equation translates the equilibrium of elastic torques within the beam. To solve it, one must also specify the boundary conditions at the two ends of the beam.

The simplest assumption is that:

$$u_1 = 0 \quad \text{in} \quad x_3 = 0 \quad \text{and} \quad L. \qquad (4.174)$$

---

[11] This is nothing other than the expression of the first principle of thermodynamics at constant temperature.

This condition does not suffice, the equation being fourth-order. One must therefore add the embedding conditions, obtained by noting that, from Eq. (4.168) and after a first integration by parts, at the two ends of the beam:

$$\frac{\partial f}{\partial u_{1,33}} \, \delta u_{1,3} = \Gamma_{\text{ext}} \, \delta u_{1,3} \quad \text{in} \quad x_3 = L, \tag{4.175}$$

$$-\frac{\partial f}{\partial u_{1,33}} \, \delta u_{1,3} = \Gamma_{\text{ext}} \, \delta u_{1,3} \quad \text{in} \quad x_3 = 0,$$

where Γ_{ext} is the external torque applied at each end.

If the beam is free to turn at both ends, implying that no external torque is applied ($\Gamma_{\text{ext}} = 0$), then $\partial f / \partial u_{1,33} = 0$, which yields explicitly, considering Eq. (4.166):

$$u_{1,33} = 0 \quad \text{in} \quad x_3 = 0 \quad \text{and} \quad L. \tag{4.176}$$

If, on the other hand, the beam is prevented from turning, $\delta u_{1,3} = 0$. This condition is achieved in particular for:

$$u_{1,3} = 0 \quad \text{in} \quad x_3 = 0 \quad \text{and} \quad L. \tag{4.177}$$

Let us now solve the differential equation (4.173) in each of these two cases.

If the beam is free to turn at its ends (Eq. (4.176)), at the buckling threshold the solution can be written as (Fig. 4.13b):

$$u_1 = u_0 \sin\left(\pi \frac{x_3}{L}\right). \tag{4.178}$$

Substituting in the differential equation (4.173) yields the force needed to induce the buckling:

$$F = F_{\text{c}} = -\pi^2 \frac{EI}{L^2}. \tag{4.179}$$

This force is negative, as expected for compression.

If the beam is embedded so that its ends cannot turn (Eq. (4.177)), the solution at the threshold is of the type (Fig. 4.13c):

$$u_1 = u_0 \left[1 - \cos\left(2\pi \frac{x_3}{L}\right)\right], \tag{4.180}$$

now yielding:

$$F = 4F_{\text{c}} = -4\pi^2 \frac{EI}{L^2}. \tag{4.181}$$

The force one must apply is larger this time, as one might expect intuitively. More precisely, it is four times larger than in the previous case, which stands to reason because the embedding amounts to buckling of a beam with free extremities but half as long.

Note that (the absolute value of) the torque one must apply at each end to prevent it from turning is:

$$C = \left| \frac{\partial f}{\partial u_{1,33}} \right| = EI \left| u_{1,33} \right| \simeq \left(\frac{2\pi}{L} \right)^2 u_0 EI, \tag{4.182}$$

where u_0 is the maximum amplitude of the deformation.

In conclusion, we emphasise that the small bending theory we have just presented does not give access to the deformation amplitude u_0 of the unstable beam (indeed, Eq. (4.173) is linear). Consequently, this theory cannot predict the nature of the bifurcation. One must then use the 'elastica theory', which can model large bending deformations. Using this theory, presented in detail in [11], it can be shown that the bifurcation is sub-critical, meaning that the beam bends strongly immediately above the threshold calculated above. In this case, the shape of the beam is no longer given by a sine or cosine function. There are however small bending states, below the absolute instability threshold, for which the sinusoidal shape (but not the amplitude) is given by Eq. (4.173) [11].

4.1.11 Flutter instability of an airplane wing

Another type of instability is the *flutter of airplane wings* (Fig. 4.15). As its name indicates, it plays an important role in aeronautics, and also in civil engineering (see below for the example of the Tacoma bridge). Although this instability is well known today,[12] this was not the case at the beginning of the twentieth century, when it provoked numerous airplane accidents. Among the most famous is that of the 'Aerodrome' (this is how the plane was baptised by its owner, Professor Samuel Langley), the wings of which broke by torsion above the Potomac river, in Washington DC, as it was making the second attempt at powered flight in the history of aviation (the first attempt having failed two months earlier).

The brothers Orville and Wilbur Wright,[13] direct competitors of Professor Langley, were more inspired and designed a system allowing the pilot to control wing warping. Wilbur had this idea after observing the flight of harriers: as they are buffeted by a gust of wind, they maintain their equilibrium by a slight twist of their wing tips. This ingenious system (among others!) allowed the Wright brothers to achieve the first powered flight with their famous *Flyer*, on a North Carolina beach, at Kitty Hawk, only a few days after the second failure of Professor Langley (Fig. 4.16). Note that the *Flyer* was a biplane, less flexible than monoplanes, and thus less sensitive to flutter. This architecture was gradually abandoned at the beginning of the 1930s, as more rigid metal skin wings began to appear.

After this brief historical presentation, let us analyse succinctly the flutter instability of an airplane wing; we will show that it is often accompanied by bend–twist deformations (Fig. 4.15). To simplify the problem as much as possible, let us replace the wing by its *'typical section'* and study its motion in a plane perpendicular to the wing. In practice, this

---

[12] Its study is now a part of aeroelasticity, a domain of applied mechanics that deals with the motion of deformable bodies in gas flows.

[13] As an anecdote, they also invented the first bicycles with rubber tyres (which made them a lot of money).

Elasticity of solids

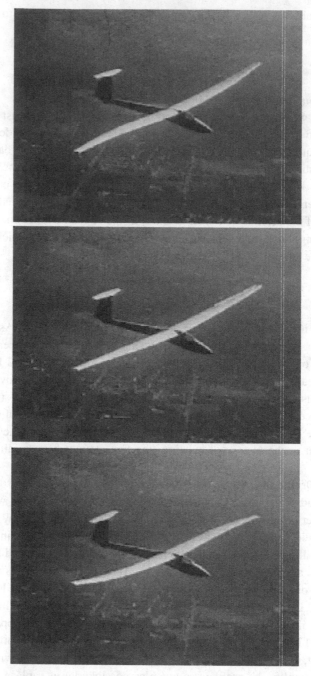

Fig. 4.15 Three snapshots of a glider experiencing flutter: the wings bend and twist out of phase (from www.onera.fr/cahierdelabo/video/hq/aleg2a.mov).

Fig. 4.16 First flight of the Wright brothers, on 17 December 1903.

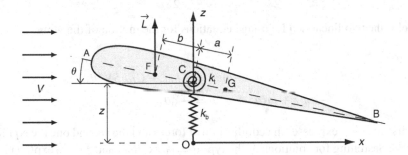

Fig. 4.17 Two-dimensional aeroelastic model of an airplane wing undergoing flutter.

profile is chosen so as to give the best agreement between theory and experiment. In the 1930s, aerodynamics specialists noticed that the section must usually be taken at 75% of the distance between the wing root and tip. This model can also be useful for interpreting wind tunnel tests, when studying the response of a wing segment immersed in a uniform air flow and flexibly attached to the walls. The profile of the typical section is sketched in Fig. 4.17. It has an axis AB and can turn by an angle θ around a point C, termed the *elastic centre*. Different from the centre of gravity G, this point can also move along the z axis (OC $= z$). To simulate the elastic twist and bend properties of the three-dimensional wing, we assume that rotation by an angle θ induces a restoring torque $-k_t\theta$ (k_t being a torsion modulus), while the drift z is compensated by a restoring force $-k_b z$ (with k_b the bending modulus).

In the absence of wind, the wing is at rest: $\theta = z = 0$.

Under the action of a wind blowing with velocity V along the x axis, it feels a (circulatory) lift force $\vec{L}$ directed along z, of amplitude $L = C_L \rho S \alpha V^2$ (where ρ is the density of air, S the wing surface, α the angle of attack and C_L the lift coefficient, constant at low incidence (see Eq. (3.167)). This force acts at a point F of the typical section, called the *aerodynamic centre*.

In the following, for simplicity's sake, we will make the (inessential) assumptions that points F and C coincide ($FC = b = 0$) and that the angle of attack α (given, strictly speaking, by $\theta - \dot{z}/V$ if we take into account the relative motion of the wing with respect to the wind) is very close to θ.

To obtain the equations of motion for the wing, let us first calculate its kinetic energy:

$$E_c = \frac{1}{2}mV_G^2 + \frac{1}{2}mr^2\dot{\theta}^2. \tag{4.183}$$

In this expression, V_G is the velocity of the centre of gravity and mr^2 the moment of inertia of the wing with respect to an axis going through G and perpendicular to the typical section (r is the radius of gyration of the wing profile). If the incidence angle θ remains small, the kinetic energy is, to second order in θ and z:

$$E_c = \frac{1}{2}m(a^2 + r^2)\dot{\theta}^2 + \frac{1}{2}m\dot{z}^2 - am\dot{z}\dot{\theta}. \tag{4.184}$$

This yields the two linearised Lagrange equations for the motion of the wing:

$$m\ddot{z} - am\ddot{\theta} = -k_b z + C_L\rho SV^2\theta,$$
$$-am\ddot{z} + m(a^2 + r^2)\ddot{\theta} = -k_t\theta. \tag{4.185}$$

The first equation expresses the equilibrium of forces and the second one the equilibrium of torques. Searching for solutions of the type $\theta = \theta_0\exp(\omega t)$ and $z = z_0\exp(\omega t)$ leads to the following characteristic equation:

$$m^2r^2\Omega^2 + m\left(A^2k_b + k_t - aC_L\rho SV^2\right)\Omega + k_bk_t = 0, \tag{4.186}$$

where $\Omega = \omega^2$ and $A^2 = a^2 + r^2$.

The wing is only stable if this equation has two real and negative roots for Ω. Its discriminant Δ must therefore be positive, and the sum of its roots negative (their product being positive). This requires:

$$\Delta = \left(A^2k_b + k_t - aC_L\rho SV^2\right)^2 - 4k_bk_tr^2 > 0 \tag{4.187}$$

and

$$A^2k_b + k_t - aC_L\rho SV^2 > 0, \tag{4.188}$$

yielding the following stability condition:

$$A^2k_b + k_t - aC_L\rho SV^2 > 2r\sqrt{k_bk_t}. \tag{4.189}$$

which defines the critical velocity V_c below which the wing vibrations are not amplified:

$$V_c = \sqrt{\frac{k_b a^2 + (r\sqrt{k_b} - \sqrt{k_t})^2}{aC_L\rho S}}.$$

(4.190)

Using Eq. (4.186), one can also calculate the eigenfrequencies of the wing in the absence of wind ($V = 0$). They are:

$$\omega = \pm i \sqrt{\frac{\omega_b^2 + \omega_t^2 \pm \sqrt{(\omega_b^2 + \omega_t^2)^2 - 4\omega_b^2\omega_t^2(1 - a^2/A^2)}}{2(1 - a^2/A^2)}},$$

(4.191)

with the typical twist $\omega_t^2 = k_t/mA^2$ and bend $\omega_b^2 = k_b/m$ eigenfrequencies of the wing when these two modes are uncoupled (this occurs when the centre of gravity coincides with the elastic centre, or $a = 0$).

On the other hand, the wing can become unstable and break for $V > V_c$.

We emphasise that, in these calculations, the twist and bend deformations are coupled. This observation is generally valid for wings (see Fig. 4.15).

The same instability was responsible for the collapse of the Tacoma Narrows bridge in the USA on 7 November 1940. The bridge collapsed after going into resonance under the action of a steady wind, blowing at the moderate speed of 60 km/h (Fig. 4.18). In this example, the main deformation mode of the bridge is a twist mode. The deformation reached an impressive amplitude immediately before the collapse, as shown in Fig. 4.18a.

To explain this instability, we will again make use of the model above, but this time considering only the twist mode. In the case of the bridge, $a = 0$ for obvious symmetry reasons. On the other hand, it is no longer reasonable (or even possible) to assume that the aerodynamic centre coincides with the elastic centre (FC $= b \neq 0$, Fig. 4.17). Under these conditions, the equation of motion for the bridge (more precisely, for its midplane) is written, taking into account the torque applied by the wind:

$$mr^2\ddot{\theta} = -k_t\theta + bC_L\rho SV^2\theta.$$

(4.192)

This equation shows immediately that the bridge becomes unstable with respect to twist if the wind velocity exceeds the critical value

$$V_c = \sqrt{\frac{k_t}{bC_L\rho S}}.$$

(4.193)

Hence, the extreme torsion flexibility of the Tacoma Narrows bridge was the main cause of its collapse.

Nowadays, civil engineers allow for the flutter instability each time they design a structure (bridge, tower, etc.) susceptible to going into resonance under the action of the wind.

Fig. 4.18 Oscillation and collapse of the Tacoma Narrows bridge, in the USA. This accident occurred in 1940, only four months after the bridge was opened to traffic. Using the amateur video from which these two frames were extracted, one could estimate that, at the oscillation peak, the height difference between the two sidewalks on either side of the bridge was about 9 m (from www.enm.bris.ac.uk/anm/tacoma/tacoma.html).

4.2 Elastodynamics of ordinary hard solids

In Chapter 2 we showed that, in the dynamic regime, the general expression for the equilibrium of forces is:

$$\rho \frac{D\vec{v}}{Dt} = \operatorname{div} \underline{\sigma} + \vec{f}. \tag{4.194}$$

In solids, $\vec{v} = \dfrac{\partial \vec{u}}{\partial t}$, where $\vec{u}$ is the displacement (no permeation).

On the other hand, the advection terms in the material derivative are completely negligible, such that the inertia force is simply $\rho \dfrac{\partial^2 \vec{u}}{\partial t^2}$. Under these conditions, the equation above becomes:

$$\rho \frac{\partial^2 \vec{u}}{\partial t^2} = \operatorname{div} \underline{\sigma} + \vec{f}, \tag{4.195}$$

yielding, for an isotropic material:

$$(\lambda + 2\mu)\overrightarrow{\operatorname{grad}} \operatorname{div} \vec{u} - \mu \, \overrightarrow{\operatorname{curl}} \operatorname{curl} \vec{u} + \vec{f} = \rho \frac{\partial^2 \vec{u}}{\partial t^2} \tag{4.196}$$

or, equivalently:

$$(\lambda + \mu)\overrightarrow{\operatorname{grad}} \operatorname{div} \vec{u} + \mu \, \Delta \vec{u} + \vec{f} = \rho \frac{\partial^2 \vec{u}}{\partial t^2}. \tag{4.197}$$

The result above is a generalisation of the Navier equation to the dynamic case.

4.2.1 Waves in an infinite medium

For simplicity's sake, we will assume that $\vec{f} = 0$ and that the elastic medium is infinite. From vector analysis we know that the displacement field can be written as the sum of the curl of a vector and the gradient of a scalar, to wit:

$$\vec{u} = \vec{u}_t + \vec{u}_l, \tag{4.198}$$

where

$$\vec{u}_t = \overrightarrow{\operatorname{curl}} \vec{A} \tag{4.199}$$

and

$$\vec{u}_l = \overrightarrow{\operatorname{grad}} \phi. \tag{4.200}$$

We can see that $\vec{u}_l$ is the irrotational component of the field:

$$\overrightarrow{\operatorname{curl}} \vec{u}_l = 0, \tag{4.201}$$

while $\vec{u}_t$, with

$$\operatorname{div} \vec{u}_t = 0, \tag{4.202}$$

is the rotational component.

From now on, we consider that the displacement is described by a plane wave with wave vector $\vec{k}$ and angular frequency ω:

$$\vec{u} = \vec{u}_0 \exp(i\omega t - i\vec{k} \cdot \vec{r}). \tag{4.203}$$

In this case, from Eq. (4.201):

$$\vec{k} \times \vec{u}_l = 0, \tag{4.204}$$

while Eq. (4.202) yields:

$$\vec{k} \cdot \vec{u}_t = 0. \tag{4.205}$$

We see then that $\vec{u}_l$ is the longitudinal component of the wave (since $\vec{u}_l \| \vec{k}$) and $\vec{u}_t$ its transverse component (because $\vec{u}_t \perp \vec{k}$). Note that longitudinal waves are accompanied by changes in volume (div $\vec{u}_l \neq 0$), unlike transverse waves, propagating at constant volume (div $\vec{u}_t = 0$).

To find the equations of motion for the components $\vec{u}_t$ and $\vec{u}_l$, let us replace $\vec{u}$ by its expression $\vec{u}_t + \vec{u}_l$ in Eq. (4.197). Using Eqs. (4.201) and (4.202), we find:

$$\ddot{\vec{u}}_l + \ddot{\vec{u}}_t = C_l^2 \, \Delta\vec{u}_l + C_t^2 \, \Delta\vec{u}_t, \tag{4.206}$$

where we set:

$$C_l^2 = \frac{\lambda + 2\mu}{\rho} = \frac{E(1 - v)}{\rho(1 + v)(1 - 2v)} \tag{4.207}$$

and

$$C_t^2 = \frac{\mu}{\rho} = \frac{E}{2\rho(1 + v)}. \tag{4.208}$$

Applying the div operator to Eq. (4.206) yields, using also Eq. (4.202):

$$\mathrm{div}\left(\ddot{\vec{u}}_l - C_l^2 \Delta\vec{u}_l\right) = 0. \tag{4.209}$$

Since the curl of the expression in brackets is also zero, as per Eq. (4.201), we infer that the longitudinal component of the wave satisfies the wave equation:[14]

$$\ddot{\vec{u}}_l - C_l^2 \Delta\vec{u}_l = 0. \tag{4.210}$$

The constant C_l is thus the propagation speed of longitudinal waves.

For the transverse component, the same method yields:

$$\ddot{\vec{u}}_t - C_t^2 \Delta\vec{u}_t = 0, \tag{4.211}$$

showing that C_t is the speed of transverse waves.

---

[14] Note that, strictly speaking, the expression in brackets is an unspecified function of time. This function is strictly zero here, since its average over the entire space must vanish, $\vec{u}_l$ describing the displacement field of a plane wave.

Since $E > 0$ and (in most cases) $0 < \nu < 1/2$, comparing the formulas (4.207) and (4.208) yields

$$C_1 > C_t\sqrt{2}. \tag{4.212}$$

4.2.2 Waves in finite medium

In this case, one must take into account the boundary conditions, usually written as:

$$\vec{u} = 0 \qquad \text{on } S_u \tag{4.213}$$

and

$$\underline{\sigma}\vec{v} = 0 \qquad \text{on } S_f. \tag{4.214}$$

The first equation describes a clamping condition on S_u and the second one a free oscillation condition on S_f.

To account for the boundary conditions, let us look for the solution in the form of stationary waves, of the type:

$$\vec{u}(\vec{r}, t) = \vec{u}'(\vec{r}) \cos(\omega t + \varphi). \tag{4.215}$$

Substituting in Eq. (4.196) yields an equation for $\vec{u}'$:

$$(\lambda + 2\mu)\overrightarrow{\text{grad}}\,\text{div}\,\vec{u}' - \mu\,\overrightarrow{\text{curl}}\,\overrightarrow{\text{curl}}\,\vec{u}' + \rho\omega^2\vec{u}' = 0. \tag{4.216}$$

One can show that this equation has solutions that obey the boundary conditions only for an *infinite and countable* set of values ω_k ($k = 1, 2, 3, \ldots$), termed *eigenfrequencies* of the system.

The general solution of Eq. (4.196) is then of the form:

$$\vec{u}(\vec{r}, t) = \sum_{k=1}^{\infty} A_k \vec{u}'_k(\vec{r}) \cos(\omega_k t + \varphi_k), \tag{4.217}$$

where the A_k and φ_k are arbitrary constants.

As an example of a system exhibiting quantified eigenfrequencies, let us search for the spectrum of longitudinal waves propagating along a rod of length L. We have already shown that, in a rod, $\sigma_{33} = B\varepsilon_{33}$ with axis 3 along the rod. One can infer that the longitudinal displacement $u_3 = u$ satisfies the equation:

$$\rho\frac{\partial^2 u}{\partial t^2} = B\frac{\partial^2 u}{\partial x_3^2}. \tag{4.218}$$

Assume that the rod is clamped at one end and free to oscillate at the other end. In these conditions:

$$u = 0 \qquad \text{at } z = 0 \tag{4.219}$$

and

$$\frac{\partial u}{\partial z} = 0 \qquad \text{at } z = L. \tag{4.220}$$

One can try solutions of the form $u = u_0 \sin(kx_3) \sin(\omega t + \alpha)$ yielding, by applying the second boundary condition:

$$\cos(kl) = 0. \tag{4.221}$$

This result proves that the wave vectors are quantified, taking values:

$$k_n = (2n + 1)\frac{\pi}{2L} \tag{4.222}$$

with integer n.

Substitution in the equation of motion (4.218) gives the corresponding frequencies:

$$\omega_n = (2n + 1)\frac{\pi}{2L}\sqrt{\frac{B}{\rho}}. \tag{4.223}$$

4.2.3 Rayleigh surface waves

In this section, we will demonstrate the presence of elastic waves, bearing the name of Rayleigh, propagating at the surface of a body without penetrating it (Rayleigh, 1945). They are important in geophysics, as we will see in the next section. We start by calculating their dispersion relation and by emphasising their essential characteristics.

We consider an elastic medium with a free surface $x_2 = 0$, occupying the half-space $x_2 \geq 0$ (Fig. 4.19). We are interested in the dispersion relation of plane waves propagating in this medium along x_1 (with $u_3 = 0$). These waves must satisfy the dynamic Navier equation (4.196) as well as the boundary condition:

$$\underline{\sigma}\vec{v} = 0 \tag{4.224}$$

at the free surface. Knowing that $\vec{v} = -\vec{x}_2$ at this surface, the condition is:

$$\sigma_{22}(x_2 = 0) = \sigma_{21}(x_2 = 0) = 0. \tag{4.225}$$

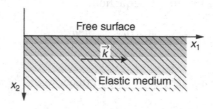

Fig. 4.19 Surface waves in a semi-infinite elastic medium.

On the other hand, we know that a wave can always be decomposed into its longitudinal and transverse parts, prompting us to search for solutions of the form:

$$\vec{u} = \vec{u}_t + \vec{u}_1 \tag{4.226}$$

with

$$\vec{u}_t = \vec{A} \exp[-ax_2 + i(kx_1 - \omega t)], \tag{4.227}$$
$$\vec{u}_1 = \vec{B} \exp[-bx_2 + i(kx_1 - \omega t)]. \tag{4.228}$$

Since, by definition, $\text{div}\,\vec{u}_t = 0$ and $\overrightarrow{\text{curl}}\,\vec{u}_1 = 0$, we must have:

$$ikA_1 = aA_2 \tag{4.229}$$

and

$$-bB_1 = B_2 ik. \tag{4.230}$$

These relations are satisfied by the choice $A_1 = -iaA$ and $A_2 = kA$, $B_1 = -ikB$ and $B_2 = bB$ with A and B real, so that the components of the displacement field can be written in the following generic form:

$$u_{t1} = aA \exp(-ax_2) \sin(kx_1 - \omega t), \tag{4.231}$$
$$u_{t2} = kA \exp(-ax_2) \cos(kx_1 - \omega t), \tag{4.232}$$
$$u_{11} = kB \exp(-bx_2) \sin(kx_1 - \omega t), \tag{4.233}$$
$$u_{12} = bB \exp(-bx_2) \cos(kx_1 - \omega t). \tag{4.234}$$

To find the dispersion relation, let us start by writing that the total displacement obeys the boundary conditions (4.225). Using the constitutive law for isotropic materials:

$$\sigma_{22} = \lambda u_{i,i} + 2\mu u_{2,2}, \tag{4.235}$$
$$\sigma_{21} = \mu(u_{1,2} + u_{2,1}), \tag{4.236}$$

we find:

$$2\mu ak A + \left[2\mu b^2 + \lambda(b^2 - k^2)\right] B = 0, \tag{4.237}$$
$$(a^2 + k^2)A + 2kbB = 0. \tag{4.238}$$

This system of homogeneous equations with unknowns A and B only has non-trivial solutions if its determinant vanishes:

$$\left[2\mu b^2 + \lambda(b^2 - k^2)\right](a^2 + k^2) - 4\mu abk^2 = 0. \tag{4.239}$$

We then write that components $\vec{u}_t$ and $\vec{u}_l$ verify the propagation equations (4.211) and (4.210), yielding the two additional equations:

$$a^2 + k^2(\chi x - 1) = 0, \tag{4.240}$$

$$b^2 + k^2(x - 1) = 0, \tag{4.241}$$

where we set:

$$x = \left(\frac{C_R}{C_t}\right)^2 \quad \text{with} \quad C_R = \frac{\omega}{k} \tag{4.242}$$

and

$$\chi = \left(\frac{C_t}{C_l}\right)^2. \tag{4.243}$$

It is then enough to calculate a and b using these two equations and to substitute into Eq. (4.239) to obtain the dispersion relation:

$$x^3 - 8x^2 + (24 - 16\chi)x - 16(1 - \chi) = 0. \tag{4.244}$$

One can easily check that, in the particular case $\lambda = \mu$ and $\chi = 1/3$, this equation has three simple solutions:

$$x = 4, \qquad x = 2 + \frac{2}{\sqrt{3}} \quad \text{and} \quad x = 2 - \frac{2}{\sqrt{3}}. \tag{4.245}$$

The first two roots must be discarded, since they give pure imaginary values for a and b, corresponding to ordinary waves. The last one, on the other hand, is suitable, since it gives positive values for a and b (the only physically acceptable ones):

$$a = \sqrt{\frac{1}{3}\left(1 + \frac{2}{\sqrt{3}}\right)} \, k = 0.847k, \tag{4.246}$$

$$b = \sqrt{\frac{2}{\sqrt{3}} - 1} \, k = 0.393k. \tag{4.247}$$

Hence, there is a *surface wave* exponentially damped over a distance comparable to the wavelength within the material (*Rayleigh evanescent wave*).

Finally, let us emphasise that:

(1) this wave propagates at a speed:

$$C_R = \sqrt{x} \, C_t \tag{4.248}$$

always *smaller than that of waves in the bulk* since only the $x < 1$ solution gives real values for the coefficients a and b, according to Eqs. (4.240) and (4.241).

(2) this wave is a *specific mixture of transverse and longitudinal waves*, with an amplitude ratio:

$$\frac{A}{B} = \frac{-2kb}{a^2 + b^2} = -2\frac{\sqrt{1-x}}{2 - \chi x}. \tag{4.249}$$

4.2.4 Waves emitted by an earthquake and seismology

As an earthquake occurs somewhere on the planet, it emits waves. These waves propagate in the Earth's crust and mantle and damp with the distance according to a $1/r$ law for bulk waves (spherical waves) and as $1/\sqrt{r}$ for surface waves. Consequently, these waves will be felt strongly close to the epicentre of the earthquake, but can also be detected (using very sensitive seismographs) anywhere on the planet, if the quake is strong enough.

This was the case for an earthquake that occurred in Peru in 1996. This quake was superficial, having taken place at a depth of about 30 km, immediately below the subduction zone, where the Nazca plate dives under the South-American plate. As shown by the record in Fig. 4.20, several wave packets arrived in succession at Grenoble (France), showing that several types of waves propagate inside the Earth at different speeds.

More specifically, one observes successively:

- a packet labelled P (*primary waves*): this corresponds to the fastest waves, therefore the first to arrive. These are the longitudinal bulk waves, with a speed given by formula (4.207). These waves take about 400 s to reach Grenoble. Their trajectory within the Earth's mantle is curved due to their propagation velocity increasing with depth (sketch (a), an effect similar to the 'mirage' observed in optics as the index, and hence the propagation speed of light, change with altitude);
- a packet labelled PP, also containing longitudinal bulk waves, after one reflection at the surface of the globe, as shown in the sketch (a);
- a packet labelled S (*secondary waves*): these are transverse waves, which are known to propagate slower than the longitudinal waves (see formula (4.212)). This packet corresponds to waves arriving directly;
- a packet labelled SS, corresponding to shear waves arriving after one reflection (sketch (b));
- a packet labelled SP, of a rather peculiar nature, arriving between the two previous ones. This stems from an S wave, converted into a P wave after reflection (sketch (c));
- the last packet, more extended and intense, corresponds to the Rayleigh surface waves. As expected, these waves arrive later than the others, since they are slower (see the relation (4.248); but they are also the most intense, and thus the most dangerous in an earthquake, being less damped than the others (indeed, their amplitude A decreases as $1/\sqrt{r}$ due to their two-dimensional nature, energy conservation imposing $A^2 H r = $ Cst, where H is the penetration distance, while for the spherical waves S and P, A decreases as $1/r$, since $A^2 r^2 = $ Cst). Finally, they last longer, being strongly dispersive due to their propagation speed depending on the depth, an effect we have completely neglected in our calculations.

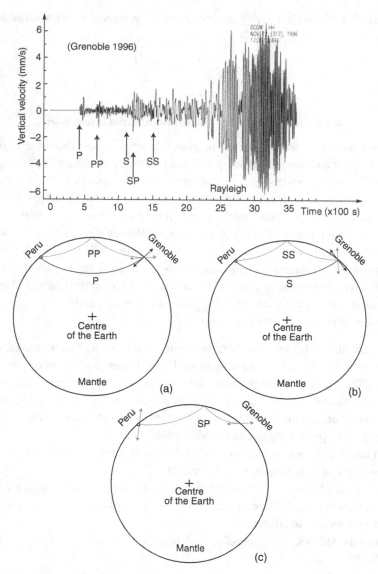

Fig. 4.20 Seismogram recorded in 1996 in Grenoble several seconds after an earthquake took place in Peru, with magnitude 7.7 on the Richter scale. This recording shows the vertical component of the displacement as a function of time. Several wave packets arrive in succession. The first ones correspond to bulk waves, either of the compression-dilation type (a) (direct waves P and waves reflected once, PP), or of the shear type (b) (direct waves S and waves reflected once, SS), or of mixed type (c) (S wave reflected once and converted into a P wave). The three diagrams drawn under the seismogram show the trajectory and the polarisation of these waves. The last packet, the most intense and long-lasting corresponds to the Rayleigh surface waves (seismogram provided by Michel Campillo, UJF, Grenoble).

4.2.5 Measuring the elastic constants: applications in geophysics

The elastic constants can be measured statically or dynamically.

Static measurements are often plagued by errors, the applied stress inducing small plastic deformations by the displacement of defects in the crystal. These displacements are reversible (contributing to the elastic response of the material) if the defects (dislocations, mainly) are trapped, either by their own internal stress field (see Section 5.2.4), or at their extremities, for instance on the nodes of the Frank network (see Section 5.1.2.7).

Dynamic measurements are more precise if performed at high enough frequency to suppress plastic deformations by 'freezing' the motion of dislocations. These studies consist of measuring the reaction of the material to the formation of (forced or resonant) elastic waves. Several independent measurements are usually needed to separate the various elastic constants involved. Thus, measuring the velocity of longitudinal and transverse waves gives access to the two Lamé coefficients, λ and μ, in an isotropic material.

Measuring the speed of waves in geology (which is the object of seismology) is also very useful, since waves help in probing the bulk mechanical properties of the Earth and hence in deducing its inner structure.

The technique consists of recording throughout the world seismograms of a localised earthquake. By correlating all of these data and trying to reproduce the trajectories of the different types of waves (see Fig. 4.20), it then becomes possible to establish an elastic seismological model of the Earth, i.e. to find the velocity of the P and S waves as a function of depth. These indirect seismological data give an image of the internal structure and composition of the Earth. Some of these results are compiled in Table 4.2.

From the table data, one can see that three regions must be distinguished inside the Earth: the crust, the mantle and the core. In each of them, the wave velocity varies almost continuously with the depth. On the other hand, strong discontinuities can be observed at the interfaces between these three regions, due to important changes in the chemical composition and the structure of the constitutive materials.[15]

Thus, in the crust, extending to a maximum depth of 40 km,[16] the main ingredient is granite (a mixture of quartz, SiO_2, and feldspars, $(K,Na)AlSi_3O_8$). The temperature at a depth of 40 km is close to 1000 K and the pressure about 1 GPa.

The mantle makes up about 80% of the total volume of the Earth, since it extends from 40 to 2900 km depth. The upper mantle is mostly composed of olivine $(Mg,Fe)_2SiO_4$, and also pyroxenes $(Mg,Fe)SiO_3$ and garnets $Mg_3Al_2[SiO_4]_3$, while the lower mantle is dominated by a compound $(Mg,Fe)SiO_3$ of perovskite structure and a $(Mg,Fe)O$ oxide, the ferropericlase. The upper mantle is divided into two regions, separated by a weak seismic discontinuity at a depth of 420 km. Another (more important) seismic discontinuity occurs at 670 km, at the limit between the upper and the lower mantle. These discontinuities are due to the structural changes of the ingredients (most importantly of olivine, transforming first into wadsleyite between 420 and 520 km, then into ringwoodite between 520 and 670 km) and to changes in the chemical composition (especially at the interface between the upper and the lower mantle). Finally, the core is essentially composed of Fe

---

[15] The limit between the crust and the mantle was discovered in 1909 by A. Mohorovičić and that between the mantle and the core by B. Gutenberg in 1913.

[16] The crust is thicker under the old continents (30–40 km) than under the oceans (5–6 km).

Table 4.2. *Seismological data*

Depth (km)	Density ρ	Pressure P (Mbar)	Compression modulus K (Mbar)	Shear modulus μ (Mbar)	C_t (km/s)	C_l (km/s)
Crust						
10	2.8	0.0003	0.54	0.35	3.54	6.00
Upper mantle						
50	3.3	0.015	1.13	0.62	4.33	7.70
100	3.35	0.025	1.25	0.63	4.34	7.90
200	3.4	0.072	1.28	0.65	4.37	7.95
420	3.55	0.141	1.76	0.82	4.81	8.97
420	3.77	0.141	2.16	0.96	5.05	9.55
570	3.95	0.199	2.35	1.11	5.30	9.85
670	4.08	0.239	2.48	1.22	5.47	10
Lower mantle						
670	4.38	0.239	3.05	1.64	6.12	10.93
1 000	4.57	0.387	3.54	1.85	6.36	11.46
1 500	4.85	0.621	4.32	2.15	6.66	12.17
2 000	5.12	0.869	5.13	2.44	6.90	12.80
2 500	5.37	1.135	5.95	2.71	7.10	13.34
2 890	5.55	1.354	6.35	2.91	7.24	13.58
Outer core						
2 890	9.91	1.354	6.58	0	0	8.15
4 000	11.32	2.461	10.31	0	0	9.54
5 150	12.14	3.289	12.77	0	0	10.26
Inner core						
5 130	12.71	3.289	13.63	1.50	3.44	11.09
6 370	13.01	3.632	14.24	1.65	3.56	11.24

1 Mbar = 100 GPa
From Dziewonski *et al.* (1975).

(at least 80%) and Ni (5–10 %) mixed with other, lighter, elements (S, C, Si, O, . . .). It is almost entirely liquid (with a viscosity of the order of 0.01 Pa s), except very close to the centre of the Earth (in the inner core) where it becomes solid again. The temperature at the surface of the inner core is estimated at 5000 K (this is the melting temperature of impure iron at 330 GPa).

4.3 Elasticity of colloidal crystals

These materials are obtained by dispersing in a liquid (the solvent) solid particles of *colloidal* size, typically between 0.1 and 1 μm (for a precise definition of the concept of colloidal particles, see Section 7.5.1). Since these particles are much larger than the atoms of an ordinary crystal and since they are immersed in a solvent, one can expect colloidal crystals to have unusual mechanic properties (Pieranski, 1983).

To show that this is indeed the case, we will study in this chapter the way they can be obtained, and then their elasticity. We will see, using simple experiments of sedimentation equilibrium and wave propagation, that they are indeed much softer than ordinary solids and that they exhibit dissipative features related to the presence of the solvent. For this reason, they can be classified as *viscoelastic solids*.

As to their behaviour under continuous shear, also very original, it will be described at the end of the next chapter, in Section 5.5, where we will see that they can melt under shear without changing the temperature.

4.3.1 General notions on colloidal crystals

One cannot understand the physics of these systems without some basic knowledge of their preparation methods and of their chemical properties. We will therefore discuss the role of the ionic strength of the solution in the formation of a colloidal crystal, and then estimate the order of magnitude of the elastic modulus.

4.3.1.1 Fabricating a colloidal crystal

The first colloidal crystals were observed in concentrated suspensions of viruses such as the TIV (Tipula Iridescent Virus). Discovered in 1957 by R. C. Williams and K. M. Smith, this virus of icosahedral structure forms in concentrated solution crystals exhibiting multiple coloured reflections (hence its name). These reflections (which can also be observed in *opals*, formed by small silica beads which have slowly settled over geological times, before adhering to each other) are simply the Bragg reflections of visible light on the lattice planes of the crystal. Since they occur in visible light, the lattice size must be in the micrometre range.

Since the mid 1970s, one could also prepare colloidal crystals from suspensions of artificially synthesised solid particles. These particles are generally spherical, with a diameter ranging from 0.1 to 1 μm, typically. Their size polydispersity does not exceed a small percentage, which is essential for achieving their crystallisation. They can be made of silica or formed by an organic polymer such as polystyrene (PS), the chemical structure of which is given in Chapter 7 (Fig. 7.2). Suspensions of polymer beads are generally termed *latex*. The experiments we will describe below have been made using PS beads.

These latex particles are obtained by polymerisation in an emulsion. The volume concentration of beads at the end of the synthesis is usually of the order of 10%. Each bead is composed of several PS macromolecules, functionalised by SO_4K groups placed at the surface of the bead, and which dissociate in contact with water (Fig. 4.21). For this reason, each bead bears a significant *negative charge* (ranging from 1000 electrons for a sphere 0.1 μm in diameter to 10^6 for a 1 μm sphere). As for the K^+ ions, they go into the solution.

Experiment shows that the suspension obtained at the end of the synthesis is always milky (due to the multiple scattering of light), a typical feature of a disordered structure. In order to obtain the nice coloured reflections of a colloidal crystal, this solution must be purified using ion exchange resins. This purification method consists of first passing the suspension over a resin that exchanges the cations (resulting from the different steps of the synthesis) against H^+ ions, then across a resin exchanging the anions for OH^- ions. The H^+ and OH^- ions then recombine giving water, H_2O, thus decreasing the ion

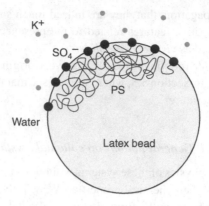

Fig. 4.21 Structure of a latex bead.

concentration in the solution. It has been observed that crystallisation occurs when the ionic strength becomes low enough. To understand this fundamental feature, a theoretical aside on the crystallisation of hard-sphere systems is in order.

4.3.1.2 *Mechanisms of formation of a colloidal crystal*

In 1950, J. G. Kirkwood, E. K. Maun and B. J. Alder suggested that a suspension of hard spheres, only interacting by steric repulsion, must exhibit an *order–disorder transition* as the volume fraction of the spheres ϕ increases. This result was confirmed in 1968 by molecular dynamics simulations (Alder *et al.*, 1968). These simulations show that the spheres form a compact crystal (with an f.c.c. or h.c.p. structure) as $\phi > 0.55$ and maintain a disordered liquid structure for $\phi < 0.5$. In the intermediate concentration range $0.5 < \phi < 0.55$, the two phases coexist, showing that this (purely entropic) transition is first order.

We will see that the Kirkwood–Alder prediction is one of the keys to explaining the formation of colloidal crystals.

To show this, let us return to the synthesis of the suspensions. We know that, at the end of the reaction, the volume fraction ϕ of latex beads is about 10%. According to the Kirkwood–Alder theory, this concentration is not sufficient to induce crystallisation; indeed, it does not occur. However, this explanation is valid only if there is no interaction between the spheres (apart from their steric repulsion), while we know that they are strongly charged. For the proposed explanation to hold (at least qualitatively), the *electrostatic interactions* between the spheres must be *screened*. This is indeed the case at the end of the synthesis, the suspension being rich in all kinds of ionic species. Debye showed that, under these conditions, each bead is shrouded in a *counterion* cloud, which screens the Coulomb interactions between beads (Fig. 4.22). The thickness of this cloud is given by the *Debye screening length* λ_D, a parameter given approximately by ([37, p. 102]):

$$\lambda_D = \sqrt{\frac{\varepsilon_0 \varepsilon k_B T}{\sum\limits_i N_i Z_i^2 e^2}}. \tag{4.250}$$

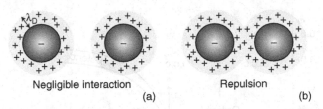

Fig. 4.22 Formation of a counterion cloud around each particle in a strong electrolyte. When the two clouds are far apart, the particles hardly feel each other, since the Coulomb interactions are screened (a); as the clouds interpenetrate, the particles repel strongly (b).

In this equation, ε is the dielectric constant of the medium (with a value of about 80 for water), e the charge of the proton (1.6×10^{-19} C) and N_i the number of ions with valence Z_i per unit volume. In pure water, at pH $= 7$, λ_D is of the order of 0.7 μm, larger than, or at least comparable to, the diameter a of the beads. On the other hand, this value is much smaller in the unpurified suspension, where the electrostatic interactions are strongly screened. Under these conditions, the behaviour of the spheres is close to the idealised Kirkwood–Alder model, explaining why they fail to crystallise.

Using these two results, we can now predict, qualitatively at least, the formation conditions of a colloidal crystal.

Assume that the ion concentration, and hence the screening length λ_D in the solution are known. Since the beads repel as their counterion clouds begin to interpenetrate, a first approximation consists in replacing them by hard spheres with an effective radius $a + \alpha\lambda_D$ larger than a (α being an adjustable parameter of order unity). Thus, everything happens as for a suspension with an effective volume fraction

$$\phi^* = \phi \left(\frac{a + \alpha\lambda_D}{a} \right)^3. \tag{4.251}$$

It is then sufficient to apply the Kirkwood–Alder theory using the effective volume fraction ϕ^* instead of the real volume fraction ϕ to predict the phase diagram of a colloidal suspension. This diagram is shown in Fig. 4.23 in the parameter plane defined by $(\phi, a/\lambda_D)$. The two separation lines of the biphase region are given by $\phi^* = 0.5$ and $\phi^* = 0.55$, respectively. This diagram can be compared to the experimental results, since the ratio a/λ_D is connected to the ion concentration through the relation (4.250). Good agreement with the data is obtained for a value $\alpha \approx 3$.

This diagram shows that, starting from any point in the disordered region, there are two ways of inducing the crystallisation: either by concentrating the solution (along path 1), or by increasing its Debye length, which amounts to purifying the solution (along path 2). In experiments, one usually adopts the second solution.

The last question to be addressed concerns the structure of the crystalline phase thus obtained. Experiment shows that it depends essentially on the ratio a/λ_D. If its value is below 1.7, the crystal has the same structure as most metals, namely b.c.c. (body-centred cubic). This result can be understood by noticing that the beads are immersed in their

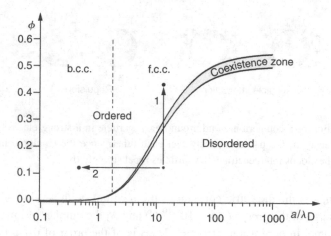

Fig. 4.23 Theoretical phase diagram calculated for $\alpha = 3$. There are two ways of inducing crystalli-sation: either by increasing its concentration (path 1), or by decreasing its ionic strength (path 2). Following path 2, one can also cross the transition line between an f.c.c. and a b.c.c. structure. This transition was observed by M. Joanicot and P. Pieranski in 1984.

large counterion clouds: the system is then close to the image of a Wigner crystal, where the atoms bathe in a cloud of completely delocalised electrons. In the opposite case, the colloidal crystal adopts a more compact structure, of the f.c.c. (face-centred cubic) or h.c.p. (hexagonal close-packed) type. These structure changes can be explained by theories more refined than that presented above, where we simply replaced the screened interactions by hard-sphere interactions. We will not go into more detail here, since they are not essential for understanding the rheological properties of colloidal crystals.

We end this section by showing, for the reader's viewing pleasure, two images taken in fluorescence confocal microscopy of a colloidal suspension of hard PMMA spheres (the chemical formula is given in Chapter 7, Fig. 7.2). The first image (Fig. 4.24a) is taken in the ordered phase, where one can observe a perfect hexagonal arrangement of the beads. The second one (Fig. 4.24b) is taken in the disordered phase and exhibits no long-range order.

After this reminder, can one predict the order of magnitude for the elastic modulus of a colloidal crystal? We answer this question in the next section.

4.3.1.3 Order of magnitude for the elastic modulus

In an ordinary crystal, the order of magnitude of the elastic modulus is obtained simply by dividing the interaction energy U between two atoms by their separation distance cubed b^3.

Let us do the same estimate for a colloidal crystal. If Z is the number of electrons on each bead, the Coulomb energy for two beads is (neglecting the influence of counterions, which is a very crude approximation):

$$U = \frac{Z^2 e^2}{4\pi \varepsilon_0 \varepsilon b}. \tag{4.252}$$

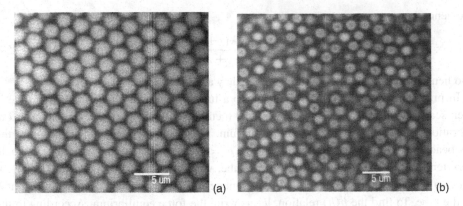

(a)   (b)

Fig. 4.24 Two confocal microscopy images of a colloidal suspension consisting of fluorescent PMMA spheres, 2.2 μm in diameter. In (a), the suspension has a hexagonal compact structure; in (b), it is in a liquid disordered state (Habdas and Weeks, 2002).

For latex beads 0.1 μm in diameter, $Z \approx 1000$. Taking a typical distance $b \approx 1$ μm yields $U \approx 10$ eV. This value is of the same order of magnitude as the interaction energy of two atoms in a crystal. This result is confirmed by indirect measurements of the *melting latent heat* of colloidal crystals, giving values of the order of 5 kcal per mole of beads (Williams *et al.*, 1976), which is comparable to the values found for ordinary crystals. Note however that, due to the size of the beads, the latent heat per unit volume is extremely low, of the order of 10^{-11} kcal/cm^3.

The interaction energy being known, one can deduce the order of magnitude of the elastic modulus:

$$E \approx \frac{U}{b^3} \approx 1 \text{ Pa.} \tag{4.253}$$

This value is about 10 orders of magnitude below that of ordinary crystals. We emphasise that this enormous difference is essentially a result of the *length scale difference*, the energy of interaction between particles being similar in the two cases.

Let us now see how to measure the elastic modulus of a colloidal crystal.

4.3.2 Measurement of the elastic modulus

In general, a cubic crystal has three elastic moduli, denoted by $\lambda_{11}, \lambda_{12}$ and λ_{66} (Section 4.1.4.2, Eq. (4.75)). Since in colloidal crystals all forces are central, one can prove that $\lambda_{12} = \lambda_{66}$, leaving only two independent moduli. In the following, we will describe two methods for measuring them.

4.3.2.1 Deformation of a crystal under the action of gravity

This is a very simple experiment, which consists of monitoring the deformation of the bead lattice under the action of gravity. This method gives access to the Young modulus, with the general expression:

$$E \approx \frac{\lambda_{66}(3\lambda_{12} + 2\lambda_{66})}{\lambda_{12} + \lambda_{66}} \tag{4.254}$$

and hence to coefficients λ_{12} and λ_{66}, since they are equal: $\lambda_{12} = \lambda_{66} = 0.4E$.

In practice, the suspension is poured into a tube several centimetres tall. The tube is then sealed to avoid any contamination, and then placed in an environment exempt from vibrations to achieve sedimentation equilibrium. This phenomenon is inevitable, since the PS beads are slightly heavier than water ($\Delta\rho = \rho_{PS} - \rho_{water} = 0.05$ g/cm^3): as such, they tend to migrate to the bottom of the tube, where their density increases slightly. It follows that, at equilibrium, the lattice parameter b varies as a function of the height h in the tube. To find the $b(h)$ relation, let us write the force equilibrium. According to the Navier equation, we should have at equilibrium:

$$E \frac{d^2 u}{dh^2} = n \, \Delta\rho v \, g. \tag{4.255}$$

In this equation, u stands for the displacement of the beads, n for the number of particles per unit volume (which varies little over the height of the column and will be considered constant in the first approximation), v is the volume of one particle and g the gravity. Integration yields immediately:

$$E \frac{\Delta b(h)}{b} = n \, \Delta\rho v \, g \, h \tag{4.256}$$

with $\Delta b(h) = b(h) - b(0)$, resulting in:

$$E = n \, \Delta\rho v \, g \, h \frac{b}{\Delta b(h)}. \tag{4.257}$$

It is then sufficient to measure the variation in the parameter $\Delta b(h)$ to find the elastic modulus. This can be done by analysing the Bragg reflections of a He-Ne laser beam at different heights. This method allowed Crandal and Williams to measure for the first time the Young modulus of a colloidal crystal and to confirm the order of magnitude given in the previous section: $E \approx 0.1$ Pa for beads 0.1 μm in diameter at a concentration $n \approx 10^{18}$ part./m^3 (Crandal and Williams, 1977).

Note the existence in this problem of a typical length, which sets the minimal height of the tubes used to observe a significant variation of the lattice parameter ($\Delta b/b \approx 0.1$). From Eq. (4.256), this length L is given by:

$$L = \frac{E}{10 \, n \, \Delta\rho v \, g}. \tag{4.258}$$

In practice, its value is of a few centimetres, which is coherent with the estimate obtained using $E \approx 0.1$ Pa, $n \approx 10^{18}$ part./m^3 and $v \approx 5 \times 10^{-22}$ m^3.

We should also point out that we neglected the Brownian motion of the particles, which is a legitimate approximation in the crystalline phase, where each particle remains trapped in the cage formed by its neighbours. The effect of Brownian motion is, however, much more important in disordered suspensions: its effect on the sedimentation equilibrium will be analysed in Section 7.5.1.3.

In the next section, we will describe another method, based on measuring the propagation speed of shear waves.

4.3.2.2 Waves in a colloidal crystal: qualitative description

In the following, we will only consider waves with a wavelength λ much larger than the lattice parameter ($\lambda \gg b$). In this limit, the colloidal crystal can be treated as a continuous medium. As for ordinary solids, two types of waves must exist in a colloidal crystal, and we will describe them qualitatively.

First, there are *longitudinal waves*, involving the compression–dilation of the bead lattice. Since the solvent is incompressible and since the density of the bead lattice changes locally, the resulting solvent flow is in the opposite direction from the motion of the beads (Fig. 4.25). This *permeation* flow across the bead lattice, behaving here as a porous medium, is strongly dissipative. Hence, we expect these waves to be strongly damped and unable to propagate over long distances. This prediction is effectively borne out by the experiments.

The *transverse waves*, depicted in Fig. 4.26, are much more interesting. In this case, the density of the bead lattice remains locally constant. Consequently, the fluid will be able to follow globally the motion of the beads, without flowing across the lattice. The absence of permeation flow predicts that these waves will be much less damped than longitudinal waves. However, the damping is not zero since the fluid is sheared, leading to dissipation.

In the next section, we will show how to calculate the dispersion relation for transverse waves.

4.3.2.3 Dispersion relation for transverse waves

For this calculation, we will use a model first proposed by T. Ohtsuki in 1978 (Ohtsuki *et al.*, 1978) and by J.-F. Joanny in 1979 (Joanny, 1979).

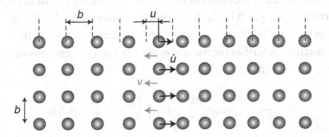

Fig. 4.25 Longitudinal wave in a colloidal crystal. The motion of the beads leads to a contrary fluid flow. Due to this permeation flow, the longitudinal waves are strongly damped and unable to propagate over long distances.

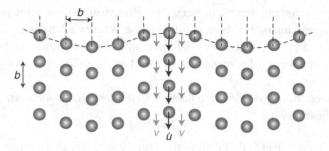

Fig. 4.26 Transverse waves. The fluid follows the motion of the beads but does not flow across the bead lattice, the density of the latter remaining constant. Hence, there is no permeation flow. The resulting dissipation is much weaker than in the case of longitudinal waves. These waves can propagate over long distances.

In this model, the crystal is considered as an elastic continuous medium immersed in a viscous background.

Two equations are needed to describe the coupled motions of the beads and the surrounding fluid.

The first one describes the *dynamics of the beads*. It is derived from the usual Navier equation of solids, supplemented by the forces describing the interaction of the beads with the fluid. Denoting by $\vec{u}$ the displacement of the beads and by $\vec{v}$ the average velocity of the fluid, we write:

$$n\,m\,\frac{\partial^2 \vec{u}}{\partial t^2} = n\,m_0\frac{\partial \vec{v}}{\partial t} + n\zeta\left(\vec{v} - \frac{\partial \vec{u}}{\partial t}\right) + \mu\,\Delta\vec{u}. \tag{4.259}$$

In this equation, m is the mass of a bead, n their number per unit volume and m_0 the mass of a quantity of solvent with the same volume as that of a bead. By comparison to the general Navier equation (4.197) one can see that we set $\operatorname{div}\vec{u} = 0$, which is legitimate for the case of transverse waves (the incompressible lattice assumption). We also introduced the shear modulus μ, which, in the case of a fine-grained polycrystal, is an average over the coefficients λ_{ij}. Finally, and more importantly, two new terms appear on the right-hand side of the equation. The first one is a kind of buoyancy resulting from the inertia of the fluid. The second term describes the friction force exerted by the fluid on the beads. ζ is thus a friction coefficient, given in the first approximation by the Stokes law: $\zeta = 6\pi\,\eta_{\text{eff}}a$, where a is the radius of a bead and η_{eff} an effective viscosity, close to that of the solvent.

The second equation describes the *motion of the fluid*. It is a generalisation of the Navier–Stokes equation, stating that:

$$\rho\frac{\partial \vec{v}}{\partial t} = -n\,m_0\frac{\partial \vec{v}}{\partial t} + n\zeta\left(\frac{\partial \vec{u}}{\partial t} - \vec{v}\right) + \eta\,\Delta\vec{v}. \tag{4.260}$$

The first two terms on the right-hand side are the inertial and viscous forces exerted by the beads on the fluid. Note that, as per the action-reaction principle, these forces are the exact opposite of the forces acting on the beads, given on the right-hand side of Eq. (4.259). In the following, we assume that the volume fraction of beads is low ($\phi = n \, m_0/\rho \ll 1$), so that the inertial force due to the beads can be neglected in this equation.

These two equations yield the dispersion relation of shear waves, by eliminating the velocity between them. Noting that, at low frequency, inertial terms are negligible compared to the viscous terms in Eq. (4.259), one has:

$$\rho\frac{\partial^2 \vec{u}}{\partial t^2} - \left(\eta + \frac{\rho\mu}{n\zeta}\right)\frac{\Delta\partial\vec{u}}{\partial t} - \mu\,\Delta\vec{u} + \frac{\eta\mu}{n\zeta}\Delta^2\vec{u} = 0. \qquad (4.261)$$

This equation can be simplified, since $\rho\mu/n\zeta \ll \eta$, as one can check using typical parameters (see the next section): $\rho = 1000 \text{ kg/m}^3$, $\mu \approx 0.1\text{--}10 \text{ Pa}$, $n \approx 10^{18} - 5 \times 10^{19} \text{ part./m}^3$, $\eta_{\text{eff}} \approx 10^{-2} \text{ Pa s}$ and $2a \approx 0.1 \text{ μm}$, resulting in $\zeta \approx 6\pi\eta_{\text{eff}}a = 10^{-8} \text{ kg s}^{-1}$.

Consider now a transverse wave of the type $\vec{u} = \vec{u}_0 \exp(i\omega t - \vec{k}\cdot\vec{r})$. Since Eq. (4.261) only holds for wave vectors $k \ll b^{-1}$, it is easily checked that the last term on the left-hand side is completely negligible compared to the penultimate one, allowing further simplification and leading in the end to the following simple form:

$$\rho\frac{\partial^2 \vec{u}}{\partial t^2} - \eta\frac{\Delta\partial\vec{u}}{\partial t} - \mu\,\Delta\vec{u} = 0. \qquad (4.262)$$

This equation yields the dispersion relation for transverse waves with a large wavelength (as compared to b):

$$k^2 = \frac{\rho\omega^2}{\mu\left(1 + \frac{i\omega\eta}{\mu}\right)}, \qquad (4.263)$$

which can also be rewritten as:

$$\omega = \pm\sqrt{\frac{\mu k^2}{\rho} - \frac{\eta^2 k^4}{4\rho^2}} + i\eta\frac{k^2}{2\rho}. \qquad (4.264)$$

The *transverse waves* are thus *dispersive* and *damped in time* due to the presence of the solvent. Their damping time is:

$$\tau = \frac{2\rho}{\eta k^2}. \qquad (4.265)$$

Expression (4.264) also shows that they become *evanescent* if their wave vector exceeds $2\sqrt{\mu\rho}/\eta$ or, equivalently, if their wavelength is below a certain threshold λ_{min} with expression:

$$\lambda_{\text{min}} = \frac{\pi\eta}{\sqrt{\mu\rho}}. \qquad (4.266)$$

In practice, λ_{min} is of the order of a few tens of micrometres.

In conclusion, this calculation shows that transverse waves with a large wavelength (in the millimetre or centimetre range) can propagate in colloidal crystals with a phase velocity

$C_t \approx \sqrt{\mu/\rho}$ (of a few cm/s), over significant distances (several mm to a few cm), since the damping is relatively low (compared to that of longitudinal waves).

In the next section, we describe an ingenious setup, developed by the team of P. Pieranski, which can be used to demonstrate these waves.

4.3.2.4 Experimental demonstration of the transverse waves and measurement of the shear modulus of a polycrystal

The experimental setup used by the Pieranski team to demonstrate the presence of transverse waves in a colloidal crystal is depicted in Fig. 4.27. It consists of a flat-bottomed tube containing the colloidal crystal. This tube is immersed in a water-filled flask serving a dual purpose: maintaining the sample at the required temperature and reducing the light refraction effects at the surface of the tube during the optical detection of crystal oscillations. The tube is vertical, guided by two ball bearings, and can turn around its axis. Its oscillations are induced by a loudspeaker. The sample is lit by the beam of a He-Ne laser. The back-scattered light creates at the surface of the spherical flask Kossel rings resulting from reflections on the various reticular planes of the crystal (similar to a powder diffraction diagram for X-rays). As the crystal vibrates, these rings start to oscillate with an amplitude

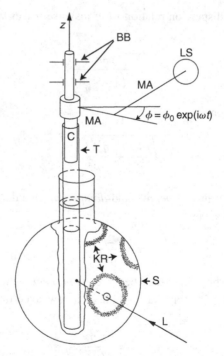

Fig. 4.27 Experimental setup. T is the tube containing the colloidal crystal; S is the surface of the flask; KR are the Kossel rings whose motion is used to detect the resonance peaks; C is a metallic cylinder fitted to the tube; BB are ball bearings; MA are metal arms; LS is the loudspeaker; finally, L shows the trajectory of the He-Ne laser (Dubois–Violette *et al.*, 1980).

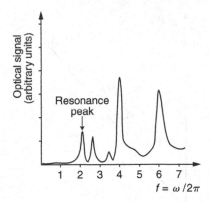

Fig. 4.28 Record of the signal output by the quadrant diode as a function of frequency. Five resonance peaks are clearly visible and two are masked by the wings of the main peaks (Joanicot, 1984b).

proportional to the displacement of the beads within the crystal. This signal (or rather its mean square amplitude) is detected using a quadrant photodiode.

The experiment consists in having the tube oscillate with a *constant amplitude* (which requires a feed-back mechanism) and of recording the optical signal given off by the sample as a function of the excitation frequency. Such a record is shown in Fig. 4.28. It exhibits several *resonance peaks*. Each peak corresponds to the selective amplification of a vibration eigenmode of the crystal due to shear.

To index these peaks properly, one must first calculate the vibration eigenmodes of the polycrystal in the tube at rest. Since the experiment is done in rotation around the z axis, the only relevant shear modes are those with components $u_r = u_z = 0$ and $u_\theta(r, z) \neq 0$. These particular modes are solutions of the dynamic equation (4.262), which is best written in cylindrical coordinates (see Appendix 4.A). They must also obey the boundary conditions on the surface of the tube (at $r = R$ and $z = 0$) and at the free surface of the colloidal crystal (at $z = H$):

$$u_\theta(r, 0) = u_\theta(R, z) = 0, \qquad (4.267)$$

$$\frac{\partial u_\theta}{\partial z}(r, H) = 0. \qquad (4.268)$$

It can be shown that the solution to this problem is given by:

$$u_\theta^{j,n}(r, z, t) = J_1\left(\mu_j \frac{r}{R}\right) \sin\left[\left(n + \frac{1}{2}\right)\pi \frac{z}{H}\right] \exp(i\omega t), \qquad (4.269)$$

where J_1 is the Bessel function of order 1, μ_j ($j = 1, 2, \ldots$) the zeros of this function (with $\mu_1 = 3.83$, $\mu_2 = 7.02$, $\mu_3 = 10.17, \ldots$) and n a positive integer ($n = 0, 1, 2, \ldots$). As to the eigenfrequencies, they are of the form:

$$\omega_{j,n} = \sqrt{\frac{\mu}{\rho}} \frac{1}{R} \sqrt{\mu_j^2 + (n + 1/2)^2 \pi^2 \alpha^2}, \qquad (4.270)$$

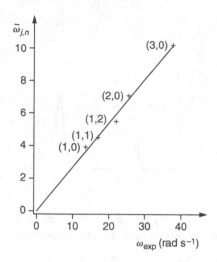

Fig. 4.29 Indexing of the resonance peaks. The value of the (j,n) indices is indicated alongside each experimental point (Joanicot, 1984b).

where $\alpha = R/H$ is the aspect ratio of the sample, with an experimental value of 0.5 (for $R = 1.75$ cm and $H = 3$ cm). Note that this choice of α is not arbitrary; it yields the lowest resonance frequencies: under these conditions, the peaks are as narrow as possible, since the dissipation is at its minimum.

This formula allows indexing the frequency of each of the resonance peaks observed. Figure 4.29 displays the rescaled frequencies $\widetilde{\omega} = \omega R \sqrt{\rho/\mu}$ as a function of the measured resonance frequencies. As expected, the graph is a straight line and its slope yields directly the elastic modulus, ρ and H being known.

This very elegant method allowed Pieranski and his collaborators to measure the shear modulus of polycrystalline colloidal crystals (Dubois–Violette *et al.*, 1980; Joanicot *et al.*, 1984a). Note that these measurements were subsequently improved, owing in particular to a new technique for sample purification using a 'continuous deionisation' procedure (Palberg *et al.*, 1994).

The values found for PS beads, 0.1 μm in diameter, range from 0.1 to 10 Pa for particle densities between 10^{18} and 5×10^{19} part./m^3.

Pieranski and Joanicot (among others) have also shown that a colloidal crystal can exhibit plastic deformation, and even melt at high shear rates. We refer the reader to the last section of the next chapter for the description of these experiments.

Various other experimental setups have been designed for studying the elasticity of colloidal crystals. We can cite, for instance, the rotating magnetic pendulum of Lindsay and Chaikin. By studying the mechanical resonances of the pendulum, these authors measured the shear modulus of ordinary colloidal crystals, and also that of colloidal glasses obtained by mixing spheres of different size and charge (Lindsay and Chaikin, 1982).

Appendix 4.A Elasticity equations in cylindrical and spherical coordinates

4.A.1 Cylindrical coordinates (r, θ, z)

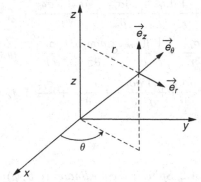

Arc length: $ds^2 = dr^2 + r^2 \, d\theta^2 + dz^2$

Scalar field: f

Gradient: $\overrightarrow{\mathrm{grad}} f = \left(\dfrac{\partial f}{\partial r}, \dfrac{1}{r} \dfrac{\partial f}{\partial \theta}, \dfrac{\partial f}{\partial z} \right)$

Laplacian: $\Delta f = \dfrac{1}{r} \dfrac{\partial}{\partial r} \left(r \dfrac{\partial f}{\partial r} \right) + \dfrac{1}{r^2} \dfrac{\partial^2 f}{\partial \theta^2} + \dfrac{\partial^2 f}{\partial z^2}$

Displacement: $\vec{u} = (u_r, u_\theta, u_z)$

Divergence (bulk dilation): $\mathrm{div}\, \vec{u} = \dfrac{1}{r} \dfrac{\partial}{\partial r}(r u_r) + \dfrac{1}{r} \dfrac{\partial u_\theta}{\partial \theta} + \dfrac{\partial u_z}{\partial z}$

Rotational:

$$\overrightarrow{\mathrm{curl}}\, \vec{u} = \left[\dfrac{1}{r} \dfrac{\partial u_z}{\partial \theta} - \dfrac{\partial u_\theta}{\partial z}, \quad \dfrac{\partial u_r}{\partial z} - \dfrac{\partial u_z}{\partial r}, \quad \dfrac{1}{r} \left(\dfrac{\partial}{\partial r}(r u_\theta) - \dfrac{\partial u_r}{\partial \theta} \right) \right]$$

Laplacian:

$$\Delta \vec{u} = \left[\Delta u_r - \dfrac{2}{r^2} \dfrac{\partial u_\theta}{\partial \theta} - \dfrac{u_r}{r^2}, \quad \Delta u_\theta - \dfrac{2}{r^2} \dfrac{\partial u_r}{\partial \theta} - \dfrac{u_\theta}{r^2}, \quad \Delta u_z \right]$$

Strain tensor:

$$\varepsilon_{rr} = \dfrac{\partial u_r}{\partial r} \qquad\qquad \varepsilon_{r\theta} = \dfrac{1}{2} \left(\dfrac{\partial u_\theta}{\partial r} - \dfrac{u_\theta}{r} + \dfrac{1}{r} \dfrac{\partial u_r}{\partial \theta} \right)$$

$$\varepsilon_{\theta\theta} = \dfrac{1}{r} \dfrac{\partial u_\theta}{\partial \theta} + \dfrac{u_r}{r} \qquad\qquad \varepsilon_{\theta z} = \dfrac{1}{2} \left(\dfrac{1}{r} \dfrac{\partial u_z}{\partial \theta} + \dfrac{\partial u_\theta}{\partial z} \right)$$

$$\varepsilon_{zz} = \dfrac{\partial u_z}{\partial z} \qquad\qquad \varepsilon_{zr} = \dfrac{1}{2} \left(\dfrac{\partial u_r}{\partial z} + \dfrac{\partial u_z}{\partial r} \right)$$

Equilibrium equations:

$$\frac{\partial \sigma_{rr}}{\partial r} + \frac{1}{r}\frac{\partial \sigma_{r\theta}}{\partial \theta} + \frac{\partial \sigma_{rz}}{\partial z} + \frac{\sigma_{rr} - \sigma_{\theta\theta}}{r} + f_r = 0$$

$$\frac{\partial \sigma_{r\theta}}{\partial r} + \frac{1}{r}\frac{\partial \sigma_{\theta\theta}}{\partial \theta} + \frac{\partial \sigma_{\theta z}}{\partial z} + 2\frac{\sigma_{r\theta}}{r} + f_\theta = 0$$

$$\frac{\partial \sigma_{rz}}{\partial r} + \frac{1}{r}\frac{\partial \sigma_{\theta z}}{\partial \theta} + \frac{\partial \sigma_{zz}}{\partial z} + \frac{\sigma_{rz}}{r} + f_z = 0$$

4.A.2 Spherical coordinates (r, θ, φ)

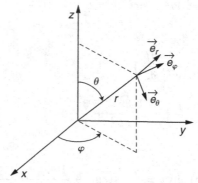

Arc element: $ds^2 = dr^2 + r^2\,d\theta^2 + r^2 \sin^2\theta\,d\varphi^2$

Scalar field: f

Gradient: $\overrightarrow{\mathrm{grad}}\, f = \left(\frac{\partial f}{\partial r}, \frac{1}{r}\frac{\partial f}{\partial \theta}, \frac{1}{r\sin\theta}\frac{\partial f}{\partial \varphi} \right)$

Laplacian: $\Delta f = \frac{1}{r^2}\frac{\partial}{\partial r}\left(r^2\frac{\partial f}{\partial r} \right) + \frac{1}{r\sin\theta}\frac{\partial}{\partial \theta}\left(\frac{\sin\theta}{r}\frac{\partial f}{\partial \theta} \right) + \frac{1}{r^2\sin^2\theta}\frac{\partial^2 f}{\partial \varphi^2}$

Displacement: $\vec{u} = (u_r, u_\theta, u_\varphi)$

Divergence (bulk dilation):

$$\mathrm{div}\,\vec{u} = \frac{1}{r^2}\frac{\partial}{\partial r}(r^2 u_r) + \frac{1}{r\sin\theta}\frac{\partial}{\partial \theta}(u_\theta \sin\theta) + \frac{1}{r\sin\theta}\frac{\partial u_\varphi}{\partial \varphi}$$

Rotational:

$$\overrightarrow{\mathrm{curl}}\,\vec{u} = \left[\frac{1}{r\sin\theta}\left(\frac{\partial}{\partial \theta}(u_\varphi \sin\theta) - \frac{\partial u_\theta}{\partial \varphi} \right),\ \frac{1}{r\sin\theta}\frac{\partial u_r}{\partial \varphi} - \frac{1}{r}\frac{\partial}{\partial r}(r u_\varphi),\ \frac{1}{r}\left(\frac{\partial}{\partial r}(r u_\theta) - \frac{\partial u_r}{\partial \theta} \right) \right]$$

Laplacian:

$$\Delta \vec{u} = \left[\Delta u_r - 2\frac{u_r}{r^2} - \frac{2}{r^2 \sin^2 \theta}\frac{\partial(u_\theta \sin \theta)}{\partial \theta} - \frac{2}{r^2 \sin \theta}\frac{\partial u_\varphi}{\partial \varphi}, \; \Delta u_\theta + \frac{2}{r^2}\frac{\partial u_r}{\partial \theta} - \frac{u_\theta}{r^2 \sin^2 \theta} \right.$$

$$\left. - \frac{2\cos \theta}{r^2 \sin^2 \theta}\frac{\partial u_\varphi}{\partial \varphi}, \; \Delta u_\varphi + \frac{2}{r^2 \sin \theta}\frac{\partial u_r}{\partial \varphi} + \frac{2\cos \theta}{r^2 \sin^2 \theta}\frac{\partial u_\theta}{\partial \varphi} - \frac{u_\varphi}{r^2 \sin^2 \theta} \right]$$

Strain tensor:

$$\varepsilon_{rr} = \frac{\partial u_r}{\partial r} \qquad\qquad \varepsilon_{r\theta} = \frac{1}{2}\left(\frac{1}{r}\frac{\partial u_r}{\partial \theta} + \frac{\partial u_\theta}{\partial r} - \frac{u_\theta}{r} \right)$$

$$\varepsilon_{\theta\theta} = \frac{1}{r}\frac{\partial u_\theta}{\partial \theta} + \frac{u_r}{r} \qquad\qquad \varepsilon_{\theta\varphi} = \frac{1}{2}\left(\frac{1}{r}\frac{\partial u_\varphi}{\partial \theta} + \frac{1}{r \sin \theta}\frac{\partial u_\theta}{\partial \varphi} - \frac{\cot \theta}{r}u_\varphi \right)$$

$$\varepsilon_{\varphi\varphi} = \frac{1}{r \sin \theta}\frac{\partial u_\varphi}{\partial \varphi} + \frac{u_\theta}{r}\cot \theta + \frac{u_r}{r} \qquad \varepsilon_{\varphi r} = \frac{1}{2}\left(\frac{1}{r \sin \theta}\frac{\partial u_r}{\partial \varphi} + \frac{\partial u_\varphi}{\partial r} - \frac{u_\varphi}{r} \right)$$

Equilibrium equations:

$$\frac{\partial \sigma_{rr}}{\partial r} + \frac{1}{r}\frac{\partial \sigma_{r\theta}}{\partial \theta} + \frac{1}{r \sin \theta}\frac{\partial \sigma_{r\varphi}}{\partial \varphi} + \frac{1}{r}(2\sigma_{rr} - \sigma_{\theta\theta} - \sigma_{\varphi\varphi} + \sigma_{r\theta}\cot \theta) + f_r = 0$$

$$\frac{\partial \sigma_{r\theta}}{\partial r} + \frac{1}{r}\frac{\partial \sigma_{\theta\theta}}{\partial \theta} + \frac{1}{r \sin \theta}\frac{\partial \sigma_{\theta\varphi}}{\partial \varphi} + \frac{1}{r}\left[3\sigma_{r\theta} + (\sigma_{\theta\theta} - \sigma_{\varphi\varphi})\cot \theta \right] + f_\theta = 0$$

$$\frac{\partial \sigma_{r\varphi}}{\partial r} + \frac{1}{r}\frac{\partial \sigma_{\theta\varphi}}{\partial \theta} + \frac{1}{r \sin \theta}\frac{\partial \sigma_{\varphi\varphi}}{\partial \varphi} + \frac{1}{r}(3\sigma_{r\varphi} + 2\sigma_{\theta\varphi}\cot \theta) + f_\varphi = 0$$

References

Alder, B. J., N. G. Hoover and D. A. Young, *J. Chem. Phys.*, **49**, 3688 (1968).

Bollmann, W., *Phys. Rev.*, **103**, 1588 (1956).

Burgers, J. M., *Proc. Kon. Ned. Acad. Wet.*, **47**, 293, 378 (1937).

Crandal, R. S. and R. Williams, *Science*, **198**, 293 (1977).

Dubois–Violette, E., P. Pieranski, F. Rothen and L. Strzelecki, *J. Phys. France*, **41**, 369 (1980).

Dziewonski, A. M., A. L. Hales and E. R. Lapwood, *Phys. Earth Planet. Int.*, **10**, 12 (1975).

Habdas, P. and E. R. Weeks, *Curr. Opin. Colloid Interface Sci.*, **7**, 196 (2002).

Hirsch, P. B., R. W. Horne and M. J. Whelan, *Phil. Mag.*, **1**, 677 (1956).

Joanicot, M., M. Jorand, P. Pieranski and F. Rothen, *J. Phys. France*, **45**, 1413 (1984).

Joanicot, M., *Elasticity, plasticity and melting of colloidal crystals under shear* (in French), Ph.D. thesis, Université Pierre et Marie Curie (1984).

Joanny, J.-F., *J. Colloid Interface Sci.*, **71**, 622 (1979).

Jouffrey, B., *Dislocations and Plastic Deformation* (in French), Yravals summer school, Sept. 1979, P. Groh, L. P. Kubin and J.-L. Martin (Eds.), Colomiers: Les éditions de physique (1980).

Lakes, R., *Science*, **235**, 1038 (1987).

Lifshitz, I. and L. Rosentsveig, *JETP*, **16**, 967 (1946).

Lindsay, H. M. and P. M. Chaikin, *J. Chem. Phys.*, **76**, 3774 (1982).

Ohtsuki, T., S. Mitacu and K. Okano, *Japan. J. Appl. Phys.*, **17**, 627 (1978).

Orowan, E., *Z. Phys.*, **89**, 634 (1934).

Palberg, T., J. Kottal, T. Loga, H. Hecht, E. Simnacher, F. Falcoz and P. Leiderer, *J. Phys. III France*, **4**, 457 (1994).

Pieranski, P., *Contemp. Phys.*, **24**, 25 (1983).

Polanyi, M., *Z. Phys.*, **98**, 660 (1934).

Rayleigh, Lord., *The Theory of Sound*, New York: Dover Publications, Inc. (1945).

Taylor, G. I., *Proc. Roy. Soc. A*, **145**, 362 (1934).

Volterra, V., *Ann. École Normale Sup. Paris*, **24**(3), 400 (1907).

Williams, R., R. S. Crandall and P. J. Wojtowicz, *Phys. Rev. Lett.*, **37**, 348 (1976).

Yamaguchi, K., *Sci. Pap. Inst. Phys. Chem. Res.*, **11**(200), 151 (1929); **11**(205), 223 (1929).

Yeganeh-Haeri, A., D. J. Weidner and J. B. Parise, *Science*, **257**, 650 (1992).

5

Defects, plasticity and fracture of solids

In the previous chapter, we discussed the limits of the theory of the elasticity of solids which, by definition, only deals with their reversible deformation. We showed the existence for each material of a *critical stress*, or *elastic limit*, above which it undergoes *irreversible deformation*. Trying to evaluate this stress for a perfect crystal, we found a value, termed the *theoretical elastic limit*, which is generally much too high: on the average, 100 to 10 000 times higher than the experimentally determined value.[1] To solve this contradiction, we recalled that crystals always have defects, in particular dislocations, that break their translational symmetry locally. We then explained qualitatively, using the image of the Mott carpet, why dislocations can propagate easily under applied stress, leading to a considerable decrease in the elastic limit of the material.

Our goal in this chapter is to delve deeper into these questions and to show that *defects* (and not only dislocations) are entirely responsible for the behaviour observed above the elastic limit. We point out straightaway that two very different types of behaviour can be observed in this regime, depending on the temperature. The solid can either *creep*, or deform slowly under the action of the applied stress or, on the contrary, it can *break* suddenly. In the first case, one speaks of *plastic deformation*, and in the second one of *brittle fracture* – a phenomenon that occurs in *almost* all materials if the temperature is below that of the *brittle to ductile transition* T_{BD}. Also note that, if the *defects of the crystal lattice* are responsible for the plastic deformation, it is mostly *microcracks* that are involved in brittle fracture, in particular in glasses, where their size starts to diverge above the elastic limit.

To discuss all these questions we divide the chapter into five main sections, the contents of which we will now present in detail.

The first one begins with some general points on crystal *defects* (Section 5.1). These can be point-like (*vacancies*, for instance), linear (these are the *dislocations* we already talked about) or surface-like (such as *walls*, or *grain boundaries*). After creating these defects topologically and giving the principle of the *Volterra construction* of dislocations, we will see how they are revealed experimentally. We will then study some of their static properties, in particular their energy and their elastic interactions, their equilibrium density and their arrangement in space (microstructure).

---

[1] Note however that for 'whiskers', very thin metal wires with almost no defects, the measured elastic limits are very close to the theoretical values.

The second section will deal almost exclusively with crystal *dislocations* and their relation to the *elastic limit*. In particular, we will discuss some of their dynamic properties, which will allow us to define their *mass* and *mobility*. We will also prove the *Peach and Koehler formula*, giving the force exerted on a dislocation by an external stress. We will see that this force is formally equivalent to the Magnus force acting on a vortex in hydrodynamics. Dislocation crossing and their multiplication processes, the best known among them being that of the '*Frank–Read mill*', will also be described. It will then be shown how all these mechanisms come into play in plasticity and can predict the correct value of the *elastic limit*. At this point, we will discuss in detail some specific problems, which are extremely important in practice, such as *grain boundary hardening* (the law of Hall and Petch), *alloy hardening* (so important for understanding the hardness of duralumin or cast iron), and *quench* or *work hardening*.

It is in the third section that we will really tackle the problem of *plastic deformation*. We will start by recalling the different regimes observed experimentally (α, β, κ and r *creep*), and continue by proving the *Orowan relation*, relating the deformation rate to the applied stress and to the *microstructure* of the crystal. We will apply this relation to the case of *Arrhenius creep*, observed at low temperature ($T < 0.5\,T_m$, where T_m is the melting temperature of the body), during which the velocity of the dislocations is governed by the thermally activated crossing of localised obstacles. We will also discuss some models for '*high-temperature*' *creep* ($T > 0.5\,T_m$), such as those of *Nabarro–Herring* and *Coble* by vacancy diffusion, or that of *Harper–Dorn* by dislocation climb, all of them Newtonian. Two examples of non-Newtonian creep (*Nabarro and Weertman models*) will also be given. We conclude that high-temperature solids belong to the category of yield stress liquids, in agreement with the classification given in Chapter 1.

The problem of *brittle fracture* will be tackled in the next part (Section 5.4). This phenomenon occurs in almost all materials at low enough temperatures, below their *ductile–brittle transition temperature*. The solid then becomes hard and brittle, as for window glass or a rock at room temperature, and exhibits almost no creep. We will prove that the *Griffith criterion*, which gives the critical stress for crack propagation, sets the elastic limit of the material in the brittle regime, and describes quite well the breaking threshold values observed, in particular for glasses. We will also give the propagation velocity of a crack as a function of its size and show that it is always below the velocity of sound.

In conclusion (Section 5.5), we will say a few words on the plasticity of *colloidal crystals*. We will see that these very soft materials can orient, or even change structure, going as far as *melting* under shear, a phenomenon never observed in ordinary solids.

5.1 General points on crystal defects

There are three types of defects in three-dimensional crystals: point-like defects, such as *vacancies* or *interstitials*, linear defects, termed *dislocations*, which are undeniably the most important in plasticity, and surface-like defects, which correspond to *walls* separating crystallites with different orientations. All these defects are responsible for the plastic

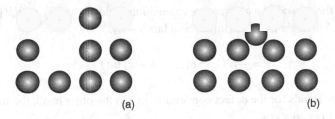

Fig. 5.1 Point-like defects in a crystal: (a) vacancy; (b) interstitial.

deformation of crystals, even if their density is generally low. This is why it is so important to study their topology and their static properties, before dealing with their dynamic properties.

We should also point out that defects (dislocations, in particular) play an important role in crystal growth for the case of faceted materials, as well as in the areas of corrosion or electronics (for instance, they can explain many of the electric properties of semiconductors or more complicated structures such as diodes).

To start with, let us describe point-like defects, which are the easiest to imagine and to study.

5.1.1 Point-like defects: vacancies, interstitials and precipitates

This generic term designates a perturbation of the crystal periodicity over a volume which is of the order of magnitude of the atomic volume Ω [12, 22].

Among the most current point-like defects, one can cite vacancies (absence of an atom at a normally occupied site, Fig. 5.1a), interstitials (excess atom occupying a normally empty site, Fig. 5.1b) or impurities (for instance, a B atom in an A crystal). These elementary defects can form aggregates (for instance, precipitates that are aggregates of B atoms in an A crystal).[2]

5.1.1.1 Equilibrium density of vacancies

Clearly, these defects have an energy cost. One might wonder why they appear. The answer is that the addition of defects increases the configuration entropy of the system, as they can be placed at different positions, and sometimes even in different ways. The equilibrium density of defects is therefore a result of the balance between the gain in entropy on the one side, and the energy loss on the other.

We will now compute the equilibrium density of vacancies. Let N be the number of atoms in the crystal and n the number of equilibrium vacancies. Since the vacancy has the symmetry of the atom it replaces, the increase in entropy is simply equal to $k_B \ln w$, where k_B is the Boltzmann constant and w the number of ways one can place the n vacancies on

---

[2] Note that in ionic crystals (such as sodium chloride), point-like defects are also present. Due to charge conservation, the vacancies can appear either as pairs with opposite signs (*Schottky defects*), or in association with an interstitial of the same ion (*Frenkel defects*) [22]. We will not consider these materials in the following.

the N sites of the crystal. An elementary calculation yields $w = C_N^n = N!/[n!(N-n)!]$, whence, using Stirling's formula $(\ln(n!) = n\ln(n) - n)$:

$$S = -Nk_B\left[c\ln c + (1-c)\ln(1-c)\right] \tag{5.1}$$

where $c = n/N$ stands for the defect concentration. On the other hand, the introduction of vacancies has an energy cost:

$$E = NcE_f \tag{5.2}$$

with E_f their individual formation energy. It follows that the free energy of the crystal

$$F = E - TS = NcE_f + Nk_BT\left[c\ln c + (1-c)\ln(1-c)\right] \tag{5.3}$$

has its minimum for a *finite* concentration of vacancies, given by the relation:

$$\frac{c}{1-c} = \exp\left(-E_f/k_BT\right). \tag{5.4}$$

If the energy E_f is much higher than k_BT, the concentration is low, reducing to:

$$c \approx \exp\left(-E_f/k_BT\right). \tag{5.5}$$

To estimate the order of magnitude for the concentration of vacancies in a crystal, one must know their formation energy. A good approximation is provided by the sublimation energy of an atom. In metals at room temperature $E_f \approx 1$ eV, yielding a concentration of the order of $\exp(-40) \approx 10^{-18}$. This value is very low, but it increases considerably on approaching the melting point due to the $1/T$ variation of the exponent. Nevertheless, the vacancy concentration never exceeds 10^{-4} in pure metals, even very close to their melting point.

From the experimental point of view, measuring the density of vacancies in metals as a function of temperature is done by *quenching*. The method consists of heating the sample to high-temperature, letting it equilibrate, and then 'quenching' it to low temperature (for instance, to 4 K, which is the boiling temperature of liquid helium). 'Quenching' the sample implies cooling it fast enough by immersion into an suitable liquid so that the density of vacancies does not have enough time to change during cooling. Since at 4 K vacancy mobility is almost null, their concentration remains unchanged over time, although the system is strongly supersaturated. One then determines their concentration by measuring the increase in volume they induce with respect to the 'perfect' crystal:

$$\frac{\Delta V}{V} \approx \frac{n\Omega}{N\Omega} = c. \tag{5.6}$$

Measuring c as a function of T gives access to E_f.

5.1.1.2 Vacancies and matter diffusion

We will see in Section 5.3, dedicated to plastic deformation, that, in spite of their low density, vacancies play an active role in high-temperature creep. It will also be shown that they are responsible for important phenomena, such as hardening of a metal after quenching (tempered steel).

These dynamic phenomena bring into play the diffusion of vacancies, which also drives matter diffusion. Indeed, one can easily conceive that, under the effect of thermal agitation, an atom changes places with a neighbouring vacancy. This exchange results in matter transport in the direction opposing the motion of the vacancy. One knows from statistical mechanics that the jump frequency v_j of an atom to the empty site of a neighbouring vacancy is given by the relation:

$$v_j = v \exp\left(-\frac{E_m}{k_B T}\right), \tag{5.7}$$

where v is an atomic vibration frequency (the Debye frequency, of the order of 10^{13} s^{-1}) and E_m a *migration energy*, equal to the height of the potential barrier that the atom must cross to pass from one site to the next (Fig. 5.2).

Let us now assume the crystal exhibits a gradient of vacancy concentration. Since the system wants to return to thermodynamic equilibrium, a flux of vacancies $\vec{J}$ (per unit surface) sets in, which we will assume proportional to their concentration gradient (*Fick's law*):

$$\vec{J} = -D_v \overrightarrow{\text{grad}}\, C \tag{5.8}$$

where $C = c/\Omega$ is the bulk concentration of vacancies (denoting by Ω the atomic volume). The phenomenological coefficient D_v is called the *diffusion coefficient of vacancies*. Its value is given by (see the demonstration in the inset):

$$D_v = v_j \frac{Z}{6} r_0^2 = v \frac{Z}{6} r_0^2 \exp\left(-\frac{E_m}{k_B T}\right), \tag{5.9}$$

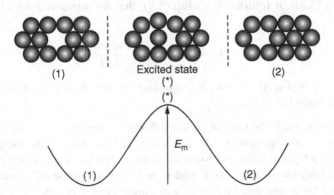

Fig. 5.2 To pass from position (1) to position (2), the atom must go through the intermediate position (*) corresponding to an excited state of the system.

where r_0 is the length of an elementary jump and Z the number of first neighbours (also called *coordinance*, with $Z = 8$ in a b.c.c. crystal, 12 in an f.c.c. crystal, etc.).

To prove the formula (5.9), let us first recall that, according to the laws of classical diffusion, the distance covered by a vacancy after a time t is given by the relation:

$$r^2 = 6D_\text{v}t. \tag{5.10}$$

On the other hand, it can be easily shown that, during a random walk (Section 7.1.4.2), the vacancy covers during time t a distance r such that (Einstein relation):

$$r^2 = nr_0^2, \tag{5.11}$$

with n the number of jumps it took during the time t. By equating these two relations, one has:

$$D_\text{v} = \frac{n}{6t}r_0^2. \tag{5.12}$$

Finally, the number of jumps taken during time t is given by:

$$n = v \times \exp\left(-\frac{E_\text{m}}{k_\text{B}T}\right) \times Z \times t, \tag{5.13}$$

where the first factor is the attempt frequency, the second one is the probability of crossing the barrier for one attempt, the third is the coordinance and the fourth is the time. Combining the last two equations yields the sought-after formula (5.9). Note that in materials with body-centred cubic (b.c.c.) or face-centred cubic (f.c.c.) structure, $(Z/6)r_0^2 = a^2$, where a is the length of the side of the cubic unit cell.

To end with, let us determine the self-diffusion coefficient of matter D_S. This transport occurs due to vacancy diffusion. Indeed, an atom can only move if it neighbours a vacancy and changes places with it. The jump frequency of an atom is thus equal to the jump frequency of a vacancy times the probability of having a vacancy nearby. This probability is nothing other than the equilibrium concentration of vacancies c calculated in the previous subsection (Eq. (5.5)). It follows, from Eq. (5.9), that for a material with b.c.c. or f.c.c. structure:

$$D_\text{S} = cD_\text{v} = va^2 \exp\left(-\frac{E_\text{m} + E_\text{f}}{k_\text{B}T}\right). \tag{5.14}$$

The parameter $E_\text{S} = E_\text{m} + E_\text{f}$ is the *self-diffusion energy*, which can be measured using the method presented in the inset.

To determine experimentally the value of the self-diffusion energy, one can measure the coefficient D_S as a function of the temperature T. This can be achieved, for instance, by covering the surface of the crystal with a layer of radioactive isotopes, which is then left to diffuse for a certain time t within the crystal heated to a temperature T (high enough to avoid 'geological' measurement times). At time t the crystal is cooled, thus freezing the concentration profile of isotopes inside. This profile can then be determined by a radioactivity measurement. By repeating the procedure several times

Table 5.1. *Parameters for vacancies in metals*

	Cu	Ag	Au	Mg	Al	Pt	W
E_f	1.28	1.06	0.95	0.79	0.69	1.50	$\approx$3.5
E_m	0.72	0.86	0.85	0.48	0.64	1.40	$\approx$1.8
va^2	0.6	0.36	0.03	$\approx$1.2	1.7	$\approx$0.33	—

Values (in eV) of the formation E_f and the migration E_m energies of vacancies [12]. The value of the prefactor is given in cm^2/s [16].

with different durations, one can obtain the evolution of the concentration profile of the radioactive tracer as a function of time, giving access to its diffusion coefficient D_S^*. It can be shown that D_S and D_S^* are not exactly identical, being connected by a relation of the type:

$$D_S^* \approx D_S \left(1 - \frac{2}{Z}\right) \sqrt{\frac{m}{m^*}} \qquad (5.15)$$

with m and m^* the atomic masses of the crystal and its isotope. This difference between the two diffusion coefficients originates in the jumps of the tracer being correlated and not following exactly a random walk, the way the crystal atoms do (in fact, tracers are discernable from the other atoms of the lattice) [12].

Table 5.1 gives a few values for the formation and migration energies of vacancies in metals, as well as the prefactor of the self-diffusion coefficient (Eq. (5.14)).

In conclusion, let us give a few orders of magnitude for D_S in solid metals, as well as in liquids and gases, for comparison. In aluminium, for instance, D_S increases enormously with temperature, from 10^{-23} cm^2/s at room temperature to 6×10^{-8} cm^2/s at the melting point ($T_m = 660.4$ °C). Even this latter value is much lower than that found in simple liquids with small molecules such as water (where $D_S \approx 10^{-5}$ cm^2/s) and in gases (where $D_S \approx 1$ cm^2/s).

5.1.2 Dislocations and Frank lattice: static properties

Dislocations are linear defects, since they break the translational order of the Bravais lattice along (not necessarily straight) lines. They are directly responsible for the plastic deformation at low temperature, as they allow the crystal planes to glide on top of each other above the ductile–brittle transition temperature (see Fig. 4.8). This is why they have been the object of extended studies for almost a century now. In this section, we will recall some of their essential properties, without going into the detail of crystal structures. On these topics, we refer the reader to specialised works [20].

To begin with, let us see how to construct a dislocation from a purely topological point of view.

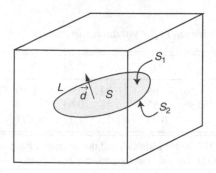

Fig. 5.3 Dislocation line L in an elastic medium.

5.1.2.1 Volterra construction

The Volterra method can be summarised thus. Consider a perfect crystal, initially with-out stress. To construct a defect line L, one performs mentally the following operations (Fig. 5.3):

(1) cut the material along a generic surface S limited by the line L;
(2) by means of an external stress, the two lips S_1 and S_2 of the cut surface are *rigidly* moved with respect to each other. In the most general case, this displacement $\vec{d}(\vec{r})$ is the sum of a translation by a vector $\vec{b}$ and a rotation about a vector $\vec{v}$ through an angle Ω, such that:

$$\vec{d}(\vec{r}) = \vec{b} + 2\sin(\Omega/2)\vec{v} \times \vec{r}; \qquad (5.16)$$

(3) if necessary, excess matter is removed or, if an empty space was created, it is filled with the perfect crystal;
(4) finally, everything is fixed back and the stress applied during step two is removed.

Obviously, this 'surgical' operation should be performed so that no scar is left (except the line) once the material has been fixed back along surfaces S_1 and S_2. *It is therefore imperative that the displacement $\vec{d}(\vec{r})$ corresponds to a symmetry operation of the perfect crystal.*

Let us now give some examples and some definitions.

If the displacement is a *pure translation* ($\vec{d}(\vec{r}) = \vec{b}$), one speaks of a *dislocation*. The vector $\vec{b}$ is called the *Burgers vector*. The dislocation is *perfect* if $\vec{b}$ is a translation vector of the Bravais lattice of the crystal. Otherwise, one speaks of *partial* or *imperfect* dislocations: these dislocations are very seldom encountered alone, since the stacking of the atomic planes one over the other has a fault over the whole cut plane, which is very costly in energy. However, perfect dislocations with Burgers vector $\vec{b}$ very often *dissociate into two partial dislocations* with Burgers vectors $\vec{b}_1$ and $\vec{b}_2$ separated by a stacking fault (with $\vec{b} = \vec{b}_1 + \vec{b}_2$) (Saada, 1966). In the following, for the sake of simplicity, we will neglect these dissociation phenomena, but it is noteworthy that they play a very important role in metallurgy. Let us also point out that three types of dislocation can be distinguished according to the orientation of the vector $\vec{b}$ with respect to the unit vector $\vec{t}$ tangent to the line. If $\vec{b}//\vec{t}$ (or $\vec{b} \perp \vec{t}$), one says that the dislocation line is of the *screw* (or *edge*) type, respectively (see Figs. 5.11 and 5.12). In all other cases, one speaks of *mixed dislocations*.

If the displacement corresponds to a *pure rotation*, a *disclination* is obtained. The energy of these defects is extremely high in solids, where they are almost never encountered. They are however very common in nematic, smectic or columnar liquid crystals. We will therefore discuss them in detail in the chapter dealing with these phases. Certain smectic phases also exhibit defects called *dispirations*, where the displacement is composed of a translation and a rotation: their original feature resides in that neither the translation nor the rotation belong to the symmetry elements of the phase. However, their composition is a symmetry element ([43], Volume 2).

5.1.2.2 Precise definition of the Burgers vector of a dislocation

In practice, it is important to indicate the norm, the direction, and also the *orientation* of the Burgers vector. This is done using the Burgers convention described in the diagram of Fig. 5.4 for an edge dislocation.

This method, applicable to all dislocations, consists of first setting an orientation $\vec{t}$ on the dislocation line (by convention, $\vec{t}$ is the unit vector tangent to the line). One then defines a closed circuit encircling the dislocation and oriented according to Maxwell's 'cork screw' rule. One associates with this circuit the corresponding circuit in the perfect crystal.

By definition, the Burgers vector of the dislocation is given by the *closure defect* of the circuit in the perfect crystal:

$$\vec{b} = \overrightarrow{QM} \qquad (5.17)$$

Note that it is also the translation vector in the Volterra process.

We emphasise that, by changing the orientation of the line, one also changes that of the Burgers vector. This caveat will take on its full meaning when calculating the Peach and Koehler elastic force exerted by an external stress field on a dislocation (Section 5.2.2).

The Burgers convention once adopted, one can easily check the following knot rule (Fig. 5.5). Let n dislocations meet at a point P and let us orient all lines either towards P, or away from P. It follows that:

$$\Sigma\vec{b}_i = 0. \qquad (5.18)$$

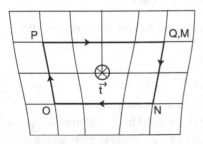 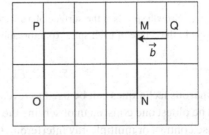

Fig. 5.4 Burgers circuit enclosing an edge dislocation. The Burgers vector is equal to the closure defect of the circuit in the perfect crystal.

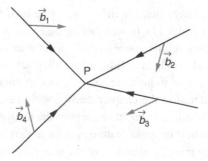

Fig. 5.5 Dislocation knot.

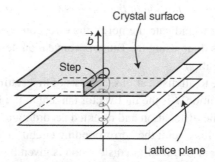

Fig. 5.6 Atomic step on a crystal surface starting at the emergence point of a screw dislocation.

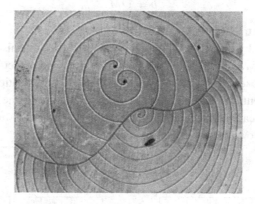

Fig. 5.7 Growth spirals at the surface of a silicon carbide crystal. A dislocation emerges at the crystal surface in the centre of each spiral (photograph from [17]).

5.1.2.3 *Experimental evidence*

Numerous techniques can be used to reveal the presence of dislocations.

The oldest one consists in observing the surface of a crystal in optical microscopy, using phase contrast or multiple-ray interferometry. If one can see atomic steps stopping abruptly on the surface, it means that screw dislocations emerge at these points. Furthermore, their Burgers vector must be equal to the height of the steps (Fig. 5.6).

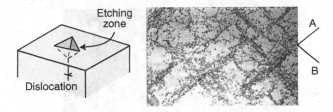

Fig. 5.8 Etching figure at the surface of a silicon monocrystal after deformation. One can see that the dislocations are grouped on two families of perpendicular lines, revealing the existence of two types of glide planes A and B [14] (photograph by A. George, École des Mines, Nancy).

It was by this technique that L. J. Griffin first observed, in 1950, the screw dislocations in a beryl crystal. He also noticed that the steps had a spiral shape centred on the emergence points of the screw dislocations (Fig. 5.7), thus demonstrating a growth mechanism peculiar to faceted crystals, and that F. C. Frank had already predicted theoretically at the time.

Another technique for revealing the emergence points of dislocations is to etch the surface of the crystal with a corrosive chemical solution, after having carefully polished it.

Experiment shows that the chemical attack occurs preferentially at the emergence points of the dislocations, leading to the formation of small pits, easily observed in optical reflection microscopy. Thus, one obtains an *etching figure* such as that shown in Fig. 5.8. This method can be used to count the dislocations and, to a certain extent, to follow their movement during plastic deformation, by successively making several etching figures.

An even more powerful visualisation technique is transmission electron microscopy (TEM). We recall that Ernst Ruska (awarded the Nobel prize for physics in 1986) built the first microscope of this type in 1931 and obtained the first images and electron diffraction patterns. However, it was only in the 1950s that this technique became fully developed and that the first thin slices of metallic samples were obtained, allowing the direct observation of dislocations (by W. Bollman in Geneva, in 1956). In practice, the thickness of these slices must not exceed a few tens of nanometres, since they must remain transparent to electrons. Nowadays, slices are obtained by ionic or chemical thinning. The photograph in Figure 5.9 shows a TEM image of a thin slice cut out of an iron sample after plastic deformation at low temperature. The dark lines one can see in this photo are the dislocations. They are visible because their core is filled with 'bad crystal', which diffracts the electrons more than the rest of the matrix, composed of good crystal (Saada, 1966). Note also that these dislocations form a regular network, called nowadays a Frank network, and to which we will return in Section 5.1.2.7.

Yet another visualisation method is X-ray topography, which relies on the fact that the dislocation core does not diffract X-rays in the same way as the rest of the crystal. It can be applied either in reflection or in transmission (Lang method). The experiment consists in moving in the X-ray beam the sample and the photographic plate used to record the diffracted beam. This technique recently prompted renewed interest due to the development of synchrotron facilities, where intense and monochromatic X-ray beams can be achieved.

Fig. 5.9 Electron microscopy image of a thin iron slice plastically deformed at low temperature (Keh and Weissmann, 1963; [15]).

Thus, the exposure times are considerably reduced and one can follow in real time the motion of dislocations during plastic deformation.

In conclusion, let us cite a decoration method that was extensively used during the 1950s (by S. Amelinckx, in particular) to study dislocations in transparent crystals such as sodium chloride, potassium chloride or potassium bromide. This method consists of doping the crystal with foreign atoms (for instance, silver atoms in potassium chloride). By an appropriate thermal treatment, one can induce the precipitation of these atoms. Experiment shows that the precipitates settle preferentially in bad crystal regions, in particular onto the cores of dislocations; it then becomes possible to decorate the dislocations, which can be observed directly in transmission optical microscopy, provided the crystal is transparent.

A nice example is shown in Fig. 5.10. This photograph shows a two-dimensional projection of the three-dimensional network formed by the dislocations within the crystal.

5.1.2.4 Elastic properties: stress field and energy

Let us begin by considering the simplest example, that of a screw dislocation. Assume it is placed at the centre of a cylinder with radius R and infinite length, of axis x_3 (Fig. 5.11). To construct the dislocation, one makes an incision along the half-plane $\theta = 0$ (with θ the polar angle) and translates the two lips of the cut by a distance b with respect to each other:

$$\vec{u}(\theta = 0) = 0 \quad \text{and} \quad \vec{u}(\theta = 2\pi) = \vec{b} \tag{5.19}$$

with $\vec{b} = b\vec{x}_3$. In this case, the atomic planes are deformed into a helicoid, as drawn in Fig. 5.6.

Fig. 5.10 Thin sample of a potassium chloride crystal examined in optical microscopy. Silver particles precipitated onto the dislocations, which are arranged in a rather regular network (Amelinckx, 1958; [15]).

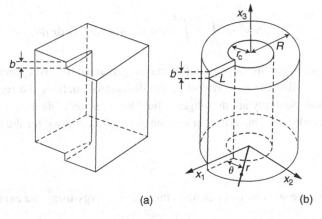

(a)   (b)

Fig. 5.11 (a) Screw dislocation in a crystal; (b) elastic distortion of a cylindrical annulus simulating the distortion induced by the screw dislocation [15].

It is easily checked that the following deformation field (Volterra, 1907):

$$u_1 = u_2 = 0 \quad \text{and} \quad u_3 = \frac{b\theta}{2\pi} = \frac{b}{2\pi}\left[\arctan\left(\frac{x_2}{x_1}\right) + k\pi\right], \tag{5.20}$$

with $k = 0$ in the first quadrant ($x_1 > 0$ and $x_2 > 0$), $k = 1$ in the second and third quadrants ($x_1 > 0$ and for any x_2) and $k = 2$ in the fourth quadrant ($x_1 > 0$ and $x_2 < 0$), satisfies the boundary conditions (5.19) as well as the Navier equation (with $\vec{f} = 0$, as we will always assume in the following).

It follows that the only non-vanishing components of the strain tensor are:

$$\varepsilon_{13} = \varepsilon_{31} = \frac{1}{2}\frac{\partial u_3}{\partial x_1} = -\frac{b}{4\pi}\frac{x_2}{x_1^2 + x_2^2}, \tag{5.21}$$

$$\varepsilon_{23} = \varepsilon_{32} = \frac{1}{2}\frac{\partial u_3}{\partial x_2} = \frac{b}{4\pi}\frac{x_1}{x_1^2 + x_2^2} \tag{5.22}$$

or, in polar coordinates:

$$\varepsilon_{\theta 3} = \varepsilon_{3\theta} = \frac{b}{4\pi r}. \tag{5.23}$$

Since $\operatorname{div} \vec{u} = 0$, one has simply $\sigma_{ij} = 2\mu\varepsilon_{ij}$, such that the stress, as well as the strain, varies as $1/r$ (with r the distance to the centre of the dislocation). Since these quantities diverge at the axis of the dislocation, a dislocation represents a *singular solution* to the equations of elasticity.

As to the energy of the dislocation per unit length $E_e(\text{screw})$, it is obtained by integrating the density of elastic free energy $f = (1/2)\sigma_{ij}\varepsilon_{ij}$ over the entire volume of the sample:

$$E_e(\text{screw}) = \int_0^{2\pi} \int_{r_c}^{R} (\sigma_{13}\varepsilon_{13} + \sigma_{23}\varepsilon_{23})\, r \, dr \, d\theta. \tag{5.24}$$

In this formula, we had to introduce a cutoff radius r_c to avoid the divergence of the energy. This radius, also called the *core radius* of the dislocation, defines the region where the equations of linear elasticity are no longer valid. One considers this to be the case as soon as the strain exceeds 5%. This criterion sets the order of magnitude for the core radius:

$$r_c \approx \frac{5b}{\pi}. \tag{5.25}$$

Once these limits are set, one can calculate the elastic energy using the expressions (5.21) and (5.22):

$$E_e(\text{screw}) = \frac{\mu b^2}{4\pi} \ln\left(\frac{R}{r_c}\right). \tag{5.26}$$

One must add to this energy the core energy of the dislocation. A good approximation is to assume that within the core the laws of elasticity remain valid, provided the strain is roughly constant and equal to 5%. This criterion leads to the following formula for the core energy:

$$E_c \approx \pi r_c^2 4\pi \left(\frac{1}{20}\right)^2 = \frac{\mu b^2}{4\pi}. \tag{5.27}$$

This contribution being very small in comparison with the elastic energy proper (a small percentage, in general) it is a good approximation to consider that

$$E_{\text{screw}} = E_e(\text{screw}) + E_c \approx E_e(\text{screw}),$$

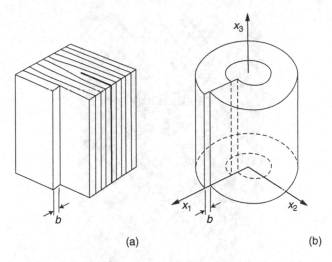

Fig. 5.12 (a) Edge dislocation in a crystal; (b) elastic distortion of a cylindrical annulus simulating the distortion produced by the edge dislocation [15].

$$\text{yielding}: \qquad E_{\text{screw}} \approx \frac{\mu b^2}{4\pi} \ln\left(\frac{R}{r_{\text{c}}}\right). \qquad (5.28)$$

This formula is important in that it shows that the energy grows as the Burgers vector squared. It is therefore preferable to create two dislocations with Burgers vectors b, rather than one with a Burgers vector $2b$ (since $(2b)^2 > b^2 + b^2$). In solids, the dislocations of small Burgers vector are thus strongly favoured. When $b = a$, the size of the unit cell of the lattice, the dislocation is called *elementary*.

One can redo the same calculations for a rectilinear edge dislocation (Fig. 5.12). In this case, the Burgers vector is perpendicular to the dislocation line. In practice, as shown by the photograph in Fig. 5.13, this amounts to adding an atomic half-plane on one side of the dislocation (plane drawn in thick line in Fig. 5.12a).

It can be shown that the deformation field obeying the Navier equation and the boundary conditions:

$$u_1(\theta = 0) = 0 \quad \text{and} \quad u_1(\theta = 2\pi) = b \qquad (5.29)$$

on the cut surface has the following form (Eshelby, 1956):

$$u_1 = \frac{b}{2\pi}\left(\theta + \frac{\sin 2\theta}{4(1-\nu)}\right),$$

$$u_2 = -\frac{b}{2\pi}\left[\frac{1-2\nu}{2(1-\nu)}\ln\left(\frac{r}{b}\right) - \frac{\sin^2\theta}{2(1-\nu)}\right], \qquad (5.30)$$

$$u_3 = 0,$$

with ν the Poisson coefficient.

3,2 Å

Fig. 5.13 High-resolution electron microscopy photograph of an edge dislocation in a germanium monocrystal [12] (photograph by A. Bourret).

Starting from these formulas, one can determine the strain and stress tensors. The stress tensor components are, in polar coordinates:

$$\sigma_{11} = -D\frac{\sin\theta(2 + \cos 2\theta)}{r},$$

$$\sigma_{22} = D\frac{\sin\theta\cos 2\theta}{r}, \qquad (5.31)$$

$$\sigma_{33} = \nu(\sigma_{11} + \sigma_{22}),$$

$$\sigma_{12} = D\frac{\cos\theta\cos 2\theta}{r},$$

where we set $D = \frac{\mu b}{2\pi(1-\nu)}$.

Once again, as in the case of the screw dislocation, we see that the stress decreases as $1/r$. However, $\operatorname{div}\vec{u} \neq 0$, meaning that the lattice is not only sheared, but also compressed or dilated depending on the position (more precisely, it is compressed on the side of the additional half-plane and dilated on the other side).

We still need to determine the elastic energy of the edge dislocation. This can be done by integrating directly the free energy density, but the calculations are tedious due to the complicated expressions for the strain and the stress. A more ingenious way is to use the work theorem proved in Section 4.1.8.1. According to this theorem:

$$2E_e(\text{edge}) = \int_{S_2} \vec{f_s} \cdot \vec{b} \, dS, \tag{5.32}$$

where $\vec{f_s}$ is the external force one must apply during the Volterra process to move the lip S_2 of the cut surface (with equation $\theta = 2\pi$) by an amount $\vec{b}$ with respect to the other lip S_1, assumed at rest (with equation $\theta = 0$). This force is:

$$\vec{f_s} = \underline{\sigma}\vec{v}, \tag{5.33}$$

where $\vec{v}$ is the unit vector $\vec{x_2}$ and $\underline{\sigma}$ the stress tensor calculated on the cut plane. Using Eqs. (5.31), one obtains successively:

$$f_{si} = \sigma_{i2}, \qquad i = 1, 2, 3 \tag{5.34}$$

then:

$$2E_e(\text{edge}) = \int_{r_c}^{R} \sigma_{12}b \, dr = Db \ln\left(\frac{R}{r_c}\right), \tag{5.35}$$

where we introduced the core radius r_c to keep the energy from diverging. Finally, replacing D by its expression yields:

$$E_e(\text{edge}) = \frac{\mu b^2}{4\pi(1-v)} \ln\left(\frac{R}{r_c}\right). \tag{5.36}$$

One must still add to this energy the core contribution, of the order of $0.1\mu b^2$. As for screw dislocations, this contribution is much smaller than the elastic energy proper and can be neglected in a first approximation ($E_{\text{edge}} \approx E_e(\text{edge})$). Note that the energy of an edge dislocation is larger than that of a screw dislocation due to the presence of the $1 - v$ factor at the denominator.

One can also determine the energy of a mixed dislocation (see inset).

In the case of a mixed dislocation, the Burgers vector makes an angle θ with the axis of the dislocation (Fig. 5.14). In this case, the deformation field is the sum of the fields produced by the edge and screw components of the Burgers vector:

$$\vec{u}_{\text{total}}(b) = \vec{u}_{\text{edge}}(b \sin\theta) + \vec{u}_{\text{screw}}(b \cos\theta). \tag{5.37}$$

Within the limits of linear elasticity, this also holds for the strain and the stress, such that:

$$\underline{\sigma}_{\text{total}}(b) = \underline{\sigma}_{\text{edge}}(b \sin\theta) + \underline{\sigma}_{\text{screw}}(b \cos\theta). \tag{5.38}$$

Also note that the tensors $\underline{\sigma}_{\text{screw}}$ and $\underline{\sigma}_{\text{edge}}$ have no non-zero component in common (in this respect, they are orthogonal) and that $\text{tr}(\underline{\sigma}_{\text{total}}) = \text{tr}(\underline{\sigma}_{\text{edge}})$. Consequently, from the general expression (4.78) of the elastic energy, one has:

$$f_{\text{total}}(b) = f_{\text{edge}}(b \sin\theta) + f_{\text{screw}}(b \cos\theta). \tag{5.39}$$

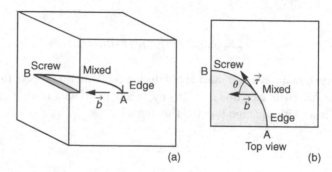

(a)   (b)

Fig. 5.14 Curved dislocation in an elastic medium. The type of the dislocation changes along the line: purely edge at its end A, it becomes mixed and then purely screw at the other end B [14].

One can therefore sum the energies of the edge and screw components and write, using Eqs. (5.26) and (5.36) and the assumption of low core energy:

$$E_{\text{mixed}} \approx E_{\text{e}}(\text{mixed}) = \frac{\mu b^2}{4\pi} \left(\cos^2\theta + \frac{\sin^2\theta}{1-\nu} \right) \ln\left(\frac{R}{r_{\text{c}}} \right). \qquad (5.40)$$

The conclusion is that the *stress field decreases as $1/r$ around a dislocation and its order of magnitude is given by $\mu b/r$. The energy per unit length of a dislocation is close to μb^2* (knowing that the logarithmic term $\ln(R/r_{\text{c}})$ is of the order of 4π), yielding μb^3 per atom if the atomic repeat distance along the dislocation is also b. In metals, this energy per atom is quite important, of the order of 5 eV. For this reason, dislocations are always *out of thermodynamic equilibrium.*

To see this, one only needs to calculate the number of loops of size $l \times l$ and with a given orientation per unit volume (for calculating the energy of a loop, see the next paragraph). J. Friedel gives [16]:

$$N \approx \frac{1}{\Omega} \exp\left(-\frac{\mu l b^2}{k_{\text{B}} T} \right), \qquad (5.41)$$

where Ω is the atomic volume.

This number is ridiculously small in metals, even for small loops of size $l = 5b$, being of the order of 10^{-400} m^{-3} at room temperature, which clearly justifies our statement.

5.1.2.5 Dislocation loop

The calculations above can be extended to the case of curved dislocations, but the general formalism is extremely heavy [18]. Fortunately, one can often avoid it by a few simple considerations. Consider the example of a dislocation forming a circular loop with diameter Φ in an isotropic material. To determine its energy, it suffices to notice that the stress field created by the dislocation in its close neighbourhood, i.e. at short distance with respect to Φ, is the same as that created by a straight dislocation tangent to the loop and with the same Burgers vector. On the other hand, one can easily check that at distances r much larger than the size of the loop the stress and strain fields go to 0 as $1/r^2$ (and no longer as $1/r$) because the contribution of each loop element is compensated by the contrary effect of the diametrally opposed portion. Due to this fast decrease, *the elastic energy of a loop in an infinite medium does not diverge*. More specifically, the elastic energy per unit loop length must be close to that of a straight dislocation in the centre of a tube with radius $R = \Phi$. As an order of magnitude, for an edge dislocation loop of diameter Φ:

$$E_{\text{loop}} \approx \pi\,\Phi \frac{\mu b^2}{4\pi(1-\nu)} \ln\left(\frac{\Phi}{r_c}\right). \tag{5.42}$$

In the next subsection, we will define a more subtle concept, that of *line tension* of a dislocation. This quantity is important, insofar as it can explain the polygonal shape of the dislocations observed in very anisotropic materials such as silicon (see, for instance, Fig. 5.40).

5.1.2.6 Line tension and polygonisation

Let us start by defining the *line tension* of a globally straight dislocation, exhibiting small undulations of wavelength λ (Fig. 5.15). We assume that the dislocation occupies the centre of a cylinder with radius R and that the material is isotropic.

For this particular case, it is convenient to separate two regions:

- a 'near field', contained within a cylinder with radius λ centred on the line ($r < \lambda$);
- a 'far field', outside this cylinder ($\lambda < r < R$).

In this latter region, the stress is the same as if the dislocation were straight, since the contributions of the line elements closer to the axis and farther away from it are rigorously

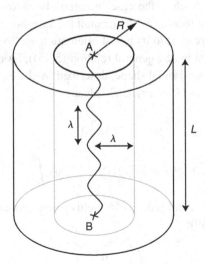

Fig. 5.15 Line tension of a sinusoidal dislocation [16].

compensated. The corresponding elastic energy is:

$$E_1 = \frac{\mu b^2}{4\pi K} L \ln\left(\frac{R}{\lambda}\right), \qquad (5.43)$$

where L is the length of the cylinder and K a factor between $1 - \nu$ and 1 which depends on the type of dislocation.

In the 'near field', the line behaves as if it were straight and contained within a tube with radius λ, but one must take into account its global lengthening dL. Hence the following contribution:

$$E_2 = \frac{\mu b^2}{4\pi K}(L + dL) \ln\left(\frac{R}{\lambda}\right). \qquad (5.44)$$

Finally, the total energy $E_1 + E_2$ differs from that of a straight dislocation, $(\mu b^2/4\pi K)L$ $\ln(R/r_c)$, by an amount $(\mu b^2/4\pi K) dL \ln(\lambda/r_c)$. This shows that a curved dislocation line is less stable than a straight line and that it tends to stiffen. Note by dE this excess energy.

By analogy with strings, one denotes as *line tension* of the dislocation the parameter:

$$\tau = \frac{dE}{dL}. \qquad (5.45)$$

In the case of a sinusoidal dislocation:

$$\tau = \frac{\mu b^2}{4\pi K} \ln\left(\frac{\lambda}{r_c}\right). \qquad (5.46)$$

This calculation takes for granted that the elastic energy of the dislocation does not depend on its orientation with respect to the crystal lattice. This assumption is not necessarily valid, as the energy of the dislocation might depend on its orientation θ with respect to a given crystallographic direction. Such is the case in materials endowed with very anisotropic elastic properties. One must then take into account the variation in dislocation energy as it turns by an angle $d\theta$ with respect to its initial orientation θ_0. To obtain the corresponding line tension (always defined by the general relation (5.45)), let us determine the increase in line energy as it takes a sinusoidal shape, described by the equation $x = A \cos kz$ with $\theta = \theta_0 + \delta\theta$, $\delta\theta = dx/dz$ and $k = 2\pi/\lambda$. This leads to:

$$dE = \tau\, dL = \int_0^L \left[E(\theta)\sqrt{1 + \delta\theta^2} - E(\theta_0) \right] dz$$

$$\approx \int_0^L [E(\theta_0 + \delta\theta) - E(\theta_0)]\, dz + E(\theta_0) \int_0^L \left(\sqrt{1 + \delta\theta^2} - 1\right) dz. \qquad (5.47)$$

Expanding $E(\theta_0 + \delta\theta)$ to second order in $\delta\theta$ yields a more general expression for the line tension (Koehler *et al.*, 1959):

$$\tau = \left(E + \frac{d^2 E}{d\theta^2} \right)_{\theta=\theta_0}. \qquad (5.48)$$

Note that in this formula one must take as the characteristic distance R the wavelength λ for a sinusoidal dislocation, the diameter Φ in the case of a loop, the mean distance between dislocations of opposite sign in the case of a Frank network, etc.

This formula is very important, as it can explain the anisotropic shape of the dislocation loops sometimes observed experimentally. Such an example in a metallic alloy is given in Fig. 5.16.

To detail this point, let us consider the example of a planar dislocation loop, of Burgers vector $\vec{b}$ contained in the plane of the loop.

Assume a shear stress is applied on the loop. We will show in Section 5.2.2 that the dislocation feels an elastic Peach and Koehler force, perpendicular to it. In the example shown in Fig. 5.17, this force is contained in the plane of the loop and tends to increase its size; of constant amplitude σb, it is opposed by the line tension force τ/R which tends to

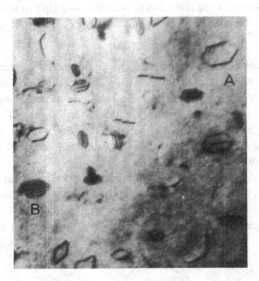

Fig. 5.16 Polygonal dislocation loops in an aluminium-magnesium alloy. Some loops, such as A, exhibit the same contrast inside and outside. They are perfect dislocation loops. Others, such as B, appear darker inside: these are imperfect dislocation loops containing a stacking fault (Westmacott et al., 1961).

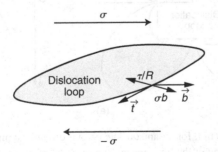

Fig. 5.17 Dislocation loop in equilibrium in a stress field.

decrease the size of the loop, where R is the local radius of curvature of the dislocation. The (unstable) equilibrium shape of the loop can thus be obtained by solving the equation:

$$\frac{\tau}{R} = \sigma b. \tag{5.49}$$

This equation is the same as the Gibbs–Thomson equation describing the shape of a seed crystal at equilibrium with its supersaturated liquid. Consequently, the shape of the loop is given by a *Wulff construction*, similar to that used to construct the equilibrium shape of a crystal as its surface stiffness $\gamma + \gamma''$ is anisotropic (see inset).

The Wulff construction is obtained as follows:

(1) Starting from an arbitrary origin, plot the polar diagram $E(\theta)$ of the line energy (*Wulff diagram*).
(2) At each point of $E(\theta)$, draw the normal to the vector radius. Up to a homothetic transformation of centre 0, the equilibrium shape of the loop is the convex envelope of these normals (the *pedal curve* of $E(\theta)$).

An example of this construction in the case of a crystal with cubic symmetry is shown in Fig. 5.18. Here, the line energy is minimal for the directions $\theta = 0$ and $\theta = \pi/2$, resulting in cusps in the Wulff diagram at these orientations. If the anisotropy is strong, as we assumed in this diagram, the equilibrium shape of the loop is square, all orientations except for 0 and $\pi/2$ disappearing from the equilibrium shape (these orientations are said to be 'missing'). In this limiting case, a straight dislocation segment (the equivalent of a facet for the equilibrium shape of a crystal) is associated with each cusp in the $E(\theta)$ diagram.

One can associate with this anisotropic equilibrium shape a polar diagram $\Gamma(\theta)$ such that, at each point of $\Gamma(\theta)$ the normal to the vector radius touches the equilibrium shape without intersecting it.

In 1962, W. W. Mullins showed that, if $E(\theta) > \Gamma(\theta)$ and if the dislocation is forced to go through two points A and B aligned along a direction θ, then this dislocation will have a polygonal shape

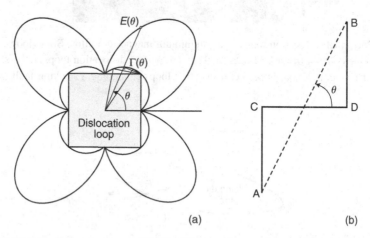

(a)   (b)

Fig. 5.18 (a) Wulff construction for an anisotropic crystal of cubic symmetry. The loop can take a square shape if the anisotropy of the line tension as a function of the angle θ is strong enough [16]; (b) polygonal shape of the dislocation when it is forced to go through points A and B.

ACDB rather than the straight one AB (Fig. 5.18b). The difference in energy between the straight and polygonal shapes is $E(\theta) - \Gamma(\theta)$ per unit length along the direction θ.

Polygonal dislocations are very rare, being encountered only in highly anisotropic materials such as silicon (or in liquid crystals such as the smectic A, see Section 8.2.9.3). In particular, they do not appear in ordinary metals, which exhibit only low anisotropy, thus justifying the approximation of isotropic elasticity used in this chapter.

To conclude, let us specify under which experimental conditions isolated dislocation loops can appear in a material. One method consists of heating the material close to its melting point, and then quenching it by immersion into an appropriate liquid. Dislocation loops appear after quenching by the local condensation of excess vacancies. This process locally decreases the chemical (or electronic) energy of the material, by recreating bonds. Another method is to irradiate the sample, for instance with neutrons or heavy ions [22]. Under these conditions, one can displace some atoms and create interstitials, which then regroup and form dislocation loops. Such effects can be important in the cores of nuclear reactors. Most notably, they affect the sheaths containing the nuclear fuel. Other nucleation mechanisms exist, e.g. the Frank and Read sources, which can generate cascades of concentric loops during plastic deformation. We will return to this mechanism in Section 5.2.6, dealing with the fundamental topic of dislocation proliferation.

5.1.2.7 Frank network

Experiment shows that, in thoroughly annealed materials exhibiting a low density of dislocations (between 10^3 and 10^8 cm per cm^3), these dislocations are arranged in a more or less regular three-dimensional network, containing essentially triple knots (Fig. 5.19) (Frank, 1955).

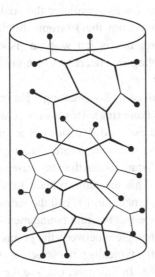

Fig. 5.19 Sketch of the Frank network in a crystal (Cottrell, 1957; [15]).

This network, named after Frank, does not modify the average orientation of the crystal lattice over long distances, containing equal numbers of dislocations of either sign. For this reason, and because the deformation fields created by two neighbouring dislocation segments, with opposite Burgers vectors, compensate over long distances, the elastic energy associated with this network is not excessive. It also follows that, when calculating the energy of a dislocation in the Frank network, the distance R is given by the typical size of the network, namely:

$$R \approx \xi = \frac{1}{\sqrt{\rho_d}} \tag{5.50}$$

with ρ_d the density of dislocations (defined as the dislocation length per unit volume).

Moreover, the Frank network is the source of a stress field inside the material, oscillating around a zero average over the typical distance ξ, with a maximum amplitude of the order of magnitude:

$$\sigma_M \approx \frac{\mu b}{2\pi (\xi/2)} = \frac{\mu b}{\pi \xi}. \tag{5.51}$$

This value, also called *internal stress*, plays an important role in plasticity, as it sets the threshold of plastic deformation (see Sections 5.2.4 and 5.3.3.2).

5.1.3 Walls

In practice, a solid sample is very seldom monocrystalline. Much more often, it is *polycrystalline*, i.e. composed of a collection of monocrystals, or *grains*, separated from one another by *walls* or *grain boundaries*.

Such an example is given in the photograph of Fig. 5.20. Here, as often happens during material preparation, the grains form after the coalescence of solid phase seeds, which nucleate at the transition from the isotropic liquid. Since there is no orientation relation between the grains, one is left with an 'isotropic' polycrystal, where the grain size decreases as the cooling rate decreases (the nucleation rate increases with the undercooling).

Polycrystals can also be obtained by hot-pressing – but in the solid phase – of a powder obtained by chemical synthesis (sintering). This is how ceramic materials are obtained.

Finally, a crystal can contain excess dislocations of a certain sign produced by plastic deformation. These dislocations often tend to form stackings, contributing to crystal hardening (Section 5.2.7). These dislocations can then organise in low-disorientation walls after appropriate thermal annealing. This mechanism leads to a polygonised structure, superposed on the Frank network, that metallurgists call the *mosaic crystal*.

Let us now discuss some properties of grain boundaries. Their nature differs according to the value of the disorientation angle θ between the grains they separate.

If this angle is small ($\theta < 20°$, typically), the structure of the boundary can be simply analysed in terms of dislocations. In this case, one speaks of *sub-grain boundaries*. The example of a tilt sub-grain boundary is shown in the diagram of Fig. 5.21.

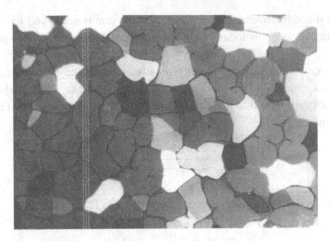

Fig. 5.20 Polycrystal of the smectic B phase, obtained by cooling from the isotropic liquid (polarised microscopy observation of a thin sample sandwiched between two glass plates). The grain size is of a few hundred micrometres.

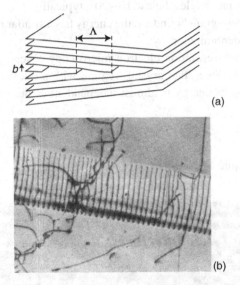

Fig. 5.21 (a) Sub-grain boundary of edge dislocations. The disorientation angle θ is directly related to the density of dislocations; (b) microscopy observation of a tilt sub-grain boundary in a thin sample of stainless steel. One can see various dislocations, in particular those of a sub-grain boundary crossed by the two surfaces of the thin slice [12] (photograph by G. Dupouy and F. Perrier).

In this case, the wall is formed by a collection of parallel edge dislocations, all of the same sign and regularly spaced. Let b be their Burgers vector and Λ the distance between them. One has immediately:

$$\tan \theta = \frac{b}{\Lambda}, \tag{5.52}$$

where θ is the disorientation angle between the two grains separated by the wall. Note that, by replacing the edge dislocations by screw dislocations, a twist sub-grain boundary is obtained.

In both these cases, the stress vanishes typically at a distance Λ from each dislocation, due to the presence of the other dislocations: one must then take $R \approx \Lambda$ when calculating the energy of dislocations within the wall (Eq. (5.36)). The surface energy of a tilt sub-grain boundary is then:

$$E_{\text{tilt}} \approx \frac{\mu b^2}{4\pi(1-\nu)\Lambda} \ln\left(\frac{\Lambda}{r_{\text{c}}}\right). \tag{5.53}$$

For small values of θ, this formula becomes, with $r_{\text{c}} \approx b$:

$$E_{\text{tilt}} \approx \frac{\mu b}{4\pi(1-\nu)}\theta|\ln\theta|. \tag{5.54}$$

The energy increases very rapidly with the angle θ. For this reason, such a model is only relevant in the case of small angles, below $10-20°$, typically.

Note that for a twist sub-grain boundary, the energy has a similar expression, but without the $1 - \nu$ factor in the denominator.

For larger angles, high-resolution electron microscopy observations (Fig. 5.22) show that the boundary resembles a 'bad crystal' region, with a thickness of the order of $2b$. Under these conditions, its energy is almost independent of the angle θ, being typically given by:

$$E_{\text{gb}} \approx 2\gamma, \tag{5.55}$$

where γ is the solid–liquid surface energy.

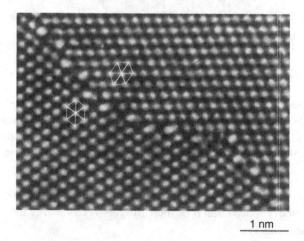

1 nm

Fig. 5.22 Grain boundary in silicon observed in high-resolution microscopy [12] (photograph by A. Bourret).

It is noteworthy that impurity atoms are mostly concentrated at the grain boundaries (and also close to the dislocation cores), since that is where they cost the least elastic energy. On the other hand, the system loses in mixing entropy; the balance between the entropy loss and the gain in elastic energy determines the ratio of impurity concentrations within the grains and at the grain boundaries [12].

One way of measuring the energy of a grain boundary consists of placing a polycrystalline sample in a temperature gradient such that it is molten at one end and solid at the other. A solid–liquid interface appears, which is crossed by grain boundaries. Optical (or electron) microscopy observation shows that the solid–liquid interface exhibits grooves at the junctions with grain boundaries. The shape and depth of the grooves depend on the temperature gradient and on the energy of the grain boundary. Such a situation in a hexagonal columnar liquid crystal is shown in the photograph of Fig. 5.23. One can clearly see two grooves of different depth, ending at two grain boundaries which also appear to have very different contrast.

Each groove is characterised by its depth h and the contact angle α at the grain boundary (Fig. 5.23b). These two parameters are accessible experimentally. One can show, using the Gibbs–Thomson law relating the temperature T_i of the solid–liquid interface to its local curvature radius R (taken as positive when the centre of curvature of the interface is in the solid phase):

$$T_i = T_m - \frac{\gamma T_m}{R\,\Delta H} \tag{5.56}$$

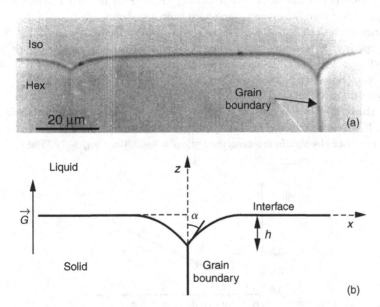

Fig. 5.23 (a) Grooves formed at the junction of two grain boundaries with the solid–liquid interface. Here, the 'solid' is a hexagonal columnar phase in homeotropic alignment. The hexagonal lattice is contained in the plane of preparation. The material is a triphenylene hexaether, known as C8HET (Oswald, 1988); (b) diagram of a groove formed at the junction of a grain boundary and the solid–isotropic liquid interface in a temperature gradient.

that the depth h is given by the following relation:

$$h = \sqrt{\frac{2\gamma T_m}{G \Delta H}} (1 - \sin \alpha), \tag{5.57}$$

where G is the temperature gradient, T_m the melting temperature and ΔH the latent heat of melting per unit volume.

On the other hand, the equilibrium of the triple point at the bottom of the groove requires:

$$E_{gb} = 2\gamma \cos \alpha. \tag{5.58}$$

The first relation yields γ and the second one, the energy of the grain boundary. Note that, by writing these two relations, we completely ignored the anisotropy in surface tension (which might not be completely negligible, reaching up to 20% in some very anisotropic materials, such as columnar phases or lamellar plastic crystals).

This method was used successfully in liquid crystals and in certain transparent plastic crystals for measuring the grain boundary energy as a function of the disorientation angle θ. The results of measurements on the columnar hexagonal phase of C8HET are shown in Fig. 5.24. Here, one encounters weakly disoriented tilt boundaries ($\theta < 20°$), such as those shown in the photograph of Fig. 5.23. This graph shows that $\theta |\ln \theta|$ is proportional to $\cos \alpha$, where α is the angle at the bottom of the groove, in agreement with the theoretical prediction (5.54).

For $\theta > 20°$, the angle α is close to 0, meaning that the energy of strongly disoriented grain boundaries is more or less constant, of the order of 2γ (Eq. (5.55)). In C8HET, $\gamma \approx 5 \times 10^{-4}$ J/m^2, yielding $E_{gb} \approx 10^{-3}$ J/m^2 for strongly disoriented grain boundaries.

Similar experiments can be performed in metals, as shown in the micrograph of Fig. 5.25. In this case, the sample is a thin bismuth film deposited by evaporation between two thin carbon layers. This sandwich is then mounted on a heating stage and observed in transmission electron microscopy. The dislocations of the sub-grain boundary can be resolved.

Unfortunately, this technique is very difficult to use. For this reason, metallurgists have rather studied the shape of grooves appearing at the free surface of a crystal where it crosses the grain boundaries. If the surface tension of the crystal γ_{air} is known, the formula (5.58) remains valid provided γ is replaced by γ_{air}. In this case, the angle α is finite, since $\gamma_{air} > 2\gamma$. The surface tension

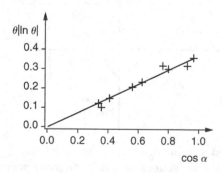

Fig. 5.24 $\theta |\ln \theta|$ as a function of $\cos \alpha$. θ is the disorientation angle across the grain boundary and α the angle at the bottom of the groove. One is dealing here with a hexagonal columnar phase (Oswald, 1988).

Table 5.2

Metal	T (°C)	γ_{air}(J/m^2)	E_{gb}(J/m^2)
Cu	850	1.64	0.59
Ag	750	1.31	0.46
Au	850	1.48	0.37
Sn	213	0.685	0.16
Fe α	1100	1.95	0.79

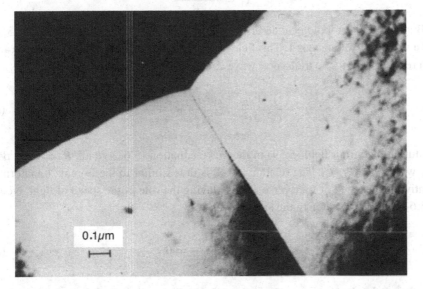

Fig. 5.25 Intersection of a weakly disoriented sub-grain boundary with the solid–liquid interface in bismuth (Woodruff, 1973) (photograph by M. E. Glicksman and C. Vold).

with air is obtained from high-temperature '*zero creep*' experiments of the Nabarro–Herring type (Section 5.3.4.1) on metal wires. More precisely, one determines the length of the wire for which the action of the surface tension exactly compensates that of weight. Values (in J/m^2) for γ_{air} and E_{gb} of metals (for strongly disoriented boundaries) are given in Table 5.2.

Note the difference in order of magnitude between these values and those given for C8HET (a factor of 1000, typically).

5.2 Dynamics of dislocations and elastic limit

In this section, we will study some dynamic properties of dislocations and discuss the origin of the elastic limit in solids.

5.2.1 Mass and vibration frequency of a dislocation

At what velocity can a dislocation propagate? Can one assign to it an effective mass?

To answer these two questions, consider a screw dislocation propagating at a constant velocity in a crystal (stationary motion).

It is easily seen that the strain field:

$$u_1 = u_2 = 0 \quad \text{and} \quad u_3 = \frac{b}{2\pi}\left[\arctan\left(\frac{x_2}{x_1^*}\right) + k\pi\right] \quad (k = 0, 1 \text{ or } 2), \tag{5.59}$$

where:

$$x_1^* = \frac{x_1 - Vt}{B} \tag{5.60}$$

with B constant, is a topological description of a screw dislocation, parallel to axis 3, moving at a velocity V along axis 1. This field must obey the dynamic Navier equation (4.197), which reduces here to the transverse wave equation (4.211), since div $\vec{u} = 0$:

$$\frac{1}{C_t^2}\frac{\partial^2 u_3}{\partial t^2} = \frac{\partial^2 u_3}{\partial x_1^2} + \frac{\partial^2 u_3}{\partial x_2^2}. \tag{5.61}$$

Substituting the strain field (5.59) in the wave equation (5.61) yields $B = (1 - \beta^2)^{1/2}$, where we set $\beta = V/C_t$. The relation (5.60) is thus similar to the Lorentz transformation in relativity, the speed of transverse waves playing the role of the speed of light. Note that Eq. (5.61) can be rewritten in the simplified form:

$$\frac{\partial^2 u_3}{\partial x_1^{*2}} + \frac{\partial^2 u_3}{\partial x_2^2} = 0, \tag{5.62}$$

by applying the Lorentz transformation (5.60), immediately showing that the strain field (5.59) is the solution to the problem.

From the strain field (5.59) one can determine the stress tensor:

$$\sigma_{13} = -\frac{\mu b B}{2\pi}\frac{x_2}{(x_1 - Vt)^2 + B^2 x_2^2}, \qquad \sigma_{23} = \frac{\mu b B}{2\pi}\frac{x_1 - Vt}{(x_1 - Vt)^2 + B^2 x_2^2}. \tag{5.63}$$

These formulas show that the shear stress σ_{13} diverges in the plane $x_1 = Vt$ for $B \to 0$, i.e. for $V \to C_t$. This result shows that *a screw dislocation cannot propagate faster than the velocity of the transverse waves C_t*.

Another interesting quantity is the kinetic energy of the dislocation. By definition, it is given by:

$$E_c = \int \frac{1}{2}\rho \dot{u}_3^2 \, dV, \tag{5.64}$$

where the integration is done over the entire sample volume. Using the expression (5.59) for the strain field, one has (in the reference system of the dislocation and taking into account the core radius of the dislocation):

$$E_c = \int_{r_c}^{R} \int_{0}^{2\pi} \frac{b^2 V^2}{4\pi^2} \frac{B^2 x_2^2}{(B^2 x_2^2 + x_1^2)^2} r \, dr \, d\theta, \tag{5.65}$$

yielding by integration:

$$E_c = E_0 \frac{1 - B^2}{2B} \tag{5.66}$$

with $E_0 = \frac{\mu b^2}{4\pi} \ln\left(\frac{R}{r_c}\right)$ the elastic energy of the dislocation at rest (see Eq. (5.26)). Once again, it becomes apparent that the kinetic energy of the dislocation diverges as its velocity approaches C_t.

One can also calculate the elastic energy and show that it is given by:

$$E_e = E_0 \frac{1 + B^2}{2B} \tag{5.67}$$

with the same divergence as $V \to C_t$.

It follows that the total energy of the dislocation has the following simple expression:

$$E_t = E_c + E_e = \frac{E_0}{B}. \tag{5.68}$$

If the velocity of the dislocation is much smaller than C_t, its total energy can be expanded as:

$$E_t = E_0 \left(1 + \frac{V^2}{2C_t^2}\right). \tag{5.69}$$

With:

$$E_t = E_0 + \frac{1}{2} m_0 V^2 \tag{5.70}$$

one finds the '*dislocation mass at rest*' m_0:

$$m_0 = \frac{E_0}{C_t^2} = \frac{\rho b^2}{4\pi} \ln\left(\frac{R}{r_c}\right). \tag{5.71}$$

Thus, the dislocation mass at rest per unit length is of the order of ρb^2, yielding per atom along the line a mass of ρb^3, i.e. more or less the mass of an atom.

One can also define the '*relativistic mass*' of a dislocation:

$$m = \frac{E_t}{C_t^2} = \frac{m_0}{B}. \tag{5.72}$$

This quantity diverges as $V \to C_t$.

Note however that, in practice, dislocations ordinarily propagate at velocities much lower than the speed of sound, such that 'relativistic' effects are almost always negligible.

Only in the case of extremely violent deformations (during an explosion, for instance) do they start to play a role.

A similar line of reasoning can be followed for an edge dislocation. As one might expect, the calculations are much more complicated, although in principle very similar. There are however a few important differences: for instance, it can be shown that the energy still diverges as the velocity of the dislocation approaches the speed of transverse waves, but this divergence is no longer 'relativistic' (in $1/B$), but in $1/B^3$ (Weertman, 1961).

One can also show that the deformation field around a screw dislocation does not couple to plane waves, so that it emits no waves as it moves at a constant velocity. Edge dislocations in stationary motion, on the other hand, emit waves and steadily lose energy by radiation [18].

To conclude this paragraph, let us determine the vibration frequency and amplitude for a dislocation segment trapped on two anchoring points, A and B. In practice, these anchoring points can be precipitates, dislocation crossings, etc. This calculation is important for its applications in certain models of plastic deformation, such as that described in Section 5.3.3.2.

To solve this problem, let us reason as for a vibrating string. Fortunately, we already know the effective mass m of the dislocation, as well as its line tension τ, equal to its line energy in an isotropic material. Consider the central line element of length dl and mass dm (Fig. 5.26). This element is subjected to an inertial force and to a line tension force proportional to τ and to the local curvature $1/R$, such that its dynamic equation is simply:

$$dm\frac{\partial^2 x}{\partial t^2} = -\frac{\tau}{R}\,dl,\qquad(5.73)$$

where x stands for the transverse displacement of element dl.

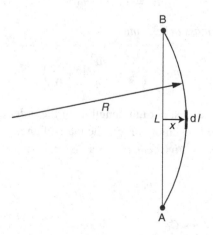

Fig. 5.26 Dislocation segment trapped at both ends; it vibrates under the action of thermal fluctuations.

Let L be the distance between anchoring points. If the vibration amplitude is small $(x \ll L)$, one has:

$$x \approx \frac{L^2}{8R}.$$ (5.74)

On the other hand, $dm = m_0\, dl \approx \rho b^2\, dl$ (neglecting relativistic effects), so that the dynamic equation (5.73) can be written as:

$$\frac{\partial^2 x}{\partial t^2} + 8\frac{C_t^2}{L^2}x = 0$$ (5.75)

with $C_t^2 = \mu/\rho$ the speed of transverse waves.

The solution to this equation is written:

$$x = x_0 \cos(2\pi\, v_L t),$$ (5.76)

with

$$v_L = \frac{\sqrt{2}}{\pi}\frac{C_t}{L} \approx \frac{C_t}{L}.$$ (5.77)

The vibration frequency of the segment is thus proportional to the reciprocal of the time needed by a transverse wave to travel the distance L between anchoring points.

As an estimate for this vibration frequency, let us take $L = 10$ μm, $\rho = 3000$ kg/m^3 and $\mu = 2.5 \times 10^{10}$ Pa. This yields $v_L \approx 1.3 \times 10^8$ s^{-1}, a relatively high value (but still much lower than the Debye frequency, of the order of 10^{13} s^{-1}).

The vibration amplitude can also be determined. For this calculation, it is enough to write that the increase in line energy as it elongates by ΔL is of the order of $k_B T$. Knowing that:

$$\Delta L \approx 2\left(\sqrt{\frac{L^2}{4} + x_0^2} - \frac{L}{2}\right) \approx \frac{2x_0^2}{L},$$ (5.78)

and that the line energy is typically μb^2 per unit length, one finally obtains the vibration amplitude:

$$x_0 \approx \sqrt{\frac{Lk_B T}{2\mu b^2}}.$$ (5.79)

With the numerical values already used above, one has $x_0 = 35$ Å, which is very small. This estimate allows us to check a posteriori that 'relativistic' effects are negligible, since the velocity of the line $v \approx v_L x_0 \approx 0.5$ m/s remains very small with respect to the speed of sound ($C_t \approx 3000$ m/s).

5.2.2 *Peach and Koehler force on a dislocation*

We will now discuss a fundamental plasticity concept. One can prove (see Appendix 5.A) that a dislocation placed in an *external stress field* $\underline{\sigma}$ experiences a *configuration force*, bearing the names of *Peach and Koehler*, given by (Peach and Koehler, 1950)[3]:

$$\vec{f} = \left(\underline{\sigma}\vec{b}\right) \times \vec{t}, \tag{5.80}$$

where $\vec{b}$ and $\vec{t}$ are respectively the Burgers vector and the unit vector tangent to the line, as defined in Section 5.1.2.2. This force resembles the Magnus force acting on a vortex in hydrodynamics, but the underlying mechanisms are very different.

To illustrate the importance of the Peach and Koehler force, let us examine the particular case of a dislocation loop contained in an elastic block under the action of an external shear stress σ (Fig. 5.27a). Suppose the Burgers vector of the dislocation is contained in the plane of the loop. The Peach and Koehler formula shows that each element of unit length of the loop experiences a force, with amplitude σb, pointing outwards or inwards depending on the sign of b. This is a *glide force*, as it is contained in the *glide plane* of the dislocation which is, by definition, the plane containing the line and its Burgers vector. These forces tend to stretch or shrink the loop, leading to plastic deformation. Note that glide is *conservative*, meaning that there is no need for mass transport by diffusion to move the dislocation. That is why this mechanism works at low temperature (i.e. far from the melting point of the material). It is also noteworthy that the shear stress exerts no force on the vertical edge dislocation. One can conclude that an applied stress does not have the same action on all the dislocations. In plasticity one must therefore make a distinction between *mobile dislocations*, which move under the action of applied forces, and the others, sometimes termed *trees*, which experience no force and remain static.

Another experimental situation is depicted in the diagram of Fig. 5.27b. In this case, the material is compressed. A loop similar to that described above would experience no force. On the other hand, an edge dislocation loop with Burgers vector perpendicular to the plane of the loop will experience a force of amplitude σb pointing inwards or outwards, depending on the sign of b. The motion now occurs in the *climb plane* of the dislocation, which is the plane containing the line and perpendicular to the Burgers vector. This motion, termed *climb*, is *not conservative*, since local matter transport must take place (by diffusion). For this reason, this process is only really effective at high-temperature, close to the melting point. Finally, we emphasise that, as in the case discussed above, there are some dislocations that experience no force: for instance, this would be the case for a screw dislocation perpendicular to the loop, which will behave as a tree.

---

[3] Note that this formula is only valid for a symmetric stress tensor. Also keep in mind that, strictly speaking, one should use in this equation the stress deviator $\underline{\sigma} - (\text{tr}\underline{\sigma}/3)\underline{I}$ instead of $\underline{\sigma}$. For simplicity's sake, we will use this formulation in the following. These two points are discussed further in Appendix 5.A.

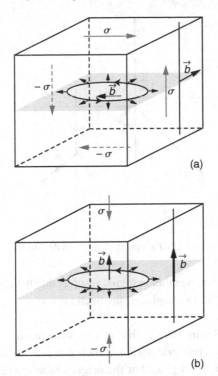

Fig. 5.27 Dislocations in an elastic block under the action of shear (a) or compression (b). A distinction must be made between mobile dislocations (here, the two loops) and trees, which do not move (the two vertical dislocations). The shaded surfaces correspond to the glide plane (a) and the climb plane (b) of the loop, respectively.

5.2.3 Interactions between dislocations

The Peach and Koehler formula also provides a way of calculating the interaction forces between dislocations [18]. These forces play a crucial role in plasticity, as we shall see below.

For simplicity's sake, let us consider the case of straight dislocations.

- The simplest situation is that of two parallel screw dislocations (Fig. 5.28). The first one, with Burgers vector b_1, coincides with the x_3 axis. The second one, with Burgers vector b_2, is at a distance r from the first. They have the same orientation ($\vec{t}_1 = \vec{t}_2 = \vec{x}_3$), while $\vec{b}_1 = b_1\vec{t}_1$ and $\vec{b}_2 = b_2\vec{t}_2$. With these conventions, dislocation 1 engenders a stress field with non-zero components $\sigma_{\theta3}^1 = \sigma_{3\theta}^1 = \mu b_1/2\pi r$, acting on dislocation 2. Using the Peach and Koehler formula, it is easily checked that dislocation 1 exerts on dislocation 2 a force:

$$\vec{f}_{1\to2} = \frac{\mu}{2\pi r}b_1b_2\vec{e}_r. \tag{5.81}$$

By virtue of the action–reaction principle, dislocation 2 exerts on dislocation 1 an opposite force. We see then that two dislocations repel if their Burgers vectors have the same

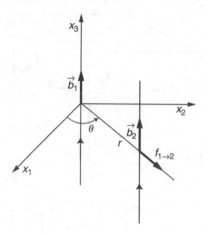

Fig. 5.28 Two parallel screw dislocations.

sign and attract otherwise. Another essential result is that the absolute value of the inter-action force between the two dislocations decreases as $1/r$, irrespective of the polar angle θ.

• Another practically relevant case is that of two straight edge dislocations, parallel to the x_3 axis (Fig. 5.29). Their orientation is such that $\vec{t}_1 = \vec{t}_2 = \vec{x}_3$. The first one, with Burgers vector $\vec{b}_1 = b_1\vec{x}_1$, is placed at the origin of the reference system. The second one, with Burgers vector $\vec{b}_2 = b_2\vec{x}_1$, is at a position (θ, r) with respect to the first. Using the Peach and Koehler formula, as well as the stress field (5.31), yields the following expression for the force exerted by 1 upon 2:

$$\vec{f}_{1\to2} = \frac{\mu}{2\pi(1-\nu)}\frac{1}{r}b_1b_2\left[f(\theta)\vec{x}_1 + g(\theta)\vec{x}_2\right], \tag{5.82}$$

with $f(\theta) = \cos\theta\cos2\theta$ and $g(\theta) = \sin\theta(2 + \cos2\theta)$.

This force is shown in Fig. 5.29a as a function of the position (θ, r) of the second dis-location. Functions $f(\theta)$ and $g(\theta)$ are plotted in Fig. 5.29b. One can see that the force is globally repulsive or attractive when b_1 and b_2 have the same sign or opposite signs, respectively. This force decreases as $1/r$, but its absolute value varies with the angle θ (for a fixed distance r), the quantity $(f(\theta)^2 + g(\theta)^2)^{1/2}$ oscillating between 1 and 1.5. Another conclusion can be drawn if dislocation 2 cannot leave its glide plane. This is the case at low temperature $(T < 0.5\,T_f)$, when molecular diffusion is negligible. Then, the glide force acting on the dislocation vanishes for $\theta = \pi/4$ (modulo $\pi/2$) and for $\theta = \pi/2$ (modulo π). Analysing the graph in Fig. 5.29a shows that the first equilibrium positions are unsta-ble, while the second ones are stable. This result entails that the tilt sub-grain boundaries described in Section 5.1.3 are energetically stable (see Fig. 5.21).

• A slightly more complicated case is that of two perpendicular dislocations. It is always possible, without loss of generality, to choose the x_1 axis parallel to dislocation 2 such that $\vec{t}_1 = \vec{x}_3$ and $\vec{t}_2 = \vec{x}_1$.

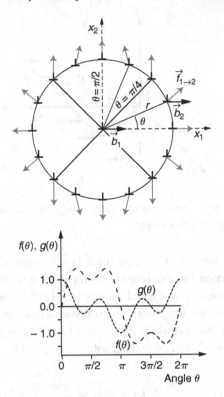

Fig. 5.29 (a) Force between two parallel edge dislocations whose Burgers vectors are parallel and of the same sign; (b) graph of the functions $f(\theta)$ and $g(\theta)$.

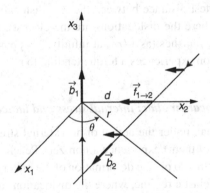

Fig. 5.30 Interaction between two perpendicular screw dislocations.

If the two dislocations are of screw type (Fig. 5.30), with $\vec{b}_1 = b_1\vec{x}_3$ and $\vec{b}_2 = b_2\vec{x}_1$, then the force per unit length exerted by dislocation 1 onto dislocation 2 is given by:

$$\vec{f}_{1\to 2} = -\frac{\mu}{2\pi r}b_1 b_2 \sin\theta\,\vec{x}_2$$

$$= -\frac{\mu d}{2\pi r^2}b_1 b_2 \vec{x}_2, \qquad (5.83)$$

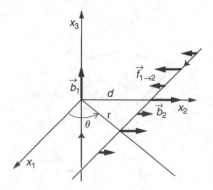

Fig. 5.31 Interaction between a screw dislocation and an edge dislocation perpendicular to it. Note that the edge dislocation experiences a torque.

where θ takes values from 0 to π. This force is not constant along dislocation 2: it goes through a maximum (in absolute value) for $\theta = \pi/2$, the point where the two dislocations are the closest ($r = d$), and vanishes (as $1/r^2$) at large distances. Note that this force is always directed along the shortest distance between the two dislocations.

• Finally, let us consider the case where dislocation 1 is of screw type (as for the previous example) and dislocation 2 is of edge type, with $\vec{t}_2 = \vec{x}_1$ and $\vec{b}_2 = b_2\vec{x}_2$ (Fig. 5.31). One then has:

$$\vec{f}_{1\to2} = \frac{\mu}{2\pi r}b_1 b_2 \cos\theta\,\vec{x}_2 = \frac{\mu\sin2\theta}{4\pi d}b_1 b_2\vec{x}_2, \tag{5.84}$$

where θ takes values between 0 and π. Once again, this force is not constant and it is directed along the shortest distance between the two dislocations. It vanishes for $\theta = \pi/2$ (i.e. for $r = d$, where the dislocations are the closest), and changes sign across this position. Finally, it vanishes (as $1/r^2$) at infinity. This particular force distribution shows that the dislocation experiences a torque tending to rotate it in its glide plane.

5.2.4 Dislocation glide: internal stress and lattice friction

We have already shown that, under the action of an external stress field, the dislocations (if not all, certainly some of them) experience a non-zero Peach and Koehler force which tends to displace them, leading to plastic deformation of the material. However, experiment shows the existence of an elastic regime, where the dislocations are practically immobile, although acted upon by a finite force. *They must therefore encounter obstacles to their propagation.*

We have already mentioned one of these obstacles, namely the internal stress field created by the dislocations themselves. Statistically, this field oscillates between σ_M and $-\sigma_M$ over a typical distance ξ equal to the average distance between dislocations. Since ξ is a macroscopic scale (several micrometres, in general), the dislocations cannot cross these maxima and propagate unless the external stress itself exceeds $\sigma_i = |\sigma_M|$ (thermal activation is not effective due to the energy scales involved). It follows that σ_i represents an

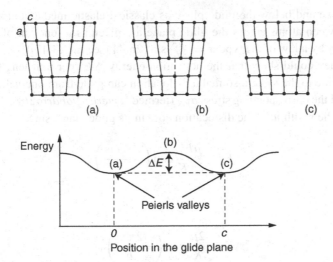

Fig. 5.32 Motion of an edge dislocation in a crystal and the corresponding energy landscape.

elasticity threshold for the material. To estimate its order of magnitude, let us consider the case of a thoroughly annealed material, with $\xi \approx 10$ μm. One then has:

$$\sigma_i \approx \frac{\mu b}{2\pi(\xi/2)} \approx 0.3\frac{\mu b}{\xi} \approx 2 \times 10^5 \text{ Pa},\qquad(5.85)$$

using $\mu \approx 2 \times 10^{10}$ Pa and $b = 3$ Å. This value is generally too small to account for the elastic limit of solids. Indeed, while it is not too far from the values measured in well-annealed monocrystals of aluminium or copper, it is no longer adequate for polycrystalline materials. It is also much too small to explain the elastic limit of ionic crystals (e.g. minerals) or that of covalent materials with directed bonds, such as diamond or silicon.

These materials must therefore raise other kinds of obstacles to the propagation of dislocations. One of them originates in the crystal lattice itself.

To understand this obstacle, let us contemplate for a moment the diagram in Fig. 5.32. It becomes apparent that, in order to glide over one atomic distance from (a) towards (c), the dislocation must go through the intermediate position (b). In general, the line energy is not the same in (b) as in (a) or (c), because the structure of the core, which is a bad crystal zone, changes during glide. Positions (a) and (c) have minimal energy and correspond to two stable equilibrium positions for the dislocation ('*Peierls valleys*'); on the other hand, the energy of the dislocation goes through a maximum in position (b). One must therefore inject into the system an additional energy ΔE if the line is to jump from one Peierls valley to the next. This implies that the force acting on the dislocation exceeds the minimal amplitude $\sigma_{PN}b$, where σ_{PN} is the stress applied in the glide plane of the dislocation.

Clearly, σ_{PN} depends on the glide plane under consideration. Thus, experiment shows that the stress σ_{PN} is lower in the dense planes than in the others. To show this, Peierls, and then Nabarro (Peierls, 1940; Nabarro, 1947; [18]), calculated ΔE and σ_{PN} by considering the media above and below the glide plane as classical elastic media, and assuming that the forces between atoms across the glide plane are given as a function of their relative displacement u by a law of the type $\sigma = \frac{\mu c}{2\pi a} \sin\left(2\pi \frac{u}{c}\right)$ (see Eq. (4.47)).

This way they could show that the activation energy ΔE (per unit length of the line) one must provide to the system so that a dislocation can glide from one valley to the next (Fig. 5.32) and the corresponding stress σ_{PN} (termed '*Peierls–Nabarro stress*') are strongly dependent on the width w of the dislocation core in its glide plane, since:

$$\Delta E \approx \frac{\mu b^2}{2\pi K} \exp\left(-\frac{2\pi w}{b}\right) \tag{5.86}$$

and

$$\sigma_{PN} \approx \frac{2\mu}{K} \exp\left(-\frac{2\pi w}{b}\right), \tag{5.87}$$

where $K = 1$ or $1 - v$, for a screw or edge dislocation, respectively.

The width, defined in Fig. 5.33 in the case of an edge dislocation, can be calculated in the framework of this model and is typically:

$$w \approx \frac{\mu b}{2\pi K \sigma_c^{theo}}, \tag{5.88}$$

where σ_c^{theo} is the theoretical elastic limit given by Eq. (4.48).

This dependence of the Peierls–Nabarro stress on the width of the dislocation can be understood qualitatively by tracing on the same sketch (Fig. 5.34) the atomic positions in the equilibrium configurations (a) and (c) defined in Fig. 5.32. When the dislocation is 'narrow' (small w) the atoms are far enough from each other that the core energy in the intermediate configuration (b) must be 'considerably' higher than in configurations (a) and (c): the stress σ_{PN} is large. Conversely, when the dislocation core is wide (large w), the atoms are almost touching, so a very small displacement is needed to go from (a) to (c). The energy of the intermediate configuration (b) must then be very close to that of configurations (a) and (c): the stress σ_{PN} is small.

One can also show that w, and hence σ_{PN}, are strongly dependent on the nature of the interatomic bonds. If they are strongly directed (as in covalent crystals, such as diamond or semiconductors) w is small ($w \approx c$) and μ is large (see Table 4.1 in Section 4.1.4.3): σ_{PN} is thus large in these materials, which are *intrinsically hard*, since it is difficult to displace the dislocations from their Peierls valleys. Ionic crystals, such as minerals, also have high Peierls–Nabarro stress values, since it is difficult to bring similarly charged ions together due to the electrostatic repulsion. Therefore, these materials are also intrinsically hard. Note that in these materials formulas (5.86) to (5.88) are not quantitatively valid, since the basic assumptions for the Peierls–Nabarro calculation are not fulfilled.

However, this is not the case for most metals, such as copper, gold or aluminium (with an f.c.c. structure), where the model applies. In these materials the bonds are only weakly

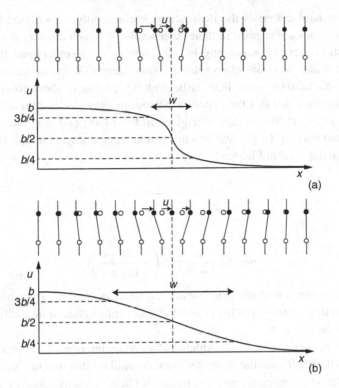

Fig. 5.33 Narrow core (a) and wide core (b) edge dislocations (from Cottrell, 1957). In each case, we represented the displacement u of the atoms in the upper row with respect to those in the lower row. The width w is defined such that $u(\pm w/2) = b/2 \mp b/4$.

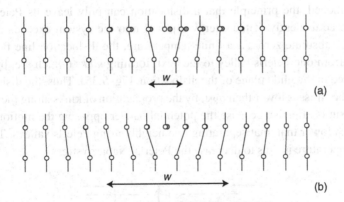

Fig. 5.34 Atom positions in the equilibrium configurations (a) and (c) defined in Fig. 5.32 (white and grey dots, respectively) when the dislocation has a narrow core (a) and a wide core (b) [12].

directed, the elastic moduli are 'low' and the dislocation cores tend to spread: these metals are *intrinsically soft* (σ_{PN} small). Note that iron (with a b.c.c. structure) represents an exception, the Peierls–Nabarro stress being important.

Note that in hard materials the dislocations become polygonised (see Fig. 5.16 and Fig. 5.40) since they are trapped in their Peierls valleys, which are very deep. This effect adds to the action of crystal anisotropy. In soft materials, on the other hand, the Peierls valleys are shallow and the dislocations seldom follow them: this further decreases the effect of the Peierls–Nabarro stress on their elastic limit (by a factor of about 1000 [16]).

It should also be noted that the structural anisotropy plays a crucial role. Indeed, let us replace σ_c^{theo} in Eq. (5.88) by the value given in Eq. (4.48) and plug the expression for w thus obtained into Eq. (5.87). We find that, for an edge dislocation with Burgers vector $b = c$, like that depicted in Fig. 5.32:

$$w \approx \frac{a}{1 - \nu} \tag{5.89}$$

and

$$\sigma_{\text{PN}} \approx \frac{2\mu}{1 - \nu} \exp\left(-\frac{2\pi}{1 - \nu}\frac{a}{c}\right). \tag{5.90}$$

This expression shows that the Peierls–Nabarro stress decreases exponentially with the structural anisotropy ratio a/c. This is why the dislocations glide more easily in the dense planes than in the others.

This anisotropy dependence also explains why certain intermetallic compounds such as $CuAl_2$ are brittle (their metallic character notwithstanding) due to a low a/c, while other materials, of lamellar structure, such as smectic B plastic crystals, soaps or graphite, flow very easily under shear parallel to the layers due to a large a/c ratio. This feature makes them usable as *solid lubricants*.

Finally, let us point out that the intrinsic hardness of a material generally decreases with temperature. Indeed, the principle that a dislocation can only leave its Peierls valley if $\sigma > \sigma_{\text{PN}}$ (or, equivalently, if the energy injected into the system exceeds ΔE) is only strictly valid at absolute zero. At a finite temperature, the dislocation line fluctuates and it can 'jump' from one Peierls valley to the next, forming *kinks* (which are, by definition, 'jogs' contained in the glide plane of the dislocation, Fig. 5.35). Thus, the dislocations can glide even if the stress is lower than σ_{PN}, by the propagation of kinks along the dislocation. This mechanism is effective because the potential barrier opposing the motion of kinks is generally much lower than that separating two neighbouring Peierls valleys. The effect of increasing temperature is thus to decrease the Peierls–Nabarro stress.

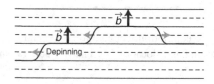

Fig. 5.35 Kinks on an edge dislocation trapped in several adjacent Peierls valleys. By propagating along the line, the kinks allow the dislocation to move from one Peierls valley to the next, even if the applied stress is below the Peierls–Nabarro threshold.

5.2.5 Dislocation crossing: forest hardening

Dislocation crossings represent a further obstacle to their propagation. This phenomenon is inevitable when a mobile dislocation encounters a *tree* of the *forest* (the set of all dislocations not acted upon by the external stress). The question is then whether the mobile dislocation will be able to cross the obstacle.

The answer depends on the Burgers vectors and on the nature of the interacting dislocations.

Let us illustrate this in the particular case of a mobile edge dislocation (with Burgers vector $\vec{b}_1$) crossing an immobile screw dislocation (with Burgers $\vec{b}_2$) (Fig. 5.36). This situation might occur in a shear experiment such as that shown in Fig. 5.27a). The topological analysis of the strain fields shows that, after crossing, each dislocation bears an edge jog parallel with and of the same magnitude as the Burgers vector of the other dislocation. Thus, in the example of Fig. 5.36, the screw dislocation bears an edge jog of length b_1, while the edge dislocation bears an edge jog, of length b_2. The order of magnitude of the jog energy is μb^3, which amounts to about 1 eV in metals at room temperature.

This energy represents a barrier that the mobile dislocation must overcome by thermal activation in order to keep on moving. However, it is much higher than the thermal energy $k_B T$, which is of the order of 1/40 eV at room temperature. Therefore, dislocation crossings constitute very serious obstacles to their propagation and, in certain situations, can be the determining factor in plastic deformation. A creep model based on thermally activated barrier crossing will be presented in Section 5.3.3.2.

In the example above, the dislocations cross, and then separate; this is not always the case, as shown in Fig. 5.37. It can happen that two dislocations with Burgers vector $\vec{b}_1$ and $\vec{b}_2$ form a junction with Burgers vector:

$$\vec{b}_3 = \vec{b}_1 + \vec{b}_2 \tag{5.91}$$

as they cross (Frank, 1955). This junction forms if energetically favourable, or more precisely if its energy, proportional to $\vec{b}_3^2 = \vec{b}_1^2 + \vec{b}_2^2 + 2\vec{b}_1 \cdot \vec{b}_2$, is lower than $\vec{b}_1^2 + \vec{b}_2^2$. It is then necessary that $\vec{b}_1 \cdot \vec{b}_2 < 0$: under these conditions, the junction is called *attractive*. Such junctions, very common in f.c.c. and b.c.c. structures, are difficult to undo and can block

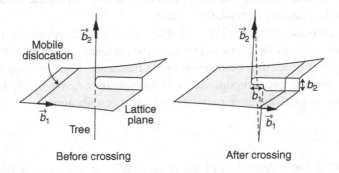

Before crossing   After crossing

Fig. 5.36 After crossing, an edge and a screw dislocation bear jogs.

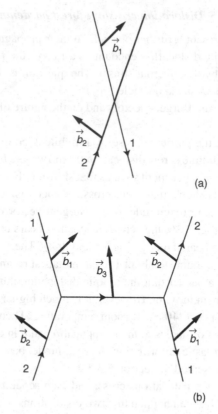

Fig. 5.37 Attractive junction between two dislocations. (a) Before junction; (b) after junction, a dislocation segment with Burgers vector $\vec{b}_3 = \vec{b}_1 + \vec{b}_2$ has formed.

the motion of mobile dislocations completely. Thus, they lead to a significant increase in the elastic limit of the material.

5.2.6 Frank–Read sources

In this section we will describe the most important mechanism for the multiplication of dislocations in the plasticity of crystalline solids.

To this end, consider a segment AB of a screw dislocation trapped at its two ends, for instance on two nodes of the Frank network. In the presence of a shear stress σ parallel to AB, the segment bends under the action of the Peach and Koehler force. This force is opposed by a restoring force proportional to its line tension τ (given by Eq. (5.48)) and to the reciprocal of its local curvature radius R, so that, at equilibrium:

$$\sigma b = \frac{\tau}{R}. \tag{5.92}$$

If τ is constant and independent of the orientation θ of the segment with respect to the crystal axes, the dislocation segment takes the shape of a circular arc, with radius (Fig. 5.38):

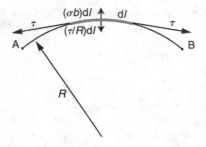

Fig. 5.38 Equilibrium shape of a dislocation segment trapped at its two ends.

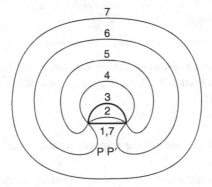

Fig. 5.39 Successive steps leading to the creation of a dislocation loop by a Frank–Read source.

$$R = \frac{\tau}{\sigma b} \approx \frac{\mu b}{\sigma},$$
(5.93)

where we took $\tau \approx \mu b^2$.

Can this equilibrium be maintained as the stress increases? Obviously, the answer is no, since the radius of curvature of the dislocation segment cannot be smaller than $L/2$, where $L = AB$ is the distance between the two anchoring points. This observation sets the critical stress value σ_{FR} above which mechanical equilibrium is broken:

$$\sigma_{FR} = \frac{2\tau}{Lb} \approx \frac{2\mu b}{L}.$$
(5.94)

Figure 5.39 shows the behaviour of a dislocation segment for $\sigma > \sigma_{FR}$. The segment keeps on bending until the two points P and P′ at the back of the curve meet. The two facing line segments can then coalesce, as they are of opposite signs: a big loop appears, which keeps on growing, while the initial segment re-forms. The process can then start again, engendering a new loop inside the first one, etc. This is a *Frank–Read source* (Frank and Read, 1950). Figure 5.40 shows an example of such a source in a silicon crystal. Note that the loops are not circular; their anisotropic shape is due to the anisotropy of

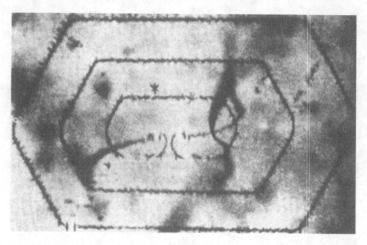

Fig. 5.40 Frank–Read source in a silicon monocrystal. The dislocations were revealed by the deco-ration technique described in Section 5.1.2.3, using copper precipitates. The photograph is taken in infrared light (Dash, 1957; [15]).

the crystal and to the dislocations getting trapped in their Peierls valleys (τ is a strongly anisotropic function of θ).

The stress σ_{FR} provides a good estimate of the elasticity threshold of a monocrystal in the case of intrinsically soft materials, such as aluminium or copper (where lattice friction is negligible).

One can draw an analogy between Frank–Read sources and the mechanism of formation of soap bubbles by blowing through a straw dipped in soapy water.

In the next section, we will discuss the case of polycrystalline materials and see how the elasticity threshold changes.

5.2.7 Grain boundary hardening

We have just seen how Frank–Read sources work under stress. These sources play a crucial role, in that they provide the crystal with new dislocations, sustaining the plastic defor-mation. Indeed, let us imagine for a moment that no new dislocations are created. In this case, the dislocations initially present in the sample will end up 'stranded' at the surface of the crystal where they will form steps, halting the plastic deformation due to lack of dislocations.

Clearly, this reasoning applies to a monocrystal. In a polycrystal, the dislocations must also cross grain boundaries before they can reach the surface of the crystal. Experiment shows that this crossing is not easy and that, in many cases, the necessary stress is much larger than that needed to activate the Frank–Read mills. Let σ_{GB} be the minimum stress one must apply to a dislocation so that it crosses the grain boundary, and let ϕ be the average grain size. Suppose that a Frank–Read source is under the action of an external stress σ larger than σ_{FR}. Under these conditions, the source will emit dislocation loops

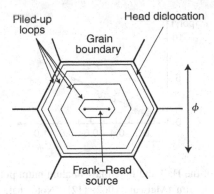

Fig. 5.41 Schematic representation of a dislocation pile-up generated by a Frank–Read source in a polycrystal grain.

that will pile up in the grain (Fig. 5.41). If the head dislocation cannot cross the boundary, the source will eventually stop, since the loops cannot pile up indefinitely (they repel each other, all having the same sign).

In theory, this occurs if the stress σ_{HD} acting on the head dislocation is below σ_{GB}. This stress is the sum of the external stress and the stress fields produced by the other dislocations. J. D. Eshelby, F. C. Frank and F. R. N. Nabarro showed in 1951 that it is proportional to the number n of dislocations in the pile-up and to the applied stress σ (Eshelby et al., 1951; [12, 16, 18]):

$$\sigma_{HD} = n\sigma. \tag{5.95}$$

This fundamental relation shows that the pile-up behaves as a *stress accumulator*.

Another result, which we will also admit without proof, is that the number of dislocations in the pile-up is proportional to the grain size ϕ and to the applied stress σ:

$$n = C\phi\sigma, \tag{5.96}$$

where C is a constant (close to $\pi(1 - v)/2\mu b$). Replacing n by its expression in Eq. (5.95) yields the head dislocation stress as a function of the average grain size ϕ and the applied stress σ:

$$\sigma_{HD} = C\phi\sigma^2. \tag{5.97}$$

This stress value is below σ_{GB} if

$$\sigma < \sqrt{\frac{\sigma_{GB}}{C\phi}}. \tag{5.98}$$

This relation defines a new stress threshold. By adding to it the value σ_c(mono) needed for dislocation glide within the monocrystal (which must exceed the Peierls–Nabarro stress

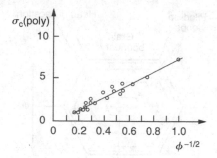

Fig. 5.42 Experimental test of the Hall and Petch law in aluminium polycrystals. The elastic limit is given in kgf/mm^2 and ϕ in μm (McLean, 1962; [12]). Note that, at the scale of the graph, σ_c(mono) ≈ 0.

σ_{PN} or the inner stress σ_i), one obtains the elastic limit of the polycrystal:

$$\sigma_c(\text{poly}) = \sigma_c(\text{mono}) + \frac{\text{Cst}}{\sqrt{\phi}}, \qquad (5.99)$$

where Cst is a constant. This is the *Hall and Petch law*, predicting that the smaller the grains, the harder a polycrystal is.

This law is well verified experimentally, as shown in Fig. 5.42. Note that grain boundary hardening is considerable, reaching a value close to $2 \times 10^{-3}\mu$ for a grain size of the order of a micrometre.

Thus, the task of the metallurgist, in order to obtain a material with the desired hardness, consists in setting the grain size as precisely as possible during material production, for instance by playing with the cooling rate during solidification of the isotropic liquid. Another trick is to make alloys, which can be even harder materials.

5.2.8 Alloy hardening

Metals are seldom used in their pure form. Thus, cast iron is an alloy of iron and carbon. Duralumin (or dural, discovered in 1904 by the German metallurgist A. Wilm) is also an alloy, composed of aluminium and copper (4%). The main advantage of alloys over pure materials is that they are much harder. In the case of dural, for instance, this hardness is obtained by quenching from a high-temperature (≈ 650 °C) down to room temperature. Experiment shows that, immediately after the quench, dural is ductile, but several hours later it becomes hard and brittle. In 1937, Guinier and Preston showed that this change in mechanical properties was due to the diffusion and precipitation of copper atoms. More specifically these atoms, after first occupying the normal sites of the aluminium crystal, cluster into flat plates, called *Guinier–Preston zones*, within which they occupy the same positions as aluminium atoms. These plates form *coherent precipitates* which are, by definition, aggregates where a certain number of solute atoms (Cu, in this case) replace the same number of solvent atoms (here, Al). There are also *incoherent precipitates*: in their

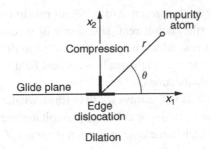

Fig. 5.43 Edge dislocation in the presence of a foreign atom (or of a coherent precipitate). The ⊥ symbol stands for the edge dislocation: the vertical bar in the middle represents the additional layer.

case, the origin of hardening is different. That is why we will discuss separately the 'coherent' and the 'incoherent' cases.

5.2.8.1 The case of coherent precipitates: summary of the Mott and Nabarro model

To understand why foreign atoms or coherent precipitates harden a material, one must analyse their elastic interactions with the dislocations (Mott and Nabarro, 1940; Mott, 1946, 1952; Nabarro, 1946; [16, 18]).

As an example, consider the case of an edge dislocation in the presence of a foreign atom placed at coordinates (θ, r). The dislocation, situated at the origin of the reference system, is parallel to the x_3 axis (Fig. 5.43). From the elastic point of view, the foreign atom behaves as a spherical cavity of radius $R(1 + \varepsilon)$ slightly different from the radius R of an atom in the perfect crystal. If the foreign atom is larger than a crystal atom, $\varepsilon > 0$; otherwise, $\varepsilon < 0$. For dural (copper atom in an aluminium matrix), $\varepsilon \approx 0.2$, which is considerable.

To understand the origin of alloy hardening, let us analyse the expression of the interaction energy E_{int} between the edge dislocation and an 'infinitely hard' cavity (in other words, with a fixed ε):

$$E_{int} = \frac{4\mu b R^3}{3r} \frac{1 + \nu}{1 - \nu} \varepsilon \sin \theta. \tag{5.100}$$

This formula (its demonstration is given in Appendix 5.B) shows that, for $\varepsilon > 0$ (the foreign atom is larger than a crystal atom), then E_{int} is positive when the atom is above the glide plane of the dislocation, and negative below. As a result, the dislocation and the foreign atom attract when this latter is below the glide plane and repel otherwise. The interaction force varies as $1/r^2$ and vanishes in the glide plane, as expected. The conclusions above must be reversed for $\varepsilon > 0$. These results show that *an attractive interaction is always possible, irrespective of the nature of the foreign atom.*

Note that the general expression of the interaction energy given in Appendix 5.B (Eq. (5B.8)) is also valid for a screw dislocation. In this case, the interaction is strictly null, since the dislocation only creates a shear stress ($\sigma_{ii}^d = 0$). However, *the screw dislocation can tilt* and acquire an edge component in order to decrease the elastic energy of the system. Once again, the interaction becomes *attractive*.

These results still apply in the presence of coherent precipitates that induce an elastic deformation of the matrix (the 'incoherent' situation will be discussed later). This is the case of Guinier–Preston zones, which behave as small edge dislocation loops. Like foreign atoms, these loops create around themselves a stress field that decreases as $1/r^2$ and interacts strongly with the dislocations.

We thus see that dislocations will always try to deform, whatever their nature, such that their interactions with foreign atoms or coherent precipitates are attractive ($E_{int} < 0$). If the latter are not mobile, which is the case at low temperature ($T \ll T_m$), they will hinder dislocation motion, with a trapping energy equal to their mutual interaction energy.

In practice, the trapping strength (at a fixed atomic concentration of solute C) depends on the size R of the aggregates and on their mutual distance Λ. This is well illustrated by the case of dural, which is ductile immediately after the quench to room temperature, when the Cu atoms still form a solid solution, but becomes hard and brittle after several hours, once the Guinier–Preston zones have appeared.

To understand the reasons for this difference in behaviour, let us compare the two situations.

In the first case (Fig. 5.44a), the copper atoms are isolated ($R \approx r_{Cu}$, with r_{Cu} the atomic radius of Cu) and very close to one another, since $\Lambda = c^{-1/3} r_{Cu} \approx 3r_{Cu}$ (in dural, where $c \approx 4\%$). As a result, the dislocations are not very sensitive to the details of the stress field created by the copper atoms, since they would need to bend over very small distances, of the order of Λ, equal here to three interatomic distances. Such distortions are impossible due to the considerable increase in line energy they would require. Since dislocations deviate very little from their straight trajectory, they do not feel the internal stress field generated by the foreign atoms and continue to glide easily ($\sin\theta$, in the interaction energy (5.100), is almost zero on the average). Dural experiences little hardening under these conditions, in agreement with experiment.

In the second case (Fig. 5.44b), the copper atoms form Guinier–Preston zones that can be identified, to a first approximation, as spherical coherent precipitates of a size R much larger than r_{Cu}. The average distance Λ between them is now:

$$\Lambda = \left(\frac{4\pi}{3r}\right)^{1/3} R \approx \left(\frac{4}{c}\right)^{1/3} R. \tag{5.101}$$

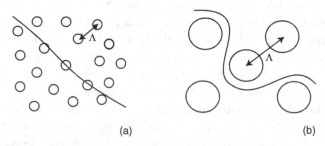

(a)   (b)

Fig. 5.44 Mott and Nabarro model: (a) solid solution; (b) Guinier–Preston zones [16].

This is much larger than the distance between solute atoms in the solid solution, so the dislocations are now able to bend to minimise their elastic interaction with the precipitates. More precisely, this occurs if the increase in line energy associated with dislocation curvature is lower than the energy of interaction with the precipitate. Using Eq. (5.100) and taking $r = \Lambda/2$, $\sin \theta = 1$ and $\nu = 0.3$ this condition reads:

$$0.5\mu b^2 \Lambda < 5 \frac{\mu b R^3}{\Lambda} \varepsilon \qquad (5.102)$$

or, from Eq. (5.101):

$$R > 0.25 b\varepsilon^{-1}c^{-2/3}. \qquad (5.103)$$

Guinier–Preston zones in dural roughly satisfy this inequality. Hence, they exert on the dislocations an oscillating trapping force (per unit length), with an absolute value:

$$F_{\text{trap}} \approx \frac{1}{\Lambda} \left| \frac{dE_{\text{int}}}{dr} \right|_{\substack{\sin\theta=1 \\ r=\Lambda/2}} \approx 10\mu b\varepsilon \left(\frac{R}{\Lambda} \right)^3 \approx 2.5\mu b\varepsilon c. \qquad (5.104)$$

This formula defines the internal stress σ_i, and thus a new elastic limit σ_c that the applied stress must exceed for the dislocations to glide over large distances. As $F_{\text{trap}} = \sigma_i b$ and $\sigma_c = \sigma_i$, one has:

$$\sigma_c \approx 2.5\mu b\varepsilon c. \qquad (5.105)$$

For dural, this calculation predicts a very high value for σ_c, of the order of $10^{-2}\mu$, in good agreement with the experimental results. Note that this alloy hardening mechanism is rather insensitive to temperature, since the distances involved in going from one equilibrium position to the next (of the order of the distance Λ between precipitates) are large compared to the atomic sizes.

It is also noteworthy that the Mott and Nabarro theory only takes into account the internal stress created by the coherent precipitates. The implicit assumption is that the dislocations, whose glide planes must necessarily contain precipitates, can easily cross the precipitates, the internal stress remaining the main obstacle to their propagation. It is then easily seen that this theory only applies if the size factor ε is large, as in the case of dural. On the other hand, this model does not explain the significant hardening of Al-Ag or Al-Zn alloys, where the size factor ε is very small. One must then account explicitly for crossing and for the short-range interactions between dislocations and the coherent Guinier–Preston zones (which are more or less spherical here) [16].

The situation changes at high-temperature, since the solute atoms, which form a solid solution, can move by diffusion. Under the action of the stress field, they can gather around the dislocations forming what is known as *Cottrell clouds* (Cottrell and Bilby, 1949; [16, 18]). These clouds have been observed by field ion microscopy, a technique able to reconstruct the three-dimensional image of the cloud around the dislocation (Fig. 5.45). These clouds can follow the motion of the dislocation if it does not move too fast, and hinder its progression. In some situations, the dislocation can escape its cloud, in which

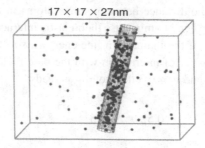

Fig. 5.45 Reconstruction of the Cottrell cloud around a dislocation in an ordered Al-Fe alloy doped with boron atoms. In this image, only the boron atoms are shown. The cylindrical envelope surrounding the dislocation is only a guide for the eye, in order to better distinguish the Cottrell cloud (Blavette *et al.*, 1999).

case it propagates much faster. This trapping–untrapping mechanism is at the origin of complex dynamic phenomena, such as the 'sawtooth creep' of Portevin–Lechatelier (Portevin and Lechatelier, 1923; [16, 18]).

5.2.8.2 *The case of incoherent precipitates: the Orowan model*

Precipitates obtained by diffusion are generally *incoherent* and replace an identical volume of the matrix by a different number of atoms. By becoming incoherent, the precipitates increase their surface energy, while releasing elastic deformation energy. This occurs if the precipitate is large enough since the second term, proportional to the volume, must exceed the first one, which is proportional to the surface. To find the minimum size above which coherence is lost, one must compare these two terms. For a coherent precipitate of size R, one can show that the elastic energy stored in the matrix is $3\mu R^3 \varepsilon^2$. For an incoherent precipitate of the same size, the surface energy is $4\pi R^2 \gamma$, where γ is the energy of the grain boundary. Coherence is lost when the first term exceeds the second one, namely for:

$$R > \frac{4\pi}{3} \frac{\gamma}{\mu \varepsilon^2}. \tag{5.106}$$

Such precipitates are formed, for instance in dural, if, after being heated at 650 °C to form a solid solution, the material is annealed for two days at 350 °C (rather than at room temperature) [16]. Under these conditions, the copper atoms no longer form Guinier–Preston zones, but incoherent Al_2Cu precipitates. For these precipitates, the condition (5.106) amounts to $R > 150$ Å since $\varepsilon \approx 0.07$, $\gamma \approx 0.5$ J/m^2 and $\mu \approx 2.7 \times 10^{10}$ Pa. This condition is verified experimentally.

What are the consequences on the elastic limit?

In the absence of coherence, there is no internal stress in the material. On the other hand, the precipitates hinder the motion of dislocations each time they cross the glide plane. More precisely, they behave as 'trees' separated by an average distance L significantly larger than Λ since:

$$2L^2 R \approx \Lambda^3. \tag{5.107}$$

This relation is obtained easily by considering a cube of unit volume with one facet parallel to the glide plane. Let us cut this cube by planes parallel to the glide plane, a distance $2R$ apart. Each precipitate intersects one of these planes. Since there are $1/2R$ planes and since each plane crosses $1/L^2$ precipitates, it follows that the number of precipitates within the cube is $(1/2R)(1/L^2)$. We obtain the formula above by noting that this number is also equal to $1/\Lambda^3$.

To determine the elastic limit σ_c in this case, let us analyse the behaviour of a dislocation as it 'collides' with the precipitates. If the applied stress is not too high, the dislocation bends locally and takes, between two consecutive anchoring points, the shape of a circular arc of radius R, given by (see Eq. (5.93)):

$$R \approx \frac{\mu b}{\sigma}. \tag{5.108}$$

As for the Frank–Read sources, the equilibrium is stable for $R > L/2$. This condition defines a new critical stress, or elastic limit σ_c:

$$\sigma_c \approx \frac{2\mu b}{L} \approx \mu\sqrt{2C}\frac{b}{R}, \tag{5.109}$$

considering that above σ_c the dislocations cross the precipitates and leave around them a dislocation loop (Fig. 5.46) (this is known as the *Orowan process*) (Orowan, 1948; [16, 18]). Note that for dural, the hardening induced by Al_2Cu precipitates is less than that obtained in the presence of Guinier–Preston zones, as in this case one can estimate $\sigma_c \approx 3 \times 10^{-3}\mu$ with $R \approx 300$ Å, which is in good agreement with the experimental values.

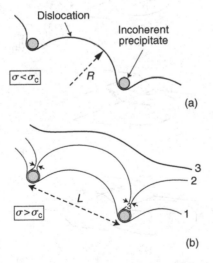

(a)

(b)

Fig. 5.46 Orowan process. (a) When the stress is not high enough, the dislocation bends without crossing the obstacles; (b) under high stress, the dislocation can cross the obstacles by 'winding' around them. After crossing, a dislocation loop is left around each of them.

5.2.9 Quench hardening

It is well known that tempered steel is very hard. In this case, the origin of the hardening is the trapping of mobile dislocations on small dislocation loops formed by vacancy condensation after the quench [16]. These loops create a strong internal stress field and act similarly to the Guinier–Preston zones discussed in the previous section.

5.2.10 Cross-slip and climb of dislocations

At low temperature $(T < 0.5\,T_m)$, the dislocations glide. This process is conservative, insofar as it involves no long-range exchange of matter.

At intermediate temperatures $(T \approx 0.5\,T_m)$, other mechanisms come into play, such as *cross-slip*, which allow dislocations to change their glide plane by crossing over to an intermediate plane (Fig. 5.47). This mechanism, which is still conservative, allows the dislocations to avoid obstacles they would not have been able to cross otherwise. It results in a decrease of the elastic limit of the material. This process can be shown to be thermally activated; it is easier in metals with the b.c.c. structure (Fe) than in those with the f.c.c. structure (Al, Cu, etc.) [16].

At even higher temperature, close to the melting point, dislocation *climb* becomes possible (see, for instance, Fig. 5.27b). This motion is not conservative, as it involves a local incoming (or outgoing) matter flux. This matter flow is made possible by the diffusion of vacancies, which are much more concentrated and mobile at these high temperatures than

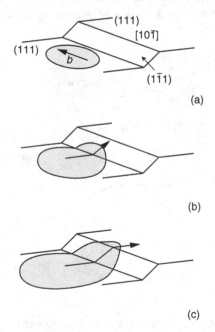

Fig. 5.47 Cross-slip of a dislocation loop in an f.c.c. crystal. In (a), the loop is in its primary plane; in (b) the screw part has undergone simple cross-slip; in (c) double cross-slip has occurred [14].

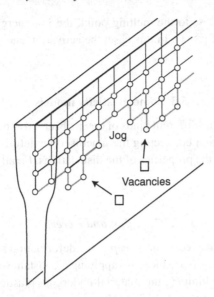

Fig. 5.48 Climb of a perfect edge dislocation line by emission or absorption of vacancies onto jogs. By vacancy absorption, the line moves upwards, while it moves downwards by emission [14].

at room temperature (interstitials might also contribute to matter transport, but they are much less frequent than vacancies due to their higher energy). The line can then leave its glide plane and 'climb' by emission or absorption of vacancies onto the jogs it contains (and which represent convenient sources or sinks for the vacancies) (Fig. 5.48).

This mechanism is made possible by the fact that, under stress, the vacancy concentration close to the dislocation core is different from the concentration at thermal equilibrium. One can show that the dislocation moves under the action of the driving Peach and Koehler force (equal to σb in the example of Fig. 5.27b) with a climb velocity given by:

$$v_c \approx 2\pi\, C_j \frac{D_S\ \sigma b^3}{l_d\ k_B T}. \tag{5.110}$$

In this formula, which is only valid under low stress ($\sigma b^3/k_B T \ll 1$), D_S is the self-diffusion coefficient (Section 5.1.1.2) and C_j is the concentration of jogs on the line, defined as the number of jogs per line atom:

$$C_j = \exp(-E_j/k_B T), \tag{5.111}$$

where E_j is the energy of a jog (a fraction of an electronvolt in metals). Finally, l_d is the typical distance over which diffusion takes place; it is also the distance from the dislocation over which vacancy concentration relaxes to its equilibrium value (in the example of Fig. 5.27b, l_d is of the order of the loop radius).

In this model, the climb velocity is proportional to the applied stress. The prefactor m defines the *mobility* of the dislocation ($v_c = m\sigma$).

At high temperature, close to the melting point, the jog energy decreases and the line can be saturated in jogs ($C_j \approx 1$). In this case, the activation energy of the mobility is that of self-diffusion.

5.3 Some plasticity models

We will start by recalling the different kinds of creep observed experimentally. We will then prove a fundamental relation connecting the macroscopic deformation to the microstructure of the material and to the properties of the dislocations. Finally, we will describe some simple plasticity models.

5.3.1 α, β, κ and r creep

Before going any further, we recall that *creep* is the deformation $\varepsilon(t)$ following the instantaneous elastic deformation ε_0 produced by applying a constant stress σ.

If σ exceeds the elastic limit σ_c, the material undergoes plastic deformation (assuming it does not break, which requires $T > T_{BD}$). Its creep curve is then strongly dependent on such parameters as temperature and stress.

Some experimental curves, obtained for very pure aluminium, are shown in Fig. 5.49. These curves show that, at a fixed stress, the creep rate is strongly temperature dependent, with a clear regime change around $0.5\,T_m$ ($T_m \approx 925$ K for aluminium).

Similar behaviour is observed in most materials. Thus, one can generally distinguish two types of creep, depending on the applied stress and the temperature.

(1) At low temperature and under low stress, $\dot{\varepsilon}$ decreases steadily, and the creep eventually stops after a certain time (Fig. 5.50). In metals, this regime is often well described by a law of the type:

$$\varepsilon_\alpha(t) = \varepsilon_0 + \alpha \ln(\gamma t + 1). \tag{5.112}$$

One speaks of *logarithmic creep*, or α *creep*.

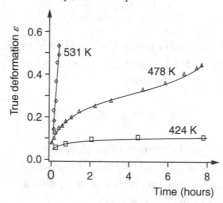

Fig. 5.49 Creep curves, at three different temperatures, for a specimen of very pure aluminium. The applied stress is the same in all three cases ($\sigma = 2.1 \times 10^7$ Pa) ([13], volume 5, p. 1625).

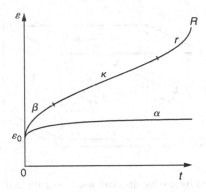

Fig. 5.50 The two types of creep [16].

(2) At high temperature, or at much higher stress values, the creep goes through a slowing down stage (*primary creep β*), before becoming linear (*secondary* or *quasi-viscous creep κ*; this phase is often followed by a new acceleration (*tertiary creep r*), followed by the *rupture* of the material *R* (Fig. 5.50).

In the β and κ regimes, the creep behaves as:

$$\varepsilon_\beta(t) = \varepsilon_0 + \beta t^m, \tag{5.113}$$

$$\varepsilon_\kappa(t) = \varepsilon_0' + \kappa t, \tag{5.114}$$

where the exponent m is often close to 1/3 (or slightly higher) (Andrade, 1910).

5.3.2 The Orowan relation

This purely geometrical relation is fundamental, as it provides a direct link between the plastic deformation rate and the properties of the dislocations present within the material (density, Burgers vector and velocity). It is therefore at the core of most plasticity models (Orowan, 1940; [19]).

To determine its expression, let us consider a monocrystalline elastic volume filled by edge dislocations of Burgers vectors b and $-b$ (Fig. 5.51). We consider equal numbers of dislocations of each sign to avoid large-scale deformations. Under the action of the shear stress, each dislocation feels a Peach and Koehler force with amplitude σb, making it glide to the right or to the left, according to the sign of its Burgers vector. Let ε be the deformation of the elastic volume ($\varepsilon = u/L_3$, where u is the displacement of the upper face with respect to the lower one and L_3 the sample thickness). At the beginning, $\varepsilon = 0$. At the instant t, each dislocation has moved by x (in one direction or the other) under the action of stress. These individual movements lead to the global deformation:

$$\varepsilon = n \frac{b}{L_3} \frac{x}{L_1}, \tag{5.115}$$

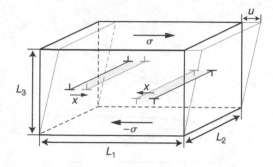

Fig. 5.51 Elastic volume filled by n edge dislocations. Under the action of shear, the dislocations glide to the right or to the left, according to the sign of their Burgers vector. These individual movements lead to the macroscopic deformation $\varepsilon = u/L_3$.

where we summed over all the deformations produced by each dislocation taken individually (equal to $(b/L_3)(x/L_1)$, since the volume deforms by b/L_3 as a dislocation moves by a distance $x = L_1$, i.e. across the whole volume).

Taking the derivative of this expression with respect to time yields immediately:

$$\dot\varepsilon = \rho_m b v, \tag{5.116}$$

where we used

$$\rho_m = \frac{n}{L_1 L_3} \tag{5.117}$$

and

$$v = \frac{dx}{dt}. \tag{5.118}$$

This is the *Orowan relation*, stating that the rate of macroscopic deformation is proportional to the density ρ_m of mobile dislocations, to their Burgers vector b and to their velocity v.

It is easily checked that this geometrical relation also applies to climb motion. It is also valid for loops, or dislocations of a generic shape (up to a geometrical prefactor between 0 and 1, which accounts for the type of the dislocation).

Obviously, one still needs to find how the dislocation velocity and density vary as a function of the applied stress and the time (if the creep process is not stationary).

5.3.3 Low-temperature creep by thermally activated obstacle crossing

Experiment shows that the creep rate often obeys an Arrhenius law of the type [16, 19]:

$$\dot\varepsilon = \dot\varepsilon_0 \exp\left(\frac{-\Delta H_0}{k_B T}\right) \exp\left(\frac{\sigma \Omega_a}{k_B T}\right). \tag{5.119}$$

In particular, this is the case for plastic smectic B crystals under shear parallel to the layers, where this law applies remarkably well. We will therefore use this example as a starting illustration; we will then discuss the consequences of this model in the much more general case of α creep in metals.

5.3.3.1 The smectic B example

Let us first recall that in this plastic crystal the elongated molecules are normal to the layers and arranged on the sites of a hexagonal lattice within the layers (see Chapter 1). The stacking is of the ABAB ... type (hexagonal close-packing, or h.c.p.).

Experiment shows that creep under shear parallel to the layers is stationary at low deformation ($\varepsilon < 1$) when the sample thickness is very small compared to its lateral size (Fig. 5.52). In Fig. 5.53 we plotted the creep rate in the 4O.8 compound under shear parallel to the layers as a function of temperature and the applied stress. Each curve, corresponding to a different temperature, is fitted with an Arrhenius law of the type (5.119) using $\Delta H_0 = 0.64$ eV and an average value for the activation volume $\Omega_a \approx 5 \times 10^{-23}$ m^3.

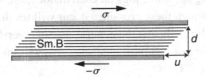

Fig. 5.52 Creep of a smectic B parallel to the layers. The deformation $\varepsilon = u/d$ is below 1. After each creep test, the sample is returned to its initial state ($u = 0$), melted in the smectic A phase, then re-crystallised. Note that the sample thickness is very small compared to its lateral size.

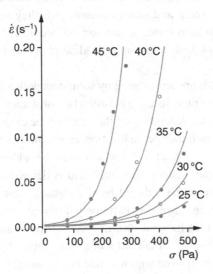

Fig. 5.53 Arrhenius creep observed under shear parallel to the layers in a smectic B plastic crystal (Oswald, 1985).

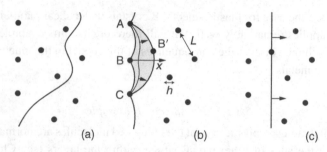

Fig. 5.54 (a) For $\sigma < \sigma_i$, the dislocations are immobile in the internal stress field; (b) when $\sigma > \sigma_i$, the dislocations stumble upon obstacles that they cross by thermal activation; (c) at high-temperature, or under very high applied field, the dislocations no longer see the obstacles. Their motion becomes athermal.

In this example, the deformation occurs by glide of the basal dislocations (whose Burgers vector is contained in the plane of the layers). As they glide, these dislocations cross 'trees', which are screw dislocations perpendicular to the layers. As we pointed out in Section 5.2.5, these crossings produce jogs on each dislocation, with an associated cost in elastic energy. The rate at which these crossings occur will set the creep dynamics.

To show this, let us start by analysing the generic case of creep by dislocation glide when it is governed by the crossing of localised obstacles. We will then return to the somewhat particular case of the smectic B.

5.3.3.2 The model

For the time being we will not concern ourselves with the nature of the obstacles (which can be trees, precipitates, etc.), and simply assume that they are *localised*, so that they can be *crossed by thermal activation*. In this section, we will show how to calculate the average velocity v of a mobile dislocation in its glide plane, and then the creep rate using the Orowan relation.

In Fig. 5.54, the obstacles are represented by solid dots with a diameter h characterising their size, while L is the average distance between two obstacles.

Three situations can occur, depending on the value of the external applied stress and on the temperature. Indeed, each mobile dislocation experiences a Peach and Koehler force induced by the external stress field σ, to which must be added the internal stress field $\sigma_i(x, T)$ created by the other dislocations (this field is defined in Section 5.2.4), as well as the local stress field $\sigma_B(x, T)$ induced by the obstacle. As shown in Fig. 5.55, the stress fields σ_i and σ_B vary in space over extremely different distances, namely the typical distance between dislocations (several µm, in general) for σ_i and the typical size of the obstacle h (ordinarily very small, of the order of a nanometre), for σ_B.

Due to this scale difference, three regimes must be distinguished, according to the value of the applied stress and the temperature of the system [19]. Let us describe them one by one.

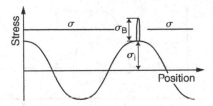

Fig. 5.55 Stress field inside the sample, as a function of the position. The internal stress $\sigma_i(x)$ (of amplitude σ_i) varies slowly, over much larger distances than the stress $\sigma_B(x)$ (with amplitude σ_B) induced by the obstacle [14].

First regime   Suppose $\sigma < \sigma_i$. The dislocations being unable to cross the maxima of $\sigma_i(x, T)$ by thermal activation (Fig. 5.54a) at any temperature, they cannot propagate over large distances. The material does not creep, the internal stress contributing to its elastic limit.

Second regime   Consider now that $\sigma_i < \sigma < \sigma_i + \sigma_B$. In this case, the dislocations can overcome the internal stress field and move over large distances. In this process, they 'stumble' upon the obstacles, that they can cross by thermal activation, as shown in Fig. 5.54b. This drawing illustrates the fact that, as a line encounters obstacles A, B and C, it is pushed to the right by the effective stress σ^*, which is the applied stress σ minus the internal stress σ_i. Under the action of this stress, the line bends. However, we know already that a dislocation arc of length L vibrates under the action of thermal activation at a typical frequency ν_L calculated in Section 5.2.1:

$$\nu_L = \frac{\sqrt{2}}{\pi} \frac{C_t}{L},$$
(5.120)

where $C_t = (\mu/\rho)^{1/2}$ is the propagation velocity of shear waves.

On the other hand, from the *theory of absolute reaction rates*, we know that the probability for the dislocation to cross the obstacle is given by the Arrhenius factor:

$$\exp\left(-\frac{\Delta G}{k_B T}\right),$$
(5.121)

where ΔG is the activation free energy of the process, which is also the energy that must be provided by the thermal agitation to cross the obstacle. This quantity is given by the free energy difference between the final state corresponding to the excited state, taken to be at thermodynamic equilibrium (the dislocation is then on top of the barrier, in position 2 shown in Fig. 5.56) and the initial equilibrium state (when the dislocation is 'at the bottom of the obstacle,' in position 1). Consequently, ΔG is given (up to a sign change) by the reversible work of the Peach and Koehler force acting on the dislocation segments AB and BC of total length $2L$, during quasi-static displacement from position 1 to position 2:

$$\Delta G = \int_1^2 \sigma_B(x, T) b 2L \, dx \quad + \int_1^2 [\sigma_i(x, T) - \sigma] b 2L \, dx.$$
(5.122)

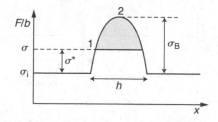

Fig. 5.56 'Force–distance' diagram and definition of the equilibrium positions 1 and 2 in the activated crossing regime. F stands for the components of the Peach and Koehler force acting on the dislocation. The shaded area represents the amount of thermal energy provided to the system during crossing [19].

If the effective stress σ^* is small compared to σ_B, this equation can be written as:

$$\Delta G \approx \Delta G_0(T) + (\sigma_i - \sigma)bA(T) = \Delta G_0(T) - \sigma^*\Omega_a(T), \qquad (5.123)$$

where $\Delta G_0(T)$ is the activation free energy of the process at zero effective stress, $A(T)$ an activation area depending on the obstacles and the distance between them ($A(T) \approx Lh$), and $\Omega_a(T) = bA(T)$ the corresponding activation volume. Note that expression (5.123) is nothing other than the series expansion of the general expression (5.122) to first order in the stress and that $\Delta G_0(T)$ can be split into:

$$\Delta G_0(T) = \Delta H_0 - T\Delta S_0, \qquad (5.124)$$

where ΔH_0 is the activation enthalpy and ΔS_0 the activation entropy, which mainly originates in the temperature variation of the shear modulus (σ_B being proportional to μ).

Finally, the frequency of obstacle jumping is given by the relation:

$$\nu_j = (\text{test frequency}) \times (\text{jump probability}) = \nu_L \exp\left(-\frac{\Delta G}{k_B T}\right), \qquad (5.125)$$

where ν_L and ΔG are given by expressions (5.120) and (5.123), respectively.

If, once obstacle B is crossed, the dislocation propagates rapidly towards the next obstacle B', which amounts to assuming that the time needed for reaching it is negligible compared to the typical crossing time $1/\nu_j$ (Fig. 5.54b), then:

$$v = \nu_j l = \nu_L l \exp\left(-\frac{\Delta G}{k_B T}\right) = \frac{\sqrt{2}}{\pi} C_t \exp\left(-\frac{\Delta G}{k_B T}\right), \qquad (5.126)$$

where l is the typical distance covered by the dislocation after each jump, given, as a first approximation, by the average distance L between obstacles (for a more refined model, see [16]).

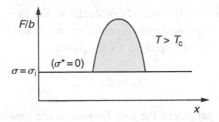

Fig. 5.57 'Force–distance' diagram at high temperature ($T > T_c$). Thermal agitation provides all the energy needed to cross the obstacle. In the stationary regime, the internal stress is balanced by the external stress applied to the system [19].

Finally, the creep rate is obtained by applying the Orowan relation and relations (5.126) and (5.123):

$$\dot{\varepsilon} = \frac{\sqrt{2}}{\pi} C_t \rho_m b \exp\left(\frac{\Delta S_0}{k_B}\right) \exp\left(-\frac{\sigma_i \Omega_a}{k_B T}\right) \exp\left(-\frac{\Delta H_0}{k_B T}\right) \exp\left(\frac{\sigma \Omega_a}{k_B T}\right). \tag{5.127}$$

This expression has indeed the expected Arrhenius form (Eq. (5.119)).

Third regime This regime occurs when the applied stress is very high, or at sufficiently high temperature.

Indeed, if $\sigma > \sigma_i + \sigma_B$, the dislocations glide and cross the obstacles 'without feeling them', at any temperature. For this reason, $\sigma_{0K} = \sigma_i + \sigma_B$ defines the critical creep stress at 0 K.

The same situation occurs when all the energy needed to cross the obstacle is provided by the thermal agitation (Fig. 5.57). This happens for temperatures above a temperature T_c given by the relation:

$$k_B T_c = \Delta G_0(T_c) \approx \sigma_B b A(T_c). \tag{5.128}$$

Under these conditions, the obstacles to glide disappear. Consequently, *other mechanisms must come into play to determine the creep rate of the sample*, and we will return to them in Section 5.3.4. However, let us point out that, if the deformation is only induced by dislocation glide, the effective stress acting on them must go to zero (otherwise, their average velocity would diverge), yielding the important relation:

$$\sigma = \sigma_i. \tag{5.129}$$

Before applying these concepts to the case of metals or minerals, let us return to the smectic B example.

5.3.3.3 Interpretation of the smectic B experiment

In smectics B, lattice friction, which could oppose the motion of basal dislocations, is completely negligible due to the very high structural anisotropy of the material ($\sigma_{PN} \approx 0$

from Eq. (5.90) since $a/c \approx 10$). The only obstacles to the motion of basal dislocations[4] are the screw dislocations that pierce the layers. Crossing leads to the appearance of a jog on each dislocation, as shown in Fig. 5.36. Consequently,

$$\Delta H_0 = E_j + E'_j, \qquad (5.130)$$

where E_j and E'_j are the energies of the jogs formed on the mobile dislocation and on the tree, respectively.

On the other hand, the width h of the obstacle is of the order of the Burgers vector b of the mobile dislocation, if it is perfect, and of the order of its dissociation width if it is dissociated into two partial dislocations separated by a stacking fault. This latter situation is quite likely in smectics B, where stacking faults cost very little energy. Let L be the width of the stacking fault ($L > b$). Hence $\Omega_a = bA \approx blL$, where l is the distance between trees.

Finally, by solving the stress field around dislocations in anisotropic elasticity, one can show that the internal stress is extremely weak in a smectic B and can be neglected (in practice, it is below 20 Pa and could not be measured). This is a result of the elastic anisotropy of the material (directly related to its structural anisotropy), whose elastic modulus in shear parallel to the layers is three to four orders of magnitude smaller than the other modules ([43], volume 2).

Gathering all these results and using Eq. (5.127), one finds that the smectic B must obey the following Arrhenius law:

$$\dot{\varepsilon} = \frac{\sqrt{2}}{\pi} C_t \rho_m b \exp\left(\frac{\Delta S_0}{k_B}\right) \exp\left(-\frac{E_j + E'_j}{k_B T}\right) \exp\left(\frac{\sigma \Omega_a}{k_B T}\right). \qquad (5.131)$$

This expression accounts very well for the observations (see Fig. 5.53). Moreover, calculating the kink energy in anisotropic elasticity using the known values of the elastic moduli yields $E_j \approx 0.1$ eV and $E'_j \approx 0.4$ eV, i.e. $\Delta H_0 \approx 0.5$ eV, close to the experimental value (0.64 eV). As to the measurement of the activation volume, it leads to a dissociation width of the basal dislocations of the order of 20–40 Å, taking as a typical distance between trees $l \approx 5-10$ μm (estimated value for the smectic A phase, found at higher temperature). Note that the presence of a high density of stacking faults in these materials was confirmed by X-ray scattering, validating the proposed model.

In conclusion, the case of the smectic B is somewhat atypical, insofar as the internal stress and lattice friction are extremely weak, but particularly instructive, providing a nice creep example where the Arrhenius law can be observed over a wide range of stress and temperature.

In metals, on the other hand, the internal stress is no longer negligible; furthermore, it can vary during the creep process due to an increase in the density of dislocations (an effect we were able to neglect in the smectic B due to the low deformations used in the

---

[4] Those contained in the plane of the layers and with a Burgers vector parallel to the layers.

experiments, see Fig. 5.52). The Arrhenius law can then lead to logarithmic creep, as we will see below.

5.3.3.4 Work hardening and α (or logarithmic) creep in metals

In metals, the internal stress is much more important than in smectics B. Experiment also shows that it increases during deformation, since the dislocations can multiply by the action of the Frank–Read sources. This effect often leads to a decrease in creep rate at the beginning of the process, which might seem paradoxical, since the number of mobile dislocations increases, and the creep rate is proportional to ρ_m according to the Orowan relation. The explanation resides in the fact that, as they multiply, the dislocations quickly start to block each other, hindering their motion. This effect also explains the increase in the elastic limit of the material after large deformation, a feature very well known by metallurgists and called *work hardening*.

To quantify this phenomenon, let us assume that the internal stress increases with the deformation ε [16]. To lowest order in the deformation, one can keep only the linear term in the series expansion, yielding:

$$\sigma_i = \sigma_i^0 + \left(\frac{d\sigma_i}{d\varepsilon}\right)_{\varepsilon_0} (\varepsilon - \varepsilon_0). \tag{5.132}$$

On the other hand, Eq. (5.131) can be written as:

$$\dot{\varepsilon} = \text{Cst} \exp\left(-\frac{\sigma_i \Omega_a}{k_B T}\right). \tag{5.133}$$

In this formula, Cst is a constant (for a fixed stress σ) of expression:

$$\text{Cst} = \frac{\sqrt{2}}{\pi} C_t \rho_m^0 b \exp\left(-\frac{\Delta G_0}{k_B T}\right) \exp\left(\frac{\sigma \Omega_a}{k_B T}\right), \tag{5.134}$$

where ρ_m was replaced by ρ_m^0, the initial dislocation density. This approximation is valid to the extent that the increase in creep rate related to the higher dislocation density is linear in the deformation, while the slowdown due to the increase of the internal stress is exponential in the deformation. Putting together Eqs. (5.132) and (5.133) and integrating the resulting equation gives a law of the type (5.112) (α or logarithmic creep, $\varepsilon_\alpha(t) = \varepsilon_0 + \alpha \ln(\gamma t + 1)$) with:

$$\alpha = \frac{k_B T}{\Omega_a \left(\dfrac{d\sigma_i}{d\varepsilon}\right)_{\varepsilon_0}} \tag{5.135}$$

and

$$\gamma = \frac{\sqrt{2} C_t \rho_m^0 b}{\pi \alpha} \exp\left(-\frac{\Delta G_0 - (\sigma - \sigma_i^0)\Omega_a}{k_B T}\right). \tag{5.136}$$

This non-stationary creep, with a decrease of the creep rate in time, is typical for *work hardening*.

5.3.4 High-temperature creep

Everybody knows one must 'strike while the iron is hot'. This old saw reminds us that it is easier to work and process a material at high temperature ($0.5\,T_m < T < T_m$) than at low temperature ($T < 0.5\,T_m$). To show this explicitly, we will examine a few models involving the diffusion of vacancies (Nabarro–Herring and Coble models) or the movement of dislocations (the models of Harper–Dorn, Nabarro and Weertman). In particular, we will see under what conditions stationary deformation (κ creep) can occur.

5.3.4.1 Nabarro–Herring creep by vacancy diffusion within a grain

This is the simplest imaginable model for high-temperature creep by vacancy diffusion (Nabarro, 1948; Herring, 1950; [16]). Consider a fine-grained crystal, with a typical grain size L. Suppose that, under the action of the applied stress, each grain experiences pure elongation. This means that some of its faces are compressed, while the others are stretched.

A situation of this type is sketched in Fig. 5.58. In this diagram, the vertical faces of the grain are compressed, while its horizontal faces are stretched. Due to these stresses, the equilibrium density of vacancies close to the compressed or stretched faces is not the same. This imbalance is due to the fact that forming a vacancy amounts to removing an atom from below the surface and bringing it to the surface. During this elementary displacement, the applied stress provides the work $\pm\sigma a^2 \times a$, whether the face is stretched ($+$ sign) or compressed ($-$ sign), where a is the interatomic distance. It follows that the energy needed for creating a vacancy becomes $E_f - \sigma a^3$ in the vicinity of a stretched face, and $E_f + \sigma a^3$ in the vicinity of a compressed face. One can then deduce the equilibrium vacancy concentrations at the two faces:

$$c^+ = c_0 \exp\left(+\frac{\sigma a^3}{k_B T}\right) \qquad \text{at the stretched faces,} \qquad (5.137)$$

$$c^- = c_0 \exp\left(-\frac{\sigma a^3}{k_B T}\right) \qquad \text{at the compressed faces,} \qquad (5.138)$$

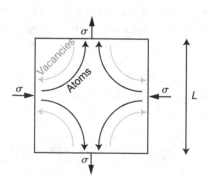

Fig. 5.58 Principle of Nabarro–Herring creep. Under the action of stress, the grain deforms due to the contrary flow of vacancies and atoms.

where c_0 is the vacancy concentration in the absence of stress (given by Eq. (5.5)). This difference in concentration creates a vacancy flow, such that the number of vacancies, and hence of atoms, N crossing from one face (with section L^2) to the other per unit time is:

$$N \approx L^2 J = L^2 D_v \frac{c^+ - c^-}{\Omega L}, \tag{5.139}$$

where $\Omega \approx a^3$ is the atomic volume. These N atoms change the grain length by

$$\delta L = N \frac{\Omega}{L^2} \tag{5.140}$$

per unit time, resulting in a creep rate:

$$\dot{\varepsilon} = \frac{N\Omega}{L^3}. \tag{5.141}$$

Using Eqs. (5.137) through (5.139), we finally obtain:

$$\dot{\varepsilon} = 2c_0 \frac{D_v}{L^2} \sinh\left(\frac{\sigma\Omega}{k_B T}\right). \tag{5.142}$$

If the argument of the hyperbolic sine is small (experimentally, this is almost always the case), this expression can be linearised to yield, using the self-diffusion coefficient D_S (Eq. (5.14)):

$$\dot{\varepsilon} \approx \frac{2D_S\Omega}{L^2 k_B T} \sigma. \tag{5.143}$$

This relation shows that *Nabarro–Herring creep is Newtonian.*

Once proven for a unique grain, this formula can be extended to the entire crystal under the assumption that the grains remain *cohesive* and that their individual deformations are additive. Then:

$$\dot{\varepsilon} \approx C \frac{D_S\Omega}{L^2 k_B T} \sigma, \tag{5.144}$$

where C is a numerical coefficient depending on the shape of the grains and on the way they glide on top of each other. More specifically, for identical spherical grains, C varies between 5 and 13 depending on whether they glide on top of each other or not [19].

In conclusion, we can see that the creep rate increases with decreasing grain size (as $1/L^2$) and with increasing temperature (because it is thermally activated, D_S increases much faster than T). This kind of creep can become dominant at high temperature for small enough grains.

Note that relation (5.144) yields the elongation viscosity of the material (defined by $\sigma = \eta_e \dot{\varepsilon}$). Using the Trouton rule for Newtonian liquids (its demonstration will be given in Section 6.3.7.1), one can obtain the 'ordinary' shear viscosity η of the material:

$$\eta = \frac{\eta_e}{3} \approx \frac{L^2 k_B T}{3C D_S \Omega}. \tag{5.145}$$

This viscosity is very sensitive to temperature (via D_S) and to grain size.

Thus, for an aluminium monocrystal of centimetre size ($L = 1$ cm), from Eq. (5.143) and the Trouton rule one has $\eta = 5 \times 10^{17}$ Pa s using $T = 400\,°C$, $D_S = 2 \times 10^{-14}\,m^2 s^{-1}$ and $\Omega = 1.6 \times 10^{-29}\,m^3$, which is very large. In this case, Nabarro–Herring creep is completely negligible.

On the other hand, for an aluminium polycrystal with grain size $L = 100\,\mu m$, using $C = 5$ in Eq. (5.145), one finds $\eta = 5 \times 10^{13}$ Pa s at $T = 400\,°C$ and $\eta = 10^{11}$ Pa s at $T = 647\,°C = 0.99\,T_m$, which becomes measurable, even though these viscosity values are still quite high.

5.3.4.2 *Coble creep by vacancy diffusion at the grain boundaries*

In the Nabarro–Herring model, matter diffuses in the bulk. However, especially at low temperature, it can happen that matter diffusion occurs preferentially along the grain boundaries (Coble, 1963). In this case, the same kind of model leads to the following result for a polycrystal:

$$\dot{\varepsilon} \approx C' \frac{D_{gb}\,\delta\Omega}{L^3 k_B T}\sigma, \tag{5.146}$$

where L is once again the grain size, δ the thickness of the grain boundaries, D_{gb} the diffusion coefficient along the grain boundaries and C' a numerical coefficient of the order of 50 [19].

This creep is also Newtonian but this time the creep rate goes as $1/L^3$. Experiment also shows that D_{gb} is thermally activated, with an activation energy lower than that of D_S. That is why Coble creep can become important at lower temperatures than those needed to observe Nabarro–Herring creep. This is the case, most notably in geology, where this type of creep seems to be relevant (see the data in Fig. 1.10 showing that the viscosity of olivine polycrystals, one of the essential components of the Earth's mantle, increases as L^3 rather than as L^2 in the Newtonian regime).

In the Nabarro–Herring and Coble models the role of dislocations was completely neglected. In fact, they often control high-temperature creep, as was already the case at low temperature. Let us give some examples.

5.3.4.3 *Creep by pure climb of isolated dislocations:* *the Harper–Dorn and Nabarro models*

The geometry is the same as for the examples above, but we further assume that inside each grain there is a high concentration of edge dislocations (Fig. 5.59), which can climb under the action of the applied stress by vacancy exchange, leading to plastic deformation.

The corresponding creep rate is given by the Orowan relation:

$$\dot{\varepsilon} \approx \frac{b v_c}{l^2}, \tag{5.147}$$

where v_c is the climb velocity of dislocations and l their average separation distance.

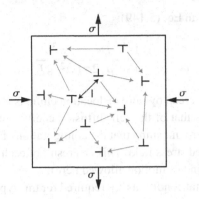

Fig. 5.59 Creep by climb of isolated dislocations. The grey arrows indicate the exchange of matter between dislocations.

If the dislocations are saturated in jogs ($C_j = 1$), their climb velocity, from Eq. (5.110) is:

$$v_c \approx 2\pi \frac{D_S \, \sigma b^3}{l_d \, k_B T}. \tag{5.148}$$

Friedel showed that, in this geometry, $l_d \approx b \ln(l/2b)$, which yields, after substitution in Eq. (5.147):

$$\dot{\varepsilon} \approx 2\pi \frac{D_S \Omega}{l^2 \ln(l/2b) k_B T} \sigma. \tag{5.149}$$

The creep is therefore Newtonian, as long as the dislocation density remains constant in time and independent of the applied stress. This type of creep was first observed in aluminium close to its melting point (Harper and Dorn, 1957) as well as in smectics B in compression normal to the layers (Oswald, 1984).

This type of creep is very seldom encountered experimentally, since the dislocation density is ordinarily stress dependent.

Indeed, since dislocation annihilation occurs at the edges or by recombination between themselves, dislocation sources are needed to maintain their density constant. If glide is forbidden (for instance, because the dislocations have too many jogs), these sources can be the Bardeen–Herring sources, which are the counterparts of Frank–Read sources for climb (Bardeen and Herring, 1952; [15,16]). On the other hand, the climb force acting on each dislocation is proportional not to σ, but rather to the effective stress $\sigma^* = \sigma - \sigma_i$, itself a function of the dislocation density, through the internal stress σ_i. Accounting for all these factors, F. R. N. Nabarro showed that, in the stationary regime, one has (Nabarro, 1967; [13, volume 5]):

$$\rho_m = \frac{1}{l^2} \approx \left(\frac{\sigma}{\mu b}\right)^2, \qquad \sigma_i \approx \frac{\mu b}{l}, \qquad \sigma^* \approx \frac{\sigma}{2\pi^2}, \tag{5.150}$$

yielding, after substitution in Eq. (5.149):

$$\dot\varepsilon \approx \frac{1}{\pi}\frac{D_S}{\ln(\mu/2\sigma)}\frac{\sigma^3 b}{\mu^2 k_B T}.\tag{5.151}$$

This formula shows that creep by pure dislocation climb is not Newtonian ($\dot\varepsilon \propto \sigma^3$) and that its activation energy is that of the self-diffusion coefficient. Note that, in this model, the effective stress is not zero, meaning that the configuration of the dislocations is never in equilibrium with the applied stress field. This is a result of each dislocation having a finite climb mobility (the only kind of motion allowed here).

Very specific experimental conditions are required for this type of creep. However, it has apparently been observed in certain alloys and oxides. It is easily recognised due to the absence of glide marks at the sample surface and to the existence of an (often very reduced) primary β creep stage.

5.3.4.4 Creep by climb-limited dislocation glide: the Weertman model

In most cases the deformation is due to dislocation glide, which, as we saw in Section 5.3.3.2, is often very easy at high-temperature, thermal agitation providing all the energy needed to cross localised obstacles, e.g. dislocation crossings. Three fundamental questions must then be answered:

(1) if glide is so easy, what are the mechanisms limiting the creep rate?
(2) under what conditions is stationary (κ) creep possible?
(3) why is this creep non-Newtonian?

To answer these three questions, let us assume that all the deformation occurs by dislocation glide and that thermal agitation provides all the energy needed to cross the obstacles to glide (such as Peierls valleys, crossings with trees, etc.). Under these conditions (and in contrast to the pure climb situation described in the previous section), the dislocations feel no viscous friction force opposing their motion in their glide plane.

In the Weertman model (Weertman, 1975; [12]; [13, volume 5]), Frank–Read dislocation sources periodically emit new dislocations in their glide planes. Consider two Frank–Read sources operating in two parallel glide planes separated by the distance d (Fig. 5.60). When the dislocations emitted by each source are close enough to each other, they start to attract (since they are of opposite sign) and 'climb' towards each other at a velocity v_c until they annihilate. During this climb process, which takes a time $t_c = d/2v_c$, the other dislocations are almost 'at rest', and their positions are such that their internal stress field exactly compensates the external field (nothing opposes their glide):

$$\sigma = \sigma_i.\tag{5.152}$$

Then comes a moment when the two dislocations annihilate; the two sources can once more operate, each one emitting a new dislocation dipole (a loop, in practice). The initial

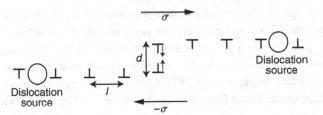

Fig. 5.60 Two dislocation sources placed in different glide planes. Each of them emits a new dislocation dipole each time annihilation occurs, after climb of the dislocations situated half way between the two sources.

distribution is restored such that, on the average, the global density of mobile dislocations ρ_m remains constant in time.

The creep rate can then be easily calculated by applying the Orowan law (Eq. (5.116)). To this end, let ΔL be the average distance covered by glide by each dislocation during the time $t_c = d/2v_c$ (annihilation time). One can immediately write the creep rate in the stationary regime:

$$\dot{\varepsilon} = 2\rho_m b \frac{\Delta L}{d} v_c. \tag{5.153}$$

This equation, coupled to Eq. (5.152), yields a prediction for the creep rate as a function of the stress and the temperature. Indeed, let l be the average distance between mobile dislocations. If no ad hoc assumption is made on the number of dislocation sources and their distribution, one must have $l \approx \Delta L \approx d$. Since

$$\rho_m = \frac{1}{l^2} \tag{5.154}$$

and, from Eq. (5.85):

$$\sigma_i \approx 0.3 \frac{\mu b}{l}, \tag{5.155}$$

one can obtain, by use of Eq. (5.152):

$$\rho_m \approx 10 \frac{\sigma^2}{\mu^2 b^2}. \tag{5.156}$$

On the other hand, we know that the climb velocity of a dislocation is proportional to the applied force σb; it can be obtained from Eq. (5.110) (using $l_d \approx b \ln(l/b) \approx 2\pi b$ and $C_j \approx 1$):

$$v_c \approx \frac{D_S}{b} \frac{\sigma b^3}{k_B T}. \tag{5.157}$$

Note that, in our case, this force is not a direct result of the external applied stress (which only induces dislocation glide), but rather of the elastic interaction between dislocation pairs of opposite sign, as they are close enough to each other, at a typical distance $d \approx l$ (Fig. 5.60). Since the attraction force between two dislocations of opposite sign is proportional to $\mu b^2/l$ (see Section 5.2.3), it is also proportional to $\sigma_i b$, from Eq. (5.155) (up to a numerical constant close to 1), and thus to σb because $\sigma = \sigma_i$ in the stationary regime. This reasoning validates expression (5.157), where σ is the applied stress.

Putting together the results of Eqs. (5.153), (5.156) and (5.157) finally yields the sought-for creep rate:

$$\dot{\varepsilon} \approx 20 D_S \frac{\sigma^3 b}{\mu^2 k_B T}. \tag{5.158}$$

This law, although resulting from a model which is clearly different from that of Nabarro, leads to the same dependence, not only with respect to stress (σ^3), but also as a function of temperature. This latter feature is however not surprising, since creep dynamics is still driven by dislocation climb, and thus by the self-diffusion coefficient. Nevertheless, the creep rates predicted by this model are much larger than in the Nabarro model.

Numerous versions of this model exist. For instance, in his initial version, Weertman took the number of dislocation sources as fixed. This assumption leads to a stress exponent of 4.5.

In a general manner, the theoretical stress exponent varies between 3 and 5, depending on the assumptions made as to the microstructure of the crystal. On the other hand, the diffusion coefficient involved can be that of self-diffusion, if the dislocations change their glide plane by climb, or that of cross-slip, if this mechanism is relevant, which often happens at lower temperatures (about $0.5 \, T_m$, see Section 5.2.10).

Thus, all models of this type predict creep rates of the form:

$$\dot{\varepsilon} = \text{Cst} \cdot \sigma^m \exp\left(-\frac{E}{k_B T}\right). \tag{5.159}$$

The exponent m ordinarily takes values between 3 and 5. As to the thermal activation energy E, this can be $E_{cs}(\sigma)$ in the case of cross-slip, a parameter that depends on the applied stress [16], or E_S, the activation energy of the self-diffusion coefficient D_S, when climb is involved, as in the example above. This type of creep law is very often observed in experiments for κ creep.

In conclusion, high-temperature stationary creep is almost never Newtonian when driven by dislocations. This is due to the fact that the number of mobile dislocations generated by the Frank–Read (or Bardeen–Herring) sources, their spatial distribution, the area they sweep and their annihilation velocity are in general dependent on σ.

Let us finally emphasise that different models can lead to similar behaviour (see, for instance, expressions (5.151) and (5.158), resulting from different mechanisms but nevertheless leading to the same exponent m and the same temperature dependence). The choice of a model and of a microscopic mechanism for describing an experiment is therefore always very delicate; almost always, complementary information on the microstructure of the crystal is needed to reach a conclusion.

5.3.4.5 *High-temperature polycrystalline solids as yield stress fluids*

In the first chapter, we classified high-temperature polycrystalline solids among yield stress fluids. We are now able to give an interpretation of this result. Indeed, we know that two mechanisms allow the material to deform.

The first one is matter diffusion across the grains (Nabarro–Herring model) or at the grain boundaries (Coble model). In both cases, the creep is Newtonian:

$$\dot{\varepsilon} = A\sigma, \tag{5.160}$$

where A is a constant depending on temperature and grain size.

The second one brings into play the cooperative motion of dislocations. In this case, the creep is strongly non-Newtonian, obeying a law of the type:

$$\dot{\varepsilon} = B\sigma^{m}, \tag{5.161}$$

where the exponent m generally takes values between 3 and 5.

Clearly, the first of these two mechanisms is predominant at low stress, while the second becomes important at high stress, the crossover between the two regimes occurring for a stress value:

$$\sigma_y = \left(\frac{A}{B}\right)^{\frac{1}{m-1}}. \tag{5.162}$$

In conclusion, a polycrystalline solid behaves as a very viscous Newtonian liquid at low stress ($\sigma < \sigma_y$), and as a strongly shear-thinning non-Newtonian liquid at high stress ($\sigma > \sigma_y$) (see Fig. 1.10). This behaviour is close to that of a yield stress fluid, such as we defined it in the first chapter, with a yield stress σ_y (see Section 1.4.1).

5.4 Crystal fracture

Two kinds of solids can be distinguished: those that are hard and brittle, such as glasses or cast iron, which break without undergoing plastic deformation, and the others, which deform plastically before breaking. In the first case, one speaks of *brittle fracture* and in the second of *ductile fracture*. Let us detail these concepts.

5.4.1 *Brittle fracture and ductile fracture*

In the graphs of Fig. 5.61, we sketch the deformation curves $\sigma = f(\varepsilon)$ during a tensile test (i.e. at a given $\dot{\varepsilon}$) for a *ductile* material (a) and a *brittle* one (b). Let σ_f be the fracture stress and σ_c the elastic limit. In the first case, the fracture occurs in the plastic regime ($\sigma_f > \sigma_c$), while in the second one, the material breaks in the elastic regime ($\sigma_f < \sigma_c$).

Experiment shows that, with the exception of metals and their alloys with f.c.c. structure (which are always ductile), each material exhibits a specific temperature, rather well defined experimentally, at which it goes from the brittle to the ductile regimes on heating:

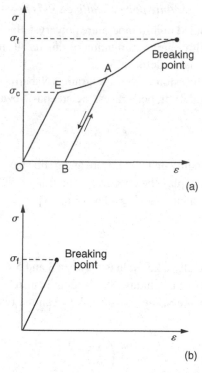

Fig. 5.61 Typical deformation curves (at a given $\dot{\varepsilon}$) for a ductile (a) and a brittle (b) material. In the first case, a permanent deformation $\varepsilon = OB$ sets in if, after having applied a stress that brings the sample to point A, relaxing this stress returns it to B. If the stress is once again applied, the sample deforms elastically until A, before undergoing a new permanent deformation. Thus, its elastic limit has increased, as a result of work hardening. The slope $d\sigma/d\varepsilon$ is called hardness. Frequently – but not always – the hardness increases, and then decreases immediately before fracture, as in the curve (a) [12].

this is the *brittle–ductile transition temperature* (T_{BD}). This specific temperature varies strongly from one material to another and is very sensitive to the type of test used (traction, shear, impact, etc.). For instance, $T_{BD} \approx 100$ K for soft steel (b.c.c.) under traction, while for molybdenum (b.c.c.), sodium chloride (f.c.c.) or tungsten (b.c.c.) in the same conditions one measures $T_{BD} \approx 300$, 650 and 750 K, respectively [12].

5.4.2 Theoretical cleavage stress

When fracture occurs along a reticular plane, one speaks of *cleavage*. This type of fracture occurs in the brittle regime, whether in monocrystals or in polycrystals, when the fracture surface cuts through the grains (*intragranular fracture*[5]).

---

[5] The fracture can also be *intergranular*. This second case is less frequent than the first one in metals with b.c.c. or h.c.p. structure: it ordinarily occurs when impurities (for instance S or P) segregate at the grain boundaries. In bulk ceramics, both types of fracture can be observed.

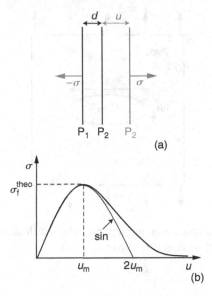

Fig. 5.62 (a) Crystal planes before and after a displacement u under the action of the stress σ; (b) shape of the stress curve $\sigma(u)$.

The theoretical cleavage stress is the stress one must apply to separate completely two adjacent reticular planes P_1 and P_2 (Fig. 5.62a), separated at rest by a distance d. Let σ be the applied stress and u the relative displacement of the two planes it produces. The graph of $\sigma(u)$ is shown in Fig. 5.62b. This curve is linear close to the origin, corresponding to the elastic deformation regime of the material. Therefore, one must have:

$$\sigma(u) = E\frac{u}{d} \quad \text{for} \quad u \ll d, \tag{5.163}$$

where E is the Young modulus of the material. As the displacement increases, one leaves the linear regime: the stress curve has an inflection, goes through a maximum σ_f^{theo} corresponding to the theoretical cleavage stress for a certain displacement u_m, then decreases and finally vanishes (Fig. 5.62b). In the interval $0, 2u_m$, the $\sigma(u)$ function can be represented by a law of the type:

$$\sigma(u) = \sigma_f^{theo} \sin\left(\frac{\pi u}{2u_m}\right). \tag{5.164}$$

This expression must reduce to Eq. (5.163) in the linear regime, yielding a first relation between u_m and σ_f^{theo}:

$$\frac{E}{d} = \frac{\pi \sigma_f^{theo}}{2u_m}. \tag{5.165}$$

To find a second relation, let us write that the work exerted by the applied stress in order to separate the two surfaces by a typical distance $2u_m$ equals the surface energy thus created,

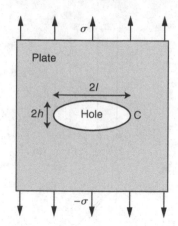

Fig. 5.63 Stressed thin plate containing an elliptical hole, which simulates a crack. The external stress is constant. The point C represents the tip of the crack.

2γ. This leads to:

$$2\gamma = \int_0^{2u_m} \sigma(u)\,\mathrm{d}u = \frac{4}{\pi} u_m \sigma_f^{theo}. \tag{5.166}$$

Comparing the two equations above, one finally obtains for the theoretical cleavage stress:

$$\sigma_f^{theo} = \sqrt{\frac{E\gamma}{d}}. \tag{5.167}$$

Experiment shows that this value is about 100 times larger than that effectively measured. One is thus faced with the same problem as in the case of the theoretical elastic limit, which is also much larger than the experimental values.

As for the elastic limit, the origin of this discrepancy can be explained by the presence of defects. One explication is that microcracks exist within the material, and they initiate the fracture (in particular in the case of glasses).

5.4.3 Fracture by propagation of cracks: the Griffith criterion

Detailed examination of material surfaces by optical or scanning electron microscopy sometimes reveals the presence of microcracks within the samples. This is the case for window glass, where their typical size is of a few tenths of a micrometre.

In 1920, Griffith showed that such a crack can become unstable under the action of stress, leading to sudden fracture of the material (Griffith, 1920; [12, 23]).

To show this, consider a thin plate pierced by an elliptical hole of length $2l$ and height $2h$ (Fig. 5.63). If the size of the plate is much larger than that of the hole, it can be shown

(Inglis, 1913) that the stress has its maximum value at point C, namely:

$$\sigma_m = \sigma \left(1 + 2\sqrt{\frac{l}{\rho}} \right), \qquad (5.168)$$

with ρ the radius of curvature at that point, given by h^2/l. For $h \ll l$, the elliptical hole is very elongated, resembling a crack. In this limit, the formula above becomes:

$$\frac{\sigma_m}{\sigma} = 2\sqrt{\frac{l}{\rho}}. \qquad (5.169)$$

The quantity σ_m/σ is termed the *stress intensity ratio*. This formula shows that the stress is strongly amplified at the tip of the crack, the amplification factor depending on the shape and length of the hole.

The estimate above yields a first criterion for crack propagation by writing that, at the tip of the crack, $\sigma_m = \sigma_f^{theo}$. This relation leads to a new expression for the cleavage stress:

$$\sigma_f = \frac{1}{2}\sqrt{\frac{\gamma E \rho}{dl}}. \qquad (5.170)$$

This expression shows that, as ρ decreases, so does the material's resilience to fracture. For a very narrow crack (in the zero-width limit), one can reasonably set $\rho \approx d$, the distance between reticular planes (also, the smallest length scale of the problem), yielding the order of magnitude (we do away with the 1/2 factor):

$$\sigma_f \approx \sqrt{\frac{\gamma E}{l}}. \qquad (5.171)$$

This is the *Griffith stress*. Obviously, its demonstration is not very rigorous, since at the tip of the crack the linear elasticity approximation is clearly unwarranted for $\rho \approx d$.

One way of avoiding this complication is by means of a global reasoning, making an energy balance over the plate. Indeed, this is the approach Griffith took in 1920. Let us follow his reasoning and assume that a crack of size $2l$ (and zero width) nucleates in the material. Due to the creation of the two free surfaces, the energy of the system increases by $4\gamma l$ (per unit length in the z direction normal to the plate). On the other hand, the surface of the two lips of the fracture being free, the stress vanishes at these surfaces. A gain in elastic energy follows, since the stress goes from 0 to σ over a distance of order l around the crack (far from the crack, the stress field remains unchanged). The density of elastic energy being given by $\sigma^2/2E$, the energy gain (per unit length along z) is roughly $4l^2\sigma^2/2E$. A more rigorous calculation, accounting for the exact stress distribution yields $\pi l^2\sigma^2/E$. By summing these two contributions, one obtains the energy variation over the entire plate due to nucleation of a crack of length l:

$$\Delta E = 4\gamma l - \frac{\pi l^2 \sigma^2}{E}. \qquad (5.172)$$

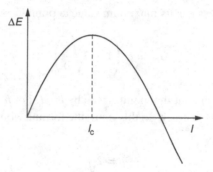

Fig. 5.64 Variation of the plate energy as a function of crack size.

We plot the function $\Delta E(l)$ in Fig. 5.64. This function goes through a maximum for

$$l = l_c = \frac{2\gamma E}{\pi \sigma^2}. \tag{5.173}$$

This means that, at a fixed stress σ, all cracks of size $l < l_c$ will collapse (admitting they can 'heal,' which requires that the two lips of the crack are not contaminated by impurities or oxidised); on the other hand, cracks of size $l > l_c$ will propagate. Thus, the value l_c defines the critical crack size for a given applied stress σ.

In fact, cracks already exist in the material. Let l_{max} be the length of the largest one. The cracks will propagate if the stress σ is larger than the cleavage stress obtained by means of formula (5.173):

$$\sigma_f = \sqrt{\frac{2\gamma E}{\pi l_{max}}} \approx \sqrt{\frac{\gamma E}{l_{max}}}. \tag{5.174}$$

In the end, we have recovered the expression for the Griffith stress, already given in Eq. (5.171).

This theory is well verified experimentally in materials such as glass, where the experiments give $\sigma_f \approx 0.2$ GPa for a Young modulus $E \approx 60$ GPa and a surface tension $\gamma \approx 0.2$ J/m^2. Using these values, the Griffith criterion yields $l_{max} \approx 0.3$ μm, which is indeed compatible with the size of the microcracks routinely observed in glass.

In the next section, we will analyse the propagation of a crack, when the fracture threshold defined by the Griffith criterion is exceeded.[6]

5.4.4 Velocity of crack propagation

Let us once again consider a thin plate under the stress σ and exhibiting a cut of size l (Fig. 5.65).

---

[6] If the stress is slightly lower than the Griffith stress, the material can still break. The fracture is then much slower, as it involves activated processes. In this case, one speaks of *delayed fracture* (Santucci, 2004).

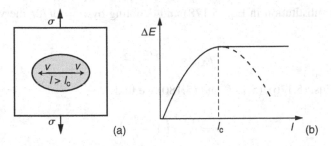

Fig. 5.65 Fracture of a plate submitted to uniaxial stress and exhibiting a cut of size l (a). The cut starts to propagate as its length reaches the value l_c. Material fracture occurs. In the dynamic regime, the total energy is conserved, as shown in diagram (b). In this regime, the stress relaxes over a surface of size l around the crack (grey area in diagram (a)). This is also the region that concentrates the kinetic energy stored by the material.

As long as $l < l_c$, with l_c the critical length defined by the Griffith criterion (5.173), the plate holds. In this quasi-static regime, its energy is given by Eq. (5.172), which can also be written as a function of l_c in the form:

$$\Delta E_{qs} = 2\gamma l_c - \frac{2\gamma}{l_c}(l - l_c)^2. \tag{5.175}$$

As l reaches the value l_c, the material breaks suddenly. The crack propagates so fast that the total energy of the plate can be considered constant in time (Fig. 5.65b):[7]

$$\Delta E_{tot} = \Delta E_{qs}(l = l_c) = 2\gamma l_c. \tag{5.176}$$

But aside from the elastic term (already calculated in Eq. (5.175)), the total energy contains the kinetic energy of the atoms contained in the region (of size l^2) where the stress relaxes, so that:

$$\Delta E_{tot} = 2\gamma l_c - \frac{2\gamma}{l_c}(l - l_c)^2 + E_{kin}. \tag{5.177}$$

As an order of magnitude:

$$E_{kin} \approx \rho l^2 \left(\frac{du}{dt}\right)^2, \tag{5.178}$$

with u the atom displacement induced by stress relaxation within an area of size l where the fracture occurs (this form requires that the propagation velocity of the crack be lower than the speed of sound, a result that we will check *ex post facto*). Consequently:

$$u \approx l\frac{\sigma}{E}, \tag{5.179}$$

---

[7] This amounts to assuming that the perturbations induced close to the crack have no time to propagate to the edges of the plate, where the external stress σ is applied. Hence, this latter does not effect any work during rapid fracture of the material, justifying the assumption of conserved total energy of the plate.

yielding, after substitution in Eq. (5.178) and denoting by $v = dl/dt$ the velocity of the crack:

$$E_{\text{kin}} \approx \rho l^2 \frac{\sigma^2}{E^2} v^2. \qquad (5.180)$$

Finally, from Eqs. (5.176), (5.177) and (5.180) we find:

$$v = v_{\text{max}} \left(1 - \frac{l_c}{l}\right), \qquad (5.181)$$

where $v_{\text{max}} \approx (E/\rho)^{1/2}$ is the maximum propagation velocity of the crack. Note that here v_{max} is identified with the speed of sound.

Remarkably, this very simplified calculation gives a good result, up to a numerical prefactor. Indeed, more sophisticated calculations also lead to the law (5.181) and show that the maximum propagation velocity is in fact given by the speed of Rayleigh waves, calculated in Section 4.2.3 (Yoffe, 1951; [23]).

This good agreement is even more surprising when we realise that the expressions proposed for the total energy and the kinetic energy are completely wrong at high velocity. Indeed, these two quantities must diverge as the propagation velocity of the crack tends to the speed of sound (the same occurs for dislocations, see Section 5.2.1), but since both exhibit exactly the same divergence, and since the velocity of the crack is given by their ratio, the errors compensate exactly.

The very simplified theory of fracture presented above is relatively accurate when describing brittle materials, in particular amorphous glasses.

On the other hand, it does not apply above the brittle–ductile transition temperature. Under these conditions, the material breaks after plastic deformation: *ductile fracture* occurs. In this regime, very relevant from a practical point of view, one must take into account the mechanisms of plastic deformation, which bring into play dislocations and their pile-ups. We will not go into these subtleties here, referring the reader to more specialised works (see [23] and also [12] for an introduction).

Instead, we will close this chapter by commenting on the case of *colloidal crystals*, already discussed at the end of the previous chapter (Section 4.3). We will show that their rheological behaviour exhibits new phenomena, not encountered in ordinary solids, such as melting under shear.

5.5 Plasticity of colloidal crystals and melting under shear

Since colloidal crystals are very soft (Section 4.3.2), they can easily be exposed to stresses much higher than their shear modulus. Obviously, such conditions are unattainable in the lab with ordinary solids. Experiments on colloidal crystals are therefore very interesting from this point of view, since they might provide, by a mere scale change, valuable information on the response of an atomic crystal to extreme shear conditions (for instance, the impact of a shell on armour plate). It seems, however, that such an extrapolation is dangerous, in view of an essential difference between a colloidal crystal and an atomic one: unlike the latter, the former is immersed in a solvent, which transmits the stresses.

Beautiful experiments on the continuous shear of colloidal crystals were performed by P. Pieranski (Pieranski, 1983), in collaboration with M. Joanicot (Joanicot, 1984). Their rheological measurements, in cylindrical geometry and under imposed stress, show that a colloidal crystal responds elastically as long as the stress remains below

$$\sigma_c \approx 2 \times 10^{-3} \, \mu. \tag{5.182}$$

This elastic limit, generally attributed to the pile-up of dislocations at the grain boundaries (Mitaku et al., 1980) (see Section 5.2.7) is extremely small due to the low value of μ. But it is also much smaller than the theoretical value $\mu/2\pi$ given by Eq. (4.48), showing that a colloidal crystal does not creep by cooperative glide of the atomic planes on top of each other, but by the propagation of dislocations, as for ordinary crystals. This assumption is validated by viscosity measurements in the plastic regime, showing that the viscosity decreases systematically at the beginning of creep, and then saturates after a certain time. These experiments also show that the build-up time of the stationary regime in the plastic domain is of the order of a minute, much longer than the viscoelastic relaxation time η/μ of the material (of the order of 10^{-2} s). This result is an additional indication that dislocations are responsible for the plastic deformation (at least for low shear rates) and their concentration increases with time until the moment their production rate balances their annihilation rate (the time scales involved depend on the grain size, which varies from 10 μm to several millimetres in the experiments cited above). Note that, if this interpretation is correct, the dislocations do not hinder each other's motion, since no work hardening is observed (at least in the materials studied by these authors, where the volume fraction of particles never exceeds 10%).

These authors also showed that, at higher shear rates, the viscosity of their samples in the stationary regime (κ creep for metallurgists) is a strongly decreasing function of the shear rate. This phenomenon can be attributed to significant *changes in the structure* of the colloidal crystal, as shown by white light observation of coloured Bragg reflections from the sheared crystal. This is shown in the photograph of Fig. 5.66, taken during shear of a colloidal crystal between two parallel planes, where three distinct areas are clearly visible:

- a central, weakly sheared, area, where the crystal deforms plastically while preserving its polycrystalline structure;
- an intermediate annulus, where the shear rate is moderate and where only three kinds of coloured reflections (yellow, red and blue) form a star. This arrangement is the mark of a significant reorientation of the crystal planes under shear;
- an outer, strongly sheared annulus, where all the reflections have disappeared. In this region, the colloidal crystal is completely destructured and behaves as a disordered suspension.

The last observation shows that *the crystal melts above a critical shear rate*, and experiment shows that this threshold is proportional to the elastic modulus μ of the crystal at

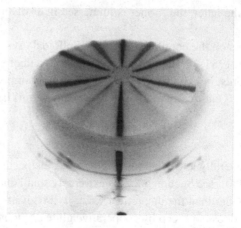

Fig. 5.66 White light observation of a colloidal crystal sheared between two parallel planes. Three regions are clearly visible: the central disc, where the material preserves its polycrystalline structure; an intermediate annulus, where the crystal is reoriented by the shear, as shown by the observation of star-like Bragg reflections; an outer annulus, where the reflections have disappeared, indicating that the crystal has melted (photo by P. Pieranski and M. Joanicot).

rest. We emphasise that this transition is not due to thermal effects, sample heating under shear remaining completely negligible due to the low viscosities involved.

Other authors have studied the behaviour of colloidal crystals under shear, in particular by light and neutron scattering. Among them, the team of N. A. Clark at Boulder showed that the structure of the crystal planes changes as they reorient under shear. More precisely, these experiments show that the particles order in hexagonal planes that glide on top of each other (Ackerson *et al.*, 1981, 1986).

More recent experiments also showed that at high shear rates (in the 'melted' regime, where all Bragg reflections have disappeared), the crystal breaks into small crystallites surrounded by the disordered liquid, their size decreasing with the shear rate (Imhof *et al.*, 1994). This interpretation explains the viscosity saturation at high shear rate in the melted phase (van der Vorst *et al.*, 1997).

In spite of all these results, it remains clear that the plasticity and *melting under shear* of colloidal crystals are complex phenomena still lacking a clear explanation, several attempts at theoretical explanation notwithstanding (Harrowell, 1990).

Finally, let us mention recent results on the plasticity of colloidal crystals in very confined environments, where one can observe numerous phase transitions not seen in the bulk (Neser *et al.*, 1997).

Appendix 5.A Demonstration of the Peach and Koehler formula

To demonstrate the Peach and Koehler formula (5.80), let us consider a dislocation loop D; we are searching for the work done by the *external forces* as the loop moves an infinitesimal distance. Let $\delta u_{i,j}$ be the variation of the strain tensor due to this displacement. In Section 4.1.3, dealing with the thermodynamics of deformation, we showed that:

$$\delta W = \int_V \sigma_{ij}\, \delta u_{i,j}\, dV. \tag{5A.1}$$

As a first approximation, the *external stress tensor*[8] does not depend on the position of the dislocation, so we can write, using the equilibrium equation $\sigma_{ij,j} = 0$:

$$\delta W = \delta \int_V \sigma_{ij} u_{i,j}\, dV = \delta \int_V (\sigma_{ij} u_i)_{,j}\, dV. \tag{5A.2}$$

In the Volterra process, the strain $\vec{u}$ is a homogeneous function exhibiting a discontinuity equal to the Burgers vector $\vec{b}$ on the cut surface S limited by the line D. One can therefore transform the volume integral above into an integral over the closed surface consisting of the two lips of the cut S_1 and S_2, connected by a tube of infinitesimal radius enclosing the dislocation line. Keeping in mind that the stress exhibits no discontinuity across the cut surface and since the Burgers vector can be extracted from the integral (being constant by definition), one has:

$$\delta W = b_i\, \delta \int_S \sigma_{ij}\, dS_j. \tag{5A.3}$$

One still needs to calculate the variation of the surface element δS_j resulting from a displacement $\delta \vec{r}$ of the line. This variation leads to a variation of the area S. Let $d\vec{l} = dl\vec{t}$, where $\vec{t}$ is the unit vector tangent to the line, oriented according to the Burgers convention. This means that, near the line, $\delta\vec{S} = -\,\delta\vec{r} \times d\vec{l}$, yielding (in index notation):

$$\delta S_j = -e_{jkl}\, \delta x_k\, dl_l = -e_{jkl}\, \delta x_k\, dl\, t_l. \tag{5A.4}$$

Finally, by substitution in Eq. (5A.3) and noting that the surface integral reduces to a line integral over the complete contour of the dislocation:

$$\delta W = \oint_D -\sigma_{ij} e_{jkl} b_i\, \delta x_k t_l\, dl. \tag{5A.5}$$

This formula can also be written in the simplified form:

$$\delta W = \oint_D f_k\, \delta x_k\, dl, \tag{5A.6}$$

where we set

$$f_k = -e_{jkl}\, \sigma_{ij}\, b_i t_l. \tag{5A.7}$$

---

[8] By definition, it is the total stress tensor minus the stress tensor due to the dislocation itself. In particular, it includes the stress created by all the other dislocations.

After rearranging the indices, one is finally left with:

$$f_i = e_{ijk}\, \sigma_{lj}\, b_l t_k \tag{5A.8}$$

or, in vector notation:

$$\vec{f} = \left(\underline{\sigma}^{\mathrm{T}} \vec{b}\right) \times \vec{t}, \tag{5A.9}$$

where $\underline{\sigma}^{\mathrm{T}}$ stands for the transpose of $\underline{\sigma}$.

This formula gives the expression for the *effective elastic force* exerted by the external stress field on the dislocation.

The calculation tells us simply that everything happens as if the dislocation line were a material line acted upon by a true force. However, this is only a convenient manner of speaking. For this reason, one speaks of an *effective configuration force*. The situation is similar to our definition of a dislocation mass, which is also a configurational effective mass.

It should be noted that no assumption was made as to the symmetry of the stress tensor: Eq. (5A.9) is therefore valid in liquid crystals, where this tensor is not symmetric.

In solids, on the other hand, the formula reduces to its ordinary form (5.80) given in the text, since $\underline{\sigma}^{\mathrm{T}} = \underline{\sigma}$ (this time, the stress tensor is symmetric).

Let us end this appendix by pointing out that this formula as we have written it can lead to erroneous results. This point was first raised by Weertman in 1965 (Weertman, 1965). Indeed, suppose $\underline{\sigma}$ is a hydrostatic pressure field. In this case, the Peach and Koehler formula predicts that an edge dislocation experiences a climb force. This result is false, the sample volume remaining unchanged as the dislocation climbs. Indeed, the applied forces perform no work, so no force acts on the dislocation. The error cause is that the appropriate thermodynamic potential for a system under the hydrostatic pressure $P = -\frac{1}{3}\mathrm{tr}\underline{\sigma}$ is not the energy W, but rather the enthalpy $W + PV$. Consequently, in this analysis $\underline{\sigma}$ must be replaced by its deviator $\underline{\sigma} - \frac{1}{3}\mathrm{tr}\underline{\sigma}\, \mathrm{I}$.

Appendix 5.B Interaction energy between a dislocation and a spherical cavity

We know that the total energy of the entire system dislocation + cavity is:

$$E_{\mathrm{tot}} = \int \frac{1}{2}\sigma_{ij}\varepsilon_{ij}\, \mathrm{d}V, \tag{5B.1}$$

where σ_{ij} and ε_{ij} are the total stress and strain fields. Here, the integration is over the entire crystal volume outside the dislocation core and the cavity. In linear elasticity the solutions can be superposed, yielding:

$$\sigma_{ij} = \sigma_{ij}^{\mathrm{d}} + \sigma_{ij}^{\mathrm{c}}, \tag{5B.2}$$

$$\varepsilon_{ij} = \varepsilon_{ij}^{\mathrm{d}} + \varepsilon_{ij}^{\mathrm{c}}, \tag{5B.3}$$

where superscripts c and d refer to the cavity and the dislocation, respectively. Substitution in Eq. (5B.1) gives:

$$E_{tot} = \int \frac{1}{2} \left(\sigma_{ij}^d \varepsilon_{ij}^d + \sigma_{ij}^c \varepsilon_{ij}^c + \sigma_{ij}^d \varepsilon_{ij}^c + \sigma_{ij}^c \varepsilon_{ij}^d \right) dV. \tag{5B.4}$$

The first two terms of this expression are the self-energy of the dislocation and of the inclusion, respectively. The two following terms correspond to the *interaction energy* between the inclusion and the dislocation:

$$E_{int} = \int \frac{1}{2} \left(\sigma_{ij}^d \varepsilon_{ij}^c + \sigma_{ij}^c \varepsilon_{ij}^d \right) dV. \tag{5B.5}$$

It can be calculated by first applying the reciprocity theorem discussed in Section 4.1.8.2. This yields:

$$E_{int} = \int \sigma_{ij}^d \varepsilon_{ij}^c \, dV. \tag{5B.6}$$

Keeping in mind that the stress tensor is symmetric and obeys the equilibrium equation $\mathrm{div}\underline{\sigma} = 0$, one then finds:

$$E_{int} = \int \sigma_{ij}^d u_{i,j}^c \, dV = \int \left(\sigma_{ij}^d u_i^c \right)_{,j} dV = \int_{\text{cavity}} \sigma_{ij}^d u_i^c v_j \, dS, \tag{5B.7}$$

where we transformed the volume integral into a surface integral over the surface of the cavity (here, $\vec{v}$ is the unit vector normal to the cavity surface and pointing inwards). Since at the scale of the cavity the stress field created by the dislocation is more or less constant, it can be extracted from the integral and then, using the spherical symmetry of the cavity, we arrive at:

$$E_{int} = \frac{1}{3} \left(\sigma_{11}^d + \sigma_{22}^d + \sigma_{33}^d \right) \int_{\text{cavity}} u_i^c v_i \, dV. \tag{5B.8}$$

The last integral is nothing other than $-\Delta V$, where ΔV is the volume variation produced by introducing the foreign atom ($\Delta V = 4\pi R^3 \varepsilon$). Using the expression of the stress tensor (5.31) calculated for an edge dislocation, one finally has:

$$E_{int} = \frac{4\mu b R^3}{3r} \frac{1+v}{1-v} \varepsilon \sin \theta. \tag{5B.9}$$

QED

References

Ackerson, B. J. and N. A. Clark, *Phys. Rev. Lett.*, **46**, 123 (1981).
Ackerson, B. J., J. B. Hayter, N. A. Clark and L. Cotter, *J. Chem. Phys.*, **84**, 2344 (1986).
Amelinckx, S., *Acta Metall.*, **6**, 34 (1958).
Andrade, E. N. da C., *Proc. Roy. Soc. A*, **84**, 1 (1910).

Bardeen, J. and C. Herring, in *Imperfections in Nearly Perfect Crystals*, W. Shockley (Ed.), New York: Wiley (1952).

Blavette, D., E. Cadel, A. Fraczkiewicz and A. Menand, *Science*, **286**, 2317 (1999).

Coble, R. L., *J. Appl. Phys.*, **34**, 1679 (1963).

Cottrell, A. H., *The Properties of Materials at High Rates of Strain*, London: Institution of Mechanical Engineers (1957).

Cottrell, A. H. and B. A. Bilby, *Proc. Phys. Soc. A*, **62**, 49 (1949).

Dash, W. C., in *Dislocations and Mechanical Properties of Crystals*, J. C. Fisher (Ed.), New York: Wiley (1957).

Eshelby, J. D., *Solid State Phys.*, **3**, 79 (1956).

Eshelby, J. D., F. C. Frank and F. R. N. Nabarro, *Phil. Mag.*, **42**, 351 (1951).

Frank, F. C., *Bristol Conference*, London: Physical Society (1955).

Frank, F. C. and W. T. Read, *Phys. Rev.*, **79**, 722 (1950).

Griffith, A. A., *Phil. Trans. Roy. Soc. Lond. A*, **221**, 163 (1920).

Harper, J. and J. E. Dorn, *Acta Metall.*, **5**, 654 (1957).

Harrowell, P., *Phys. Rev. A*, **42**, 3427 (1990).

Herring, C., *J. Appl. Phys.*, **21**, 437 (1950).

Imhof, A., A. van Blaaderen, G. Maret, J. Mellela and J. K. G. Dhont, *J. Chem. Phys.*, **100**, 2170 (1994).

Inglis, C. E., *Trans. Inst. Naval Archit.*, **55**, 219 (1913).

Joanicot, M., Ph.D. Thesis, *Elasticity, plasticity and melting of colloidal crystals under shear* (in French), Université Pierre et Marie Curie (1984).

Keh, A. S. and S. Weissmann, in *Electron Microscopy and Strength of Crystals*, G. Thomas and J. Washburn (Eds.), New York: Interscience (1963) p. 231.

Koehler, J. S. and G. deWit, *Phys. Rev.*, **116**, 1121 (1959).

McLean, D., *Mechanical Properties of Metals*, New York: Wiley (1962).

Mitaku, S., T. Ohtsuki and K. Okano, *Jap. J. Appl. Phys.*, **19**, 439 (1980).

Mott, N. F. and F. R. N. Nabarro, *Proc. Phys. Soc.*, **52**, 86 (1940).

Mott, N. F., *J. Inst. Met.*, **72**, 367 (1946).

Mott, N. F., *Phil. Mag.*, **43**, 1151 (1952).

Nabarro, F. R. N., *Proc. Phys. Soc.*, **58**, 699 (1946).

Nabarro, F. R. N., *Proc. Phys. Soc.*, **59**, 256 (1947).

Nabarro, F. R. N., *Bristol Conference*, London: Physical Society (1948).

Nabarro, F. R. N., *Phil. Mag.*, **16**, 231 (1967).

Neser, S., C. Bechinger, P. Leiderer and T. Palberg, *Phys. Rev. Lett.*, **79**, 2348 (1997).

Orowan, E., *Proc. Phys. Soc.*, **52**, 8 (1940).

Orowan, E., *Symposium on Internal Stresses in Metals*, London: The Institute of Metals (1948) p. 451.

Oswald, P., *J. Physique Lett. France*, **45**, L1037 (1984).

Oswald, P., *J. Physique France*, **46**, 1255 (1985).

Oswald, P., *J. Physique France*, **49**, 1083 (1988).

Peach, M. and J. S. Koehler, *Phys. Rev.*, **80**, 436 (1950).

Peierls, R., *Proc. Phys. Soc. Lond.*, **52**, 34 (1940).

Pieranski, P., *Contemp. Phys.*, **24**, 25 (1983).

Portevin, A. and F. Lechatelier, *C. R. Acad. Sci. Paris*, **176**, 507 (1923).

Saada, G., *Microscopie électronique des lames minces cristallines*, Paris: Masson & Cie Éditeurs (1966).

Santucci, S., *Croissance lente thermiquement activée et piégeage d'une fissure dans les matériaux structurés à une échelle mésoscopique : expériences et modèles*, Ph.D. Thesis, École Normale Supérieure de Lyon (2004).

van der Vorst, B., D. van den Ende, H. J. J. Aelmans and J. Mellema, *Phys. Rev. E*, **56**, 3119 (1997).

Volterra, V., *Ann. École Normale Sup. Paris*, **24**(3), 400 (1907).

Weertman, J., in *Response of Metals to High Velocity Deformation*, P. G. Schewmon and V. F. Zackay (Eds.), New York: Interscience (1961) p. 205.

Weertman, J., *Phil. Mag.*, **11**, 1217 (1965).

Weertman, J., *Rate Processes in Plastic Deformation of Materials*, Proceedings J. E. Dorn Symposium, J. C. M. Li and A. K. Mukherjee (Eds.), ASM 4 (1975) p. 315.

Westmacott, K. H., R. S. Barnes, D. Hull and R. E. Smallman, *Phil. Mag.*, **6**, 929 (1961).

Woodruff, D. P., *The Solid-Liquid Interface*, Cambridge: Cambridge University Press (1973).

Yoffe, E. H., *Phil. Mag.*, **42**, 739 (1951).

6

Rheology of isotropic viscoelastic materials: macroscopic aspects

After having studied simple liquids and solids, we will now be concerned with *viscoelastic* and *isotropic* fluids and solids. The former category being the most commonly encountered, we will concentrate upon it, giving only a few general indications on viscoelastic solids. Such materials do exist in nature, however most of them are nowadays synthesised by the chemical industry. We can cite as examples polymer materials, either molten (silicone oils) or in reticulated form (rubber, plastics), emulsions or suspensions (paint, printing ink, drilling mud), surfactant solutions (dish detergent, shampoo), as well as a sizable fraction of biological liquids (blood, tree sap, etc.). This list, far from being exhaustive, highlights the importance and the ever-increasing role played by these materials in our day-to-day lives.

This chapter aims to define the most general concepts and mathematical tools used in describing their rheological behaviour at the macroscopic scale, without going into the detail of the microscopic structure. For a microscopic treatment, we refer the reader to the next chapter, where we describe model systems such as emulsions, polymer melts or surfactant solutions. Let us now present the contents of the chapter.

We begin by recalling the concepts of *linear* and *nonlinear regimes* (Section 6.1). As we already mentioned several times in Chapter 1, this distinction is essential, since the flow properties are strongly dependent on the regimes under study.

We then approach the problems specific to *linear viscoelasticity* and *oscillating flow* (Section 6.2), beginning with viscoelastic liquids and the *Maxwell model* with one relaxation time; we continue by extending this model to the case of materials exhibiting a (finite or infinite) distribution of relaxation times. Next, we give the general definition of the *stress relaxation modulus* under shear $G(t)$, to which can be associated (in Fourier representation under oscillating shear) the *complex modulus* $G^*(\omega)$, consisting of a real part describing the elastic response of the material (*elastic modulus* $G'(\omega)$) and the imaginary part, due to the viscous response (*loss modulus* $G''(\omega)$). We also give the definitions of the memory function $m(t)$, of the compliance $J(t)$, as well as of the transient viscosities $\eta^+(t)$ and $\eta^-(t)$ and of the complex viscosity $\eta^*(\omega)$. These laws will then be generalised to the case of compressible materials, where the *relaxation modulus for uniform compression or dilation* $K(t)$ can no longer be considered infinite. We also give a brief description of *extensional flow* (of substantial practical importance, as exemplified in Fig. 1.22), and define a time-dependent *Young modulus* $E(t)$. Then comes the description of viscoelastic solids. We will see that the *Kelvin–Voigt model* is better adapted to their description than the generalised Maxwell model, although they are formally equivalent. Finally, we will discuss the experimental techniques for measuring the linear viscoelasticity functions using traditional rheometers, or more sophisticated ones, such as the piezorheometer or the magnetic bead rheometer.

The chapter continues with the study of *continuous flow*, which, as we will see, requires the framework of *nonlinear viscoelasticity* (Section 6.3). We will start by analysing the case of shear flow. We will see that the viscosity generally depends on the shear rate, and that finite normal stress differences (traditionally denoted by N_1 and N_2) can appear, with unordinary consequences. As an example, we will discuss the Weissenberg effect, whereby a liquid climbs up a rotating rod, and also the swelling of the free surface of a liquid flowing in a tilted trough. We will then analyse the stability of simple Couette flow, and present some results relating to Poiseuille flow. We will also discuss the methods for measuring the viscosity and the two normal stress differences using a cone-plate rheometer. The section ends with the description of elongational flow and of an experimental setup that can be used to measure the corresponding viscosity η_e.

We will then approach the difficult problem of calculating the viscometric functions (η, η_e, N_1 and N_2) and their relations with the linear viscoelasticity functions (G', G'') (Section 6.4). We will see that kinematic nonlinearities must be taken into account in the constitutive equations, even in the linear regime, when continuous flow is considered. In this respect, we will discuss several quasi-linear models, among them the *convected Maxwell model* and the *Oldroyd-B and Lodge models*.

We will see, however, that these models have shortcomings (for instance, they predict a viscosity divergence at high elongation rate) that must be palliated by accounting for the nonlinearities of the material. Consequently, we will describe several nonlinear models, extending those mentioned above; among them, the *Oldroyd 8-constant model*, those of *Giesekus* or *Phan-Thien and Tanner*, as well as the *FENE* or *KBKZ* models. We will see that these models provide (relatively accurate) predictions for the viscometric functions, as well as for the functions of linear viscoelasticity. We will also give two empirical laws, very useful in practice, which apply in particular to polymer melts, in the linear and nonlinear regimes, and connect the viscosity η and the first normal stress difference N_1, measured at a shear rate $\dot\gamma$, to the complex modulus G^*, measured in linear viscoelasticity at the frequency $\omega = \dot\gamma$ (the *Cox and Merz rule* and the *Laun rule*).

We will then discuss the *time–temperature superposition principle*. This principle is also widely employed in the study of polymer systems: it allows one to extrapolate measurements performed at a certain temperature and over a certain frequency range to measurements taken at a different temperature in a different frequency range (Section 6.5).

Finally, we will say a few words about the *Darcy law* and about the *Saffman–Taylor instability* in shear–thinning fluids (Section 6.6).

We emphasise that throughout this chapter (except for Section 6.2.3) we will assume all materials to be *incompressible*.

6.1 Linear and nonlinear regimes

Since the concepts of linear and nonlinear regimes are essential in rheology, we will define them right away.

As we have already pointed out, the objects composing the structure of a viscoelastic fluid are deformed under stress. In a polymer melt or solution, these objects are the macro-molecules themselves; in an emulsion, they are the liquid droplets, etc. If the stress is released suddenly, these objects retrieve their equilibrium configuration over a typical time defining the *viscoelastic relaxation time* τ. This time plays an important role in rheology and sets the limit between the *linear* and the *nonlinear* regimes.

More specifically, we will say that the system is in the *linear regime* if the deformation of the structural objects is proportional to the applied stress; this occurs for moderate stress values. In the *nonlinear regime*, under stronger applied stress, the deformations are no longer proportional to the stress and begin to saturate, corresponding to the appearance of nonlinear terms in the constitutive law of the material.

In more detail:

- In the case of oscillating motion, the system remains in the linear regimes as long as the imposed strain γ is small, i.e. below a certain limiting value γ_c (of the order of 1 in emulsions or in polymer melts, provided the system is far enough from the glass transition temperature, in which case γ_c can be reduced to a small percentage, see Section 1.5.1.1). Also note that γ_c tends to decrease with increasing frequency.
- Under continuous shear, the relevant parameter is no longer the imposed strain γ, which can be much larger than 1, but rather the shear rate $\dot{\gamma}$. To fix the ideas, let us consider the case of an emulsion or of a dilute polymer solution. If $\dot{\gamma}$, which also sets the 'collision frequency' between objects (droplets or polymer molecules), is small compared to γ_c/τ (where τ is the terminal relaxation time, defined as the longest typical relaxation time of the material), then the objects have enough time to relax between two successive collisions and their deformation remains small: thus, the system remains locally in the linear regime, exhibiting essentially Newtonian behaviour under continuous shear and moderate normal stress values (see Chapter 1). On the other hand, the objects are strongly deformed for shear rates above γ_c/τ. The system crosses over to the nonlinear regime, characterised by a clearly non-Newtonian behaviour (the shear viscosity now depends on $\dot{\gamma}$) and by the presence of significant normal stress values, which can sometimes exceed by far the viscous shear stress.

These results are schematically summarised in Fig. 6.1, showing in the plane of the experimental parameters ((ω, γ) plane for oscillating shear and ($\dot{\gamma}, \gamma$) plane for continuous shear) the regions corresponding to the linear and nonlinear regimes (white and cross-hatched regions, respectively).

6.2 Linear viscoelasticity and oscillating flow

In this section, we only consider the *linear regime*, namely a flow regime where the *local constitutive law of the material is linear* (the object deformation is small).

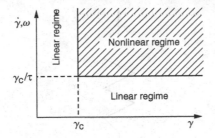

Fig. 6.1 Linear and nonlinear regimes, under oscillating or continuous shear. The boundaries between the two are rather fuzzy and vary considerably from one material to the next (this concerns in particular the value of γ_c, which can be much smaller than 1).

Furthermore, we assume that the *strain γ of the material remains small* at all times ($\gamma \ll 1$), such that the stress remains proportional to γ or to its time derivative $\dot{\gamma}$ (*linear response*).[1]

Both these conditions, not just the first one, are required to remain in the domain of *linear viscoelasticity*, to which this section is limited.

Obviously, oscillating flows can fulfil both conditions, as long as their amplitude remains small. We will therefore give them particular attention.

This discussion is built upon the *Maxwell model*, which is the simplest one for describing the behaviour of a viscoelastic liquid, and on the *Kelvin–Voigt model*, better adapted to the case of viscoelastic solids.

6.2.1 Maxwell model (viscoelastic liquids)

First proposed by J. C. Maxwell in 1867, this model contains the essential ingredients of viscoelasticity. For simplicity's sake, we will begin with the textbook case of *simple shear strain* in a material described by a *single relaxation time*. Note that, although rather rare, such materials do exist. In particular, this is the case for certain lyotropic mixtures that form giant micelles, also known as 'living polymers' (see the next chapter).

6.2.1.1 Single relaxation time model

In this case, the material behaves as a spring connected in series with a viscous damper, or dashpot. This assembly is depicted schematically in Fig. 6.2.

In this model the strains, and hence the strain rates, add up, yielding:

$$\dot{\gamma} = \dot{\gamma}_E + \dot{\gamma}_v, \tag{6.1}$$

where γ_E and γ_v are the strain of the spring and the dashpot, respectively, under the action of the applied stress. As the spring models a perfect Hookean solid, we have:

$$\sigma = G\gamma_E, \tag{6.2}$$

where G is the shear elastic modulus.

---

[1] Indeed, we will see in Section 6.4 that, even in the linear regime, kinematic nonlinearities are bound to appear in the constitutive laws of materials for large values of the strain γ.

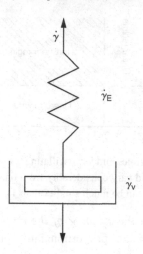

Fig. 6.2 Single relaxation time Maxwell model (a spring connected in series with a damper).

As to the dashpot, behaving as an ideal fluid with viscosity η, it obeys the Newton law:

$$\sigma = \eta \dot{\gamma}_v. \tag{6.3}$$

These three equations immediately yield the *Maxwell equation*, relating the stress to the total strain:

$$\dot{\gamma} = \frac{\dot{\sigma}}{G} + \frac{\sigma}{\eta}. \tag{6.4}$$

Multiplication by η gives:

$$\sigma + \tau \dot{\sigma} = \eta \dot{\gamma}, \tag{6.5}$$

where τ is the *Maxwell relaxation time*:

$$\tau = \frac{\eta}{G}. \tag{6.6}$$

When the system is suddenly sheared (starting at a time $t = 0$) at a constant shear rate $\dot{\gamma}$, the solution to Eq. (6.5) reads:

$$\sigma(t) = \eta \dot{\gamma} \left[1 - \exp\left(-\frac{t}{\tau}\right) \right]. \tag{6.7}$$

This is again the typical behaviour described in the first chapter (see Fig. 1.11).

It is easily checked that the general solution to the Maxwell equation can be written under the integral form:

$$\sigma(t) = \frac{\eta}{\tau} \int_{-\infty}^{t} \exp\left[-(t - t')/\tau\right] \dot{\gamma}(t') \, dt'. \tag{6.8}$$

There are numerous materials that cannot be described by a simple Maxwell model. One possible generalisation consists of adding a second Maxwell element, connected in parallel with the first one.

6.2.1.2 Burgers model and Jeffrey model

We now consider a new assembly, composed of two 'spring–dashpot' combinations connected in parallel (Fig. 6.3).

In this case, the constitutive equations are as follows:

$$\dot{\gamma} = \frac{\dot{\sigma}_1}{G_1} + \frac{\sigma_1}{\eta_1}, \tag{6.9}$$

$$\dot{\gamma} = \frac{\dot{\sigma}_2}{G_2} + \frac{\sigma_2}{\eta_2}. \tag{6.10}$$

Knowing that the total stress is:

$$\sigma = \sigma_1 + \sigma_2, \tag{6.11}$$

one can easily verify that they satisfy the following equation, which generalises the Maxwell equation (6.5):

$$(\eta_1 + \eta_2)\dot{\gamma} + (\eta_1\tau_2 + \eta_2\tau_1)\ddot{\gamma} = \sigma + (\tau_1 + \tau_2)\dot{\sigma} + (\tau_1\tau_2)\ddot{\sigma}. \tag{6.12}$$

This is the *Burgers equation*. This (also linear) equation now involves the second time derivatives of the stress σ and the strain γ.

A simplified (but extremely useful) version of this equation is that of Jeffrey. In this case, one of the Maxwell elements is simply a dashpot. Setting $\eta_1 = \eta_P$, $G_1 = G_P$ and $\tau_1 = \tau_P$, as well as $\eta_2 = \eta_s$, $G_2 = \infty$ and $\tau_2 = 0$ yields the *Jeffrey equation*:

$$(\eta_P + \eta_s)\dot{\gamma} + \eta_s\tau_P\ddot{\gamma} = \sigma + \tau_P\dot{\sigma}, \tag{6.13}$$

which can also be written as:

$$\eta_0(\dot{\gamma} + \tau_R\ddot{\gamma}) = \sigma + \tau_P\dot{\sigma}, \tag{6.14}$$

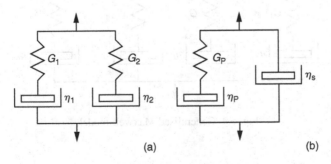

Fig. 6.3 (a) Burgers model; (b) Jeffrey model.

with $\eta_0 = \eta_P + \eta_s$ and $\tau_R = \tau_P \eta_s / \eta_0$. Two characteristic times appear here. The first one, τ_P, is the *viscoelastic relaxation time* of the particles. The other one, τ_R, is called the *retardation time*, for reasons that will become apparent in Section 6.2.1.9. Finally, η_0 gives the continuous shear viscosity in the stationary regime, as will be generally shown in Section 6.2.1.7.

Dilute emulsions of one incompressible viscous liquid in another, or dilute suspensions of hard spheres in a viscous liquid, are well described by this type of model. Under these examples, the solitary dashpot represents the host fluid (or the solvent, hence the subscript s), while the other Maxwell element describes the suspended objects (or particles, hence the P).

6.2.1.3 Model involving a finite distribution of relaxation times

Connecting in parallel a finite number of 'spring–dashpot' Maxwell elements, each characterised by the elastic modulus G_i and the viscosity η_i, yields the *generalised Maxwell model*, displayed schematically in Fig. 6.4.

In this case, each Maxwell element follows an equation of the type:

$$\dot{\gamma} = \frac{\dot{\sigma}_i}{G_i} + \frac{\sigma_i}{\eta_i} \quad (i = 1, \dots n), \tag{6.15}$$

so the total stress can be obtained in integral form:

$$\sigma(t) = \sum_1^n \sigma_i = \sum_1^n \frac{\eta_i}{\tau_i} \int_{-\infty}^t \exp\left[-(t - t')/\tau_i\right] \dot{\gamma}(t') \, dt', \tag{6.16}$$

where $\tau_i = \eta_i / G_i$ is the relaxation time associated with the ith Maxwell element.[2]

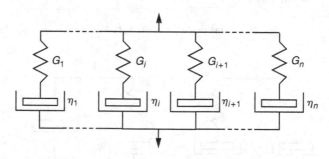

Fig. 6.4 Generalised Maxwell model.

---

[2] In the following, we will denote by τ_{min} and τ_{max} the shortest and longest relaxation times, respectively, provided that these two times can be defined. τ_{max} is called the 'terminal relaxation time'.

One can show that, in this case, the stress and the strain satisfy an equation of the type [24]:

$$\left(1 + \alpha_1 \frac{\partial}{\partial t} + \alpha_2 \frac{\partial^2}{\partial t^2} + \cdots + \alpha_n \frac{\partial^n}{\partial t^n}\right) \sigma = \left(1 + \beta_1 \frac{\partial}{\partial t} + \beta_2 \frac{\partial^2}{\partial t^2} + \cdots + \beta_n \frac{\partial^n}{\partial t^n}\right) \gamma.$$

(6.17)

This is the general differential equation of linear viscoelasticity.

6.2.1.4 Model involving a continuous distribution of relaxation times: definition of the relaxation modulus under shear $G(t)$

This model can be generalised to the case of a continuous distribution of relaxation times. Let $H(\tau)$ be the *distribution function (or spectrum) of relaxation times*, such that $H(\tau)\,d\tau$ accounts for the contribution to the total viscosity of all Maxwell elements with relaxation times contained between τ and $\tau + d\tau$. Equation (6.16) becomes:

$$\sigma(t) = \int_0^{+\infty} \frac{H(\tau)}{\tau} \left[\int_{-\infty}^t \exp\left[-(t - t')/\tau\right] \dot{\gamma}(t')\,dt'\right] d\tau.$$

(6.18)

By introducing the *shear stress relaxation modulus* $G(t)$, defined as:

$$G(t) = \int_0^{+\infty} H(\tau) \exp(-t/\tau) \frac{d\tau}{\tau},$$

(6.19)

one can rewrite the stress $\sigma(t)$ in the following general form:

$$\sigma(t) = \int_{-\infty}^t G(t - t')\,\dot{\gamma}(t')\,dt'.$$

(6.20)

Let us now see how to measure the modulus $G(t)$.

6.2.1.5 Measuring the relaxation modulus $G(t)$: stress relaxation after sudden deformation

To determine the modulus $G(t)$, one starts by applying to the (initially relaxed) sample a sudden deformation γ_0:

$$\gamma(t < 0) = 0 \quad \text{and} \quad \gamma(t > 0) = \gamma_0 = \text{Cst.}$$

(6.21)

'Sudden' in this context means that the deformation is applied over a time much shorter than the shortest time of the Maxwell distribution τ_{min}. Under these conditions, one can write:

$$\dot{\gamma}(t) = \gamma_0\,\delta(t),$$

(6.22)

where $\delta(t)$ is the Dirac function. From Eq. (6.20), one has:

$$\sigma(t) = \int_{-\infty}^t \gamma_0\,G(t - t')\delta(t')\,dt',$$

(6.23)

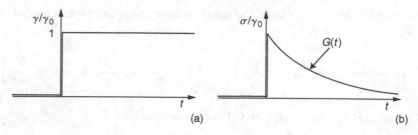

Fig. 6.5 Stress relaxation after sudden deformation. (a) Strain curve; (b) stress curve.

yielding

$$\sigma(t < 0) = 0 \quad \text{and} \quad \sigma(t > 0) = \gamma_0 G(t). \tag{6.24}$$

It is then sufficient to measure the stress relaxation curve to find $G(t)$ (Fig. 6.5).

This experiment justifies the name given to $G(t)$, that of *shear stress relaxation modulus*. It is easily checked that, in the case of the discrete model, one has:

$$G(t) = \sum_{1}^{n} G_i \exp\left(-\frac{t}{\tau_i}\right). \tag{6.25}$$

This expression shows that the instantaneous response of the material is that of an elastic solid of shear modulus

$$G(0) = \sum_{1}^{n} G_i. \tag{6.26}$$

Another essential feature of $G(t)$ is that it tends towards zero exponentially as t tends to infinity. We will use this property in the following.

6.2.1.6 Definition of the memory function $m(t)$

Our goal here is to relate the stress measured at a certain moment to the past deformations applied to the sample. This can be done by integrating by parts Eq. (6.20), resulting in:

$$\sigma(t) = \gamma(t)G(0) - \gamma(-\infty)G(+\infty) + \int_{-\infty}^{t} \dot{G}(t - t')\gamma(t')\,\mathrm{d}t'. \tag{6.27}$$

On the other hand:

$$\int_{-\infty}^{t} -\dot{G}(t - t')\gamma(t)\,\mathrm{d}t' = G(0)\gamma(t) - G(+\infty)\gamma(t). \tag{6.28}$$

Using these two equations, knowing that $G(+\infty) = 0$, yields:

$$\sigma(t) = \int_{-\infty}^{t} \dot{G}(t - t')\left[\gamma(t') - \gamma(t)\right]\mathrm{d}t'. \tag{6.29}$$

Introducing the *memory function* $m(t)$, defined by:

$$m(t) = -\dot{G}(t), \tag{6.30}$$

this formula becomes:

$$\sigma(t) = \int_{-\infty}^{t} m(t - t')\gamma(t, t')\,dt', \tag{6.31}$$

where $\gamma(t, t') = \gamma(t) - \gamma(t')$ is the strain at time t relative to that measured at time t'.

This integral formula provides a connection between the stress measured at time t to the strain undergone by the sample in the past ($t' < t$). This contribution is weighted by the $m(t - t')$ function, characterising the memory of the sample, hence the name. Note that this function has a maximum at time $t' = t$ and vanishes as $t' \to -\infty$. This behaviour reflects the time-limited memory of the sample.

6.2.1.7 Stress relaxation after the sudden cessation of a stationary flow: transient viscosity $\eta^-(t)$

Let us now assume that the sample is sheared at a constant shear rate $\dot{\gamma}_0$ (lower than $1/\tau_{max}$, to remain in the linear regime). At the time $t = 0$, the shear is suddenly stopped (over a time much shorter than τ_{min}) (Fig. 6.6). To calculate the stress, let us use the general formula (6.20). With the variable change $u = t - t'$, this formula becomes:

$$\sigma(t) = \int_0^{+\infty} G(u)\dot{\gamma}(t - u)\,du. \tag{6.32}$$

It follows that, before the shear is removed (at times $t < 0$), the stress is given by:

$$\sigma(t) = \eta_0\dot{\gamma}_0, \tag{6.33}$$

where η_0 is the stationary viscosity under continuous shear, given by:

$$\eta_0 = \int_0^{+\infty} G(u)\,du \tag{6.34}$$

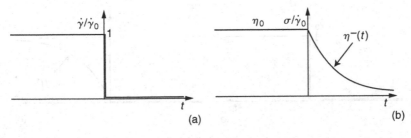

(a)   (b)

Fig. 6.6 Stress relaxation after the sudden cessation of flow. (a) Shear rate as a function of time; (b) stress as a function of time.

in the case of the continuous model and by:

$$\eta_0 = \sum_1^n \eta_i \tag{6.35}$$

for the discrete model.

On the other hand, immediately after the shear has stopped ($t > 0$), the stress relaxes (Fig. 6.6) being given by (using the notations of Bird *et al.*, [28]):

$$\sigma(t) = \eta^-(t)\dot\gamma_0, \tag{6.36}$$

where

$$\eta^-(t) = \int_t^{+\infty} G(u)\, du. \tag{6.37}$$

Thus, measuring the *transient viscosity* $\eta^-(t)$ can in principle yield the relaxation modulus $G(t)$ (since $G(t) = -\,d\eta^-/dt$).

Note that in the discrete model:

$$\eta^-(t) = \sum_1^n \eta_i \exp\left(-\frac{t}{\tau_i}\right). \tag{6.38}$$

6.2.1.8 Stress growth after the sudden application of a constant shear rate: transient viscosity $\eta^+(t)$

One can also suddenly apply, starting from the moment $t = 0$, a constant shear rate $\dot\gamma_0$ to an initially relaxed sample (Fig. 6.7).

In this case, formula (6.32) shows that the stress is zero for $t < 0$, while for $t > 0$ it increases with time before saturating, following the law:

$$\sigma(t) = \eta^+(t)\dot\gamma_0 \tag{6.39}$$

with, for the continuous model:

$$\eta^+(t) = \int_0^t G(u)\, du \tag{6.40}$$

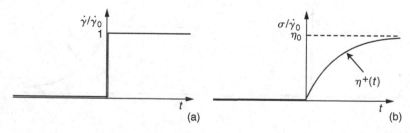

Fig. 6.7 Stress growth after the sudden application of shear. (a) Shear rate as a function of time; (b) stress as a function of time.

and for the discrete model:

$$\eta^+(t) = \sum_1^n \eta_i \left[1 - \exp\left(-\frac{t}{\tau_i}\right)\right].$$ (6.41)

This time, the transient viscosity $\eta^+(t)$ tends towards the viscosity η_0 given by the formulas (6.34) or (6.35) (Fig. 6.7).

Note that $G(t) = d\eta^+/dt = -d\eta^-/dt$.

It is also instructive to calculate the elastic energy stored by the fluid during the transient regime. To this end, consider the general diagram of Fig. 6.4. When the system is in its stationary state, each Maxwell element bears a stress:

$$\sigma_i = \eta_i \dot{\gamma}_0$$ (6.42)

and is deformed by an amount:

$$\gamma_{Ei} = \frac{\sigma_i}{G_i} = \frac{\eta_i}{G_i} \dot{\gamma}_0.$$ (6.43)

The total elastic energy stored in the fluid is obtained by summing over the contributions of all the springs:

$$E_{elas} = \sum_1^n \frac{1}{2} G_i \gamma_{Ei}^2 = \frac{1}{2} \left(\sum_1^n \frac{\eta_i^2}{G_i}\right) \dot{\gamma}_0^2.$$ (6.44)

We have thus recovered the expression (1.34), stated without proof in the first chapter for a Maxwell fluid with a single relaxation time.

Note that, in the continuous case, the expression above is written as:[3]

$$E_{elas} = \frac{1}{2} \left(\int_0^{+\infty} u G(u)\, du\right) \dot{\gamma}_0^2.$$ (6.45)

Also note that at the instant $t = 0$, the elastic energy stored by the fluid is zero. The importance of this remark will be made clear in the next subsection.

6.2.1.9 Creep after the sudden application of a stress: creep compliance $J(t)$

The *creep compliance* $J(t)$ is defined by the general relation:

$$\gamma(t) = \int_{-\infty}^t J(t - t')\dot{\sigma}(t')\, dt'.$$ (6.46)

This function can be measured experimentally by imposing suddenly, at the moment $t = 0$, a constant stress σ_0 (Fig. 6.8). Assuming that $\gamma = 0$ for $t < 0$, using the equation above one finds that, for $t > 0$:

$$\gamma(t) = J(t)\sigma_0.$$ (6.47)

---

[3] It suffices to replace $G(u)$ by its expression (6.25) in Eq. (6.45), immediately retrieving Eq. (6.44).

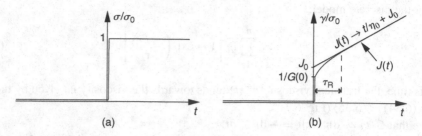

Fig. 6.8 Deformation after the sudden application of stress. (a) Stress as a function of time; (b) deformation (normalised by σ_0) as a function of time (or creep compliance). In the case of the Jeffrey model, the asymptotic regime is reached after a time τ_R (known as the retardation time).

The relation between $J(t)$ and $G(t)$ is not obvious, but one can at least define its limits. Indeed, one knows that at long times $\dot{\gamma}$ tends to σ_0/η_0. It results that

$$J(t) \to t/\eta_0 + J_0 \quad \text{as} \quad t \to \infty, \tag{6.48}$$

where η_0 is the 'total' viscosity, given by:

$$\eta_0 = \sum_1^n \eta_i = \int_0^{+\infty} G(u)\, du \tag{6.49}$$

and J_0 is a constant (termed *equilibrium compliance*). One can show ([28], volume 1) that it is expressed by:

$$J_0 = \sum_1^n \frac{G_i \tau_i^2}{\eta_0^2} = \frac{\int_0^{+\infty} u G(u)\, du}{\left(\int_0^{+\infty} G(u)\, du \right)^2}. \tag{6.50}$$

One also knows that, at the instant $t = 0$, the system undergoes elastic deformation, by an amount $\gamma(0) = \sigma_0 / \sum_1^n G_i$. This provides the value of the initial time compliance:

$$J(0) = 1 \bigg/ \sum_1^n G_i = 1/G(0). \tag{6.51}$$

Note that $J_0 \neq J(0)$ in general, except in the case of the Maxwell model with a single relaxation time, where:

$$J(t) = \frac{1}{G} + \frac{t}{\eta} \tag{6.52}$$

and

$$J_0 = J(0) = \frac{1}{G}. \tag{6.53}$$

On the other hand, for a Jeffrey fluid, with rheological behaviour described by Eq. (6.14), the expression for the creep compliance $J(t)$ is easily found by using Eq. (6.50):

$$J(t) = \frac{t}{\eta_0} - \frac{G_P \tau_P^2}{\eta_0^2} \left[\exp\left(-\frac{t}{\tau_R}\right) - 1 \right], \tag{6.54}$$

where

$$J(0) = 0 \quad \text{and} \quad J_0 = \frac{G_P \tau_P^2}{\eta_0^2}. \tag{6.55}$$

As expected, $J_0 \neq J(0)$. As an aside, the *retardation time* τ_R is the time after which the creep compliance, starting at the initial value $J(0)$, reaches its asymptotic $J_0 + t/\eta_0$ behaviour (Fig. 6.8b).

These two examples show clearly that the wider the distribution of relaxation times, the larger the difference between $J(0)$ and J_0. To quantify the width of the distribution, one can define the two following typical times:

$$\tau_n = \eta_0 J(0) = \frac{\eta_0}{G(0)} = \frac{\sum \eta_i}{\sum G_i} \tag{6.56}$$

and

$$\tau_w = \eta_0 J_0 = \frac{\sum \eta_i^2 / G_i}{\sum \eta_i}. \tag{6.57}$$

The first one (a number-average time) is characteristic of the short times, while the second one (which is a weight-average time) is typical of the long times. The ratio τ_w/τ_n (always ≥ 1), given by

$$\frac{\tau_w}{\tau_n} = \frac{J_0}{J(0)}, \tag{6.58}$$

provides a good indication as to the distribution width.

Note that in a Jeffrey fluid, this ratio is infinite since the shortest time τ_n (set by the solvent, with infinite rigidity) is zero.

One can also prove that the functions G and J are connected by the integral relation:

$$\int_0^t G(u) J(t - u) \, du = t, \tag{6.59}$$

from which it follows that $J(t)G(t) \leq 1$ [29].

It is also worthwhile to provide and compare the expressions for the elastic energy stored by the fluid in the stationary regime and at the initial time.

In the stationary regime, the formulas (6.44) and (6.45) are directly applicable. Knowing that the strain rate tends towards $\dot{\gamma}_0 = \sigma_0/\eta_0$ as $t \to \infty$, one has:

$$E_{\text{elas}}^{\infty} = \frac{1}{2}\left(\sum_{1}^{n}\frac{G_i\tau_i^2}{\eta_0^2}\right)\sigma_0^2 = \frac{1}{2}\frac{\int_0^{+\infty} uG(u)\,du}{\left(\int_0^{+\infty} G(u)\,du\right)^2}\sigma_0^2 = \frac{1}{2}J_0\sigma_0^2, \tag{6.60}$$

where the ∞ superscript emphasises that this is the energy stored during the entire transient regime (namely, after a time much longer than all relaxation times of the fluid).

This energy is different from the elastic energy stored in the fluid at the time $t = 0$. Indeed, each spring deforms suddenly by an amount $\gamma(0) = \sigma_0/\sum G_i$ as the stress is applied suddenly, yielding the initial elastic energy:

$$E_{\text{elas}}^{0} = \sum_{1}^{n}\frac{1}{2}G_i\gamma(0)^2 = \frac{1}{2}\frac{\sigma_0^2}{\sum G_i} = \frac{1}{2}\frac{\sigma_0^2}{G(0)} = \frac{1}{2}J(0)\sigma_0^2. \tag{6.61}$$

Since $J_0 \geq J(0)$ always, it follows that $E_{\text{elas}}^{\infty} \geq E_{\text{elas}}^{0}$.

Finally, note that the ratio τ_w/τ_n, characterising the width of the distribution of relaxation times, can also be written as:

$$\frac{\tau_w}{\tau_n} = \frac{E_{\text{elas}}^{\infty}}{E_{\text{elas}}^{0}} \geq 1. \tag{6.62}$$

6.2.2 Case of oscillating shear at the frequency $f = \omega/2\pi$

Oscillating shear of a fluid is easily achieved experimentally, using a cone-plate rheometer or a piezoelectric setup. Widely employed by physicists, this type of flow provides a simple way of probing the viscoelastic properties of a material in the linear regime.

6.2.2.1 Complex shear modulus $G^*(\omega)$

Consider a sample undergoing sinusoidal deformation, with an amplitude γ_0 and angular frequency ω:

$$\gamma(t) = \gamma_0 \exp(i\omega t). \tag{6.63}$$

If γ_0 remains small enough (in practice $\gamma_0 \ll 1$), the stress stays proportional to γ_0 (linear viscoelasticity) and the general equation (6.20) holds, yielding:

$$\sigma(t) = i\omega\gamma_0 \int_{-\infty}^{t} G(t - t')\exp(i\omega t')\,dt'. \tag{6.64}$$

After the variable change $u = t - t'$, one has:

$$\sigma(t) = i\omega\gamma_0 \exp(i\omega t)\int_0^{\infty} G(u)\exp(-i\omega u)\,du, \tag{6.65}$$

which can also be written as:

$$\sigma(t) = G^*(\omega)\gamma(t), \tag{6.66}$$

where G^* is the *complex shear modulus*, connected to the relaxation modulus $G(t)$ by the following expression:

$$G^* = i\omega \int_0^\infty G(u)\exp(-i\omega u)\,du. \tag{6.67}$$

One ordinarily sets

$$G^* = G' + iG'', \tag{6.68}$$

where G' is the *elastic modulus* (also termed the *storage modulus*) and G'', the *loss modulus*.

In real notation, one can see that for $\gamma(t) = \gamma_0\cos(\omega t)$:

$$\sigma(t) = G'\gamma_0\cos(\omega t) - G''\gamma_0\sin(\omega t) = G'\gamma(t) + \frac{G''}{\omega}\dot{\gamma}(t). \tag{6.69}$$

It follows that G' represents the stress component that is *in phase* with the deformation (*elastic response*) and G'', the stress component that is *in quadrature advance* over the deformation (*viscous response*).

Also note that, from Eq. (6.67):

$$G' = \omega \int_0^\infty G(u)\sin(\omega u)\,du \tag{6.70}$$

and

$$G'' = \omega \int_0^\infty G(u)\cos(\omega u)\,du. \tag{6.71}$$

It is very instructive to calculate G' and G'' in the case of the discrete Maxwell model (Fig. 6.4). In this case, from Eqs. (6.25) and (6.67) one has:

$$G^*(\omega) = i\omega \sum_1^n \int_0^\infty \frac{\eta_i}{\tau_i}\exp\left[-u\left(\frac{1}{\tau_i}+i\omega\right)\right]du = \sum_1^n \frac{i\omega\eta_i}{1+i\omega\tau_i}. \tag{6.72}$$

This results in:

$$G' = \sum_1^n \frac{\eta_i\tau_i\omega^2}{1+\omega^2\tau_i^2} = \sum_1^n \frac{G_i\tau_i^2\omega^2}{1+\omega^2\tau_i^2}, \tag{6.73}$$

$$G'' = \sum_1^n \frac{\eta_i\omega}{1+\omega^2\tau_i^2} = \sum_1^n \frac{G_i\tau_i\omega}{1+\omega^2\tau_i^2}. \tag{6.74}$$

In Fig. 6.9a, we plot G'/G and G''/G as a function of the scaled angular frequency $\omega\tau$ for a Maxwell model with a single relaxation time ($\tau = \eta/G$). Note that $G' \to G$ as

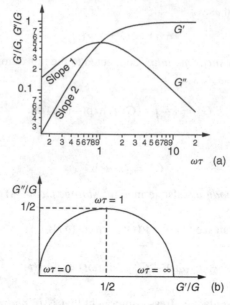

Fig. 6.9 (a) G' and G'' as a function of $\omega\tau$ in the case of a viscoelastic liquid described by the Maxwell model with one relaxation time; (b) corresponding Cole–Cole diagram.

$\omega \to \infty$ and it behaves as $G\tau^2\omega^2 = (\eta^2/G)\omega^2$ for $\omega \to 0$. As to G'', it varies as $G/\omega\tau$ for $\omega \to \infty$ and as $G\tau\omega = \eta\omega$ for $\omega \to 0$.[4] Finally, note that $G' = G'' = 1/2$ for $\omega\tau = 1$.

It is also instructive to plot the Cole–Cole diagram $G''/G = f(G'/G)$. For a *Maxwell fluid* with a unique relaxation time, it is easily checked that:

$$(G''/G)^2 + (G'/G - 1/2)^2 = 1/4. \tag{6.75}$$

The representative plot is thus a semi-circle with radius $1/2$, centred at $(1/2, 0)$ (Fig. 6.9b).

To complete this discussion, let us also plot the same graphs (Fig. 6.10) for a *Jeffrey fluid* (Fig. 6.3b). In this case, the elastic and loss moduli are, respectively:

$$\frac{G'}{G_P} = \frac{(\omega\tau_P)^2}{1 + (\omega\tau_P)^2}, \tag{6.76}$$

$$\frac{G''}{G_P} = r\omega\tau_P + \frac{\omega\tau_P}{1 + (\omega\tau_P)^2}, \tag{6.77}$$

where $\tau_P = \eta_P/G_P$ describes the 'particles' of the fluid, which are responsible for its viscoelasticity, while $r = \eta_s/\eta_P$ is the ratio of the solvent viscosity to the viscosity excess due to the particles (knowing that at zero frequency the fluid has a viscosity $\eta_0 = \eta_P + \eta_s$).

This type of model applies in the first approximation to emulsions or suspensions.

---

[4] We emphasise that, for a Newtonian liquid (corresponding to the $G \to +\infty$ and $\tau \to 0$ limit), $G' = 0$ and $G'' = \eta\omega$ for all frequency values.

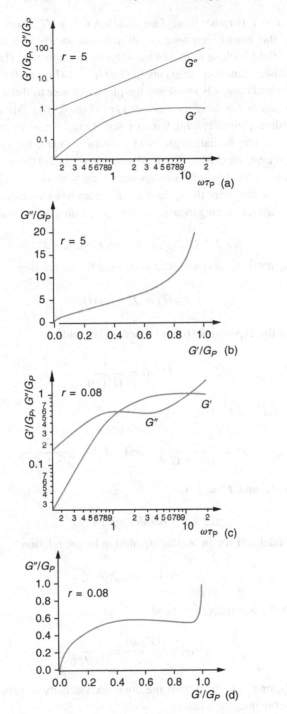

Fig. 6.10 G' and G'' as a function of $\omega\tau$ and corresponding Cole–Cole diagram for a Jeffrey fluid with $r = 5$ (a) and (b), and $r = 0.08$ (c) and (d).

In the dilute case, r is rather large. One then has $G'' \gg G'$, since the viscous response dominates over the elastic response at all frequencies (Fig. 6.10a). The rheological behaviour of the fluid is close to that of the solvent, as expected. The Cole–Cole diagram is a strictly increasing function, diverging in $G'/G_P = 1$ (Fig. 6.10b).

In the concentrated case, r is small and the plots are close to those of a simple Maxwell fluid at low frequency (as long as $\omega\tau_P < 1/r$) (Fig. 6.10c). More precisely, the fluid behaves as an ordinary viscous fluid, with a viscosity $\eta_0 = \eta_s + \eta_P$, at very low frequency ($\omega\tau_P \ll 1$). In the intermediate regimes ($1 < \omega\tau_P < 1/r$), the fluid exhibits typically viscoelastic behaviour, the elastic response being stronger than the viscous one. Finally, the fluid behaves as its solvent, which is an ordinary fluid with viscosity η_s, at high frequency ($\omega\tau_P \gg 1/r$). Note that here the Cole–Cole diagram has two extrema that move closer together as r increases, reducing to an inflexion point with a horizontal tangent for $r = 1/8$.

6.2.2.2 Complex shear compliance $J^*(\omega)$

For an oscillating motion, this parameter is defined by the relation:

$$\gamma(t) = J^*(\omega)\sigma(t). \tag{6.78}$$

Comparison with the expression (6.66) immediately yields:

$$J^*(\omega) = \frac{1}{G^*(\omega)} \tag{6.79}$$

so that, with $J^* = J' - iJ''$:

$$J' = \frac{G'}{G'^2 + G''^2} \quad \text{and} \quad J'' = \frac{G''}{G'^2 + G''^2}. \tag{6.80}$$

Note that $J' \neq 1/G'$ and $J'' \neq 1/G''$.

6.2.2.3 Complex viscosity $\eta^*(\omega)$

This parameter is defined for an oscillating motion by the relation:

$$\sigma(t) = \eta^*(\omega)\dot{\gamma}(t). \tag{6.81}$$

Comparison with the formulas (6.66) and (6.78) gives:

$$\eta^*(\omega) = \frac{G^*(\omega)}{i\omega} = \frac{1}{i\omega J^*(\omega)}. \tag{6.82}$$

Using Eq. (6.67), one can check that the complex viscosity is related to the relaxation modulus $G(t)$ by the integral relation:

$$\eta^*(\omega) = \int_0^\infty G(u)\exp(-i\omega u)\,du. \tag{6.83}$$

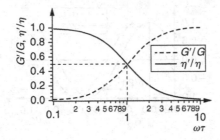

Fig. 6.11 Elastic modulus G' and dynamic viscosity η' as a function of ω for the Maxwell model with a single relaxation time.

One usually sets

$$\eta^* = \eta' - i\eta''. \tag{6.84}$$

The parameter η', known as the *dynamic viscosity*, corresponds to the ordinary viscosity. It is related to the loss modulus G'' and to the relaxation modulus $G(t)$ by the following relation:

$$\eta' = \frac{G''}{\omega} = \int_0^\infty G(u) \cos(\omega u) \, du. \tag{6.85}$$

The other coefficient η'' has no particular name (to our knowledge). It is connected to the elastic modulus G' and to the relaxation modulus $G(t)$ by the relation:

$$\eta'' = \frac{G'}{\omega} = \int_0^\infty G(u) \sin(\omega u) \, du. \tag{6.86}$$

In Fig. 6.11, we plot the evolution of the viscosity η' and the elastic modulus G' as a function of the angular velocity ω for a Maxwell model with a single relaxation time.

A Jeffrey fluid would exhibit the same type of behaviour, except for the viscosity tending towards the solvent viscosity at high frequency, instead of zero.

6.2.2.4 Loss angle $\delta(\omega)$

Another way of describing oscillating flow is by defining the loss angle δ, such that:

$$\sigma(t) = \sigma_0 \exp i(\omega t + \delta). \tag{6.87}$$

With this definition, the angle δ represents the phase advance of the stress with respect to the strain (Fig. 6.12). It is easily checked that, in the general case:

$$\tan \delta = \frac{G''}{G'}. \tag{6.88}$$

For a *Maxwell fluid* with a single relaxation time, this formula yields:

$$\tan \delta = \frac{1}{\tau \omega}. \tag{6.89}$$

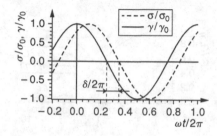

Fig. 6.12 Stress and strain as a function of time in oscillating shear: definition of the loss angle δ.

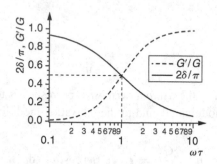

Fig. 6.13 Loss angle δ and elastic modulus G' in the case of the Maxwell model with a single relaxation time.

In Fig. 6.13, we plot the loss angle δ and the elastic modulus G' as a function of ω for this model.

In a *Jeffrey fluid*, it is easy to verify that the loss angle tends towards $\pi/2$ at low and high frequency (where the fluid exhibits purely Newtonian behaviour) and has a minimum of $\delta_{\min} = \arctan\{2r(r+1)^{1/2}\}$ at the scaled frequency $(\omega\tau_P)_{\min} = (r+1)/r^{1/2}$, where $r = \eta_s/\eta_P$.

6.2.2.5 Energy balance

As the sample deformation goes from 0 to γ_0, the system stores, per unit volume, a certain amount of elastic energy E_{elast} given by:

$$E_{\text{elast}} = \int_0^{\gamma_0} \sigma' \, d\gamma = \int_0^{\gamma_0} G'\gamma \, d\gamma = G'\frac{\gamma_0^2}{2}. \tag{6.90}$$

This energy is recovered as the deformation returns from γ_0 to 0, and so on.

One can also calculate the energy dissipated during a complete cycle:

$$E_{\text{diss}} = \int_0^{2\pi/\omega} \sigma'' \frac{d\gamma}{dt} \, dt = \int_0^{2\pi/\omega} (-G''\gamma_0 \sin \omega t)(-\gamma_0\omega \sin \omega t) \, dt$$

$$= \pi G''\gamma_0^2. \tag{6.91}$$

This yields the general relation connecting the loss angle to the ratio between the dissipated energy and the stored elastic energy:[5]

$$\frac{E_{\text{diss}}}{E_{\text{elast}}} = 2\pi \frac{G''}{G'} = 2\pi \tan \delta. \tag{6.92}$$

6.2.3 Complete expression of the stress tensor: definition of the shear modulus $G(t)$ and the compression modulus $K(t)$

So far, we have always assumed that the material was undergoing simple shear. To describe a more general deformation, one must introduce the tensor γ_{ij} defined as:

$$\gamma_{ij} = \frac{\partial u_i}{\partial x_j} + \frac{\partial u_j}{\partial x_i}, \tag{6.93}$$

identical, up to a prefactor of 2, to the strain tensor ε_{ij} already defined in Chapter 4 on elasticity ($\gamma_{ij} = 2\varepsilon_{ij}$). We have also shown that, for an isotropic elastic solid, the most general linear relation between the stress tensor σ_{ij} and the tensor γ_{ij} is of the type (Hooke's law):

$$\sigma_{ij} = \frac{K}{2}\text{tr}\,\underline{\gamma}\,\delta_{ij} + \mu \left(\gamma_{ij} - \frac{1}{3}\text{tr}\,\underline{\gamma}\,\delta_{ij} \right), \tag{6.94}$$

where K is the compression modulus and μ the shear modulus (denoted by G in viscoelastic materials).

This will lead us to generalising the expression (6.20) connecting the strain to the deformation in the linear regime.

6.2.3.1 General constitutive law for an isotropic viscoelastic material

Formula (6.20), giving the constitutive law for a viscoelastic material under shear can be easily generalised along the lines of the constitutive law (6.94) for an isotropic elastic solid, trivially leading to:

$$\sigma_{ij}(t) = \int_{-\infty}^{t} \left[G(t-t') \left(\dot{\gamma}_{ij}(t') - \frac{1}{3}\dot{\gamma}_{kk}(t')\delta_{ij} \right) + \frac{1}{2}K(t-t')\dot{\gamma}_{kk}(t')\delta_{ij} \right] dt'. \tag{6.95}$$

In this expression, the modulus $G(t)$ is the *shear stress relaxation modulus* and $K(t)$ is the *relaxation modulus for uniform compression or dilatation stress*.

This equation can also be written as:

$$\sigma_{ij}(t) = -P(t)\delta_{ij} + \int_{-\infty}^{t} \left[G(t-t') \left(\dot{\gamma}_{ij}(t') - \frac{1}{3}\dot{\gamma}_{kk}(t')\delta_{ij} \right) \right] dt', \tag{6.96}$$

---

[5] This relation is at the origin of the term 'loss angle'.

where $P(t)$ is the *hydrostatic pressure*, i.e. the thermodynamic variable coupled to the bulk dilation:

$$P(t) = -\int_{-\infty}^{t} \frac{1}{2} K(t - t') \dot{\gamma}_{kk}(t') \, dt'. \tag{6.97}$$

6.2.3.2 The case of oscillating motion at the angular frequency ω

In the case of oscillating motion, the expression above can be rewritten as:

$$\sigma_{ij}(t) = G^*(\omega) \left(\gamma_{ij}(t) - \frac{1}{3} \gamma_{kk}(t) \delta_{ij} \right) + \frac{1}{2} K^*(\omega) \gamma_{kk}(t) \delta_{ij}, \tag{6.98}$$

where G^* and K^* are the *complex shear and compression moduli*. Accounting for the pressure, this equation becomes:

$$\sigma_{ij}(t) = -P(t) \delta_{ij} + G^*(\omega) \left(\gamma_{ij}(t) - \frac{1}{3} \gamma_{kk}(t) \delta_{ij} \right), \tag{6.99}$$

with

$$P(t) = -\frac{1}{2} K^*(\omega) \gamma_{kk}(t). \tag{6.100}$$

6.2.3.3 The case of simple extensional flow

This type of flow is very important in practice, as it occurs each time a rod, fibre or film of viscoelastic material is placed under tension.

The basic geometry is shown in Fig. 6.14. A cubic element of material undergoes simple extension by γ_{11} along direction 1 under the action of the aplied stress σ_T. If the edges are free, the distortions γ_{22} and γ_{33} are equal, and of opposite sign with respect to γ_{11}. Under these conditions, the stress tensor reads, from Eq. (6.95):

$$\sigma_{11}(t) = \int_{-\infty}^{t} \left\{ \frac{2}{3} G(t - t') \left[\dot{\gamma}_{11}(t') - \dot{\gamma}_{22}(t') \right] + \frac{1}{2} K(t - t') \left[\dot{\gamma}_{11}(t') + 2\dot{\gamma}_{22}(t') \right] \right\} dt'$$
$$= \sigma_T(t) - P_a, \tag{6.101}$$

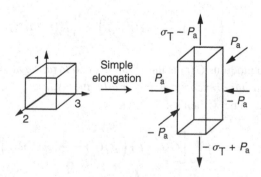

Fig. 6.14 Simple extensional flow.

with

$$\sigma_{22}(t) = \sigma_{33}(t) = -P_a$$

$$= \int_{-\infty}^{t} \left\{ \frac{1}{3} G(t - t') \left[\dot{\gamma}_{22}(t') - \dot{\gamma}_{11}(t') \right] + \frac{1}{2} K(t - t') \left[\dot{\gamma}_{11}(t') + 2\dot{\gamma}_{22}(t') \right] \right\} dt',$$

$$(6.102)$$

where P_a is the ambient pressure. Eliminating $K(t)$ between these equations yields:

$$\sigma_{11}(t) = \int_{-\infty}^{t} G(t - t')\dot{\gamma}_{11}(t') \, dt' - \int_{-\infty}^{t} G(t - t')\dot{\gamma}_{22}(t') \, dt' - P_a. \qquad (6.103)$$

Let us now assume that, at the time $t = 0$, the sample is deformed by the constant strain $\varepsilon = \gamma_{11}/2$. Since the material is viscoelastic, the measured stress $\sigma_T(t)$ depends on the time, and ultimately relaxes towards zero. The stress law is obtained by setting $\dot{\gamma}_{11}(t') = \gamma_{11}\delta(t')$ in the equation above, giving immediately:

$$\sigma_T(t) = \gamma_{11} \left[G(t) + \int_{-\infty}^{t} G(t - t')\dot{v}(t') \, dt' \right], \qquad (6.104)$$

where we introduced, by analogy with ordinary solids, the Poisson coefficient (here, it is time dependent):

$$v(t) = -\frac{\gamma_{22}(t)}{\gamma_{11}} = \frac{1 - \Delta V(t)/\varepsilon}{2}. \qquad (6.105)$$

In this expression, $\Delta V = (\gamma_{11} + \gamma_{22} + \gamma_{33})/2$ represents the bulk dilation. This equation can be rewritten in the form:

$$\sigma_T(t) = \varepsilon \, E(t), \qquad (6.106)$$

where (as for solids) $E(t)$ stands for the *Young modulus* of the material, also depending on the time:

$$E(t) = 2 \left[G(t) + \int_{-\infty}^{t} G(t - t')\dot{v}(t') \, dt' \right]. \qquad (6.107)$$

6.2.3.4 The incompressible case

In many viscoelastic materials, the compression modulus is much larger than the shear modulus ($K \gg G$). In particular, this is the case for polymer liquids, where their values are (at least) two orders of magnitude apart, over a very wide frequency range.

Under these conditions, the bulk dilation γ_{kk} is close to zero and the viscoelastic fluid can be considered as incompressible (div $\bar{v} = 0$). Equations (6.96) and (6.99) simplify to the following expressions:

$$\sigma_{ij}(t) = -P(t)\delta_{ij} + \int_{-\infty}^{t} G(t - t')\dot{\gamma}_{ij}(t') \, dt' \qquad (6.108)$$

in the case of generic flow, and

$$\sigma_{ij}(t) = -P(t)\delta_{ij} + G^*(\omega)\gamma_{ij}(t) \tag{6.109}$$

for oscillating flow at the angular frequency ω.

The Young modulus also reduces to a particularly simple expression in the *incompressible limit*, since the Poisson coefficient takes the value 1/2:

$$E(t) = 3G(t). \tag{6.110}$$

This expression is easily found from Eq. (6.107), where one must take $\dot{v}(t') = \delta(t')/2$ since $v(t' < 0) = 0$ and $v(t' > 0) = 1/2$. The ratio $E(t)/G(t)$ of the traction resistance to the shear resistance was first introduced by F. T. Trouton. It takes the value 3 in the previous case (see also Section 6.3.7).

In conclusion, it appears that elongational flow provides essentially the same information as simple shear flow. We emphasise however that this conclusion only holds in the linear regime, and only if the Poisson coefficient remains close to 1/2 (which requires working at a sufficiently low frequency so that the material can be considered incompressible).

Things change during continuous deformation, where the nonlinearities of the material become important. In this case, the value of the Trouton ratio is larger than 3.

6.2.4 Kelvin–Voigt model (viscoelastic solids)

Natural rubber, used for producing car tyres, is the archetype of viscoelastic solids. Although it can withstand reversibly a deformation of up to 200%, it cannot give back this entire energy after deformation, due to the internal friction.

The simplest way of modelling this viscoelastic behaviour is by connecting in parallel a spring and a dashpot (Fig. 6.15). This is the *Kelvin model* (or *Voigt model* according to some authors). The spring, endowed with a stiffness G, describes the elasticity of the material; the dashpot, with a viscosity η, accounts for the internal friction and dissipation.

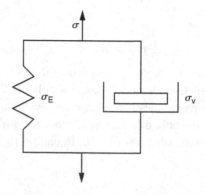

Fig. 6.15 Kelvin model with a single relaxation time (a spring connected in parallel with a dashpot).

While in the Maxwell model the deformations of the two elements are added, since they are connected in series, here it is the stresses that are summed, the elements being in parallel. One must then have:

$$\sigma = \sigma_E + \sigma_v \tag{6.111}$$

where σ is the applied stress. In terms of the deformation γ, this equation reads:

$$\sigma = G\gamma + \eta\dot{\gamma}, \tag{6.112}$$

where the viscoelastic relaxation time $\tau = \eta/G$ appears explicitly. This parameter is the typical time needed by the material to deform after the sudden application (or cessation) of stress (Fig. 6.16).

Like the Maxwell model, the Kelvin model can be extended to materials exhibiting several relaxation times.

It is also noteworthy that the simple Kelvin model is formally equivalent to the Burgers model, where the first Maxwell element is a spring of stiffness G in series with a dashpot of infinite viscosity, while the second Maxwell element is a dashpot with viscosity η in series with a spring of infinite stiffness.

Consequently, most of the concepts defined above, such as the flow compliance $J(t)$, the complex modulus G^* or the complex viscosity η^*, can be used to describe viscoelastic solids.

For instance, in the case of a Kelvin solid with a single relaxation time (Figs. 6.16b and 6.17) one has:

$$J(t) = \frac{1}{G}\left[1 - \exp\left(-\frac{t}{\tau}\right)\right]. \tag{6.113}$$

This formula reminds us that the time τ defines the crossover between the viscous regime ($t < \tau$) and the elastic regime ($t > \tau$).

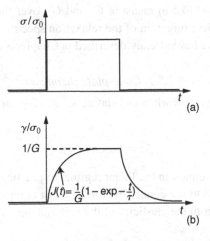

Fig. 6.16 Deformation curve after the application and the sudden cessation of stress in a viscoelastic solid. (a) Stress as a function of time; (b) deformation as a function of time.

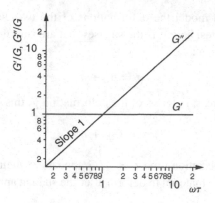

Fig. 6.17 G' and G'' as a function of $\omega\tau$ for a viscoelastic solid described by the Kelvin model with a single relaxation time.

One can also calculate

$$G' = G \quad \text{and} \quad G'' = \eta\omega \tag{6.114}$$

and

$$\eta' = \eta \quad \text{and} \quad \eta'' = \frac{G}{\omega}, \tag{6.115}$$

showing that, at the angular frequency $\omega = 1/\tau$, the elastic and viscous effects have an equal contribution, since $G' = G'' = G$ or, equivalently, $\eta' = \eta'' = \eta$.

6.2.5 Measuring the linear viscoelasticity functions $G'(\omega)$ and $G''(\omega)$

Experimentally, one would like to measure G' and G'' over the widest frequency range possible, for an optimal determination of the relaxation spectrum of the material. One can use several devices that we have already described in Chapter 3 on hydrodynamics.

6.2.5.1 Cone-plate rheometer

This setup (see Fig. 3.47) can work in oscillating shear. Assume that the rotation angle Ω of the cone is imposed:

$$\Omega = \Omega_0 \cos(\omega t), \tag{6.116}$$

with Ω_0 small enough to remain in the linear regime. In practice, one measures the torque Γ that needs to be applied to the cone. This torque has a component that is in phase with the excitation, resulting from the elasticity of the material, and a component in quadrature to it, due to the viscous response:

$$\Gamma = \Gamma' \cos(\omega t) - \Gamma'' \sin(\omega t). \tag{6.117}$$

Practically, the torques Γ' and Γ'' are measured using a lock-in amplifier. Their experimental values then give the sought-for coefficients G' and G''. Indeed, one has from Eq. (3.398), proven in Chapter 3 and still valid here in the linear regimes provided one adopts the complex notation:

$$\Gamma = \frac{2\pi}{3\theta_0} \eta^* R^3 \Omega_0 i\omega \exp(i\omega t), \qquad (6.118)$$

where $\eta^* = G^*/i\omega$ is the complex viscosity of the sample, θ_0 the angle of the cone and R its radius. Taking the real part of this expression and comparing it to Eq. (6.117), identifying the terms in $\cos(\omega t)$ and $\sin(\omega t)$ yields:

$$G' = \frac{3\theta_0}{2\pi R^3} \frac{\Gamma'}{\Omega_0} \qquad (6.119)$$

and

$$G'' = \frac{3\theta_0}{2\pi R^3} \frac{\Gamma''}{\Omega_0}. \qquad (6.120)$$

In the plate-plate geometry, it is easy to verify that these formulas become, according to Eq. (3.399):

$$G' = \frac{2h}{\pi R^4} \frac{\Gamma'}{\Omega_0}, \qquad (6.121)$$

$$G'' = \frac{2h}{\pi R^4} \frac{\Gamma''}{\Omega_0}, \qquad (6.122)$$

where h is the gap between the plates.

Note that the loss angle is given here by the expression:

$$\tan \delta = \frac{\Gamma''}{\Gamma'}. \qquad (6.123)$$

Commercial rheometers cannot reach high frequencies, being limited to a few tens of hertz (in practice, $\omega < 200$ Hz).

6.2.5.2 Piezoelectric acoustic rheometer

This rheometer (see Fig. 3.49) has the advantage of reaching much higher frequencies (in the acoustic range) than the setup described above, since its mass can be considerably smaller. For instance, the setup developed by J.-F. Palierne works in the compression–dilation mode (it can also work in shear mode using a different polarisation of the ceramic elements, but this deformation mode is less sensitive, due to the weaker signal), and has a frequency range 0.1 Hz $< f = \omega/2\pi < 10^4$ Hz.

For an incompressible viscoelastic fluid, the velocity and stress fields are calculated in exactly the same way as for a viscous fluid, provided the real viscosity is replaced by the complex viscosity η^*. Experimentally, one imposes on the sample a sinusoidal deformation:

$$\varepsilon = \varepsilon_0 \cos(\omega t) \tag{6.124}$$

and one measures the force:

$$F = F' \cos(\omega t) - F'' \sin(\omega t) \tag{6.125}$$

one must apply to the upper plate in order to deform the sample.

In complex notation, this force (which is also equal to the force exerted by the fluid onto the lower plate) is given by Eq. (3.421):

$$F = \frac{3\pi R^4}{2h^2} \eta^* i\omega\varepsilon_0 \exp(i\omega t) = \frac{3\pi R^4}{2h^2} G^* \varepsilon_0 \exp(i\omega t), \tag{6.126}$$

yielding immediately, by taking the real part of this expression and comparing the $\cos(\omega t)$ and $\sin(\omega t)$ terms:

$$G' = \frac{2h^2}{3\pi R^4} \frac{F'}{\varepsilon_0} \tag{6.127}$$

and

$$G'' = \frac{2h^2}{3\pi R^4} \frac{F''}{\varepsilon_0}. \tag{6.128}$$

This time, the loss angle δ is given by the equation:

$$\tan \delta = \frac{F''}{F'}. \tag{6.129}$$

6.2.5.3 Capillary rheometer

The viscoelastic properties of an isotropic fluid in the linear regimes can also be measured using the Poiseuille flow in a capillary tube. One possible setup is shown in Fig. 6.18.[6]

This device induces an oscillating flow within the capillary by the vibrations of the membrane closing the reservoir on the left. Let u be its displacement:

$$u = u_0 \cos(\omega t). \tag{6.130}$$

This leads to the (volume) flow rate:

$$Q(t) = -\omega A u_0 \sin(\omega t), \tag{6.131}$$

where A is a constant of the setup, with surface units, that we will consider as known. Experimentally, one measures the pressure difference $\Delta P(t) = P(t) - P_a$ between the inside of the reservoir and the atmosphere using a pressure gauge:

$$\Delta P(t) = \Delta P' \cos(\omega t) - \Delta P'' \sin(\omega t). \tag{6.132}$$

---

[6] This is a simplified version of a capillary rheometer developed at the Physics Laboratory of the ENS of Lyon by H. Gayvallet.

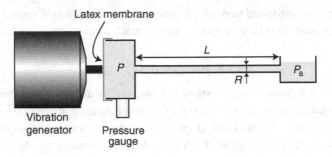

Fig. 6.18 Capillary rheometer.

The relation between ΔP and Q can be found using the Poiseuille law, proved in the case of viscous fluids (Eq. (3.330)), provided the real viscosity is replaced by the complex viscosity. In complex notation, this yields:

$$\Delta P = \frac{8L}{\pi R^4} \eta^* Q,$$ (6.133)

where L is the length of the capillary and R its radius. Taking the real part of this expression gives:

$$G' = \frac{\pi R^4}{8AL} \frac{\Delta P'}{u_0}$$ (6.134)

and

$$G'' = \frac{\pi R^4}{8AL} \frac{\Delta P''}{u_0}.$$ (6.135)

Note that, in the linear regime, the velocity profile within the capillary is still parabolic. Finally, the loss angle is given by the equation:

$$\tan \delta = \frac{\Delta P''}{\Delta P'}.$$ (6.136)

6.2.5.4 Magnetic bead rheometer

In the linear regime, the force acting on a bead immersed in a fluid is given by the Stokes law, with the complex viscosity replacing the real viscosity. In a magnetic bead rheometer (Adam *et al.*, 1984), a small magnetic bead is immersed in the fluid under study, itself contained in a hermetically closed and temperature-controlled container (to keep the sample from drying if it contains a solvent). The bead, which feels the action of gravity, is kept in levitation by applying a magnetic force, produced by a controlled current running through a coil. An optical system is used to detect the position of the bead with extreme precision. The complex viscosity of the fluid is obtained by measuring the force to be applied on the bead (using the coil) so that it remains immobile as the container is set in motion. This motion can be constant-speed translation, sinusoidal oscillation or a sudden displacement of a given amplitude. This type of rheometer can reach frequencies of up to 1 kHz.

6.2.5.5 Dynamic light scattering

A much more recent technique consists of determining by means of dynamic light scattering the mean square displacement $\langle \Delta r^2(t) \rangle$ of colloidal particles dispersed in the viscoelastic medium under study.[7] This method gives access to the complex viscosity of the fluid over a much wider frequency range (from Hz to MHz) than using the classical rheometers described above (Mason and Weitz, 1995).

6.3 Nonlinear viscoelasticity and continuous flow

At the beginning of this chapter we emphasised the existence of nonlinear effects, particularly important under continuous shear for $\dot{\gamma} > \gamma_c/\tau$. The origin of these effects is elastic (and kinematic, as we will show in Section 6.4); their manifestations are mostly related to the appearance of normal stresses leading to unusual effects, such as fluid climbing up a rotating rod (Fig. 1.20), swelling as it exits a nozzle (Fig. 1.21) or the formation of a tubeless siphon (Fig. 1.22).

In this section, dealing mainly with Couette and Poiseuille flow, as well as with elongational flow, we will try to give a phenomenological explanation of these features.

6.3.1 Expression of the stress tensor under simple shear

Let us assume that the viscoelastic fluid is sheared in the (x_1, x_2) plane. By convention, the velocity is taken along the x_1 axis and the velocity gradient along the x_2 axis. The flow is invariant along the x_3 axis (the neutral axis) (Fig. 6.19).

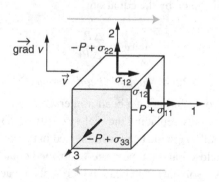

Fig. 6.19 General form of the stress tensor in the case of simple shear flow.

---

[7] In practice, the self-correlation function of the light scattered by the particles across the sample is measured in the multiple scattering regime. Analysing this function yields the mean square displacement of the particles.

Let us write the total stress tensor in the general form:[8]

$$\underline{\sigma} = \begin{pmatrix} -P + \sigma_{11}^E & \sigma_{12} & \sigma_{13} \\ \sigma_{12} & -P + \sigma_{22}^E & \sigma_{23} \\ \sigma_{13} & \sigma_{23} & -P + \sigma_{33}^E \end{pmatrix}, \tag{6.137}$$

where P is the hydrostatic pressure.

By symmetry, the forces acting on the faces 1 and 2 (perpendicular to the axes x_1 and x_2, respectively) of a matter cube must be contained in the symmetry plane of the flow. Hence:

$$(\underline{\sigma}\vec{x}_1)_3 = \sigma_{13} = 0 \tag{6.138}$$

and

$$(\underline{\sigma}\vec{x}_2)_3 = \sigma_{23} = 0. \tag{6.139}$$

The most general form of the tensor $\underline{\sigma}$ for the simple shear flow under consideration is thus:

$$\underline{\sigma} = \begin{pmatrix} -P + \sigma_{11}^E & \sigma_{12} & 0 \\ \sigma_{12} & -P + \sigma_{22}^E & 0 \\ 0 & 0 & -P + \sigma_{33}^E \end{pmatrix}. \tag{6.140}$$

In the following, to avoid unnecessarily complicating the notation, we will set $\sigma_{ii}^E = \sigma_{ii}$ ($i = 1, 2, 3$) (with the understanding that the total normal stress along the i axis is $-P + \sigma_{ii}$).

We now assume that the fluid is contained between two parallel plates and forms an infinite ribbon along direction 3, with a width L much larger than its thickness d (Fig. 6.20). Under these conditions, we will neglect the edge effects and set $\partial/\partial x_3 = \partial/\partial x_1 = 0$. In the absence of external bulk forces, the equation of momentum conservation results in:

$$\frac{\partial \sigma_{12}}{\partial x_2} = 0, \tag{6.141}$$

$$-\frac{\partial P}{\partial x_2} + \frac{\partial \sigma_{22}}{\partial x_2} = 0. \tag{6.142}$$

At the vertical boundaries of the sample, the normal stress balance requires:

$$-P + \sigma_{11} = -P_a, \tag{6.143}$$

where P_a is the ambient pressure.

---

[8] Note that, in the case of simple shear, the viscous stress elements σ_{11}^V, σ_{22}^V and σ_{33}^V are zero.

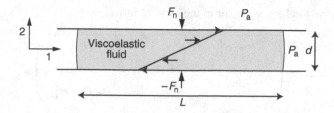

Fig. 6.20 Couette flow in a viscoelastic fluid. The fluid exerts a normal force on the plates. This is the Weissenberg effect.

Equation (6.142) shows that the fluid exerts on the upper plate a constant normal force, given by $P - \sigma_{22}$. In order to maintain the plate at the required height, one must therefore push down on it with the surface force:

$$F_n = -P + \sigma_{22} - (-P_a), \tag{6.144}$$

which, by use of Eq. (6.143), can be rewritten as:

$$F_n = \sigma_{22} - \sigma_{11}. \tag{6.145}$$

Up to a sign change, this force is equal to the *first normal stress difference*. This brings us to the following definitions.

6.3.2 *First and second normal stress differences N_1 and N_2: normal stress coefficients Ψ_1 and Ψ_2*

Generally, using the notation in Fig. 6.19, one defines the '*first normal stress difference*' as the quantity:

$$N_1 = \sigma_{11} - \sigma_{22} \tag{6.146}$$

and the '*second normal stress difference*' as the quantity:

$$N_2 = \sigma_{22} - \sigma_{33}. \tag{6.147}$$

As we have already stated in the first chapter, and as shown in the section above, the normal stresses are the origin of peculiar effects, such as the Weissenberg effect responsible for the increase of the gap between plates under simple shear (Fig. 6.20), or the climb of liquid along a rotating rod dipped into it (see Fig. 1.20). On symmetry grounds, these phenomena must be independent of the sign of the velocity, and hence of the sign of the shear rate, in agreement with the experiments. It follows that, in the most general fashion, the normal stresses are necessarily even functions of the shear rate $\dot{\gamma}$. In the following, we will suppose that they can be expanded in a power series of $\dot{\gamma}$; the lowest-order non-zero terms in this expansion will therefore be proportional to $\dot\gamma^2$.[9] The same holds for their differences N_1 and N_2, ordinarily written in the form:

$$N_1 = \Psi_1 \dot\gamma^2 \tag{6.148}$$

and

$$N_2 = \Psi_2 \dot\gamma^2. \tag{6.149}$$

The *viscometric functions* Ψ_1 and Ψ_2 (which are also even functions of $\dot\gamma$) are the *normal stress coefficients*.

As we have pointed out repeatedly, the normal stress differences are due to an accumulation of elastic energy in the fluid, most often due to the deformation of the 'particles' in the fluid under shear. These elastic effects are weak at low shear rate in the linear regime (where they do exist, obviously!) but can become very important in the nonlinear regime, where N_1 can sometimes exceed by an order of magnitude the value of the shear stress that engenders it. Intuitively, one would also expect the deformations of the objects to be much more important in the shear plane (the 1, 2 plane, by convention) than in a perpendicular plane. Consequently (the absolute value of) N_2 must be much lower than N_1. This is indeed observed for most materials. Also, one almost always has $N_1 > 0$ and $N_2 < 0$.

Let us now analyse two types of flow where the normal stresses play a direct role.

6.3.3 Some experimental manifestations of normal stresses

The two experiments we will now describe are easy to perform. The first one illustrates the existence of N_1 and the second one, that of N_2.

6.3.3.1 Climb of a viscoelastic fluid along a rotating rod

This phenomenon, illustrated by the photograph in Fig. 1.20, is mainly due to the first normal stress difference N_1.

To show this, let us assume that the surface of the viscoelastic liquid is covered by a lid that prevents the fluid from climbing and which is pierced in the centre to allow the rod to pass. We also assume that the liquid slips on the surface of the lid without friction (Fig. 6.21). In this particular geometry, the velocity only has an orthoradial component v_θ and $\partial/\partial z = \partial/\partial\theta = 0$, such that in the stationary regime the equations of motion are written:

$$-\rho \frac{v_\theta^2}{r} = -\frac{dP}{dr} + \frac{1}{r}\frac{d}{dr}(r\sigma_{rr}) - \frac{\sigma_{\theta\theta}}{r} \quad \text{(motion along } r\text{),} \tag{6.150}$$

$$0 = \frac{1}{r^2}\frac{d}{dr}(r^2\sigma_{r\theta}) \quad \text{(motion along } \theta\text{).} \tag{6.151}$$

[9] This result is not general, as the normal stress differences can be proportional to $|\dot\gamma|$ in some special materials such as certain bicontinuous emulsions [27]. A similar dependence is also observed in liquid crystals (see Section 8.1.7.4). Nevertheless it is worth noting that in liquid crystals normal stresses are of viscous – and not elastic – origin and are due to their structural anisotropy.

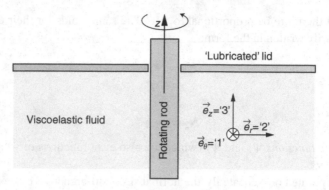

Fig. 6.21 Simplified geometry for studying the origin of the Weissenberg effect shown in Fig. 1.20. The surface of the lid is considered lubricated, such that the viscoelastic fluid slips over it without friction.

The fluid exerts on the lid the surface force $P - \sigma_{zz}$. To find its profile along r, let us write Eq. (6.150) in the form:

$$\frac{d}{dr}(P - \sigma_{zz}) = \frac{d}{dr}(\sigma_{rr} - \sigma_{zz}) + \frac{\sigma_{rr} - \sigma_{\theta\theta}}{r} + \rho\frac{v_\theta^2}{r} \tag{6.152}$$

or, equivalently:

$$\frac{d}{d\ln r}(P - \sigma_{zz}) = -2\sigma_{r\theta}\frac{d(\sigma_{rr} - \sigma_{zz})}{d(\sigma_{r\theta})} + (\sigma_{rr} - \sigma_{\theta\theta}) + \rho v_\theta^2, \tag{6.153}$$

knowing that, from Eq. (6.151):

$$\frac{d\sigma_{r\theta}}{dr} = -\frac{2}{r}\sigma_{r\theta}. \tag{6.154}$$

With the notation of Section 6.3.1, the θ direction corresponds to direction 1, the r direction to direction 2 and the z direction to direction 3 so that finally Eq. (6.153) can be rewritten as:

$$\frac{d}{d\ln r}(P - \sigma_{33}) = -2\sigma_{21}\frac{dN_2}{d\sigma_{21}} - N_1 + \rho v_1^2, \tag{6.155}$$

where we introduced again the two normal stress differences N_1 and N_2 defined in the section above.

If the liquid is Newtonian, $N_1 = N_2 = 0$. Hence, the normal stress $P - \sigma_{33}$ exerted by the fluid on the lid increases upon moving away from the z axis. However, this quantity must tend towards the ambient pressure far away from the rotation axis. We can thus conclude that the fluid exerts a stress $P - \sigma_{33}$ that is lower than the ambient pressure close to the axis. Under the action of the centrifugal force, a Newtonian liquid will therefore present a 'dip' in the vicinity of the rotating rod if the lid is removed.

If the fluid is viscoelastic, the normal stresses N_1 and N_2 no longer vanish; on the contrary, they quickly dominate over the centrifugal force. Since N_2 is generally much lower than N_1 (in the absolute value), we can neglect its contribution and write:

$$\frac{d}{d\ln r}(P - \sigma_{33}) \approx -N_1. \tag{6.156}$$

Therefore, if $N_1 > 0$, the stress $P - \sigma_{33}$ decreases on moving away from the axis. Since it tends towards P_a far away from the axis, it must increase – and take values larger than P_a – on approaching the rotation axis. That is why the liquid starts climbing up the rod if the lid is removed.

Since we now understand why the liquid tends to climb up the rod, let us search for the shape of the free surface once the lid is removed. In this case, one has to solve the equation of motion along the vertical direction. Taking the z axis to point upwards, this equation is written:

$$\frac{\partial}{\partial z}(P - \sigma_{zz}) = -\rho g, \tag{6.157}$$

where g is the gravitational acceleration. If the effect of the second normal stress difference N_2 is negligible, the equation of motion along r becomes explicitly, according to Eq. (6.155) (after going back to cylindrical coordinates):

$$r\frac{\partial}{\partial r}(P - \sigma_{zz}) = -N_1 + \rho v_\theta^2. \tag{6.158}$$

As to the equation of motion along the orthoradial direction, it is given by Eq. (6.151).

To solve these three equations, we assume that the fluid is weakly sheared (linear regime). In this limit, $\sigma_{r\theta} = \eta_0\dot\gamma$ and $N_1 = \Psi_1\dot\gamma^2$, where the viscosity η_0 and the normal stress coefficient Ψ_1 can be taken as constants. Furthermore, knowing that $\dot\gamma = \frac{\partial v_\theta}{\partial r} - \frac{v_\theta}{r}$, formula (6.151) yields an equation for v_θ:

$$r^2\frac{\partial^2 v_\theta}{\partial r^2} + r\frac{\partial v_\theta}{\partial r} - v_\theta = 0. \tag{6.159}$$

This equation has the obvious solution:

$$v_\theta = \frac{\omega R^2}{r}, \tag{6.160}$$

where R is the rod radius and ω its rotation rate. As for the stress field, it must satisfy the boundary conditions $\sigma_{rz} = \sigma_{\theta z} = 0$ and $P - \sigma_{zz} = P_a$ at the free surface, given by $z = h(r)$ (assuming that its normal remains close to the z axis). The first two conditions are automatically satisfied by the velocity field given in Eq. (6.160). As for the stress $P - \sigma_{zz}$, it is obtained by integrating Eqs. (6.157) and (6.158), which yields:

$$P - \sigma_{zz} = \Psi_1\frac{\omega^2 R^4}{r^4} - \frac{\rho}{2}\frac{\omega^2 R^4}{r^2} - \rho g z + \text{Cst}. \tag{6.161}$$

At the free surface (described by the equation $z = h(r)$, with $h = 0$ far away from the rod) one must have $P - \sigma_{zz} = P_a$. From the equation above, this condition gives $\mathrm{Cst} = P_a$ and:

$$h(r) = \frac{\omega^2}{\rho g} \left(\frac{\Psi_1 R^4}{r^4} - \frac{\rho}{2} \frac{R^4}{r^2} \right). \tag{6.162}$$

This approximate formula (which also neglects the capillary effects) shows that the liquid climbs along the rod for $\Psi_1 > \frac{\rho}{2} R^2$. In the limit of $\Psi_1 \gg \frac{\rho}{2} R^2$, the elastic effects dominate over the effect of the centrifugal force. Then, the liquid climbs along the rod up to a height $h(r) = \frac{\Psi_1 \omega^2}{\rho g}$, independent of the rod radius and proportional to the normal stress coefficient Ψ_1, providing a simple method for measuring it. These predictions are borne out rather well by the experiments (Beavers and Joseph, 1975).

6.3.3.2 Swelling of the free surface of a viscoelastic fluid flowing down a tilted trough

The experiment we are about to describe allows a direct demonstration of the second normal stress difference. Proposed by Tanner in 1970, it consists simply in letting the fluid flow down a tilted trough under the action of gravity. Experiment shows that, if the fluid is Newtonian, its surface remains planar; on the other hand, the surface swells if the fluid is viscoelastic. This phenomenon is shown in Fig. 6.22, presenting a front view of the trough in the case of a Newtonian fluid (the photo on the left), and of a viscoelastic one (the photo on the right).

To explain this phenomenon and calculate the height h by which the surface bulges, let us write the equations of motion in the (x, y, z) reference frame. The z axis is along the trough, the x axis is perpendicular to it and the y axis makes an angle β with the vertical, where β is the tilt angle of the trough with respect to the horizontal (Fig. 6.23).

Fig. 6.22 Tanner experiment. In the image on the left, the fluid is Newtonian. The free surface remains planar. In the image on the right, the fluid is viscoelastic. The free surface of the fluid bulges upwards (Tanner, 1970; [28, volume 1]).

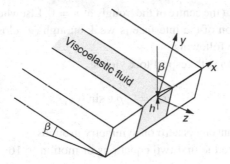

Fig. 6.23 Schematic representation of the Tanner setup. Definition of the x, y and z axes and of the height h by which the free surface of the fluid swells.

To do this calculation, we will assume that the trough is deep enough to neglect the effect of the bottom on the flow. Under these conditions, the only non-zero component of the velocity is along z and only depends (in the first approximation) on the x coordinate:

$$\vec{v} = (0,\, 0,\, v_z(x)). \tag{6.163}$$

Hence, the stress components σ_{ij} (due to the local shear) only depend on the x coordinate, unlike the pressure P, which is in principle a function of the x, y and z coordinates. Since the fluid is only sheared in the (x, z) plane, $\sigma_{xy} = \sigma_{yz} = 0$. One can then deduce the equations of motion:

$$0 = -\frac{\partial P}{\partial x} + \frac{d\sigma_{xx}}{dx} \qquad \text{(motion along } x\text{)}, \tag{6.164}$$

$$0 = -\frac{\partial P}{\partial y} - \rho\, g \cos \beta \qquad \text{(motion along } y\text{)}, \tag{6.165}$$

$$0 = -\frac{\partial P}{\partial z} + \frac{d\sigma_{xz}}{dx} + \rho\, g \sin \beta \qquad \text{(motion along } z\text{)}, \tag{6.166}$$

where g is the acceleration of gravity.

Taking the z derivative of these three equations yields $\partial^2 P/\partial z^2 = \partial^2 P/\partial x \partial z = \partial^2 P/\partial y \partial z = 0$, showing that $\partial P/\partial z = \text{Cst}$. To show that the value of this constant is in fact zero, let us write that, at the free surface of the fluid $\underline{\sigma} \vec{v} = 0$, where $\vec{v}$ is the normal to the free surface (we neglect here the surface tension). It follows that:

$$(-P + \sigma_{xx})v_x = -P_a v_x, \tag{6.167}$$

$$(-P + \sigma_{yy})v_y = -P_a v_y, \tag{6.168}$$

$$\sigma_{xz} v_x = 0, \tag{6.169}$$

where P_a is the ambient pressure. From the first two equations, $\partial P/\partial z = 0$ at the free surface. Since $\partial P/\partial z = \text{Cst}$ everywhere in the fluid, one has $\text{Cst} = 0$. Note that Eq. (6.169) is only rigorously valid at the centre of the trough, at $x = 0$. Elsewhere it holds more or less, provided the deformation of the interface is weak enough (v_x close to 0), an assumption that we will make in the following.

Integrating Eq. (6.166) with respect to x yields:

$$\sigma_{xz} = -\rho g \sin \beta \, x, \qquad (6.170)$$

where the integration constant is zero on symmetry grounds.

One can also integrate the first two equations of motion (6.164) and (6.165), resulting in:

$$P = \sigma_{xx} - \rho g \cos \beta \, y + C, \qquad (6.171)$$

where C is a constant. This equation can also be written as:

$$P - \sigma_{yy} = \sigma_{xx} - \sigma_{yy} - \rho g \cos \beta \, y + C. \qquad (6.172)$$

To determine the swelling height h, we will use the boundary conditions at the free surface. From Eq. (6.168), at the top of the free surface, at $x = 0$ and $y = h$:

$$P - \sigma_{yy} = P_a. \qquad (6.173)$$

Thus, from Eq. (6.172), we deduce that at this point:

$$P_a = \sigma_{xx} - \sigma_{yy} - \rho g \cos \beta \, h + C, \qquad (6.174)$$

whence:

$$h = \frac{\sigma_{xx} - \sigma_{yy} + C - P_a}{\rho g \cos \beta}. \qquad (6.175)$$

However, the shear rate is zero at the centre of the trough, by symmetry. One therefore has here $\sigma_{xx} = \sigma_{yy} = 0$, finally leading to:

$$h = \frac{C - P_a}{\rho g \cos \beta}. \qquad (6.176)$$

To find the constant C, let us rewrite Eq. (6.172) at the edge of the trough, at $x = \pm W/2$ and $y = 0$. Knowing that here $P - \sigma_{yy} = P_a$, one has:

$$P_a = (\sigma_{xx} - \sigma_{yy})_{\text{edge}} + C \qquad (6.177)$$

and, substituting for C its expression in Eq. (6.176):

$$h = \frac{-(\sigma_{xx} - \sigma_{yy})_{\text{edge}}}{\rho g \cos \beta}. \qquad (6.178)$$

Returning now to the standard notations of Section 6.3.1, one can see that $N_2 = \sigma_{xx} - \sigma_{yy}$, since the x axis corresponds to axis 2, the y axis to axis 3 and the z axis to axis 1. Finally, this leads to:

$$h = \frac{-(N_2)_{edge}}{\rho g \cos \beta}, \tag{6.179}$$

where $(N_2)_{edge}$ is the second normal stress difference corresponding to the shear stress $\sigma_{12} = \rho g \sin \beta \, W/2$ exerted by the fluid onto the walls of the trough (given by Eq. (6.170)).

This formula shows that, for a Newtonian fluid, the surface of the liquid remains planar ($h = 0$ since $N_2 = 0$).

If the fluid is viscoelastic, the surface bulges, in general, meaning that $N_2 < 0$. This experiment could in principle be used to measure N_2, but it is not very precise. Moreover, the range over which the shear stress σ_{12} varies is not very large.

In the next section, we return to the case of simple shear flow described in Fig. 6.20 and analyse the velocity profile and its stability.

6.3.4 Velocity profile and stability of Couette flow

The geometry is that shown in Fig. 6.20. To find the velocity profile in the permanent regime, let us assume that the viscosity η depends on the shear rate $\dot{\gamma}_{12}$. Under these conditions, Eq. (6.141) becomes, setting $\dot{\gamma}_{12} = \dot{\gamma}$:

$$\frac{\partial \eta(\dot{\gamma})\dot{\gamma}}{\partial x_2} = 0 \tag{6.180}$$

and yields, by expanding this expression:

$$\left(\eta + \dot{\gamma}\frac{\partial \eta}{\partial \dot{\gamma}}\right)\frac{\partial \dot{\gamma}}{\partial x_2} = 0. \tag{6.181}$$

If the fluid does not slip at the walls, the boundary conditions read as follows:

$$v_1 = \pm\frac{V}{2} \quad \text{at} \quad x_2 = \pm\frac{d}{2}. \tag{6.182}$$

The solution to this problem is a linear velocity profile:

$$v_1 = V\frac{x_2}{d} \tag{6.183}$$

with $\dot{\gamma} = V/d = \dot{\gamma}_0$.

Is the flow stable? To answer this question, let us analyse its *linear stability*. Consider the appearance of a velocity fluctuation $w_1(x_2, t)$, of infinitesimal amplitude:

$$v_1 = \dot{\gamma}_0 x_2 + w_1(x_2, t). \tag{6.184}$$

The local shear rate becomes:

$$\dot{\gamma} = \dot{\gamma}_0 + \frac{\partial w_1}{\partial x_2}. \tag{6.185}$$

Since the flow is no longer stationary, one must take into account the inertial term. The equation of motion itself becomes:

$$\rho \frac{\partial v_1}{\partial t} = \left(\eta + \dot{\gamma} \frac{\partial \eta}{\partial \dot{\gamma}} \right) \frac{\partial \dot{\gamma}}{\partial x_2}, \tag{6.186}$$

so that, only preserving the terms of first order in the perturbation:

$$\rho \frac{\partial w_1}{\partial t} = \left[\eta(\dot{\gamma}_0) + \dot{\gamma}_0 \left(\frac{\partial \eta}{\partial \dot{\gamma}} \right)_{\dot{\gamma}_0} \right] \frac{\partial^2 w_1}{\partial x_2^2}. \tag{6.187}$$

Setting $w_1 = w_{10} \exp(\omega t + ikx_2)$ one obtains, after substitution in the previous equation, the dispersion equation:

$$\omega = -\frac{k^2}{\rho} \left[\eta(\dot{\gamma}_0) + \dot{\gamma}_0 \left(\frac{\partial \eta}{\partial \dot{\gamma}} \right)_{\dot{\gamma}_0} \right], \tag{6.188}$$

where ω is the growth rate for a perturbation of wave vector k. The flow is unstable for $\omega > 0$, leading to the *instability condition*:

$$\eta(\dot{\gamma}_0) + \dot{\gamma}_0 \left(\frac{\partial \eta}{\partial \dot{\gamma}} \right)_{\dot{\gamma}_0} < 0, \tag{6.189}$$

which can also be expressed, as a function of the *stress*, in the form:

$$\frac{\partial \sigma_{12}}{\partial \dot{\gamma}} < 0. \tag{6.190}$$

Conversely, *the flow is stable* if the *viscosity* satisfies the following condition:

$$\dot{\gamma} \frac{\partial \ln(\eta)}{\partial \dot{\gamma}} > -1. \tag{6.191}$$

If the fluid obeys an Ostwald–de Waehle law ($\eta = m\dot{\gamma}^{n-1}$, see Eq. (1.4)), the flow will be linearly stable only if $n > 0$.

Note that these results apply equally well to non-Newtonian viscous fluids and to viscoelastic fluids (which are also non-Newtonian in general) in the permanent flow regime.

In conclusion, we can see that *flow in shear-thickening fluids is always stable*; on the other hand, *in shear-thinning fluids the flow becomes unstable if the viscosity decreases faster than the reciprocal of the local shear rate*.

In the following section, we will analyse Poiseuille flow.

6.3.5 Poiseuille flow

We will now consider the flow of a viscoelastic fluid (or of a viscous non-Newtonian fluid) in a tube sustaining a pressure difference between its extremities. We assume the tube to be much longer than its diameter, in order to neglect end effects. Let R be its radius, L its length and $\Delta P = P(0) - P(L)$ the imposed pressure loss. Working in cylindrical coordinates (r, θ, z), we are searching for the simplest solution, of the form $v_z = v_z(r)$, $v_\theta = 0$, $v_r = 0$ and $\mathbb{P} = \mathbb{P}(z)$. Here, $\mathbb{P}$ stands for the pressure P in the case of a simple viscous fluid, or for the combination $P - \sigma_{zz}$ if it is viscoelastic (keeping in mind that σ_{zz} is the elastic component of the stress tensor). The equation of motion along z is:

$$0 = -\frac{d\mathbb{P}}{dz} + \frac{1}{r}\frac{d}{dr}(r\sigma_{rz}), \tag{6.192}$$

where $d\mathbb{P}/dz$ is the imposed pressure gradient $-\Delta P/L$. This is simply a constant, so the equation is immediately integrated, yielding:

$$\sigma_{rz} = -\frac{\Delta P r}{2L} + \frac{C}{r}. \tag{6.193}$$

Since the stress cannot diverge at the axis, the integration constant C must be zero, finally leading to:

$$\sigma_{rz} = -\sigma_R\frac{r}{R}, \tag{6.194}$$

denoting by σ_R the absolute value of the shear stress at the walls of the capillary ($\sigma_R = R\Delta P/2L$). Thus, the shear stress varies linearly from the wall (where it reaches its maximum absolute value) towards the centre of the tube, where it falls to zero. To find the velocity profile, we need the constitutive law of the material. Let us assume it is given by a power law $\eta = m\dot\gamma^{n-1}$. In this expression, $\dot\gamma$ must be positive. We should therefore take $\dot\gamma = -dv_z/dr$, resulting in the shear stress:

$$\sigma_{rz} = \eta\frac{dv_z}{dr} = -m\left(-\frac{dv_z}{dr}\right)^n. \tag{6.195}$$

We deduce the differential equation for v_z:

$$m\left(-\frac{dv_z}{dr}\right)^n = \sigma_R\frac{r}{R} \tag{6.196}$$

and then, by integration:

$$v_z = \left(\frac{\sigma_R}{m}\right)^{1/n}\frac{R}{(1/n)+1}\left[1 - \left(\frac{r}{R}\right)^{(1/n)+1}\right], \tag{6.197}$$

assuming that the velocity vanishes at the wall. We recall that this boundary condition is not always satisfied in complex fluids, as they can exhibit wall slippage (see Section 3.8.5). In particular, this occurs for certain polymers with high molecular mass.

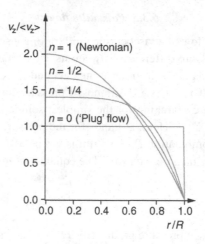

Fig. 6.24 Some velocity profiles of Poiseuille flow for different values of the exponent n.

Calculating the flow rate gives:

$$Q = \frac{\pi R^3}{(1/n) + 3} \left(\frac{\Delta P R}{2mL} \right)^{1/n}.$$ (6.198)

Some velocity profiles (normalised by the average velocity $\langle v_z \rangle = Q/\pi R^2$) are shown in Fig. 6.24 for different values of n. For $n = 1$, one finds the typical Poiseuille parabolic velocity profile. For smaller n values, the profile flattens at the centre ('plug' flow).

In fact, these profiles are not very accurate, except in the Newtonian case, because the viscosity no longer follows the power law at low shear rates, where it reaches a constant value η_0. This creates a problem at the centre of the tube, where the shear rate necessarily vanishes. One can however show that the flow rate formula (6.198) remains valid if the wall stress σ_R is large enough (more precisely, if $\sigma_R \gg \eta_0 \dot{\gamma}_0$, where η_0 is the viscosity at 'zero' shear rate and $\dot{\gamma}_0$ the shear rate at which the fluid starts to exhibit shear-thinning behaviour).

Another interesting case is that of yield stress fluids. The simplest among them follow the Bingham law given in Chapter 1 (Eqs. (1.7) and (1.8)). In this particular case, one can assume that the fluid does not flow as long as $|\sigma| < \sigma_y$ and behaves as a Newtonian fluid with a viscosity η_p for $|\sigma| > \sigma_y$ (σ_y being the yield stress, taken as positive by convention). We will now show that a Bingham fluid exhibits 'plug' flow if the imposed pressure gradient is large enough ($\Delta P/L > \Delta P_{\min}/L$).

To calculate the velocity profile, we assume that the fluid flows close to the walls ($r_0 < r < R$) and moves as a solid block ($v_z = v_0 = \mathrm{Cst}$) at the centre of the tube ($0 < r < r_0$) (Fig. 6.25). Since $\mathrm{div}\,\underline{\sigma}_{\mathrm{total}} = 0$ in the absence of external forces, Eqs. (6.192) and (6.193) still apply. It follows, from the Bingham law (1.8), that in the region $r_0 < r < R$, the velocity is given by the solution of the following equation:

$$\sigma_{rz} = -\frac{\Delta P r}{2L} + \frac{C}{r} = -\sigma_y + \eta_p \frac{dv_z}{dr} \quad (r_0 < r < R).$$ (6.199)

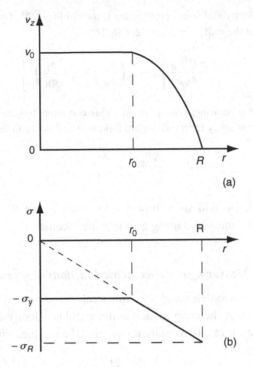

Fig. 6.25 'Plug' flow in a Bingham fluid. (a) Velocity profile; (b) stress profile. Assuming that the stress is constant in the 'plug' part of the flow amounts to considering that the fluid behaves elastically below the yield stress. In fact, we have shown in Chapter 1 that most yield stress fluids exhibit a very high viscosity at low stress. If this is the case, the shear stress must vanish in the central region (dashed line in the graph). On the other hand, the velocity profile remains almost unchanged.

One can easily check that $C = 0$ in the stationary regimes by writing the equilibrium of the forces acting on the fluid contained within a tube slice of length L : $\sigma_R L 2\pi R = \Delta P \pi R^2$ (where σ_R is again the absolute value of the shear stress at the tube wall). The equation above can thus be integrated to yield, assuming that the velocity goes to zero at the walls of the tube:

$$v_z = \frac{\Delta P(R^2 - r^2)}{4\eta_p L} - \frac{\sigma_y}{\eta_p}(R - r) \qquad (r_0 < r < R). \tag{6.200}$$

The radius r_0 is obtained by writing that $\sigma_{rz}(r = r_0) = -\sigma_y$:

$$r_0 = 2L \frac{\sigma_y}{\Delta P}. \tag{6.201}$$

Finally, the stress and the velocity are constant in the central part of the flow, and their respective values are:

$$\sigma_{rz} = -\sigma_y \quad (0 < r < r_0), \tag{6.202}$$

$$v_z = \frac{\Delta P(R - r_0)^2}{4\eta_p L} \quad (0 < r < r_0). \tag{6.203}$$

The ('plug' type) velocity and stress profiles are shown in Fig. 6.25. One can also calculate the flow rate as a function of the wall stress $\sigma_R = \Delta P R / 2L$:

$$Q = \frac{\pi R^3 \sigma_R}{4\eta_p} \left[1 - \frac{4}{3} \left(\frac{\sigma_y}{\sigma_R} \right) + \frac{1}{3} \left(\frac{\sigma_y}{\sigma_R} \right)^4 \right]. \tag{6.204}$$

As expected, the flow rate vanishes for $\sigma_R = \sigma_y$. This condition sets the value of the minimum pressure gradient one must apply for the Bingham fluid to start flowing in the tube:

$$\frac{\Delta P_{\min}}{L} = \frac{2\sigma_y}{R}. \tag{6.205}$$

In the next section, we will show how the viscosity η and the normal stress differences N_1 and N_2 can be measured using a cone-plate rheometer.

6.3.6 Measuring the viscometric functions η, Ψ_1 and Ψ_2

The geometry of the rheometer and the spherical coordinate system are recalled in Fig. 6.26. We assume that the cone rotates at the angular velocity Ω. If the angle of the cone θ_0 is small, the shear rate is constant throughout the sample, with a value:

$$\dot{\gamma} = \frac{\Omega}{\theta_0}. \tag{6.206}$$

Let Γ be the torque one must apply to turn the cone. It can be calculated as in the case of a Newtonian fluid (Eq. (3.398)). Measuring it yields the viscosity of the viscoelastic fluid under simple shear:

$$\eta = \frac{3\theta_0}{2\pi} \frac{\Gamma}{R^3 \Omega}. \tag{6.207}$$

The viscoelastic fluid also exerts on the surfaces of the plate and the cone normal stresses, which can be measured using pressure gauges.

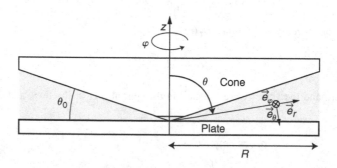

Fig. 6.26 Cone-plate rheometer.

To calculate these stresses (acting on the plate, for instance), let us begin by giving the expression for the stress tensor in spherical coordinates (r, θ, φ):

$$\underline{\sigma} = \begin{pmatrix} -P + \sigma_{rr} & 0 & 0 \\ 0 & -P + \sigma_{\theta\theta} & \sigma_{\theta\varphi} \\ 0 & \sigma_{\theta\varphi} & -P + \sigma_{\varphi\varphi} \end{pmatrix} \qquad (6.208)$$

and by writing the equation of motion along the r direction:

$$0 = -\frac{\partial P}{\partial r} + \frac{1}{r^2}\frac{\partial}{\partial r}(r^2\sigma_{rr}) - \frac{\sigma_{\theta\theta} + \sigma_{\varphi\varphi}}{r} + \frac{\rho v_\varphi^2}{r}. \qquad (6.209)$$

The last term of this expression represents the centrifugal force; it is in general negligible, so we will discard it in the following.

The lower plate feels the joint action of the fluid and the atmosphere, resulting in the normal stress $(P_a - P + \sigma_{\theta\theta})\vec{z}$. To determine this quantity, multiply the previous equation by r. Since the shear rate does not depend on r, it follows that $\partial\sigma_{rr}/\partial r = \partial\sigma_{\theta\theta}/\partial r = 0$, finally leading to:

$$\frac{\partial}{\partial \ln r}(P_a - P + \sigma_{\theta\theta}) = -\frac{1}{r}\frac{\partial}{\partial r}(r^2\sigma_{rr}) + \sigma_{\theta\theta} + \sigma_{\varphi\varphi} = -2\sigma_{rr} + \sigma_{\theta\theta} + \sigma_{\varphi\varphi}. \qquad (6.210)$$

Returning to the notation of Section 6.3.1 and to the definitions of N_1, and N_2 (the φ direction corresponds to direction 1, the θ direction to direction 2 and the r direction to direction 3), one can easily see that this equation can be rewritten as:

$$\frac{\partial}{\partial \ln r}(P_a - P + \sigma_{\theta\theta}) = N_1 + 2N_2 \qquad (6.211)$$

and, after integration:

$$P_a - P + \sigma_{\theta\theta} = (N_1 + 2N_2)\ln\left(\frac{r}{R}\right) + \text{Cst.} \qquad (6.212)$$

The constant is obtained by writing the equilibrium of the normal stresses at the free surface, considered spherical, with a radius R (we neglect here the capillary forces due to surface tension):

$$\underline{\sigma}\vec{e}_r = -P_a\vec{e}_r. \qquad (6.213)$$

This boundary condition yields:

$$(-P + \sigma_{rr})(r = R) = -P_a, \qquad (6.214)$$

whence:

$$(P_a - P + \sigma_{\theta\theta})(r = R) = \sigma_{\theta\theta} - \sigma_{rr} = N_2. \qquad (6.215)$$

We deduce that $Cst = N_2$, so that Eq. (6.212) can be rewritten in the final form:

$$P_a - P + \sigma_{\theta\theta} = (N_1 + 2N_2)\ln\left(\frac{r}{R}\right) + N_2. \tag{6.216}$$

This equation shows that it is enough to measure the radial profile of the normal stress $P_a - P + \sigma_{\theta\theta}$ and fit it with the theoretical law above to obtain the values of N_1 and N_2 at the chosen shear rate.

One can also calculate the total force F exerted by the fluid onto the plate in the z direction. This force (responsible for the Weissenberg effect described in Section 1.5.7) is much easier to measure than the radial distribution of normal stresses. By definition, it is:

$$F = \int_0^R (P_a - P + \sigma_{\theta\theta})2\pi r \, dr \tag{6.217}$$

or, using Eq. (6.216):

$$F = \int_0^R \left[(N_1 + 2N_2)\ln\left(\frac{r}{R}\right) + N_2\right]2\pi r \, dr = -\frac{\pi}{2}N_1 R^2. \tag{6.218}$$

One can see that the second normal stress difference 'miraculously' vanishes from the final result. Thus, the measure of the total force directly yields the first normal stress difference N_1:

$$N_1 = -\frac{2F}{\pi R^2}. \tag{6.219}$$

However, in order to find N_2 one still needs to measure the complete profile of the normal force.

6.3.7 Elongational flow

We have already mentioned this type of flow at the end of our presentation of linear viscoelasticity, where we generalised the concept of the Young modulus (widely used for solids), showing that its value is three times that of the shear modulus when the compressibility effects are negligible (or, equivalently, when the Poisson coefficient remains very close to its asymptotic 1/2 value). We will now deal with continuous flow of the same type.

6.3.7.1 Definition of the elongational viscosity

Let us first consider simple elongational flow (Fig. 6.14). Two situations can occur.

At low drawing velocity, the deformation of the objects (or particles) is low and they store very little elastic energy. The system is in the *linear regime*, as defined at the beginning of the chapter. Under these conditions, the fluid behaves as an incompressible

Newtonian fluid, with the same viscosity η as under continuous shear. The velocity field is then given by the following equations, valid for an incompressible Newtonian fluid:

$$v_1 = \dot{\varepsilon}\, x_1,$$
$$v_2 = -\frac{1}{2}\dot{\varepsilon}\, x_2, \tag{6.220}$$
$$v_3 = -\frac{1}{2}\dot{\varepsilon}\, x_3,$$

so that the total stress tensor is given in this case by:

$$\begin{pmatrix} -P + 2\eta\dot{\varepsilon} & 0 & 0 \\ 0 & -P - \eta\dot{\varepsilon} & 0 \\ 0 & 0 & -P - \eta\dot{\varepsilon} \end{pmatrix}. \tag{6.221}$$

In the *nonlinear regime*, at high elongation, the most general form of this tensor (compatible with the flow symmetry) is:

$$\begin{pmatrix} -P + \sigma_{11} & 0 & 0 \\ 0 & -P + \sigma_{22} & 0 \\ 0 & 0 & -P + \sigma_{33} \end{pmatrix}, \tag{6.222}$$

where, in the case of simple elongation, $\sigma_{22} = \sigma_{33}$. The disappearance of the non-diagonal terms of this tensor is due to the invariance of flow under $180°$ rotations about the axes 1, 2 and 3.

In the stationary regime, the *elongational viscosity* describes the resistance opposed by the fluid to elongation. It is defined in the following manner:

$$\eta_e \equiv \frac{\sigma_{11} - \sigma_{22}}{\dot{\varepsilon}}. \tag{6.223}$$

In a Newtonian fluid, or in a viscoelastic fluid at moderate elongation rates (rigorously speaking, in the $\dot{\varepsilon} \to 0$ limit), this viscosity is given by 3η, as shown by Eq. (6.221). Under these two limiting cases, the value of the Trouton ratio defined in Section 6.2.3.4 is 3.

On the other hand, experiment shows that η_e does not vary like the shear viscosity η in the nonlinear regime; the former generally tends to decrease, while the latter can sometimes exhibit considerable increase (of a few orders of magnitude). Experimental values will be presented in the next chapter.

6.3.7.2 *Experimental measurements*

The elongational viscosity is not very easy to measure. A classical experimental setup, the 'opposed jet rheometer', is shown in Fig. 6.27. This device contains two pipes of radius R,

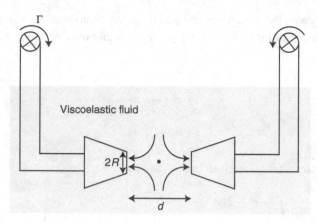

Fig. 6.27 'Opposed jet rheometer' used for measuring the elongational viscosity in the permanent regime.

a distance d apart, immersed in a container filled with the liquid under study. As the liquid is sucked into the pipes at a constant rate Q, a stationary elongational flow develops, with a stagnation point in the centre. The pipes experience a force proportional to the elongational viscosity of the fluid. This force is determined by measuring the torque Γ that must be applied to the arms supporting the pipes in order to keep them at a constant distance d. The device is calibrated using Newtonian viscous fluids where the elongational viscosity is well known, since $\eta_e = 3\eta$.

We will now see how to connect the measurements performed under oscillating flow (in the linear viscoelasticity regime) to those done in continuous flow (belonging, by their very nature, to the realm of nonlinear viscoelasticity).

6.4 Calculation of the viscometric functions and their relation with the functions of linear viscoelasticity

Appearances notwithstanding, this is a highly non-trivial problem. Indeed, we will show that, *even in the linear regime, the constitutive equations of a viscoelastic fluid are nonlinear*. This nonlinear character of the equations is of *kinematic origin* and stems from the *objectivity principle*, which states that *a constitutive equation must not depend on the chosen frame of reference* (in particular, this frame of reference need not be Galilean, in contrast with the frame of reference employed to write the equations of motion).

We will first demonstrate these features starting from the linear Maxwell model. We will see that this model gives erroneous results under continuous shear and that it must be modified. To this end, we will introduce a new tensor, the *Finger tensor*, which generalises the ordinary definition of the strain tensor to motions of large amplitude. Starting from this tensor, we will construct a new constitutive equation, generalising the previous one. Known as the *convected Maxwell equation*, it can be applied somewhat successfully to the study of continuous flow and can be used to obtain the viscometric functions. We will see that in this equation the time derivative is replaced by a *convected time derivative, accounting for the rotations of the fluid elements*.

We will then extend these results to the case of Jeffrey fluids (Oldroyd-A or B models), and then to the case of generalised Maxwell fluids. We will then see that it is not enough to make use of the convective derivatives or to add viscoelastic relaxation times to the constitutive equations in order to explain the non-Newtonian character of these fluids; new terms must be added, translating their *nonlinear behaviour at the local level*. These insights will lead us to the *phenomenological Oldroyd 8-constant model*, and then to the *Giesekus and FENE models* that – in contrast with the previous model – are built on a microscopic basis (provided by the *dumbbell model* described in the next chapter). We will also mention a different family of models (including the Phan-Thien and Tanner model and the KBKZ model) built upon an alternative microscopic model, that of a *transient network of elastic junctions*. We will see that all these models predict a shear-thinning behaviour, which is indeed observed in most viscoelastic fluids.

6.4.1 The shortcomings of the Maxwell model

For the time being, we will assume that the viscoelastic material is described by a unique relaxation time τ (with $\tau = \eta/G$). In the linear viscoelasticity regime, it is therefore described by the Maxwell equation (6.5); in tensor notation, this equation is written:

$$\underline{\sigma} + \tau \underline{\dot{\sigma}} = \eta \underline{\dot{\gamma}}. \tag{6.224}$$

Suppose now that the material undergoes *solid rotation* by an angle θ about the x_3 axis (Fig. 6.28). If the rotation takes place over a time much shorter than τ, the material must respond elastically. Its 'instantaneous' response is therefore given by the equation:

$$\underline{\sigma} = G\underline{\gamma}, \tag{6.225}$$

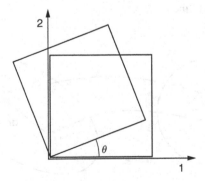

Fig. 6.28 Solid rotation of a fluid.

yielding, for a generic rotation angle θ (not necessarily small):

$$\underline{\sigma} = G \begin{pmatrix} 2(\cos\theta - 1) & 0 & 0 \\ 0 & 2(\cos\theta - 1) & 0 \\ 0 & 0 & 0 \end{pmatrix}. \tag{6.226}$$

This result is unacceptable, since one should have $\underline{\sigma} = 0$ for all values of the angle θ.

This model must therefore be improved, as it is incapable of describing the flow as soon as the deformations become important. In particular, this is the case for continuous shear, where $\gamma \to \infty$.

These considerations lead to a generalisation of the concept of deformation tensor.

6.4.2 Convective transport and Finger tensor

Consider a vector $\mathrm{d}\vec{r}'$ connecting two neighbouring points M_1' and M_2' in the fluid at the instant t'. Let us follow these two points along their motion; they will be found at M_1 and M_2, respectively, at the instant t, defining a new vector $\mathrm{d}\vec{r}$ (Fig. 6.29).

The application $\mathrm{d}\vec{r}' \to \mathrm{d}\vec{r}$ is linear and regular. It defines the invertible tensor $\underline{F}$ (we identify here the tensor and its matrix, written in an orthonormal reference frame) such that:

$$\mathrm{d}\vec{r} = \underline{F}(t, t') \, \mathrm{d}\vec{r}'. \tag{6.227}$$

The components of this tensor are:

$$F_{ij} = \frac{\partial x_i}{\partial x_j'}. \tag{6.228}$$

It measures the displacement at time t with respect to the position at time t'.

We mention the trivial relation:

$$\underline{F}(t, t) = \underline{I}, \tag{6.229}$$

where $\underline{I}$ is the unit matrix. Note that, in the literature, $\underline{F}$ is sometimes defined as the inverse of the tensor we have just defined.

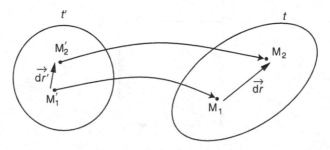

Fig. 6.29 Convective transport of a vector.

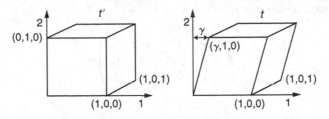

Fig. 6.30 Simple shear flow.

Consider now an elementary volume shaped, for instance, as a cube of unit size at time t'. Under the action of flow, this cube is deformed and has a new form at time t. It so happens that this transformation can always be analysed as the composition of a solid rotation and a stretching deformation. More specifically, it can be shown that the tensor $\underline{F}$ can be *uniquely* decomposed as (*polar decomposition theorem* [30]):

$$\underline{F} = \underline{V}\,\underline{R}, \tag{6.230}$$

where $\underline{R}$ represents a rotation matrix (such that $\underline{R}^{-1} = \underline{R}^{\mathrm{T}}$ and $\det(\underline{R}) = 1$) and $\underline{V}$ a positive definite symmetric matrix (namely, for any vector $\vec{a}$, the scalar quantity $\vec{a} \cdot \underline{V}\vec{a}$ is positive). An example is given in the inset below.

To illustrate the polar decomposition theorem, consider the simple shear deformation in Fig. 6.30. In this case, one has $x_1 = x_1' + \gamma(t, t')x_2'$, $x_2 = x_2'$ and $x_3 = x_3'$, yielding, from Eq. (6.228):

$$\underline{F} = \begin{pmatrix} 1 & \gamma & 0 \\ 0 & 1 & 0 \\ 0 & 0 & 1 \end{pmatrix}. \tag{6.231}$$

Calculating the matrices $\underline{R}$ and $\underline{V}$ leads to the following result:

$$\underline{R} = \begin{pmatrix} \dfrac{2}{\sqrt{4+\gamma^2}} & \dfrac{\gamma}{\sqrt{4+\gamma^2}} & 0 \\ -\dfrac{\gamma}{\sqrt{4+\gamma^2}} & \dfrac{2}{\sqrt{4+\gamma^2}} & 0 \\ 0 & 0 & 1 \end{pmatrix} \tag{6.232}$$

and

$$\underline{V} = \begin{pmatrix} \dfrac{2+\gamma^2}{\sqrt{4+\gamma^2}} & \dfrac{\gamma}{\sqrt{4+\gamma^2}} & 0 \\ \dfrac{\gamma}{\sqrt{4+\gamma^2}} & \dfrac{2}{\sqrt{4+\gamma^2}} & 0 \\ 0 & 0 & 1 \end{pmatrix} \tag{6.233}$$

yielding, for γ much smaller than 1:

$$\underline{R} \approx \begin{pmatrix} 1 & \gamma/2 & 0 \\ -\gamma/2 & 1 & 0 \\ 0 & 0 & 1 \end{pmatrix} \tag{6.234}$$

and

$$\underline{V} \approx \begin{pmatrix} 1 & \gamma/2 & 0 \\ \gamma/2 & 1 & 0 \\ 0 & 0 & 1 \end{pmatrix}. \tag{6.235}$$

Under this form, one can see clearly that the motion is decomposed into a rotation by the angle $-\gamma/2$ about axis 3, followed by stretching along the bisectrix of axes 1 and 2 (Fig. 6.31).

In the following, we will use the general decomposition of motion, defined by Eq. (6.230), to describe the action of a generic flow on the fluid.

Before we start, let us return to the tensor $\underline{F}$: since, by definition, it accounts for the rotation of the fluid elements, it cannot be used directly in the constitutive laws of the material (which must obey the *objectivity principle*, namely the independence of the reference frame). To eliminate the effect of rotations, the simplest choice is the following combination:

$$\underline{B} = \underline{F}\,\underline{F}^{\mathrm{T}} = \underline{V}^2. \tag{6.236}$$

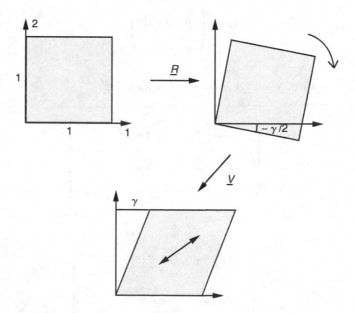

Fig. 6.31 Decomposition of a simple shear motion into a rotation followed by stretching.

This is the *Green tensor*, which can be used to build the *Finger (deformation) tensor*, given by [30]:

$$\underline{H} = \frac{1}{2}(\underline{B} - \underline{I}) = \frac{1}{2}(\underline{F}\,\underline{F}^T - \underline{I}) = \frac{1}{2}(\underline{V}^2 - \underline{I}). \tag{6.237}$$

It is easily checked that this tensor reduces to the ordinary strain tensor $\underline{\varepsilon}$ in the case of small deformations. This result is shown explicitly for two examples in the inset below.

In the case of simple shear (Figs. 6.30 and 6.31), one obtains directly from Eq. (6.231):

$$\underline{B} = \begin{pmatrix} \gamma^2 + 1 & \gamma & 0 \\ \gamma & 1 & 0 \\ 0 & 0 & 1 \end{pmatrix} \tag{6.238}$$

and then:

$$\underline{H} = \begin{pmatrix} \gamma^2/2 & \gamma/2 & 0 \\ \gamma/2 & 0 & 0 \\ 0 & 0 & 0 \end{pmatrix}. \tag{6.239}$$

Thus, this tensor does reduce to the $\underline{\varepsilon}$ tensor in the case of small deformations:

$$\underline{\varepsilon} = \begin{pmatrix} 0 & \gamma/2 & 0 \\ \gamma/2 & 0 & 0 \\ 0 & 0 & 0 \end{pmatrix}. \tag{6.240}$$

On the other hand, in the case of elongational flow described in Fig. 6.32:

$$\underline{F} = \begin{pmatrix} \lambda_1 & 0 & 0 \\ 0 & \lambda_2 & 0 \\ 0 & 0 & \lambda_3 \end{pmatrix} \tag{6.241}$$

yielding, from the definition (6.237):

$$\underline{H} = \begin{pmatrix} (\lambda_1^2 - 1)/2 & 0 & 0 \\ 0 & (\lambda_2^2 - 1)/2 & 0 \\ 0 & 0 & (\lambda_3^2 - 1)/2 \end{pmatrix}. \tag{6.242}$$

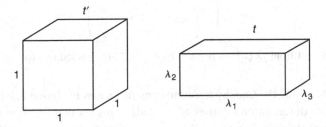

Fig. 6.32 Elongational deformation.

Note that, for a simple elongation, $\lambda_2 = \lambda_3$. If the material is also incompressible, then $\lambda_1\lambda_2\lambda_3 = 1$, whence $\lambda_2 = \lambda_3 = \lambda_1^{-1/2}$.

As in the previous case, it is easy to check that this tensor reduces to the ordinary strain tensor $\underline{\varepsilon}$ in the case of small perturbations:

$$\underline{\varepsilon} = \begin{pmatrix} \lambda_1 - 1 & 0 & 0 \\ 0 & \lambda_2 - 1 & 0 \\ 0 & 0 & \lambda_3 - 1 \end{pmatrix}. \tag{6.243}$$

Before generalising the Maxwell equation, let us introduce the Cauchy tensor. The latter, as the Finger tensor, vanishes in the case of solid rotation and tends towards $\underline{\varepsilon}$ in the small deformation limit.

6.4.3 Cauchy tensor

The *Cauchy (deformation) tensor* is defined as [30]:

$$\underline{G} = \frac{1}{2}(\underline{I} - \underline{B}^{-1}), \tag{6.244}$$

where $\underline{B}^{-1} = (\underline{F}^{\mathrm{T}})^{-1}\underline{F}^{-1} = (\underline{F}^{-1})^{\mathrm{T}}\underline{F}^{-1}$ is the inverse tensor of $\underline{B}$. Note that:

$$F_{ij}^{-1} = \frac{\partial x_i'}{\partial x_j} \tag{6.245}$$

from Eq. (6.228).

One can also prove in the general case that $\underline{G}$ tends to $\underline{\varepsilon}$ for small deformations ([28, volume 1]). To see this, let us examine a particular case.

For simple shear flow (Fig. 6.30), one has from Eq. (6.238):

$$\underline{B}^{-1} = \begin{pmatrix} 1 & -\gamma & 0 \\ -\gamma & \gamma^2 + 1 & 0 \\ 0 & 0 & 1 \end{pmatrix}, \tag{6.246}$$

whence

$$\underline{G} = \begin{pmatrix} 0 & \gamma/2 & 0 \\ \gamma/2 & \gamma^2/2 & 0 \\ 0 & 0 & 0 \end{pmatrix}. \tag{6.247}$$

Clearly, $\underline{G}$ is different from $\underline{H}$ (which is given by Eq. (6.239)), but reduces to $\underline{\varepsilon}$ in the small deformation limit.

It turns out that using the Cauchy tensor to construct a new evolution law for the material in the continuous deformation regimes often leads to results that are in worse agreement with the experimental data than those obtained by using the Finger tensor. That is why we will mostly use the latter in the following.

6.4.4 Generalisation of the Maxwell equation and calculation of the viscometric functions using the Finger tensor

6.4.4.1 The convected Maxwell equation

In Section 6.2.1.1 we have shown that, in the small deformation limit, a viscoelastic material undergoing simple shear can be described by the following constitutive law:

$$\sigma(t) = G \int_{-\infty}^{t} \exp\left[-(t - t')/\tau\right] \dot{\gamma}(t') \, dt'. \tag{6.248}$$

This law (which is the integral form of the Maxwell differential equation (6.224)) shows that each deformation increment $\dot{\gamma}(t') \, dt'$ that occurred in the past, between the moments t' and $t' + dt'$, entails at the instant t (taken as the present time) a stress:

$$d\sigma = G \exp\left[-(t - t')/\tau\right] \dot{\gamma}(t') \, dt', \tag{6.249}$$

as weak as the instant t' is far from the instant t (a result one might call the *principle of memory loss at long times*). From this point of view, the exponential factor $\exp\left[-(t - t')/\tau\right]$ represents the weight of the deformation that had occurred at the instant t'. Note that this weight becomes negligible for deformation having occurred before $t - \tau$. The time τ (or *viscoelastic relaxation time*) can thus be seen as quantifying the *memory* of the material (see Chapter 1).

Let us now see how Eq. (6.248) can be generalised to the case of large amplitude flow. Before that, it is convenient to rewrite this equation under the following equivalent form, obtained after integrating by parts:

$$\sigma(t) = G \int_{-\infty}^{t} \frac{2}{\tau} \exp\left[-(t - t')/\tau\right] \varepsilon(t, t') \, dt', \tag{6.250}$$

where we employed the strain $\varepsilon(t, t')$ occurring between the instants t' and t:

$$\varepsilon(t, t') = \frac{1}{2}\left[\gamma(t) - \gamma(t')\right] = \frac{1}{2} \int_{t'}^{t} \dot{\gamma}(t'') \, dt''. \tag{6.251}$$

With the notation in Fig. 6.30, we can see that $\varepsilon = \varepsilon_{12}$. It is therefore very tempting to generalise Eq. (6.250) to the tensor case by replacing $\underline{\varepsilon}$ with the Finger tensor $\underline{H}$ defined above (we recall that it reduces to the $\underline{\varepsilon}$ tensor in the small deformation limit). In this way, one obtains a new integral formulation of the Maxwell equation for incompressible viscoelastic fluids (known as the *Lodge equation*), which has the advantage of being invariant under changes in the reference frame (since the effect of rotations has been eliminated, see Section 6.4.1):

$$\underline{\sigma}(t) = G \int_{-\infty}^{t} \frac{2}{\tau} \exp\left[-(t - t')/\tau\right] \underline{H}(t, t') \, dt'. \tag{6.252}$$

In order to return to the differential formulation, one simply needs to derivate this equation with respect to the time t. This calculation gives:

$$\dot{\underline{\sigma}} = \frac{2G}{\tau}\underline{H}(t,t) - G\int_{-\infty}^{t}\frac{2}{\tau^{2}}\exp\left[-(t-t')/\tau\right]\underline{H}(t,t')\,dt'$$

$$+ G\int_{-\infty}^{t}\frac{2}{\tau}\exp\left[-(t-t')/\tau\right]\underline{\dot{H}}(t,t')\,dt'. \tag{6.253}$$

The first term on the right-hand side is zero, since $\underline{H}(t,t)=0$ (obviously, no deformation can occur between two identical moments). The second term is given by $-\frac{1}{\tau}\underline{\sigma}$, from Eq. (6.252). To evaluate the third term, one must first calculate $\underline{\dot{H}}(t,t')$. From the definition of $\underline{H}$ (Eq. (6.237)), one has:

$$\underline{\dot{H}} = \frac{1}{2}\underline{\dot{B}} = \frac{1}{2}\left[\underline{\dot{F}}\,\underline{F}^{\mathrm{T}} + \underline{F}\,\underline{\dot{F}}^{\mathrm{T}}\right]. \tag{6.254}$$

In addition, Eq. (6.228) yields:

$$\dot{F}_{ij} = \frac{\partial \dot{x}_{i}}{\partial x'_{j}} = \frac{\partial v_{i}}{\partial x_{k}}\frac{\partial x_{k}}{\partial x'_{j}} = (\vec{\nabla}\vec{v})_{ik}F_{kj} = (\vec{\nabla}\vec{v}\,\underline{F})_{ij}, \tag{6.255}$$

where $\vec{v} = \dot{\vec{r}}$ is the velocity of the fluid and $\vec{\nabla}\vec{v}$ the velocity gradient tensor, with components:[10]

$$(\vec{\nabla}\vec{v})_{ij} = \frac{\partial v_{i}}{\partial x_{j}}. \tag{6.256}$$

It follows that:

$$\underline{\dot{H}} = \frac{1}{2}\left[\left(\vec{\nabla}\vec{v}\,\underline{F}\right)\underline{F}^{\mathrm{T}} + \underline{F}\left(\vec{\nabla}\vec{v}\,\underline{F}\right)^{\mathrm{T}}\right] = \frac{1}{2}\left[\vec{\nabla}\vec{v}\left(\underline{F}\,\underline{F}^{\mathrm{T}}\right) + \left(\underline{F}\,\underline{F}^{\mathrm{T}}\right)\vec{\nabla}\vec{v}^{\mathrm{T}}\right]$$

$$= \frac{1}{2}\left[\left(\vec{\nabla}\vec{v}\,\underline{B} + \underline{B}\,\vec{\nabla}\vec{v}^{\mathrm{T}}\right)\right] = \vec{\nabla}\vec{v}\,\underline{H} + \underline{H}\,\vec{\nabla}\vec{v}^{\mathrm{T}} + \underline{A}, \tag{6.257}$$

where $\underline{A}$ is the strain rate tensor ($\dot{\varepsilon}$ in the small deformation limit), generally defined as:

$$\underline{A} = \frac{1}{2}\left(\vec{\nabla}\vec{v} + \vec{\nabla}\vec{v}^{\mathrm{T}}\right). \tag{6.258}$$

The third term on the right-hand side of Eq. (6.253) is therefore, using Eqs. (6.252) and (6.257):

$$G\int_{-\infty}^{t}\frac{2}{\tau}\exp\left[-(t-t')/\tau\right]\underline{\dot{H}}(t,t')\,dt' = \vec{\nabla}\vec{v}\,\underline{\sigma} + \underline{\sigma}\vec{\nabla}\vec{v}^{\mathrm{T}} + 2G\underline{A}. \tag{6.259}$$

---

[10] Note that this tensor is sometimes defined as its transpose.

Putting together all these three terms, and keeping in mind that $\eta = \tau G$, leads to the following differential equation:

$$\tau \dot{\underline{\sigma}} = -\underline{\sigma} + \tau \vec{\nabla}\vec{v}\,\underline{\sigma} + \tau \underline{\sigma}\,\vec{\nabla}\vec{v}^{\mathrm{T}} + 2\eta\,\underline{A}, \tag{6.260}$$

where $\dot{}$ designates the material derivative introduced in Chapter 2.

This equation can also be written in the following compact form:

$$\tau \overset{\triangledown}{\underline{\sigma}} + \underline{\sigma} = 2\eta\,\underline{A}, \tag{6.261}$$

where the symbol $^{\triangledown}$ stands for the *upper-convected* (or *contravariant*) *time derivative* defined as:

$$\overset{\triangledown}{\underline{\sigma}} = \dot{\underline{\sigma}} - \vec{\nabla}\vec{v}\,\underline{\sigma} - \underline{\sigma}\,\vec{\nabla}\vec{v}^{\mathrm{T}}. \tag{6.262}$$

This new equation, extending the Maxwell equation of linear viscoelasticity to the generic flow case, is the *upper-convected* (or *contravariant*) *Maxwell equation*.

Note that we could also have used the Cauchy tensor $\underline{G}$ (see Section 6.4.3) instead of the Finger tensor $\underline{H}$. This would have led to the following equation:

$$\tau \overset{\triangle}{\underline{\sigma}} + \underline{\sigma} = 2\eta\,\underline{A}, \tag{6.263}$$

where this time the symbol $^{\triangle}$ stands for the *lower-convected* (or *covariant*) *time derivative*, defined as:

$$\overset{\triangle}{\underline{\sigma}} = \dot{\underline{\sigma}} + \vec{\nabla}\vec{v}^{\mathrm{T}}\underline{\sigma} + \underline{\sigma}\,\vec{\nabla}\vec{v}. \tag{6.264}$$

This *lower-convected* (or *covariant*) *Maxwell equation* is very seldom used, as for most fluids,[11] and in particular for polymer melts or solutions, its rheological predictions are in poor agreement with the experiments (see the next section).

More generally, we could have replaced $\underline{H}$ in the Lodge equation (6.252) by a linear combination of the Finger and Cauchy tensors, of the type $(1-u)\underline{H}+u\underline{G}$, with $0 \leq u \leq 1$. This is legitimate, since this combination also reduces to $\underline{\varepsilon}$ in the small deformation regime. With this choice, the constitutive equation for the stress becomes:

$$\tau \overset{\square}{\underline{\sigma}} + \underline{\sigma} = 2\eta\underline{A}, \tag{6.265}$$

where

$$\overset{\square}{\underline{\sigma}} = (1-u)\,\overset{\triangledown}{\underline{\sigma}} + u\,\overset{\triangle}{\underline{\sigma}}. \tag{6.266}$$

The time derivative $^{\square}$ is known as the Gordon–Schowalter derivative.[12]

---

[11] One notable exception is that of suspensions of discoidal particles (clays).

[12] In the case $u = 1/2$, one obtains the corotational (or Jaumann) derivative, of expression $\overset{\circ}{\underline{\sigma}} = \dot{\underline{\sigma}} - \underline{\Omega}\underline{\sigma} - \underline{\sigma}\underline{\Omega}^{\mathrm{T}}$, where $\underline{\Omega} = \frac{1}{2}(\vec{\nabla}\vec{v} - \vec{\nabla}\vec{v}^{\mathrm{T}})$ is the local rotation rate tensor already introduced in Chapter 3.

Clearly, one can choose among countless possibilities for the convected time derivative. For simplicity's sake, in the following we will use almost exclusively the *contravariant* version of the *convected Maxwell equation*. At this point, it is very important to note that this new equation, although nonlinear due to the presence of terms containing the velocity gradient squared, absolutely does not account for a possible nonlinear behaviour of the material that might come into play, at the level of the 'particles', for high shear rates, with $\dot{\gamma} > 1/\tau$. Hence, this equation only holds in the *linear regime*, for shear rates $\dot{\gamma} < 1/\tau$ (being understood that, in this case, the nonlinearities of the model are of purely kinematic origin).

The upper-convected Maxwell equation is nevertheless very useful, since it predicts the existence of a finite first normal stress difference, as well as the non-trivial behaviour of the elongational viscosity in the continuous regime.

We will show in the next chapter that, in the case of polymers, the microscopic dumbbell model (where each molecule is represented by two beads connected by an infinitely extensible spring[13]) and more generally the Rouse model (where the dumbbell is replaced by a chain of beads connected by infinitely extensible springs) 'naturally' lead to the upper-convected Maxwell equation (6.261).

6.4.4.2 Calculating the viscosity $\eta(\dot{\gamma})$ and the normal stress coefficients $\Psi_1(\dot{\gamma})$ and $\Psi_2(\dot{\gamma})$ under continuous shear

In this section, we will use the *upper-convected* (or *contravariant*) *Maxwell equation*. Similar calculations can be performed in the covariant version, on condition to replace the Finger tensor by the Cauchy tensor, or the contravariant derivative $\overset{\triangledown}{}$ by the covariant derivative $\overset{\triangle}{}$.

The easiest way of analysing a shear flow, such as that represented in Fig. 6.30, is to use the integral Lodge equation (6.252). In theory, the stationary regime is reached after an infinite time; in practice, this time only needs to be much longer than τ. Let $\dot{\gamma}$ be the imposed shear rate (assumed constant). If the flow started at $t' = -\infty$, the cumulative deformation between any instant t' and the present t is $\gamma(t, t') = \dot{\gamma}(t - t')$. Since, from Eq. (6.239), $H_{12}(t, t') = \gamma(t, t')/2$, one can determine, using Eq. (6.252), the viscosity in the stationary regime:

$$\eta(\dot{\gamma}) \equiv \frac{\sigma_{12}}{\dot{\gamma}} = \frac{G}{\dot{\gamma}} \int_{-\infty}^{t} \frac{2}{\tau} \exp\left[-(t - t')/\tau\right] H_{12}(t, t') \, dt'$$

$$= \frac{G}{\dot{\gamma}} \int_{-\infty}^{t} \frac{1}{\tau} \exp\left[-(t - t')/\tau\right] \dot{\gamma}(t - t') \, dt' = G\tau, \tag{6.267}$$

resulting in

$$\eta(\dot{\gamma}) = \eta. \tag{6.268}$$

This model predicts that the viscosity measured under continuous shear is the same as that measured in linear viscoelasticity.

---

[13] This model will be described in detail in Appendix 7.A.

In the same way, to calculate the normal stresses one can replace $H_{11}(t, t')$, $H_{22}(t, t')$ and $H_{33}(t, t')$ by their respective expressions in the integral equation (6.252). Since, from Eq. (6.239):

$$H_{11}(t, t') = \frac{1}{2}\dot{\gamma}^2(t' - t)^2, \qquad H_{22}(t, t') = H_{33}(t, t') = 0, \qquad (6.269)$$

one finds:

$$\sigma_{11} = 2G\tau^2\dot{\gamma}^2, \qquad \sigma_{22} = \sigma_{33} = 0. \qquad (6.270)$$

This model leads to a finite first normal stress difference:

$$N_1 = \sigma_{11} - \sigma_{22} = 2G\tau^2\dot{\gamma}^2 = 2\eta\tau\dot{\gamma}^2, \qquad (6.271)$$

but the second normal stress difference vanishes:

$$N_2 = \sigma_{22} - \sigma_{33} = 0 \qquad (6.272)$$

or, equivalently, the first normal stress coefficient is constant:

$$\Psi_1 = 2\eta\tau \qquad (6.273)$$

and the second normal stress coefficient is zero:

$$\Psi_2 = 0. \qquad (6.274)$$

This behaviour is qualitatively correct. In particular, measurements performed on polymer melts or solutions confirm the validity of Eq. (6.271) in the $\dot{\gamma} \to 0$ limit and show that the second normal stress difference is typically one order of magnitude weaker than the first one.

Redoing the same calculations in the covariant version yields the same results for the viscosity and for the first normal stress coefficient, but this time with $\Psi_2 = -\Psi_1$. If the sign of Ψ_2 is in agreement with the experiments, its order of magnitude is generally off the mark, since very often (in particular for polymers) $\Psi_2 \approx -0.1\Psi_1$. That is why this model is seldom used.[14]

Note that the calculations above are only valid in the stationary regime. One can easily extend them to the case of a deformation switched on suddenly at $t = 0$. From Eq. (6.239) one finds that, in the case of simple shear, the Finger tensor has the following finite components (at times $t > 0$):

$$H_{11}(t, t' < 0) = \dot{\gamma}^2 t^2/2 \quad \text{and} \quad H_{11}(t, t' > 0) = \dot{\gamma}^2(t - t')^2/2, \qquad (6.275a)$$

$$H_{12}(t, t' < 0) = \dot{\gamma}t/2 \quad \text{and} \quad H_{12}(t, t' > 0) = \dot{\gamma}(t - t')/2. \qquad (6.275b)$$

---

[14] Note that using the generalised Gordon and Schowalter derivative (6.265) leads to $\Psi_2/\Psi_1 = u - 1$ ($0 \le u \le 1$).

These formulas directly yield the shear stress in the transient regimes from the Lodge equation (6.252):

$$\sigma_{12}(t > 0) = G \int_{-\infty}^{t} \frac{2}{\tau} \exp\left(-\frac{t-t'}{\tau}\right) H_{12}(t, t') dt'. \tag{6.276a}$$

Replacing $H_{12}(t, t')$ by its expression (6.275b) leads to:

$$\sigma_{12}(t > 0) = \eta \dot{\gamma} \left[1 - \exp\left(-\frac{t}{\tau}\right)\right]. \tag{6.276b}$$

One finds the expected behaviour (see Fig. 1.11 in Chapter 1).

In the same way, using Eq. (6.275b), one can determine the variation of the normal stress N_1 during the transient regime:

$$N_1(t) = 2G\tau^2\dot{\gamma}^2 \left[1 - \left(1 + \frac{t}{\tau}\right)\exp\left(-\frac{t}{\tau}\right)\right]. \tag{6.277}$$

6.4.4.3 Calculating the elongational viscosity

The upper-convected Maxwell model also predicts the behaviour of the elongational viscosity η_e for simple stretching at a constant rate (see Fig. 6.32). From Eq. (6.223), by definition:

$$\eta_e = \frac{\sigma_{11} - \sigma_{22}}{\dot{\varepsilon}}. \tag{6.278}$$

If the elongation rate $\dot{\varepsilon} = \frac{1}{\lambda_1}\frac{d\lambda_1}{dt}$ is constant, then $\lambda_1 = \exp\left[\dot{\varepsilon}(t - t')\right]$. On the other hand, from Eq. (6.242) one has, for an incompressible fluid:

$$H_{11} = \frac{1}{2}\left(\lambda_1^2 - 1\right), \qquad H_{22} = \frac{1}{2}\left(\frac{1}{\lambda_1} - 1\right). \tag{6.279}$$

Hence, from the Lodge equation (6.252), the elongational viscosity is given by:

$$\eta_e = \frac{G}{2\dot{\varepsilon}} \int_{-\infty}^{t} \frac{2}{\tau} \exp\left[(t' - t)/\tau\right] \left\{\exp\left[2\dot{\varepsilon}(t - t')\right] - \exp\left[-\dot{\varepsilon}(t - t')\right]\right\} dt', \tag{6.280}$$

yielding, after calculations:[15]

$$\eta_e = \tau G \left[\frac{2}{1 - 2\tau\dot{\varepsilon}} + \frac{1}{1 + \tau\dot{\varepsilon}}\right]. \tag{6.281}$$

The formula above shows that, in the $\dot{\varepsilon} \to 0$ limit, $\eta_e \to 3\eta$ (the *Trouton rule* for Newtonian fluids). Above all, however, it shows that the elongational viscosity increases

---

[15] We emphasise, once again, that this formula only applies in the permanent regime. In order to calculate the viscosity $\eta_e(t)$ in the transient regimes after the sudden onset at $t = 0$ of a constant elongation rate $\dot{\varepsilon}$, the $\exp\left[2\dot{\varepsilon}(t - t')\right] - \exp\left[-\dot{\varepsilon}(t - t')\right]$ term in integral (6.280) should be preserved unchanged over the time interval $0 < t' < t$ and replaced by $\exp(2\dot{\varepsilon}t) - \exp(-\dot{\varepsilon}t)$ in the interval $-\infty < t' < 0$.

with the elongation rate $\dot{\varepsilon}$ and diverges for $\dot{\varepsilon} \to 1/2\tau$. Such an increase is indeed observed in experiments, but not the divergence, although the experiments often show a sudden increase in η_e for $\dot{\varepsilon} \to 1/2\tau$. We will see in Section 6.4.6 that this divergence can be suppressed by taking into account the nonlinearities in the local constitutive law of the material. In polymers, for instance, which are fairly well described by the *dumbbell model*, the nonlinearities show up when including either the local anisotropy of the friction force acting on each bead (Giesekus model), or the finite extensibility of the springs (FENE model). We will present a phenomenological description of these models in Section 6.4.6, and their microscopic analysis in Appendices 7.A and 7.B.

In conclusion, the *upper-convected Maxwell model makes qualitatively correct predictions for the linear regime* (i.e. for $\dot{\gamma} < 1/\tau$, or $\dot{\varepsilon} < 1/2\tau$). It shows directly that the *appearance of normal stresses and the increase in elongational viscosity are of elastic origin* (these phenomena disappear in the Newtonian case, as $G \to \infty$ and $\tau = 0$). On the other hand, it does not account for the generally observed shear-thinning behaviour of the viscosity $\eta(\dot{\gamma})$ and of the normal stress coefficients $\Psi_1(\dot{\gamma})$ and $\Psi_2(\dot{\gamma})$.

Before approaching this problem in the framework of more complex models allowing for the local nonlinearities of the material, let us extend these calculations to the case of a Jeffrey fluid.

6.4.5 *The convected Jeffrey models (or 'Oldroyd-A or -B')*

Many systems, among them polymer solutions or suspensions of rigid or deformable particles (emulsions) behave, in the dilute regime, as Jeffrey fluids (see Fig. 6.3b). The material is described by three constants, instead of two: the viscosity η_P and the elastic modulus G_P associated with the deformable particles in the mixture (polymer coils, liquid droplets in suspension, etc.) and the viscosity η_s of the solvent (taken as Newtonian) in which these particles bathe.

In this model, the total stress $\underline{\sigma}$ is the sum of the stresses due to the particles and the solvent:

$$\underline{\sigma} = \underline{\sigma}_P + \underline{\sigma}_s. \tag{6.282}$$

At length scales much larger than the typical distance between particles, the hydrodynamic velocity of the particles is equal to that of the solvent molecules. One can thus write for the particles the following constitutive equation:

$$\tau_P \, \overset{\triangledown}{\underline{\sigma}}_P + \underline{\sigma}_P = 2\eta_P \, \underline{A}, \tag{6.283}$$

while for the solvent one has simply:

$$\underline{\sigma}_s = 2\eta_s \, \underline{A}, \tag{6.284}$$

where $\underline{A}$ is the strain rate tensor (identical for the particles and the solvent). Replacing $\underline{\sigma}_P$ by its expression $\underline{\sigma} - 2\eta_s \underline{A}$ in Eq. (6.283) yields an equation for $\underline{\sigma}$:

$$\tau_P \overset{\triangledown}{\underline{\sigma}} + \underline{\sigma} = (2\eta_P + 2\eta_s) \underline{A} + 2\tau_P \eta_s \overset{\triangledown}{\underline{A}}. \tag{6.285}$$

Introducing the viscosity $\eta_0 = \eta_P + \eta_s$ and setting $\tau_R = \tau_P \eta_s/(\eta_P + \eta_s) = \tau_P \eta_s/\eta_0$ (*retardation time*) leads to a new form for this equation:

$$\tau_P \overset{\triangledown}{\underline{\sigma}} + \underline{\sigma} = 2\eta_0 \left(\underline{A} + \tau_R \overset{\triangledown}{\underline{A}} \right). \tag{6.286}$$

This is the *upper-convected Jeffrey equation* (or *Oldroyd-B*), which is nothing other than a generalisation of the upper-convected Maxwell equation to the case of a Jeffrey fluid.

Obviously, the lower-convected Maxwell equation can be generalised in the same way. This leads to the *Oldroyd-A equation*, which is similar to Eq. (6.286), except the contravariant derivative is replaced by the covariant derivative:

$$\tau_P \overset{\triangle}{\underline{\sigma}} + \underline{\sigma} = 2\eta_0 \left(\underline{A} + \tau_R \overset{\triangle}{\underline{A}} \right). \tag{6.287}$$

This equation is not very useful, as it often leads to incorrect rheological predictions.

It is noteworthy that the Oldroyd-B or -A equations can be obtained from the Jeffrey equation (6.13), established in the framework of linear viscoelasticity, by replacing $\dot{\gamma}/2$ with $\underline{A}$ and the time derivative $\dot{}$ by the convected time derivatives, contravariant $\overset{\triangledown}{}$ or covariant $\overset{\triangle}{}$.

Also note that both these new equations are nonlinear, whereas the constitutive law of the material remains locally linear. Thus, they are only valid in the linear regimes ($\dot{\gamma} < 1/\tau_P$), like the convected Maxwell equations.

To conclude this paragraph, let us determine the viscometric functions $\eta(\dot{\gamma})$, $\Psi_1(\dot{\gamma})$, $\Psi_2(\dot{\gamma})$ and $\eta_e(\dot{\varepsilon})$ in the framework of the *Oldroyd-B model*. The calculations are easily done using the integral formulation of the constitutive equation (see Eq. (6.252)):

$$\underline{\sigma}(t) = 2\eta_s \underline{A} + G_P \int_{-\infty}^t \frac{2}{\tau_P} \exp\left[-(t - t')/\tau_P\right] \underline{H}(t, t') \, dt' \tag{6.288}$$

and lead to the following results:

$$\eta(\dot{\gamma}) = \eta_0 = \eta_s + \eta_P, \tag{6.289}$$

$$\Psi_1(\dot{\gamma}) = 2\eta_P \tau_P, \tag{6.290}$$

$$\Psi_2(\dot{\gamma}) = 0, \tag{6.291}$$

$$\eta_e(\dot{\varepsilon}) = 3\eta_s + \eta_P \left[\frac{2}{1 - 2\tau_P \dot{\varepsilon}} + \frac{1}{1 + \tau_P \dot{\varepsilon}} \right]. \tag{6.292}$$

We can see that this model, like the previous ones, fails to predict the shear-thinning behaviour of the fluid. One might hope that adding a second characteristic time, or even a discrete or continuous distribution of relaxation times, would change things. It is however obvious from the integral formulation that this procedure does not change the final result. Indeed, it is easily checked that, in the general case, one has:

$$\eta(\dot{\gamma}) = \eta_0 = \eta_s + \sum \eta_i = \eta_s + \int_0^\infty G(u)\, du, \tag{6.293}$$

$$\Psi_1(\dot{\gamma}) = 2 \sum \eta_i \tau_i = 2 \int_0^\infty u\, G(u)\, du, \tag{6.294}$$

$$\Psi_2(\dot{\gamma}) = 0 \tag{6.295}$$

and

$$\eta_e(\dot{\varepsilon}) = 3\eta_s + \sum \left[\eta_i \left(\frac{2}{1 - 2\tau_i\dot{\varepsilon}} + \frac{1}{1 + \tau_i\dot{\varepsilon}} \right) \right]. \tag{6.296}$$

Note that the Oldroyd-A model leads to the same results for the viscosity and for the first normal stress coefficient, but it predicts $\Psi_2 = -\Psi_1$, which is generally in disagreement with the experiments.

In conclusion, *the nonlinearities of the material must necessarily be included in its local constitutive law in order to explain its rheological properties in the nonlinear regime.* We will show this explicitly, by discussing two phenomenological models that contain new nonlinear terms. For simplicity's sake, we assume that the fluid behaves as a Jeffrey fluid in linear viscoelasticity (with a single characteristic Maxwell time τ_P).

6.4.6 Generalisation to the nonlinear regimes of the convected Jeffrey models (or Oldroyd-A or -B)

6.4.6.1 The Oldroyd 8-constant model

Oldroyd noticed that the convected Jeffrey equations do not contain all generally allowed terms that are quadratic in the velocity gradient. He therefore proposed to complete these equations by all terms invariant under a reference change that contain the product of $\underline{\sigma}$ with $\underline{A}$ and of $\underline{A}$ with itself. This procedure leads to the following generic equation (Oldroyd, 1961; [28, volume 1]):

$$\underline{\sigma} + \tau_1 \overset{\triangledown}{\underline{\sigma}} + \tau_3 \left(\underline{A}\,\underline{\sigma} + \underline{\sigma}\,\underline{A} \right) + \tau_5 \left(\text{tr}\,\underline{\sigma} \right) \underline{A} + \tau_6\, \text{tr} \left(\underline{\sigma}\,\underline{A} \right) \underline{I} =$$

$$2\eta_0 \left[\underline{A} + \tau_2 \overset{\triangledown}{\underline{A}} + 2\tau_4 \underline{A}\,\underline{A} + \tau_7\, \text{tr} \left(\underline{A}\,\underline{A} \right) \underline{I} \right], \tag{6.297}$$

where we recognise classical invariants such as $\text{tr}\,\underline{\sigma}$, $\text{tr}(\underline{\sigma}\,\underline{A})$ or $\text{tr}(\underline{A}\,\underline{A})$. This equation contains the Oldroyd-B equation (6.286), as one can check immediately by taking $\tau_1 = \tau_P$, $\tau_2 = \tau_R$ and $\tau_i = 0$ for $(i = 3, 4, \ldots, 7)$. Less obviously, one can also retrieve the

Oldroyd-A equation (6.287) using $\tau_1 = \tau_P$, $\tau_2 = \tau_R$, $\tau_3 = 2\tau_1$, $\tau_4 = 2\tau_2$ and $\tau_i = 0$ ($i = 5, 6, 7$).

This model leads to more realistic behaviour of the viscometric functions, as we show in the inset below. In particular, it predicts their shear-thinning features and avoids the divergence of the elongational viscosity.

Let us first calculate the viscometric functions in the case of simple shear (Fig. 6.30). In the stationary regime, the stress tensor has the following general form:

$$\underline{\underline{\sigma}} = \begin{pmatrix} \sigma_{11} & \sigma_{12} & 0 \\ \sigma_{12} & \sigma_{22} & 0 \\ 0 & 0 & \sigma_{33} \end{pmatrix} \tag{6.298}$$

while

$$\underline{\underline{A}} = \begin{pmatrix} 0 & \dfrac{\dot{\gamma}}{2} & 0 \\ \dfrac{\dot{\gamma}}{2} & 0 & 0 \\ 0 & 0 & 0 \end{pmatrix} \quad \text{and} \quad \vec{\nabla}\vec{v} = \begin{pmatrix} 0 & \dot{\gamma} & 0 \\ 0 & 0 & 0 \\ 0 & 0 & 0 \end{pmatrix}. \tag{6.299}$$

Substituting in Eq. (6.297) yields a system of four equations with the four unknowns σ_{11}, σ_{22}, σ_{33} and σ_{12}, to be solved as a function of the shear rate $\dot{\gamma}$. The calculation leads to the following results:

$$\frac{\eta(\dot{\gamma})}{\eta_0} = \frac{1 + \sigma_2 \dot{\gamma}^2}{1 + \sigma_1 \dot{\gamma}^2}, \tag{6.300}$$

$$\frac{\Psi_1(\dot{\gamma})}{\eta_0(\tau_1 - \tau_2)} = \left(\frac{\eta(\dot{\gamma})}{\eta_0} - \frac{\tau_2}{\tau_1} \right) \frac{\tau_1}{\tau_1 - \tau_2}, \tag{6.301}$$

$$\frac{\Psi_2(\dot{\gamma})}{\eta_0 \tau_1} = -\frac{\Psi_1(\dot{\gamma})}{2\eta_0\tau_1} + \frac{(\tau_1 - \tau_3)}{\tau_1}\frac{\eta(\dot{\gamma})}{\eta_0} - \frac{\tau_2 - \tau_4}{\tau_1}, \tag{6.302}$$

with $\sigma_i = \tau_i(\tau_3 + \tau_5) + \tau_{i+2} + 2(\tau_1 - \tau_3 - \tau_5) + \tau_{i+5}(\tau_1 - \tau_3 - \frac{3}{2}\tau_5)$.

One can also determine the elongational viscosity:

$$\frac{\eta_e(\dot{\varepsilon})}{3\eta_0} = \frac{1 - (\tau_2 - \tau_4)\dot{\varepsilon} + \left(\dfrac{3}{2}\tau_5 - \tau_1 + \tau_3\right)(2\tau_2 - 2\tau_4 - 3\tau_7)\dot{\varepsilon}^2}{1 - (\tau_1 - \tau_3)\dot{\varepsilon} + \left(\dfrac{3}{2}\tau_5 - \tau_1 + \tau_3\right)(2\tau_1 - 2\tau_3 - 3\tau_6)\dot{\varepsilon}^2}. \tag{6.303}$$

All these functions are plotted in the graphs of Fig. 6.33 for $\tau_3 = \tau_4 = \tau_6 = \tau_7 = 0$ (simplified Oldroyd 4-constant model). As stated, the fluid is shear-thinning and its elongational viscosity no longer diverges.

This model is still not very satisfactory, although it is better than those presented above (it does however contain eight adjustable parameters!). Indeed, it cannot describe the very significant decrease in viscosity observed for certain polymer melts at very high shear

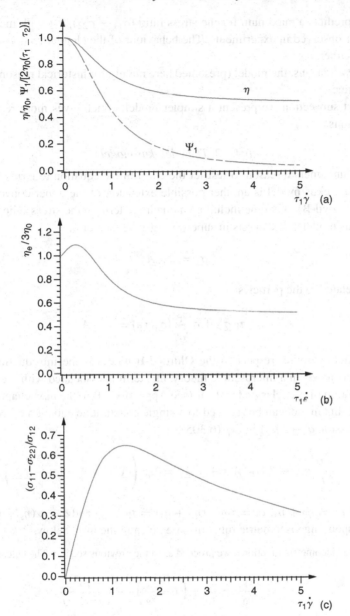

Fig. 6.33 Predictions of the Oldroyd 4-constant model for $\tau_2/\tau_1 = 0.5$ and $\tau_5/\tau_1 = 1.2$. (a) Viscosity, first normal stress coefficient and stress ratio as a function of the shear rate (note that, for these values, $\Psi_2 = 0$); (b) elongational viscosity as a function of the elongation rate. It no longer diverges at $\dot{\varepsilon} = \tau_1/2$, and decreases at high shear rates; (c) ratio of the first normal stress difference to the shear stress as a function of shear rate. This function exhibits a maximum, never observed in experiments, thus showing the limits of the model.

rate; it also predicts a maximum for the stress ratio $(\sigma_{11} - \sigma_{22})/\sigma_{12}$ with increasing $\dot{\gamma}$, a feature never observed in experiments. The behaviour of the elongational viscosity is not too realistic either.

For all these reasons, the model (presented here mainly for historical reasons) is seldom used in practice.

In the next subsection we present a simpler model, which gives more realistic viscometric functions.

6.4.6.2 The Giesekus model

There is no fundamental reason to restrict the constitutive equation to terms linear in the stress. The Giesekus model is another possible extension of the upper-convected Jeffrey model (or Oldroyd-B), this time including a quadratic term in the stress (Giesekus, 1982; [26–28]). This model also consists in superposing a solvent term:

$$\underline{\sigma}_s = 2\eta_s \underline{A} \qquad (6.304)$$

to the term related to the particles:

$$\underline{\sigma}_P + \tau_P \overset{\triangledown}{\underline{\sigma}}_P + \alpha \frac{\tau_P}{\eta_P} \left(\underline{\sigma}_P \underline{\sigma}_P \right) = 2\eta_P \underline{A}. \qquad (6.305)$$

The only difference with respect to the Oldroyd-B model is the introduction of a new unitless parameter α. In a microscopic theory, this term is associated with the anisotropic Brownian motion of particles under shear (see Appendix 7.B to the next chapter).

As before, this model can be reduced to a single constitutive equation by replacing $\underline{\sigma}_P$ with its expression $\underline{\sigma} - 2\eta_s \underline{A}$ in Eq. (6.305):

$$\underline{\sigma} + \tau_P \overset{\triangledown}{\underline{\sigma}} + a \frac{\tau_P}{\eta_0} \underline{\sigma}\,\underline{\sigma} - 2a\tau_R \left(\underline{A}\,\underline{\sigma} + \underline{\sigma}\,\underline{A} \right) = 2\eta_0 \left(\underline{A} + \tau_R \overset{\triangledown}{\underline{A}} - 4a \frac{\tau_R^2}{\tau_P} \overset{\triangledown}{\underline{A}}\,\overset{\triangledown}{\underline{A}} \right), \qquad (6.306)$$

where we set $\eta_0 = \eta_P + \eta_s$, $\tau_R = \tau_P \eta_s/(\eta_P + \eta_s) = \tau_P \eta_s/\eta_0$ and $a = \alpha(\eta_s + \eta_P)/\eta_P$.

The corresponding viscometric functions are given in the inset below.[16]

To calculate the viscometric functions, we proceed as in the previous section. The calculation yields:

$$\frac{\eta(\dot{\gamma})}{\eta_0} = \frac{\tau_R}{\tau_P} + \left(1 - \frac{\tau_R}{\tau_P} \right) \frac{(1-f)^2}{1 + (1-2\alpha)f}, \qquad (6.307)$$

$$\frac{\Psi_1(\dot{\gamma})}{2\eta_0(\tau_P - \tau_R)} = \frac{f(1-\alpha f)}{(\tau_P \dot{\gamma})^2 \alpha(1-f)}, \qquad (6.308)$$

$$\frac{\Psi_2(\dot{\gamma})}{\eta_0(\tau_P - \tau_R)} = \frac{-f}{(\tau_P \dot{\gamma})^2} \qquad (6.309)$$

[16] In practice, the Giesekus model (like that of Oldroyd, incidentally) cannot fit correctly the time-dependent experimental results. One possible improvement consists in introducing a spectrum of relaxation times τ_{Pi}, and writing for each corresponding Maxwell mode an equation analogous to Eq. (6.305), where σ_P is replaced by σ_{Pi}, τ_P by τ_{Pi} and η_P by η_{Pi}. In this case, the total stress due to the particles is obtained by summing over all stress components associated with these modes: $\sigma_P = \sum_i \sigma_{Pi}$.

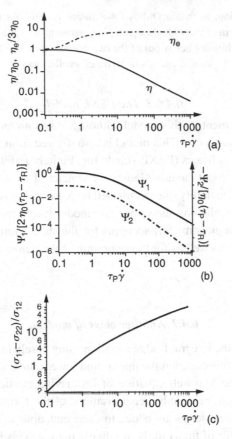

Fig. 6.34 Predictions of the Giesekus model for the following parameter values: $\tau_R/\tau_P = 0.001$ and $\alpha = 0.1$. (a) Viscosity under simple shear and elongational viscosity; (b) first and second normal stress coefficients; (c) stress ratio as a function of the shear rate.

and

$$\frac{\eta_e(\dot\varepsilon)}{3\eta_0} = \frac{\tau_R}{\tau_P} + \left(1 - \frac{\tau_R}{\tau_P}\right)\frac{1}{6\alpha}\left[3 + \frac{1}{\tau_P\dot\varepsilon}\left\{\left[1 - 4(1-2\alpha)\tau_P\dot\varepsilon + 4\tau_P^2\dot\varepsilon^2\right]^{1/2}\right.\right.$$

$$\left.\left. - \left[1 + 2(1-2\alpha)\tau_P\dot\varepsilon + \tau_P^2\dot\varepsilon^2\right]^{1/2}\right\}\right], \qquad (6.310)$$

where we set:

$$f = \frac{1-\chi}{1+(1-2\alpha)\chi} \quad \text{with} \quad \chi^2 = \frac{\left[1 + 16\alpha(1-\alpha)(\tau_P\dot\gamma)^2\right]^{1/2} - 1}{8\alpha(1-\alpha)(\tau_P\dot\gamma)^2}. \qquad (6.311)$$

Plotting the viscometric functions shows that this model only gives realistic results in the range $0 < \alpha < 1/2$.

Some typical curves are given in Fig. 6.34. They exhibit a strong decrease of the viscosity under simple shear, as well as an elongational viscosity that increases, and then saturates at strong shear rates (instead of decreasing, as in the Oldroyd 4-constant model). As to the stress ratio ($\sigma_{11} - \sigma_{22})/\sigma_{12}$, it also tends to increase with shear rate, in agreement with the observations. Finally, this model predicts the shear-thinning behaviour of the two normal stress coefficients, with a positive Ψ_1 coefficient and a negative Ψ_2 coefficient, about 10 times smaller than the former.

6.4.6.3 The FENE model

It is important that we mention this model, although it has no analytical differential formulation, unlike the ones above. This model is also derived from the dumbbell model for polymers. As its name indicates (FENE stands for 'Finitely Extensible Nonlinear Elastic spring'), it takes into account the non-Gaussian character of the springs in the dumbbells, related to their finite extensibility (the details will be presented in Appendix 7.B of the next chapter). Numerical calculations show that this model (more realistic than the Oldroyd 8-constant phenomenological model) accounts for the shear-thinning behaviour of polymers at high shear rates, as well as for the saturation of their elongational viscosity at high elongation rate.

6.4.7 Another class of models

Above, we constructed the integral Lodge equation using purely kinematic considerations. We then deduced from this equation the upper- and lower-convected differential Maxwell equations (extending the Maxwell equation of linear viscoelasticity), and then those of Oldroyd-A and B, which are simply generalisations of the former to Jeffrey fluids. We then showed how nonlinear terms are added to these equations in order to better describe the rheological behaviour of fluids in the nonlinear regime. This led us to discussing the Oldroyd 8-constant model, as well as the Giesekus and FENE models. We also emphasised that all these models (except for the Oldroyd 8-constant one, which is purely phenomenological) are based on the dumbbell model (see Appendices 7.A and 7.B to the next chapter).

In fact, one can find in the literature other models, built on the concept of a 'transient network'. This concept, inspired by the theory of rubber elasticity,[17] was put forward by Green and Tobolsky in 1946 to describe the behaviour of polymer melts (Green and Tobolsky, 1946).

It is based on the idea that the response of the polymer to large amplitude deformations is that of a 'transient network of elastic junctions'. Assuming that the junctions deform affinely until the moment they break,[18] that they all have the same probability of breaking, irrespective of the deformation value, and that they reform in an equilibrium configuration as soon as they come undone, Green and Tobolsky showed that the stress still obeys the Lodge equation (6.252), where the viscoelastic relaxation time corresponds to the lifetime of a junction. This model was later independently improved by Lodge and Yamamoto,

---

[17] The elasticity of completely reticulated rubbers will be described in the next chapter.
[18] This means that the deformation of the network coincides with that of the continuum.

taking into account the possible existence of multiple relaxation processes. This approach leads to the following 'generalised Lodge equation':

$$\underline{\sigma}(t) = \int_{-\infty}^{t} 2m(t - t')\underline{H}(t, t')dt', \tag{6.312}$$

where $m(t - t')$ is a memory function of the type:

$$m(t - t') = \sum_{i} \frac{G_i}{\tau_i} \exp\left(-\frac{t - t'}{\tau_i}\right). \tag{6.313}$$

We recall that, in differential form, this equation leads to an upper-convected Maxwell equation for each of the components $\underline{\sigma}_i$ of the total stress $\underline{\sigma}$.

Consequently, the rheological predictions of this model suffer from the same shortcomings as those of the infinitely extensible Hookean dumbbell model, namely the shear viscosity is constant and the elongational viscosity diverges at high elongation rates.

To palliate these defects, two modifications are generally proposed.

(1) The first one consists of introducing a non-affine deformation of the junction network with respect to the continuum (which implies junction slipping) using the Gordon–Schowalter derivative $^{\square}$ with $u \neq 0$ and 1 (Eq. 6.265).
(2) The second one resides in assuming that the creation and breaking rates of the junctions are constant, but that they depend on the deformation of the segments between junctions by an invariant of the stress tensor, such as its trace.

This type of approach leads to the differential equation of Phan-Thien and Tanner (Phan-Thien and Tanner, 1977):

$$\underline{\sigma}_i \exp\left[\xi \frac{\tau_i}{\eta_i} \text{tr}(\underline{\sigma}_i)\right] + \tau \overset{\square}{\underline{\sigma}}_i = 2\eta_i \underline{A} \quad \text{with} \quad \underline{\sigma} = \sum_i \underline{\sigma}_i, \tag{6.314}$$

where u (in the definition of the Gordon–Schowalter derivative $^{\square}$) and ξ are two adjustable parameters. This equation is another possible generalisation of the convected Maxwell equations.

In the same vein, one can also modify the Lodge equation (6.312). A classical generalisation is the *integral KBKZ equation* (Kaye, 1962; Bernstein *et al.*, 1963)

$$\underline{\sigma}(t) = \int_{-\infty}^{t} 2m(t - t')\left[\frac{\partial U}{\partial I_1}\underline{H}(t, t') + \frac{\partial U}{\partial I_2}\underline{G}(t, t')\right]dt'. \tag{6.315}$$

In this model, the weight function measuring the influence of past deformation is the product of the memory function (describing the disappearance of network junctions with time) with functions describing the influence of the deformation on the disappearance of the network through the first two fundamental invariants $I_1 = \text{tr}(\underline{B})$ and $I_2 = \frac{1}{2}\left[(\text{tr}(\underline{B}))^2 - \text{tr}(\underline{B}^2)\right]$ of the Green tensor $\underline{B}$ (Eq. 6.236). $U(I_1, I_2)$ is termed the *stress energy function.* Note that the Lodge model is recovered by taking $U = I_1$. We should also point out that the U function does not depend explicitly on the third fundamental invariant of the Green tensor, $\det(\underline{B})$ (see Section 4.1.1.5). This is legitimate for an incompressible fluid, because in this limit $\det(\underline{B}) = 1$. In this case, one can also show that $I_2 = \mathrm{tr}(\underline{B}^{-1})$.[19]

Various expressions for the $U(I_1, I_2)$ function have been proposed in the literature. Some are derived from microscopic models, such as the Doi–Edwards reptation model, which we will develop in the next chapter and which applies to entangled polymers. In the first approximation, this model gives (Currie, 1982; [27, 28]):

$$U(I_1, I_2) \approx 5\ln(J - 1) \quad \text{with} \quad J = I_1 + 2(I_2 + 13/4)^{1/2}. \tag{6.316}$$

As for the memory function, it is given by Eq. (6.313) taking for the relaxation times τ_i the values given by the reptation model (see Eq. (7.111) in the next chapter).

The Phan-Thien and Tanner or KBKZ models give reasonable predictions of the stationary or time-dependent regimes for most viscoelastic fluids, provided the parameters are well chosen.

To end this section, we deem it useful to present two empirical formulas relating the viscometric functions measured at a given shear rate $\dot{\gamma}$ (even in the nonlinear regime) to the linear viscoelasticity functions measured at the pulsation $\omega = \dot{\gamma}$.

6.4.8 Empirical formulas of Cox and Merz and of Laun

In 1958, Cox and Merz were searching for a connection between the viscosity $\eta(\dot{\gamma})$ measured under continuous shear and the complex viscosity $\eta^*(\omega)$ measured under oscillating flow. Comparing the experimental data obtained by these two methods, they deduced the following empirical formula (Cox and Merz, 1958):

$$\eta(\dot{\gamma}) = \left| \eta^*(\omega) \right|_{\omega=\dot{\gamma}} = \sqrt{\eta'^2 + (G'/\omega)^2} \,\Big|_{\omega=\dot{\gamma}}. \tag{6.317}$$

This formula, known as the *Cox and Merz rule*, works particularly well for polymer melts in the linear regimes (as we justified in the previous section in the $\omega \to 0$ limit), and also in the nonlinear regime, as shown in Fig. 6.35. This result is more surprising (and, to our knowledge, not yet well understood).

Another formula of the same type is that of *Laun*, connecting the first normal stress difference or, equivalently, the first viscometric function $\Psi_1(\dot{\gamma})$ (see Eq. (6.148)), to the complex viscosity $\eta^*(\omega)$ (Laun, 1986):

$$\Psi_1(\dot{\gamma}) = \frac{2\eta''(\omega)}{\omega} \left[1 + \left(\frac{\eta''}{\eta'} \right)^2 \right]^{0.7} \Big|_{\omega=\dot{\gamma}}. \tag{6.318}$$

---

[19] As a concrete example, one has $I_1 = I_2 = 3 + \dot{\gamma}^2(t - t')^2$ in the case of a simple shear flow in the permanent regime. This result is easily obtained from Eqs. (6.238) and (6.246).

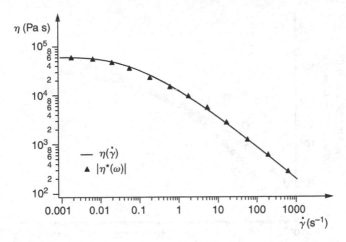

Fig. 6.35 Experimental comparison between the viscosity $\eta(\dot{\gamma})$ (solid line) and the absolute value of the complex viscosity $|\eta^*(\omega)|$ measured at $\omega = \dot{\gamma}$ (triangles) in a low-density polyethylene melt. The two curves are perfectly superimposed, in agreement with the Cox and Merz rule (Retting and Laun, 1991).

This formula also applies to polymers, in the linear and nonlinear regimes.

6.5 'Time temperature' superposition principle

Experiment shows that one can sometimes superimpose the $G'(\omega)$ and $G''(\omega)$ plots measured at different temperatures simply by shifting them along the frequency axis (Fig. 6.36a). In this case, one can define a unitless parameter a_T, depending only on the temperature (and the material under consideration), such that all G' and G'' plots fall on two master curves (given for a reference temperature T_0), provided they are plotted as a function of the reduced frequency $a_T\omega$. This is the 'time–temperature' superposition principle. The fluids exhibiting this feature are said to be 'thermo-rheologically simple'. Note that, at the reference temperature, $a_T = 1$ by definition.

In practice, most polymer melts and concentrated polymer solutions are of this type, provided they form a homogeneous system. On the other hand, the dilute solutions do not always follow this law, so that it must be used carefully.

From the experimental point of view, this principle is very useful as it can be used, at a given temperature, to 'artificially' extend the frequency range of the rheometer used. Indeed, this is equivalent to cooling the material to gain access to frequency values above the maximum working frequency of the rheometer, and to heating it in order to go below its minimum working frequency. This result follows from the observation that the typical relaxation frequencies of the material are most often thermally activated, all increasing with temperature in the same way. Note that, in this case, a_T varies as an exponential function of the reciprocal temperature and can be written as:

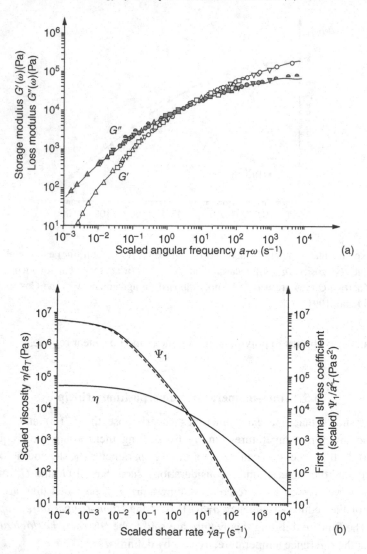

Fig. 6.36 (a) Elastic and loss moduli of a polymer melt (polyethylene) plotted as a function of the reduced angular frequency $a_T\omega$ (the reference temperature is 150 °C, meaning that $a_{T=150\,°C} = 1$); (b) reduced viscosity and reduced first normal stress coefficient plotted as a function of the reduced shear rate $\dot{\gamma}a_T$ (Laun, 1978).

$$a_T = C \exp\left(\frac{E}{k_B T}\right), \tag{6.319}$$

where C is a constant and E an activation energy per molecule.

It turns out that this formula does not apply close to the glass transition temperature, where the temperature dependence is stronger. One must then use a more general

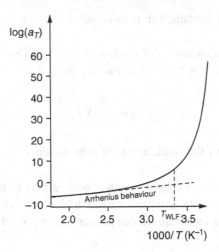

Fig. 6.37 Variation of the temperature coefficient a_T as a function of the reciprocal temperature, calculated using $A = 20$, $B = 75$ K and $T_v = 300$ K. At high temperature, Arrhenius-type behaviour is retrieved.

formula, known as the WLF equation after the names of the authors who first proposed it (Williams, Landel and Ferry, 1955):

$$a_T = \exp\left[\frac{-A(T - T_{\mathrm{WLF}})}{B + (T - T_{\mathrm{WLF}})}\right]. \tag{6.320}$$

In this equation, T_{WLF} is a temperature (ordinarily close to, but larger than, the glass transition temperature T_g) and A and B are two material constants (ranging from 6 to 34 for the former and from 20 to 130 K for the latter, according to the experimental work of Ferry, who investigated a large number of polymers [29]). The shape of this function is shown in Fig. 6.37.

Experiment also shows that the superposition principle applies to nonlinear viscometric functions such as the viscosity η under continuous shear, or the first normal stress coefficient Ψ_1. In this case, one must plot the scaled parameters η/a_T and Ψ_1/a_T^2 as a function of the reduced shear rate $\dot{\gamma}a_T$, using the same shift factor as for the linear viscoelasticity functions G' and G''. Note that these particular combinations are obtained naturally from the Cox and Merz and Laun rules, stating that $\eta = G''/\omega$ and $\Psi_1 = 2G'/\omega^2$ in the $\omega = \dot{\gamma} \to 0$ limit. It is, however, surprising that this superposition principle also applies in the nonlinear regime.

These results are illustrated in the graph of Fig. 6.36b in the case of a polymer melt (polyethylene).

6.6 Saffman–Taylor finger in complex fluids

In Chapter 3, dealing with the hydrodynamics of Newtonian fluids, we described an instability that leads to fingering as a less viscous fluid (air, in practice) replaces a more viscous liquid (oil, for instance) contained in a rectangular and elongated Hele–Shaw cell (see Fig. 3.38).

In this experiment, the dynamics of the fluid is described by the Darcy law of porous media written, in the case of a Newtonian fluid, in the form (see Eq. (3.342)):

$$v = -\frac{b^2}{12\eta}\nabla P, \tag{6.321}$$

where v is the velocity of the fluid, averaged over the thickness b of the cell, and η its viscosity.

We showed that a finger, of relative width λ with respect to the width L of the channel, is formed after a transient regime during which all other fingers are eliminated. From dimensional considerations we deduced that the parameter controlling this instability is the dimensionless number:

$$\frac{1}{B} = 12\left(\frac{L}{b}\right)^2 \frac{\eta U}{\gamma}, \tag{6.322}$$

where γ is the surface tension between the fluid and air and U is the velocity of the finger. Finally, we showed that λ is indeed selected as a function of the unitless parameter $1/B$ if the fluid is Newtonian, and that it (approximately) tends towards 1/2 as $1/B \to \infty$, in agreement with the two-dimensional theory.

How do these results change if the viscous fluid is, for instance, shear-thinning?[20] An element of response was provided recently by A. Lindner, D. Bonn and J. Meunier from the experimental point of view (Lindner *et al.*, 2000), and by E. Corvera-Poiré and M. Ben Amar from the theoretical point of view (Corvera-Poiré and Ben Amar, 1998).

In the following, we will generalise the Darcy law to the case of shear-thinning viscous fluids. We will then describe the main features of the Saffman–Taylor fingers observed in these fluids. Finally, we will give a few indications as to the morphologies of the fingers obtained in the more complex case of viscoelastic or yield-stress fluids.

6.6.1 Darcy law in a shear-thinning viscous fluid

In this section, we assume that the fluid obeys the Ostwald–de Waehle law:

$$\eta(\dot{\gamma}) = m\dot{\gamma}^{n-1}. \tag{6.323}$$

Under these conditions, the Poiseuille flow along the cell thickness can be solved exactly, showing that the velocity profile is given by the following formula (similar to Eq. (6.197)):

$$v_z = \frac{n}{n+1}\left(\frac{|\nabla P|}{m}\right)^{1/n}\left[\left(\frac{b}{2}\right)^{1+1/n} - \left|\frac{b}{2} - z\right|^{1+1/n}\right], \tag{6.324}$$

---

[20] We assume here that the elasticity of the fluid can be neglected.

where z is the coordinate perpendicular to the plates, situated at positions $z = 0$ and $z = b$. This formula yields the average flow velocity:

$$v = \frac{b}{2(2 + 1/n)} \left(\frac{b}{2m}\right)^{1/n} |\nabla P|^{1/n} \tag{6.325}$$

as a function of the applied pressure gradient. This is the *non-Newtonian Darcy law*, which extends the law (6.321).

This law is often written in the following *approximate form*:

$$v = -\frac{b^2}{12\eta(\dot{\gamma})}\nabla P, \tag{6.326}$$

where $\dot{\gamma}$ is the *average shear rate* within the cell:

$$\dot{\gamma} = \frac{v}{b} \frac{2(1 + 2n)}{1 + n}, \tag{6.327}$$

which can be determined from Eq. (6.324). It is easily checked that Eq. (6.326) is close to Eq. (6.325) for a weakly shear-thinning fluid (typically, for $0.65 < n < 1$). More precisely, these two laws give exactly the same dependence of the average velocity on the pressure gradient ($v \propto |\nabla P|^{1/n}$), with very similar prefactors, as long as n is not too far from 1 (the prefactor is identical for $n = 1$).

Let us now see how the Saffman–Taylor finger is selected in the case of a shear-thinning fluid.

6.6.2 Selection of a Saffman–Taylor finger in a shear-thinning viscous fluid

It turns out that the selection rule stating that $\lambda \to 1/2$ at high velocity is rather fragile.

Indeed, we have already mentioned in Chapter 3 that, even in the purely Newtonian case, the wetting effects of the liquid on the plates can lead to a decrease in the asymptotic value of λ. This effect remains small however, as λ is only reduced by a small percentage with respect to the theoretical value of 1/2 given by the two-dimensional model (see Fig. 3.45).

On the other hand, a much more spectacular decrease in λ can be obtained by stretching a wire within the cell (Zocchi *et al.*, 1987) (Fig. 6.38). Experiment shows that the fingers are then much more stable than in the wireless case, and that λ tends towards a plateau value close to 0.3 with increasing velocity. The same result is observed if one of the plates is scratched lengthwise (Rabaud *et al.*, 1988). These experiments clearly show *the stabilising effect of anisotropy induced by an external perturbation* (here, the wire or the scratch).

However, the *anisotropy can also be due to the fluid itself, provided it is non-Newtonian*. This was demonstrated by A. Lindner, D. Bonn and J. Meunier in the case of a shear-thinning viscous fluid.

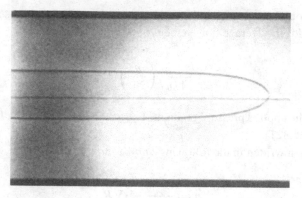

Fig. 6.38 Saffman–Taylor finger in a Hele–Shaw cell containing a stretched 13 μm thick tungsten wire (Zocchi *et al.*, 1987).

Fig. 6.39 Contours of a few Saffman–Taylor fingers obtained with a 2000 ppm Xanthane solution. For the thinnest finger, $L = 2$ cm and $\lambda = 0.56$; for the largest one, $L = 8$ cm and $\lambda = 0.38$; for the intermediate finger, $L = 4$ cm and $\lambda = 0.49$. The solid lines are the best fits with the theoretical Saffman–Taylor form (see Eq. (3.384)) calculated by neglecting the surface tension (Lindner *et al.*, 2000).

Using a mixture of water and Xanthane,[21] they obtained stable fingers, much thinner than in the Newtonian case (Fig. 6.39). Surprisingly, their shape is still very well described by the Saffman–Taylor law without surface tension (Eq. (3.384)), irrespective of their width.

These authors also measured λ as a function of the new dimensionless parameter:

$$\frac{1}{B} = 12 \left(\frac{L}{b}\right)^2 \frac{U m \, [\dot{\gamma}(U)]^{n-1}}{\alpha}, \tag{6.328}$$

where α is the surface tension, while $\dot{\gamma}(U)$ is given by Eq. (6.327). By using this definition, all the measurements performed in cells of various width or thickness collapse onto master curves, which only depend on n (Fig. 6.40). These curves show that, the more shear-thinning the liquid is (namely, the smaller n is), the thinner the fingers are. They also show

---

[21] This is a water-soluble *stiff polymer*. We will see in the next chapter that, in this case, the final solution is not too elastic, since the polymer molecules deform very little under shear. They do however orient under flow, which leads to shear-thinning behaviour.

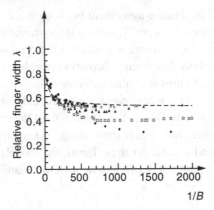

Fig. 6.40 Relative width of the fingers as a function of $1/B$ for water (diamonds) and for aqueous Xanthane solutions at 50 ppm (open triangles), 100 ppm (solid squares), 500 ppm (solid triangles), 1000 ppm (empty squares) and 1750 ppm (solid dots) (Lindner *et al.*, 2000).

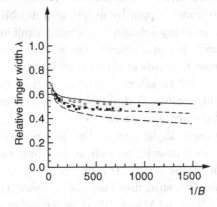

Fig. 6.41 Theoretical predictions for the finger width (given by the model of Corvera-Poiré and Ben Amar) and their comparison with the experimental results of Lindner. From top to bottom, the theoretical curves correspond to $n = 1$ (Saffman–Taylor), $n = 0.8$ and $n = 0.5$. The empty squares correspond to a 100 ppm Xanthane solution ($n = 0.82$) and the solid dots to a 500 ppm Xanthane solution ($n = 0.61$) (Lindner *et al.*, 2000).

that, for each value of n, λ saturates towards a plateau value, with decreasing values as the fluid is more shear-thinning.

Trying to explain these results, E. Corvera-Poiré and M. Ben Amar solved the problem of finger selection for a fluid obeying the effective Darcy law given by Eq. (6.326). The problem is then much more complicated than in the Newtonian case, since the pressure field driving the instability no longer follows the Laplace equation. Nevertheless, these authors show that one can retrieve the Laplacian character of the equations by means of a conformal transformation, finally allowing their resolution (Corvera-Poiré and Ben Amar, 1998). By pursuing this analysis, they obtain a numerical solution for λ as a function of $1/B$ when the exponent n remains close to 1 (weakly shear-thinning fluid). Their results are shown in Fig. 6.41, also displaying the results of the measurements performed by A. Lindner.

This figure shows good qualitative agreement between theory and experiment, at least as far as the general tendency is concerned, namely the systematic decrease in finger width with respect to the Newtonian case, more important as the fluid is more shear-thinning.

However, the theory predicts that λ tends (slowly) towards 0 for $1/B \to \infty$, while the experiment clearly shows a finite-value plateau in the same conditions. This discrepancy is more marked when the fluid is more shear-thinning. A fairly obvious explanation is that the fluid does not follow the Ostwald–de Waehle law at all the shear rates investigated, in particular at low and high shear rate, where the viscosity is known to approach two constant (but different) values, η_0 and η_∞. Under these limits, the models clearly overestimate the shear-thinning character of the fluid, leading to an underestimated value of λ.

6.6.3 Saffman–Taylor fingers in viscoelastic or yield stress liquids

We conclude this chapter by noting that Saffman–Taylor fingers were also observed in quasi-Newtonian viscoelastic liquids (Lindner *et al.*, 2002). These fluids are prepared by diluting in water a few hundred ppm by weight of a flexible polymer, such as PEO (polyethylene oxide). The resulting solutions are elastic, exhibiting strong normal stress differences N_1 under shear,[22] but quasi-Newtonian, as their viscosities show very little decrease with the shear rate. Experiment shows that, in this case, the relative width of the fingers is always above 1/2 (it saturates at a value close to 0.7, typically for $1/B$ larger than 100). This result, completely different from that observed with shear-thinning Xanthane solutions, is not yet well understood.

Experiments on the Saffman–Taylor fingers have also been performed in yield-stress fluids, such as clays or polymer gels, as well as in foams (which exhibit behaviour closer to that of an ideal plastic solid). The fingers are then very different, forming fractal branched or fracture-like structures (see, for instance, (van Damme *et al.*, 1988) for experiments on clays, (Zhao and Maher, 1993) for those done using polymer gels and (Park and Durian, 1994) for experiments involving foams).

References

Adam, M., M. Delsanti, P. Pieranski and R. B. Meyer, *Revue de Phys. Appl.*, **19**, 253 (1984).
Beavers, G. S. and D. D. Joseph, *J. Fluid Mech.*, **69**, 475 (1975).
Bernstein, B., E. A. Kearsley and L. J. Zapas, *Trans. Soc. Rheol.*, **7**, 391 (1963).
Corvera Poiré, E. and M. Ben Amar, *Phys. Rev. Lett.*, **81**, 2048 (1998).
Cox, W. P. and E. H. Merz, *J. Polym. Sci.*, **28**, 619 (1958).
Currie, P. K., *J. Non-Newtonian Fluid. Mech.*, **11**, 53 (1982).
van Damme, H., E. Alsac, C. Laroche and L. Gatineau, *Europhys. Lett.*, **5**, 25 (1988).
Giesekus, H., *J. Non-Newtonian Fluid Mech.*, **11**, 69 (1982); *Rheol. Acta.*, **21**, 366 (1982).
Green, M. S. and A. V. Tobolsky, *J. Chem. Phys.*, **14**, 80 (1946).
Kaye, A., *Non-Newtonian Flow in Incompressible Fluids*, College of Aeronautics, Cranford, UK, Note No. 134 (1962).

---

[22] This is due to the molecules deforming very easily under shear and storing considerable elastic energy.

Laun, H. M., *Rheol. Acta*, **17**, 1 (1978).

Laun, H. M., *J. Rheol.*, **30**, 459 (1986).

Lindner, A., Ph.D. Thesis, *The Saffman–Taylor instability in complex fluids: relation between rheological properties and pattern formation* (in French), Université de Paris VI (2000).

Lindner, A., D. Bonn and J. Meunier, *Phys. of Fluids*, **12**, 256 (2000).

Lindner, A., D. Bonn, E. Corvera-Poiré, M. Ben Amar and J. Meunier, *J. Fluid Mech.*, **469**, 237 (2002).

Mason, T. G. and D. A. Weitz, *Phys. Rev. Lett.*, **74**, 1250 (1995).

Oldroyd, J. G., *Rheol. Acta*, **1**, 337 (1961).

Park, S. S. and D. J. Durian, *Phys. Rev. Lett.*, **72**, 3347 (1994).

Phan-Thien, N. and R. I. Tanner, *J. Non-Newtonian Fluid Mech.*, **2**, 353 (1977).

Rabaud, M., Y. Couder and N. Gerard, *Phys. Rev. A*, **37**, 935 (1988).

Retting, W. and H. M. Laun, *Kunststoff-Physik*, Munich: Hanser–Verlag (1991) p. 138.

Tanner, R. I., *Trans. Soc. Rheol.*, **14**, 483 (1970).

Williams, M. L., R. F. Landel and J. D. Ferry, *J. Am. Chem. Soc.*, **77**, 3701 (1955).

Zhao, H. and J. V. Maher, *Phys. Rev. E*, **47**, 4278 (1993).

Zocchi, G., B. E. Shaw, A. Libchaber and L. P. Kadanoff, *Phys. Rev. A*, **36**, 1894 (1987).

7

Rheology of isotropic viscoelastic materials: examples and microscopic theories

In the previous chapter, we recalled the general tools one can use to study the rheological properties of viscoelastic isotropic materials. For the *linear regime* of 'small' deformations, we introduced the *linear viscoelasticity functions* $G(t)$ and G^*. These two parameters, known as the *relaxation modulus* and the *complex modulus*, respectively, are equivalent since G^* is the Laplace transform of $G(t)$. We also defined the *viscometric functions*, used to describe the material under continuous shear. First among them is the *viscosity* $\eta(\dot{\gamma})$: denoted by η_0 in the linear regime (as $\dot{\gamma} \to 0$), it generally varies with the shear rate $\dot{\gamma}$ in the nonlinear regime. The two other functions (that will be only briefly discussed in this chapter, in spite of their considerable practical importance) are the *first and second normal stress differences* N_1 and N_2. At the same time, we described the *phenomenological Maxwell model*, associated with a *(discrete or continuous) spectrum of relaxation times*. This model gives access to the linear viscoelasticity functions $G(t)$ and G^*, as well as to the viscosity. However, the calculations become extremely complicated for large deformations, where one must deal with both kinematic nonlinearities (convected derivatives) and intrinsic nonlinearities due to the constitutive laws of the material itself (when it is strongly deformed). For this reason, we will mostly be concerned with the linear regime, and only provide a few results on the nonlinear regime.

In this chapter we will show, starting from concrete examples, how the rheological functions can be deduced from the microscopic theories. To this end, we will study four large families of materials.

The first one is that of polymer melts (Section 7.1). After giving some concrete examples, we recall some of their essential rheological properties. This will lead to distinguishing two large classes of polymers: those of low molecular mass, where the chains are independent, and those of high molecular mass, where the chains are entangled. After a brief reminder on polymer theory, we will describe two models: that of *Rouse*, applicable to low molecular mass polymers; and that of *Doi and Edwards*, which extends the former model to high molecular mass polymers by taking into account the *entanglements* and chain *reptation* – a concept discovered by *de Gennes*. In particular, we will see that these two models can be used to calculate the spectrum of Maxwell relaxation times, and to predict the existence of a *rubber plateau* over a frequency range that becomes wider as the molecular mass of the polymer increases.

The second family to be studied is that of *elastomers*, among them natural rubber (Section 7.2). We will see how they are produced by the *chemical or physical reticulation* of polymer chains, and then we will establish the general expression of their *law of elastic behaviour* as a function of the density of *reticulation knots*. We will see that they are *viscoelastic solids*, much softer than ordinary solids, since their elasticity is of *entropic*, rather than enthalpic, origin.

The third class of materials to be discussed is that of *polymer solutions* (Section 7.3). We will see that these systems are used in practice to increase the viscosity of solutions. We will show that this *shear-thickening* effect occurs in the *semi-dilute regime*, where the chains are entangled. This is the regime we will mainly describe in this section. In particular, we will show the existence – as in melts – of universal behaviour resulting from fundamental concepts such as those of *excluded volume* or '*blob*' – another concept invented by de Gennes.

We will then tackle the problem of *giant micelles*, formed in lyotropic mixtures (Section 7.4). We will see that these micelles resemble very long polymer chains in solution, with an *extremely large size distribution*, predicting very complicated rheological behaviour. However, this is not at all the case, especially in semi-dilute solution, their behaviour being astonishingly simple, of the purely Maxwell type. We will explain this 'anomaly' using the framework of the *Cates model*, based on the fact that the micelles can *break* and *recombine* (which is why they are sometimes called '*living polymers*', like certain biological polymers).

The last big family of materials we will describe is that of *dispersions* (Section 7.5). These can be *suspensions of solid particles*, *emulsions*, when a liquid is dispersed within another one, or *macromolecular solutions* (polymers in dilute solution).

In the first case, we will start with a historical note on how *Faraday* prepared and characterised the first colloidal gold suspension. We will then define the *Péclet number*, quantifying the importance of the *Brownian motion* relative to the *hydrodynamic effects*. This will allow us to define the concept of *colloidal suspension*. We will then say a few words on the topic of *sedimentation*, before approaching suspension viscosity proper. We will show how to calculate the viscosity of a suspension of spheres, all of the same size, first in the dilute regime (*Einstein formula*), and then at higher concentration, as the particles encounter each other and interact. On this topic, we will describe two *effective medium* theories, yielding simple predictions for the viscosity in the concentrated regime (the theories of *Arrhenius* and of *Ball and Richmond*). We will also give some experimental results on the linear viscoelasticity functions, highlighting the importance of the Péclet number. In conclusion, we will discuss the problem of *rod suspensions*. We will see that the *rotational Brownian motion*, which plays no role in the case of spheres, becomes very important here, even in the dilute regime, where strongly *shear-thinning* behaviour is to be expected.

In the case of emulsions, we will first discuss their formation mechanisms. We will see that they can be obtained either by chemical or by mechanical methods, by vigorously stirring the two fluids. This will bring up the *Taylor model*, which predicts the size of the droplets as a function of the shear rate during stirring. We will then give the *Taylor formula*, extending that of Einstein to the case of emulsions, then the *Pal formula*, predicting the viscosity of concentrated emulsions in the framework of an effective medium approach, similar to that of Arrhenius. Finally, we will discuss the viscoelasticity of emulsions. On this topic, we will only present the calculation of the complex modulus G^* in the simplified case where the two liquids (assumed to be Newtonian) have the same viscosity (a particular case of the *general Oldroyd model*, where the fluids can have different viscosities).

To conclude this section on dispersions, we will discuss the case of *dilute macromolecular solutions*. We will describe briefly the *Zimm model*, which extends the Rouse model by including the effect of *hydrodynamic interactions* between monomers. We will see that this model can predict the viscoelastic properties of the solutions, in good agreement with the experiment.

7.1 Polymer melts

We have already mentioned (on several occasions) the viscoelastic properties of polymer melts, in particular in Chapters 1 and 6. On the other hand, we have not yet explained their rheology in terms of their microscopic structure. In this section, our goal is precisely to establish this connection and to show that their behaviour under shear can be described using several *simple and universal concepts*, such as that of *statistical coils*, of *entanglements* (already mentioned somewhat elliptically in Chapter 1), and of *tubes* or *reptation* (a concept invented by de Gennes in 1978). This will lead to discussion of two complementary theories: that of Rouse, describing the behaviour of polymer chains at short times, and that of Doi and Edwards, dealing with their long-time reptation. We will mainly be concerned with their linear viscoelastic properties, which are much more easily treated theoretically than the nonlinear properties, on which we will not say much in this section. Before delving into the theory, let us give a general introduction to polymers, their chemistry and classification, as well as to their most striking rheological properties.

7.1.1 General concepts on polymers

We recall that, by definition, a polymer is a substance consisting of long molecules, formed by the repetition of the same pattern, comprising one or more elementary units (the *monomers*). The average number of such units is called the *degree of polymerisation*. If this number is large (in the thousands), one speaks of *'high' polymers*; the molecule can then reach a length of a few tens of micrometres (and sometimes even more, as in the case of DNA), which is quite considerable. If the number is only in the range of hundreds, or less, the result is an *oligomer*: the molecule is then so short that it can hardly be seen as a polymer at all.

Homopolymers contain a single type of monomers ([A-A-...-A-A]), while *copolymers* are formed of at least two different types of monomers. If the pattern sequence is random ([A-A-B-A-...-A-B-B-B]) one speaks of *statistical copolymers*. There are also *sequence* (or *block*) *copolymers*, such as the *diblock* [A-A-A-A-...-B-B-B-B], associating a sequence of A monomers to another sequence of B monomers. In this case, the material exhibits a heterogeneous structure, combining the properties of monomers A and B. We should also mention here *alternating copolymers* ([A-B-A-B-...-A-B-A-B]) which are generally difficult to synthesise.

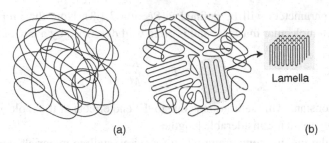

(a)   (b)

Fig. 7.1 Amorphous (a) or semi-crystalline (b) structure of a linear polymer. In the crystallised regions, the molecule folds back upon itself more or less regularly and forms lamellas.

From the structural point of view, there are three main types of polymers (see Fig. 7.3 in the next section):

- *linear* polymers, consisting of long chains of monomers, attached to each other by covalent chemical bonds;
- *ramified* (or *branched*) polymers, formed by a *main chain* to which are grafted shorter *side chains*;
- *reticulated* polymers, where the chains are attached to each other forming *crosslinks*. These crosslinks connect all the chains, forming a *disordered three-dimensional network*.

In the industry, linear and branched polymers are used to produce *solids* or *viscoplastic elastomers*. These two types of materials have different *structures*. The former always have an *amorphous* (i.e. disordered) structure, for all temperatures (Fig. 7.1a). The latter, can have a *semi-crystalline* structure (Fig. 7.1b) over a certain temperature range. In this case, some fraction of the polymer (generally between 40 and 70% by mass) crystallises locally, while the remainder stays amorphous: this phenomenon occurs in numerous polymers, such as polyethylene (to be discussed in more detail in the next section).

For all these materials, there exists a temperature above which they behave as viscoelastic liquids, making them easy to process. In materials with amorphous structure, this temperature is that of the *glass transition*, T_g. In semi-crystalline polymers, on the other hand, this is the *melting* temperature T_m of their crystalline part, knowing that the glass transition temperature of their amorphous part T_g is generally lower than T_m. This leads to a marked difference in behaviour between materials with amorphous structure and thermoplastic elastomers, since the former become tough and brittle below their glass transition temperature, while the latter start by behaving as elastomers (which are soft viscoelastic solids) immediately below their melting temperature T_m and only become tough and brittle below the glass transition temperature T_g of their amorphous part.

We recall that there is no jump in latent heat (or in density) across the glass transition. Thus, it is not a first-order phase transition. On the other hand, this transition is marked by a *considerable slowing down of the molecular motions* (in particular, of monomer rotation around their chemical bonds), leading to a break in the slope of the specific heat when plotted against the temperature, as well as to a sharp increase in viscosity (at 'zero' shear rate) and in the elastic modulus (on the rubber plateau) when the temperature drops below

T_g (these two parameters will be defined in Section 7.1.3). Note that, for polymers, T_g saturates at high molecular mass M, and is often well described by a law of the type:

$$T_g = T_g^\infty - \frac{K}{M}, \qquad (7.1)$$

where K is a constant. This saturation is due to the end effects of the molecules becoming negligible owing to their considerable length.

In contrast, the melting temperature T_m of semi-crystalline materials corresponds to a genuine first-order phase transition (with the associated latent heat), above which the crystalline lamellas melt. Since, in general, $T_g < T_m$, the polymer behaves as a viscoelastic liquid with completely amorphous structure above T_m.[1]

As to reticulated polymers, they form *elastomers*. If the crosslinks at the level of the reticulation points are very strong (as in the case of natural rubber, where the crosslinks are chemical covalent bonds), they cannot be broken by heating; indeed, the polymer burns before it can ever melt. The number of reticulation points can also increase by heating, so that the material hardens: it is then termed *thermosetting* (this is the case for some resins). If, on the other hand, the reticulation points can become undone on heating (as can happen with *physical crosslinks*, obtained, for instance, by absorbing the chains on solid micropar-ticles, such as carbon black), the material melts above a certain temperature: it should then be classified in the family of *thermoplastic elastomers* (together with semi-crystalline materials, where the crystalline lamellas play the same role as the microparticles).

Finally, let us emphasise that a polymer is not in general a pure body, being formed by *macromolecules* of various lengths, or even with different chemical compositions. To characterise it, one should therefore use statistical parameters, such as the *average molecular mass*, the *average chemical composition*, the *degree of polymerisation* and the *polydisper-sity index* (giving the mass dispersion around the average mass).

7.1.2 Some examples of polymers

Although they can be found in nature (leather, natural rubber, cellulose, proteins, ... and DNA molecules), the overwhelming majority of polymers are artificially synthesised, either by polycondensation, or by a chain polymerisation process, which can be of the radical, anionic or cationic type.

Among the very first polymers synthesised at the beginning of the twentieth century was cellulose nitrate. Obtained by the reaction of cotton with nitric acid, it is a powerful explosive, used to replace gunpowder in the ammunition employed during the First World War. It was also used, owing to its thermoplastic properties, to produce smooth fibres with a silky sheen, as well as billiard balls (which occasionally exploded!) and film stock (which often caught fire in contact with the arc lamps in the projectors, thus destroying quite a few cinemas!). As a result, this material was quickly replaced in photographic uses with

---

[1] We emphasise that semi-crystalline polymers are not transparent below T_m (rare exceptions notwithstanding), since the optical indices of the amorphous and crystalline parts are generally different (resulting in strong light scattering).

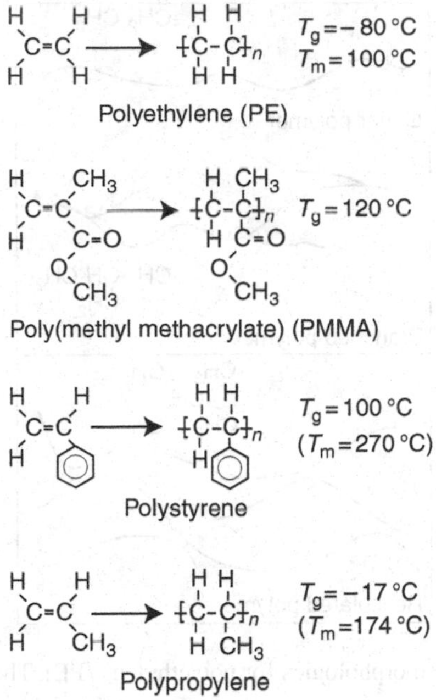

$$\begin{array}{c} \text{H} \quad \text{H} \\ \diagdown \!\!\! \diagup \\ \text{C=C} \\ \diagup \!\!\! \diagdown \\ \text{H} \quad \text{H} \end{array} \longrightarrow \begin{array}{c} \text{H} \ \text{H} \\ +\!\!\!\overset{|}{\text{C}}\!\!-\!\!\overset{|}{\text{C}}\!\!+_{n} \\ \overset{|}{\text{H}} \ \overset{|}{\text{H}} \end{array} \quad \begin{array}{l} T_g = -80\,°\text{C} \\ T_m = 100\,°\text{C} \end{array}$$

Polyethylene (PE)

$$\begin{array}{c} \text{H} \quad \text{CH}_3 \\ \diagdown \!\!\! \diagup \\ \text{C-C} \\ \diagup \!\!\! \diagdown \\ \text{H} \quad \text{C=O} \\ \qquad \overset{|}{\text{O}} \\ \qquad \diagdown \!\! \text{CH}_3 \end{array} \longrightarrow \begin{array}{c} \text{H} \ \text{CH}_3 \\ +\!\!\!\overset{|}{\text{C}}\!\!-\!\!\overset{|}{\text{C}}\!\!+_{n} \\ \overset{|}{\text{H}} \ \overset{|}{\text{C=O}} \\ \quad \overset{|}{\text{O}} \\ \quad \diagdown \!\! \text{CH}_3 \end{array} \quad T_g = 120\,°\text{C}$$

Poly(methyl methacrylate) (PMMA)

$$\begin{array}{c} \text{H} \quad \text{H} \\ \diagdown \!\!\! \diagup \\ \text{C=C} \\ \diagup \!\!\! \diagdown \\ \text{H} \quad \bigcirc \end{array} \longrightarrow \begin{array}{c} \text{H} \ \text{H} \\ +\!\!\!\overset{|}{\text{C}}\!\!-\!\!\overset{|}{\text{C}}\!\!+ \\ \overset{|}{\text{H}}\bigcirc \end{array} \quad \begin{array}{l} T_g = 100\,°\text{C} \\ (T_m = 270\,°\text{C}) \end{array}$$

Polystyrene

$$\begin{array}{c} \text{H} \quad \text{H} \\ \diagdown \!\!\! \diagup \\ \text{C=C} \\ \diagup \!\!\! \diagdown \\ \text{H} \quad \text{CH}_3 \end{array} \longrightarrow \begin{array}{c} \text{H} \ \text{H} \\ +\!\!\!\overset{|}{\text{C}}\!\!-\!\!\overset{|}{\text{C}}\!\!+_{n} \\ \overset{|}{\text{H}} \ \text{CH}_3 \end{array} \quad \begin{array}{l} T_g = -17\,°\text{C} \\ (T_m = 174\,°\text{C}) \end{array}$$

Polypropylene

Fig. 7.2 Chemical formulas of some vinyl-based polymers and their starting molecules. In each case, we give the typical values of the glass transition temperature T_g and of the melting temperature T_m (the latter is only defined if the material has a semi-crystalline form; that is why it is sometimes enclosed in parentheses).

the much less flammable cellulose acetate,[2] then by cellulose xanthate, which can also be drawn into nice smooth fibres ('rayon'). Note, however, that cellulose nitrate continues to be the main ingredient of nail polish.

Another synthetic polymer, with a much simpler chemical structure, is polyethylene (PE), produced on an industrial scale from the early 1930s. This polymer belongs to the family of *vinyl polymers*, as its monomer derives from ethylene, a molecule containing a carbon–carbon double bond. As shown in Fig. 7.2, its chemical formula is extremely simple, corresponding to a regular sequence of CH_2 groups. In contrast, its structure depends on the type of synthesis employed. As such, one must distinguish three types of PE, composed of linear, branched or reticulated chains (Fig. 7.3).

It is also noteworthy that PE is semi-crystalline at room temperature (the length of the ordered segments in the lamellas depicted in Fig. 7.1 being of the order of 100 Å) and that it melts between 80 and 120 °C, depending on its molecular mass (given here in g/mol) which can be low, typically between 20 000 and 40 000 (it is then branched, and denoted

---

[2] However, this material can also catch fire, as shown by the famous disaster of the German dirigible Hindenburg, on 6 May 1937 not far from New York City. The skin of this dirigible was made of cellulose acetate and filled with hydrogen. For a long time, it was believed that the hydrogen was at the origin of this catastrophe, which is erroneous since other dirigibles, filled with helium, burned in the same way.

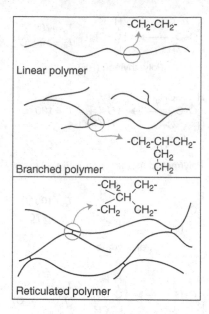

Fig. 7.3 Three possible morphologies for polyethylene (PE). The last case is very rare.

as LDPE, from 'Low-Density PE') or high, from 200 000 to 500 000 (it is then linear and labelled as HDPE, or 'High-Density PE'). As to the glass transition temperature of its amorphous part, it is much lower (about $-80\,°C$).

In practice, PE is used in its semi-crystalline state. LDPE (less expensive) is used for producing packaging film (the plastic bags distributed in supermarkets), while HDPE, which is more resistant (and more expensive), is used for fabricating everyday items such as gas jerricans, swimming pool toys, certain types of shoes, etc. There also exists linear PE with a very high molecular mass (several millions), denoted by UHMWPE (for 'Ultra-High-Molecular-Weight PE'), that one can use to draw extremely resistant fibres, which can be used in bullet-proof vests (like Kevlar). One can also produce large UHMWPE slabs, sometimes used to replace the ice in skating rinks.

Three other classical examples of vinyl polymers are shown in Fig. 7.2.

The first one is poly(methyl methacrylate) (PMMA) (Fig. 7.2). Commercially available since the early 1930s under the name Plexiglas, this amorphous polymer is perfectly transparent (even more so than glass) at room temperature. Its molecular mass generally ranges from 50 000 to 1 million, and its glass transition temperature is close to $120\,°C$. It is mainly used for producing safety glass. Very thick panes can be manufactured (up to 30 cm) and used for building large aquariums, such as the one in Atlanta, holding several million litres of water.

The two other examples we chose are polystyrene and polypropylene. These are also 'classical' systems, with uncountable applications. For each of them, we give the glass transition temperature and the melting temperature (Fig. 7.2). The latter is parenthesised, since it cannot be defined for amorphous polymers. This is the case when the polymer

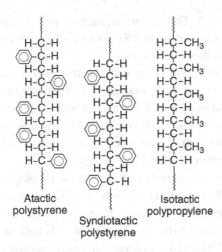

Fig. 7.4 Polymer tacticity. It depends on the polymerisation method employed.

is in its *atactic* form, namely when the phenyl groups (for polystyrene) or methyl groups (for polypropylene) are randomly placed on one side or the other of the chain (Fig. 7.4a). On the contrary, polystyrene is strongly crystalline when the phenyl groups are regularly arranged in an alternating pattern on either side of the chain: it is then termed *syndiotactic* (Fig. 7.4b). As to polypropylene, it is semi-crystalline when *isotactic*, namely when the methyl groups are all on the same side of the chain (Fig. 7.4c).

Polystyrene is tough at room temperature, with a high glass transition temperature (of the order of 100 °C). It is used for manufacturing various kinds of containers (cases for computers and household appliances), as well as in expanded form, for packaging and insulation. It is transparent when amorphous (in its atactic form).

At room temperature, polypropylene is softer than polystyrene. In its isotactic semi-crystalline form (in which case it can be classified as a thermoplastic elastomer) it is used in the production of food containers that resist boiling water (since it only melts above 170 °C). In fibre form, it is also used for floor coatings.

Obviously, many other polymers exist, such as the PVC used by plumbers due to its resistance to water and fire (under the action of heat it gives off chlorine, hindering combustion), the polycarbonates used for manufacturing the CDs in our computers, etc. It is impossible to give here an exhaustive list, which would be much too long.

Let us nevertheless point out that all the examples presented above belong to the family of *organic polymers*, where the main chain is composed essentially of carbon atoms. However, one can also have *inorganic polymers*, containing no carbon atoms in their main chain. Silicones, which should be more aptly called polysiloxanes since their main chain does not contain silicone groups (as the chemists believed in the beginning), belong to this category. An example was given in Chapter 1 (see Fig. 1.14). That polymer was PDMS (or polydimethylsiloxane), a molecule obtained by the polymerisation of octamethylcyclotetrasiloxane. This material has an amorphous structure, and its glass transition temperature is very low ($T_g \approx -130\,°C$). This feature is due to a very flexible main chain, since the angle of the Si–O bonds can vary easily. Silicones can also be reticulated (by irradiation, for instance) yielding excellent elastomers.

In conclusion, we should point out that the well-known modelling paste, Silly Putty, that we talked about in Chapter 1, is obtained by mixing PDMS and boric acid, $B(OH)_3$. The role of the boric acid is to induce a transient reticulation of the material over short distances, increasing its elastic modulus on the rubber plateau, but without preventing it from flowing over very long times. This mixture can thus be classified among viscoelastic fluids, as already discussed in Chapter 1.

In the following, we will concern ourselves with the rheological behaviour of polymer *melts*. This implies that the temperature is above the glass transition temperature for completely amorphous systems, or above the melting temperature of their crystalline fraction for semi-crystalline ones. Under these conditions, they behave as viscoelastic fluids. Our discussion will be limited to the case of linear polymers, requiring the simplest theoretical treatment. Before beginning, let us recall some of their rheological properties.

7.1.3 Rheological behaviour of polymer melts

Although their chemical structures are extremely diverse, (linear) polymers exhibit in their molten state universal features, some of them being revealed by their rheological behaviour. This is what we intend to show in the following.

7.1.3.1 Flow viscosity η_0 as a function of the molecular mass M

Figure 7.5 shows, in logarithmic scale, the variation of the shear viscosity η_0 in the linear regime (i.e. at low shear rates) for different polymer melts as a function of the molecular mass M.

These curves, plotted with arbitrary vertical shifts to avoid superpositions, show that, for all the polymers studied, the viscosity starts by increasing linearly with the molecular mass, followed by a much steeper regime, described by a power law with a constant exponent, close to 3.3.

This change in slope occurs at a certain value M_e of the molecular mass, which depends on the material under consideration. This behaviour has an obvious universal character, which one can summarise as:

$$\eta_0 \propto M \qquad (M < M_e),$$
$$\eta_0 \propto M^{3.3} \qquad (M > M_e). \tag{7.2}$$

In particular, this crucial observation shows that a short polymer flows much more easily than a long one. This effect is generally interpreted in terms of *entanglements* since shorter chains form independent coils, while longer ones tend to entangle and form knots (thus increasing their mutual friction and therefore their viscosity, Fig. 7.6).

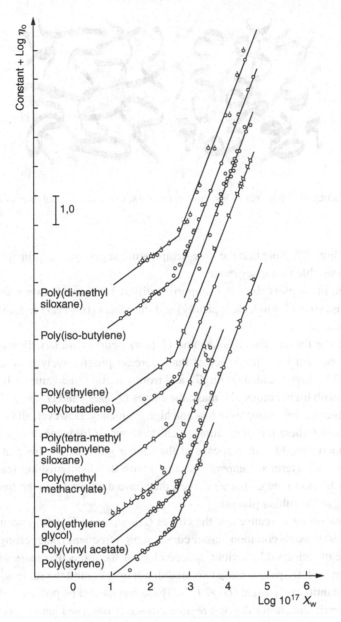

Fig. 7.5 Viscosity as a function of the molecular mass for various polymers (Berry and Fox, 1970) (here, the parameter X_w is proportional to the molecular mass).

7.1.3.2 Elastic and loss moduli G' and G''

Another experiment that illustrates the difference between the two regimes described above consists in measuring the elastic modulus G' and the loss modulus G'' as a function of frequency for polymers of different molecular mass. Such data, measured for linear PE,

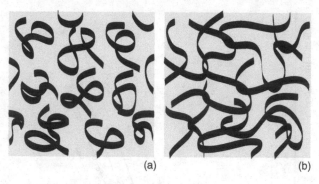

$$(a) \hspace{5cm} (b)$$

Fig. 7.6 (a) Unentangled polymer of low molecular mass; (b) entangled polymer of high molecular mass.

are shown in Fig. 7.7. Note that the time–temperature superposition principle was used to extend the accessible frequency range.

Once again, these plots show that important differences exist between samples of low molecular mass (from L9 to L16, typically) and the others (from L37 to L18), with higher molecular mass.

Indeed, for the former, the $G'(\omega a_T)$ and $G''(\omega a_T)$ curves are strictly increasing functions with slopes (in logarithmic scale) that decrease progressively from 2 (at low frequency) to 1/2 (at high frequency) for G', and from 1 to the same value of 1/2 for G''.

In samples with higher molecular mass, the curves $G'(\omega a_T)$ and $G''(\omega a_T)$ have the same slopes as for the previous samples at low and high frequency. However, all $G'(\omega a_T)$ curves exhibit a plateau (where the slope almost vanishes) over a frequency range that becomes wider as M increases. One then speaks of the *rubber plateau*. On this plateau, G' takes a specific value G_p (termed '*plateau modulus*') with the distinguishing feature of being insensitive to M. As to the $G''(\omega a_T)$ curves, they have a minimum in the frequency range corresponding to the rubber plateau.

It is very important to realise that the curves $G'(\omega a_T)$ and $G''(\omega a_T)$ seem to converge (for all samples) towards common master curves at high frequency, suggesting the presence in this regime of universal behaviour, independent of the molecular mass M. Even more surprisingly, at high frequency these two moduli not only exhibit the same $\omega^{1/2}$ behaviour, but are also quantitatively equal ($G' \approx G''$). These results will be proven in the following.

Finally, experiment shows that the regime crossover observed here occurs at the same molecular mass M_e (termed '*entanglement mass*') as revealed by the previous experiment.

It is also worthwhile comparing the moduli G' and G'' over a wider frequency range than shown above (although it does span nine decades in frequency).

Such an example is shown in Fig. 7.8, where the two moduli are plotted on the same graph as a function of frequency. The material is a statistical styrene-butadiene copolymer (with 26% styrene), with the chemical formula $[CH_2–CH(C_6H_5)]$ $[CH_2–CH=CH–CH_2]$ and an average molecular mass of 170 000. Note that this amorphous copolymer behaves as a homopolymer above its glass transition temperature.

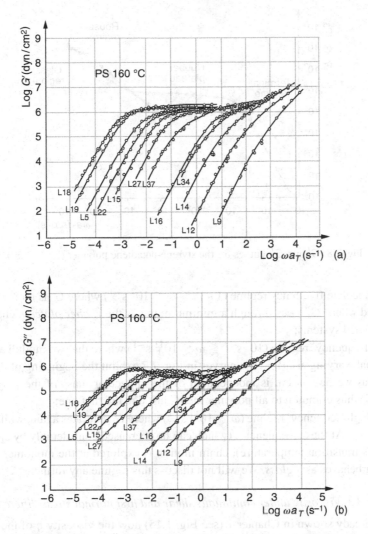

Fig. 7.7 Elastic G' (a) and loss G'' (b) moduli as a function of the scaled frequency, measured in samples of linear PE with different molecular mass. This parameter varies from 8 900 for sample L9 to 580 000 for sample L18 (Onogi *et al.*, 1970).

In this experiment, the frequency range covered is impressive (more than 14 decades!). Obviously, this is only possible by superposing, using the WLF equation (see Eq. (6.320)) data measured between $-40\,°C$ (temperature close to T_g) and $125\,°C$. The master curves are given here for a reference temperature of $75\,°C$. They show clearly the existence of four distinct regimes (three of them should already be familiar):

- a 'low frequency' regime ($\omega \lesssim 1\,s^{-1}$) where G' varies as ω^2 and G'' as ω; this is the *terminal zone* common to all polymer melts;

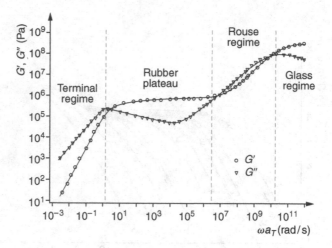

Fig. 7.8 G' and G'' curves for the styrene-butadiene polymer ([36], p. 50).

- an intermediate frequency regime ($1\ \mathrm{s}^{-1} \lesssim \omega \lesssim 10^7\ \mathrm{s}^{-1}$) where G' is more or less constant and where G'' goes through a minimum: this is the *rubber plateau*, characteristic of entangled systems;
- a 'high frequency' regime ($10^7\ \mathrm{s}^{-1} \lesssim \omega \lesssim 10^{11}\ \mathrm{s}^{-1}$) where the two moduli are more or less equal, varying as $\omega^{1/2}$. The curves in Fig. 7.7 exhibit the beginning of this regime, which, as we have seen, is independent of the molecular mass of the polymer. This regime is thus common to all polymer melts, just like the first one;
- a 'very high frequency' regime ($\omega \gtrsim 10^{11}\ \mathrm{s}^{-1}$) where G' saturates again, while G'' starts to decrease. At these frequencies (which can be decreased considerably by approaching the glass transition temperature), chain motion is explored at the monomer scale. The polymer behaves as a *glass*. We will not discuss this regime any further.

7.1.3.3 Viscosity under continuous shear and first normal stress difference

We have already shown in Chapter 1 (see Fig. 1.15) how the viscosity η of the Rhodorsil silicone oils (essentially composed of PDMS) varies as a function of the shear rate $\dot{\gamma}$ in the stationary flow regime.

These plots show that for each molecular mass (in the present case, it is always larger than the entanglement mass M_e) there exists a shear rate range ($\dot{\gamma} < \dot{\gamma}_c$) where the viscosity does not change, taking the value η_0. In this regime, the oil behaves as a *Newtonian viscous fluid*. As $\dot{\gamma} > \dot{\gamma}_c$, the viscosity starts to decrease and the oil becomes *shear thinning*. As $\dot{\gamma} \gg \dot{\gamma}_c$, its viscosity follows the Ostwald–de Waehle law (Eq. (1.4)), varying as $1/\dot{\gamma}^{1-n}$, with $n \approx 0.25$.

We should make two comments on these plots:

- first, the shear rate $\dot{\gamma}_c$ above which the oil becomes non-Newtonian is inversely proportional to the molecular mass M (here, with $M\dot{\gamma}_c \approx 5 \times 10^6\ \mathrm{g\,s^{-1}\,mol^{-1}}$ for molecular masses between 1000 and 2.5 million);

- second, all plots seem to converge towards a common master curve at very high shear rate, for all molecular mass values.

Comparing these results with linear viscoelasticity measurements is very interesting. Indeed, experiment shows that in polymer melts in general, and silicone oils in particular, the Cox and Merz rule is very well verified (see Section 6.4.8). This implies that the viscosity $\eta'(\omega)$ measured at the frequency ω (and connected to the loss modulus G'' by the relation $\eta' = G''/\omega$ according to Eq. (6.85)) equals (with a very good approximation) the viscosity $\eta(\dot\gamma)$ measured under continuous shear at the shear rate $\dot\gamma = \omega$. In this case, it follows very clearly that the break in slope at $\dot\gamma_c$ for the viscosity curve $\eta(\dot\gamma)$ coincides with the first maximum in the $G''(\omega)$ curve, which has the same shape as in Fig. 7.8. However, we have seen that this maximum marks the beginning of the rubber plateau, where the entanglements start to hinder the molecular motion. We conclude that the critical shear rate $\dot\gamma_c$, which sets the beginning of the nonlinear regime under continuous shear, is that above which the molecules start to disentangle and align substantially under the action of shear.

To summarise this discussion graphically, we have drawn in Fig. 7.9 a schematic representation of the two linear viscoelasticity functions G' and G'' as a function of the frequency, as well as the two viscometric functions η and Ψ_1 as a function of shear rate. The shear stress $\sigma = \eta\dot\gamma$ and the first normal stress difference $N_1 = \Psi_1\dot\gamma^2$, which are the experimentally accessible quantities, are also shown. All calculations were performed assuming that the polymer is characterised by a unique Maxwell time τ close to the edge of the rubber plateau, itself described by the elastic modulus G_p (obviously, the existence of this plateau requires that the polymer is entangled and that its molecular mass is larger than M_e). We have also applied the Cox and Merz (Eq. (6.317)) and Laun (Eq. (6.318)) formulas to calculate the viscometric functions.

These curves show clearly that the nonlinear regime under continuous shear begins as the shear rate exceeds $1/\tau$ (implicitly assuming that $\dot\gamma_c \simeq 1$). This value also sets the frequency at which the G' and G'' curves cross. Note that in the case of polymers, even for monodisperse systems, the relaxation is not described by a single viscoelastic time, but by a rather complicated distribution (calculable, for instance, in the framework of the de Gennes–Doi–Edwards reptation model discussed in Section 7.1.7, see Eq. (7.111)). This becomes immediately apparent by comparing the real viscosity curves, decreasing as $1/\dot\gamma^{0.75}$ in the strongly nonlinear regime (Fig. 1.15), while the Maxwell model with a single relaxation time and the application of the Cox and Merz formula (Eq. (6.317)) lead to an unrealistic $1/\dot\gamma$ variation (in this case, the shear flow is marginally stable, as we showed in Section 6.3.4). On the other hand, the time τ that we introduced has a precise physical meaning, being the *longest time* of this distribution. Also known as the *terminal relaxation time* (it coincides with the reptation time τ_{rep} in the model cited above, see Eq. (7.101)), it plays a crucial role in experiments, as it sets the width of the *terminal zone* in linear viscoelasticity (corresponding to frequencies $\omega < 1/\tau$, for which G' varies as ω^2 and G'' as ω), as well as the beginning of the nonlinear regime under continuous shear (assuming, once again, that $\dot\gamma_c \approx 1$, which is indeed the case for polymer melts, such as silicone oils, far from their glass transition temperature).

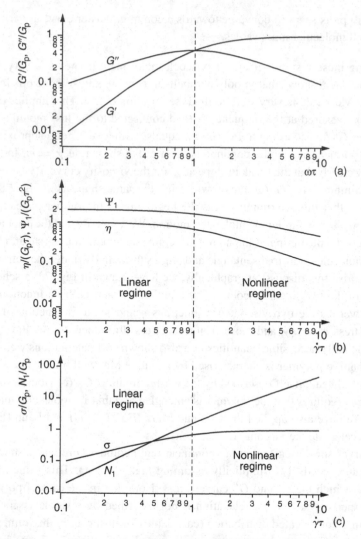

Fig. 7.9 (a) Shape of the $G'(\omega)$ and $G''(\omega)$ curves for a Maxwell fluid with plateau modulus G_p and viscoelastic relaxation time τ; (b) comparison with the $\eta(\dot{\gamma})$ and $\Psi_1(\dot{\gamma})$ curves calculated using the Cox and Merz and Laun formulas. The last plot (c) shows the shear stress and the first normal stress difference N_1.

7.1.3.4 *Microscopic origin of the first normal stress difference N_1: qualitative explanation*

Before delving into the theoretical models, let us try to give a qualitative explanation for the microscopic origin of the normal stress difference N_1.

We have already seen that the simplest experiment one can perform in order to demonstrate the existence of N_1 consists in shearing a polymer slab between two parallel plates. In this geometry (represented in Fig. 7.10, as well as in Figs. 6.20 and 1.18), we have

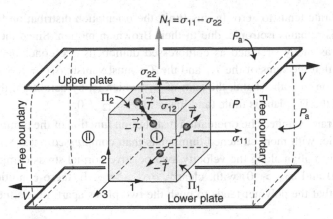

Fig. 7.10 Dumbbell model providing an explanation for the presence of a normal force. This force appears due to the molecules being both stretched and aligned along the flow direction.

shown (see Section 6.3.1) that the force per unit surface F exerted by the fluid normal to the upper plate is of elastic origin (related to the diagonal elements of the stress tensor) with expression:

$$F = P - \sigma_{22} - P_{\mathrm{a}} = \sigma_{11} - \sigma_{22} \equiv N_1, \tag{7.3}$$

where N_1 is, by definition, the first normal stress difference.

To achieve a physical comprehension of the origin of the elastic stresses, let us imagine each polymer molecule as two small beads connected by a spring (this is the 'dumbbell' model already mentioned in Section 1.5.6, which we will analyse in detail in Appendices 7.A and 7.B to this chapter in relation with the more general Rouse model described in Section 7.1.6). When the sample is sheared, the dumbbells tend to elongate or to shorten, depending on their orientation (Fig. 7.10). As a result, two tension forces of opposite sign $\vec{T}$ and $-\vec{T}$ act on each of their extremities. Consider now a cube with unit volume inside the fluid. Label by (I) the fluid domain within the cube and by (II) the one outside. We showed in Chapter 2 that the stress σ_{11} is the component along the 1 axis of the force exerted by part (II) onto part (I) across the face Π_1 of the cube; in the same way, the stress σ_{22} is the component along the 2 axis of the force exerted by part (II) onto part (I) across the face Π_2 of the cube. Consider at present all the dumbbells 'i' (or 'j') that intersect faces Π_1 (or Π_2, respectively) of the cube and denote by $\vec{T}_i$ (or $\vec{T}_j$) the tension forces acting on the outside beads of these dumbbells. We then have, by the definition of the stress components:

$$\sigma_{11} = \sum_i T_{i1}, \tag{7.4}$$

$$\sigma_{22} = \sum_j T_{j2}. \tag{7.5}$$

One can then imagine two limiting situations:

(1) the shear rate tends to zero. In this limit, the orientation distribution function of the dumbbells remains isotropic, due to their Brownian motion. Since there are (on the average) as many stretched as compressed dumbbells, we conclude, on symmetry grounds, that the sums of the T_{i1} and the T_{j2} must vanish: $\sigma_{11} = \sigma_{22} = 0$, such that $N_1 = 0$; on the other hand, the sum of T_{j1} which is, by definition, the shear stress acting on the Π_2 plane is finite (and linear in $\dot{\gamma}$): $\sigma_{12} \neq 0$;

(2) the shear rate is finite. The orientation distribution function of the dumbbells becomes anisotropic, with more stretched dumbbells than compressed ones. Since the dumbbells tend to align along the velocity, two positive normal stress components appear ($\sigma_{11} > 0$ and $\sigma_{22} > 0$) with, clearly, $\sigma_{11} > \sigma_{22}$. In these conditions $N_1 > 0$, meaning that the polymer tends to push the two plates apart: this is the Weissenberg effect.

It is very important to note here the role of the boundary conditions. Indeed, due to the stress balance at the free vertical edges, one must have $P_a = P - \sigma_{11}$. If the shear rate is very strong, all the dumbbells tend to lie flat: σ_{11} becomes very large and σ_{22} very small, so the force exerted by the fluid on the upper plate is essentially due to the pressure term: $F \approx P - P_a = \sigma_{11}$.

This said, one can conclude that the appearance of normal stresses is generally related to *two coupled effects*:

(1) the *elongation of the particles* (here, the polymer molecules) under shear (an effect proportional to $\dot{\gamma}$ in the linear regime);

(2) their *tendency to align* along the velocity (this effect is also proportional to $\dot{\gamma}$ and translates the competition between Brownian motion and hydrodynamic effects).

It follows that, for polymers (where the molecules are deformable), the normal stresses and, consequently, their differences N_1 and N_2, must vary as $\dot{\gamma}^2$.

Finally, note that the change in molecular length being inversely proportional to the stiffness, the normal stresses (and hence the stored elastic energy) are more important when the molecules deform easily, i.e. when they are long and flexible. This result is in agreement with the experiments and with the upper-convected Maxwell model, which predicts that N_1 is proportional to $\eta^2 \dot{\gamma}^2 / G$ in the linear deformation regime (see Sections 6.4.4.2 and Section 7.1.6.10).

In the following, we will show how the behaviour of monodisperse polymer melts can be modelled starting from a microscopic approach in the *linear regime*. Two models will allow us to calculate explicitly the spectrum of relaxation times and to reproduce the shape of the experimental curves shown in Figs. 7.7 and 7.8: that of Rouse, deriving from the dumbbell model and only applying to unentangled polymers, and that of Doi and Edwards, which extends the Rouse model to the case of entangled polymers by taking into account the reptation of chains across the topological knots formed by the entanglements.

Before that, let us give a reminder on polymer chains.

7.1.4 Reminder on the theory of polymers

A polymeric material is characterised by the presence of chains. These chains can exhibit different types of interactions, such as entropic, electrostatic (van der Waals forces), hydro-dynamic,[3] excluded volume, or of topological origin (entanglements). In polymer melts (and their very concentrated solutions), all these interactions are present, but they are not all of equal importance. Thus, experiment shows that the hydrodynamic interactions (which are very important in dilute polymer solutions, as we will see later) are negligible in melts. Even more surprisingly (and it took the theorists some time to be convinced), the attractive van der Waals forces between neighbouring chains are compensated over very short dis-tances by the repulsive, excluded volume forces acting between the monomers of the same chain. Hence, in polymer melts, the chains are almost *ideal* and behave as a *random walk*. In the next section, we will describe the statistical properties (entropic in origin) of such a chain.

7.1.4.1 The random walk model and the statistical Kuhn segment

For simplicity's sake, let us consider the case of a homopolymer, where each chain con-sists of a regular sequence of monomers. The monomers are connected to one another by more or less flexible chemical bonds. For this reason, each type of chain has a typical min-imal distance such that two monomers, at least this distance apart, can assume unrestricted relative orientations.

It is then possible to describe any polymer chain as a *random walk* composed of rigid segments in succession, with completely arbitrary relative orientations. These chain links are the *statistical Kuhn segments*. In flexible polymers (such as those described above), these segments are generally a few monomers long. Note that this length is equal (up to a factor of 2 due only to the definitions) to the *persistence length* of the chain. As the name shows, this parameter indicates the distance below which the chain can be seen as stiff: above this value, it becomes flexible. In this representation, the chain looks as in Fig. 7.11.

In the following, to simplify even further, we will identify the statistical Kuhn segment with the monomer (note that this is only a convenient choice of terms, with no consequence on the physics of the problem). We denote its length by b and we assume that each chain contains N monomers. The nth monomer of the chain is represented by its direction vector $\vec{r}_n$ and the radius vectors of its two extremities are denoted by $\vec{R}_{n-1}$ and $\vec{R}_n$, such that:

$$\vec{r}_n = \vec{R}_n - \vec{R}_{n-1} \qquad (n = 1, 2 \dots N). \qquad (7.6)$$

Within the random walk model, the statistical distribution function of the vectors $\vec{r}_n$ is simply given by the equation:

$$\Psi\left(\{\vec{r}_n\}\right) = \prod_{n=1}^{N} \psi\left(\vec{r}_n\right), \qquad (7.7)$$

---

[3] This is the case for polymers dissolved in solvent.

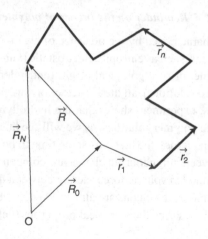

Fig. 7.11 Schematic representation of a polymer chain according to the random walk model. Each segment represents a statistical Kuhn segment.

where $\psi(\vec{r}_n)$ stands for the probability that the vector $\vec{r}_n$ (with a constant length, given by b) has a certain orientation:

$$\psi(\vec{r}_n) = \frac{1}{4\pi b^2} \delta(|\vec{r}_n| - b).$$

(7.8)

Note that this function obeys the following normalisation condition: $\int_0^\infty \psi(r)\, 4\pi r^2 dr = 1$.

7.1.4.2 The end-to-end vector $\vec{R}$: mean square value and distribution function

We will define the *end-to-end vector* $\vec{R}$ of the chain as the vector:

$$\vec{R} = \vec{R}_N - \vec{R}_0 = \sum_{n=1}^{N} \vec{r}_n.$$

(7.9)

Its mean square value $\overline{R}$ (*known as the Flory radius* [33]) is very easily obtained by writing that:

$$\overline{R}^2 = \left\langle \vec{R}^2 \right\rangle = \left\langle \vec{R}_N - \vec{R}_0 \right\rangle^2 = \sum_{n=1}^{N} \left\langle \vec{r}_n^2 \right\rangle + 2 \sum_{n>m} \left\langle \vec{r}_n \cdot \vec{r}_m \right\rangle.$$

(7.10)

Since there is no orientation correlation between the monomers and since their length is fixed, one has $\langle \vec{r}_n \cdot \vec{r}_m \rangle = 0$ for $n \neq m$ and $\langle \vec{r}_n^2 \rangle = b^2$ giving, by substitution into the formula above:

$$\overline{R} = \sqrt{N}\, b.$$

(7.11)

As to the distribution function $\Phi(\vec{R}, N)$ of the end-to-end vector, giving the probability $\Phi \, d\vec{R}$ for the extremity of the $\vec{R}$ vector to be found within the volume element $d\vec{R}$, it is given by the equation:

$$\Phi(\vec{R}, N) = \int d\vec{r}_1 \int d\vec{r}_2 \ldots \int d\vec{r}_N \delta\left(\vec{R} - \sum_{n=1}^{N} \vec{r}_n\right) \Psi(\{\vec{r}_n\}). \tag{7.12}$$

This integral can be easily calculated using the closure relation: $\delta(\vec{r}) = (2\pi)^{-3} \int d\vec{k} \exp(i\vec{k} \cdot \vec{r})$, yielding:

$$\Phi(\vec{R}, N) = \frac{1}{(2\pi)^3} \int d\vec{k} \exp(i\vec{k} \cdot \vec{R}) \left(\frac{\sin(kb)}{kb}\right)^N. \tag{7.13}$$

In the large N limit, this results in:

$$\left(\frac{\sin(kb)}{kb}\right)^N \approx \exp\left(-\frac{Nk^2b^2}{6}\right) \tag{7.14}$$

and finally, after substitution into Eq. (7.13):

$$\Phi(\vec{R}, N) = \left(\frac{3}{2\pi Nb^2}\right)^{3/2} \exp\left(-\frac{3\vec{R}^2}{2Nb^2}\right). \tag{7.15}$$

This formula shows that *the distribution function of the end-to-end vector $\vec{R}$ is a Gaussian*, centred in 0 and with a width $\sqrt{3Nb}/3$.

7.1.4.3 The Gaussian chain equivalent model

The simplest analytical model that yields the distribution function above[4] and leads to correct rheological predictions (see below for the Rouse model), is that of the *Gaussian chain*. This chain is modelled as a series of springs, with a null rest length, connecting fictitious beads (Fig. 7.12).

In this case, the potential energy of the chain is the sum of the elastic energies of the springs:

$$U(\{\vec{r}_n\}) = \frac{3k_BT}{2b^2} \sum_{n=1}^{N} \left(\vec{R}_n - \vec{R}_{n-1}\right)^2 = \frac{3k_BT}{2b^2} \sum_{n=1}^{N} \vec{r}_n^2. \tag{7.16}$$

Using the Boltzmann distribution, one can then obtain the following distribution function for the chain segments:

$$\Psi(\{\vec{r}_n\}) = \left(\frac{3}{2\pi b^2}\right)^{3N/2} \exp\left[-\sum_{n=1}^{N} \frac{3\vec{r}_n^2}{2b^2}\right]. \tag{7.17}$$

---

[4] which, although not exact, provides a very good description of the statistical properties of the polymer chain.

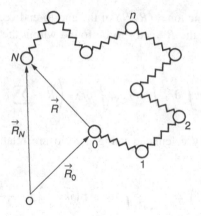

Fig. 7.12 The Gaussian chain model.

This expression and Eq. (7.12) (where $\vec{R} - \sum_{n=1}^{N} \vec{r}_n$ is replaced by $\vec{R}_n - \vec{R}_m - \sum_{i=n+1}^{m} \vec{r}_i$) yield the distribution function for the vector $\vec{R}_n - \vec{R}_m$ connecting two arbitrary monomers along the chain:

$$\Phi\left(\vec{R}_n - \vec{R}_m, n - m\right) = \left[\frac{3}{2\pi b^2 |n - m|}\right]^{3/2} \exp\left[-\frac{3(R_n - R_m)^2}{2b^2 |n - m|}\right]. \qquad (7.18)$$

As expected, this expression gives back (for $n = 0$ and $m = N$) the distribution function of the end-to-end vector of the entire chain (Eq. (7.15)).

In this model, the polymer forms a *Gaussian statistical coil at all scales*.

7.1.4.4 Gyration radius and its determination by radiation scattering

In practice, one does not measure $\vec{R}$ directly, but rather the gyration radius R_g of the molecule, defined by the relation:

$$R_g^2 = \frac{1}{2N^2} \sum_{n,m=1}^{N} \left\langle \left(\vec{R}_n - \vec{R}_m\right)^2 \right\rangle = \frac{1}{N} \sum_{n=1}^{N} \left\langle \left(\vec{R}_n - \vec{R}_G\right)^2 \right\rangle, \qquad (7.19)$$

where $\vec{R}_G$ is the centre of mass of the chain:

$$\vec{R}_G = \frac{1}{N} \sum_{n=1}^{N} \vec{R}_n. \qquad (7.20)$$

However, $\left\langle \left(\vec{R}_n - \vec{R}_m\right)^2 \right\rangle = |n - m| b^2$ owing to the Gaussian statistics of the chain at all scales (see Eq. (7.18)), so that substitution in Eq. (7.19) yields:

$$R_g^2 = \frac{1}{2N^2} \sum_{n=1}^{N} \sum_{m=1}^{N} |n - m| b^2. \qquad (7.21)$$

Going to the continuous limit results in:

$$R_g^2 = \frac{1}{2N^2} \int_0^N dn \int_0^N dm |n - m| b^2 = \frac{N}{6} b^2, \tag{7.22}$$

so that, using Eq. (7.11):

$$R_g = \frac{\overline{R}}{\sqrt{6}}. \tag{7.23}$$

Without going into the details, let us recall that neutron scattering was used to measure the gyration radius in polymer melts, which was indeed shown to vary as the square root of the molecular mass, itself proportional to N:[5]

$$R_g \propto \sqrt{M}. \tag{7.24}$$

This experimental result is the *first direct evidence for the ideal behaviour of the chain and its Gaussian statistics in melts.*

These basic concepts having been presented, we should now establish a microscopic expression for the stress tensor, taking into account the statistical coil configuration of the chain.

7.1.5 Microscopic expression of the stress tensor

To derive such an expression, let us consider a cube of side L and a plane (P), perpendicular to the j axis and cutting the cube into two parts, (I) and (II) (Fig. 7.13). By definition, the stress is the component along i of the force per unit surface exerted by part (II) onto part (I).[6]

To calculate this force, consider all polymer chains that cross this plane at least once. Let $\vec{r}_p$ be a segment of one of these chains that cut the plane. According to the model of the Gaussian chain (and to Eq. (7.16)), in order to extend this segment by a length $\vec{r}_p$, one must exert on its side (II) end a force with an i component equal to κr_{pi}, where

$$\kappa = \frac{3k_B T}{b^2} \tag{7.25}$$

is its *Brownian tension* (entropic in origin). Obviously, part (I) exerts on the other end of the segment an opposite force.

Since this spring cuts the plane (P) with a probability r_{pj}/L, it contributes thus to the average stress an amount:

$$\frac{1}{L^2} \kappa r_{pi} \frac{r_{pj}}{L}, \tag{7.26}$$

---

[5] If m is the mass of a Kuhn segment and N_A the Avogadro number, then $M = N_A m N$.
[6] Note that we are only interested here in the viscoelastic part of the stress tensor directly associated with the polymer chains (the σ_P part in the Oldroyd-B model described in Section 6.4.5).

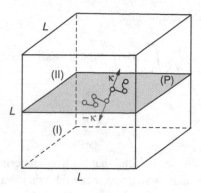

Fig. 7.13 Principle of stress calculation.

where the $1/L^2$ factor signifies that the stress is a force per unit surface. The stress is obtained by summing over all segments of a chain, and then over all chains. In practice, this amounts to summing over all segments of a chain, taking the statistical average (implicitly assuming that the chains are in local thermodynamic equilibrium), and then multiplying by the number of molecules contained in the volume L^3. This yields:

$$\sigma_{ij} = \frac{c}{N} L^3 \sum_{p=1}^{N} \left\langle \frac{1}{L^3} \kappa r_{pi} r_{pj} \right\rangle = \frac{c}{N} \frac{3k_B T}{b^2} \sum_{p=1}^{N} \langle r_{pi} r_{pj} \rangle, \tag{7.27}$$

where c is the number of Kuhn segments per unit volume. Since N is large, one can consider that the radius vector $\vec{R}_n$ is a continuous function of n, such that $r_{pi} = \left(\frac{\partial R_{ni}}{\partial n} \right)_{n=p}$. The discrete sum is then transformed into an integral, and the formula above becomes:

$$\sigma_{ij} = \frac{3k_B T}{b^2} \frac{c}{N} \int_0^N \left\langle \frac{\partial R_{ni}}{\partial n} \frac{\partial R_{nj}}{\partial n} \right\rangle dn. \tag{7.28}$$

Let us now present the Rouse model, the simplest one that can describe the dynamics of a polymer chain.

7.1.6 The Rouse model for unentangled polymers

For the moment, we assume that the chains are short enough that they do not entangle with one another. Experimentally, this is the case when the molecular mass remains below the entanglement mass M_e, defined in the experimental section. We emphasise that, in practice, M_e typically corresponds to a molecular length of a hundred monomers (or, more precisely, Kuhn segments).

7.1.6.1 Langevin equation for the motion of the chain in a quiescent fluid

The model chosen by P. E. Rouse to represent the polymer coil is that of the Gaussian chain (Fig. 7.12) (Rouse, 1953; [32]). In this representation, the nth bead of the chain is connected to its neighbours (the $n - 1$th and $n + 1$th beads) by two springs of tension $\kappa = 3k_BT/b^2$, exerting on it two elastic forces. The medium also exerts on each bead a viscous friction force proportional to its velocity $d\vec{R}_n/dt$, as well as a random force $\vec{f}_n$ due to the thermal agitation and the collisions with the other molecules (*stochastic Langevin force*). The beads at the two ends of the chain have $n = 1$ and $n = N$, respectively. In these conditions, the dynamic equation for an intermediate nth bead (neglecting its inertia) is given by:

$$\zeta \frac{d\vec{R}_n}{dt} = -\kappa \left(2\vec{R}_n - \vec{R}_{n+1} - \vec{R}_{n-1} \right) + \vec{f}_n, \tag{7.29}$$

where ζ is a phenomenological friction coefficient (equal to $6\pi\eta R$ for a bead with radius R moving in a fluid with viscosity η). For the beads at each end, the equation is:

$$\zeta \frac{d\vec{R}_1}{dt} = -\kappa \left(\vec{R}_1 - \vec{R}_2 \right) + \vec{f}_1 \quad \text{and} \quad \zeta \frac{d\vec{R}_N}{dt} = -\kappa \left(\vec{R}_N - \vec{R}_{N-1} \right) + \vec{f}_N. \tag{7.30}$$

The random forces $\vec{f}_n$ average to zero over time:

$$\left\langle \vec{f}_n(t) \right\rangle = 0. \tag{7.31}$$

One can also show using the fluctuation-dissipation theorem (and we will admit this result in the following) that their time correlation function is given by the following expression [32]:

$$\left\langle f_{n\alpha}(t) \, f_{m\beta}(t') \right\rangle = 2\zeta k_B T \, \delta_{nm} \delta_{\alpha\beta} \, \delta(t - t'). \tag{7.32}$$

This expression shows that there is no time correlation between the force experienced by a bead at a moment t and the force it feels at a different instant t' (the $\delta(t - t')$ factor). At the same instant, the forces acting on two different beads are also independent (hence the δ_{nm} factor). The same applies to the three components of the force acting on a particle ($\delta_{\alpha\beta}$ factor). Finally, since this force is due to thermal agitation, it is proportional to $k_B T$.

In practice, it is more convenient to work in a continuous representation (see the end of the previous section). One can also introduce two fictitious beads with radius vectors $\vec{R}_0 = \vec{R}_1$ and $\vec{R}_{N+1} = \vec{R}_N$, such that the two end equations (7.30) are contained in the following general equation:

$$\zeta \frac{\partial \vec{R}_n}{\partial t} = \kappa \frac{\partial^2 \vec{R}_n}{\partial n^2} + \vec{f}_n \tag{7.33}$$

with $\frac{\partial \vec{R}_n}{\partial n} = 0$ in $n = 0$ and $n = N$.

Note that, in the continuous representation, the Kronecker symbol δ_{nm} must be replaced by the $\delta(n-m))$ function, so that Eq. (7.32) takes the form:

$$\langle f_{n\alpha}(t)\, f_{m\beta}(t')\rangle = 2\zeta k_B T\, \delta(n-m)\delta_{\alpha\beta}\,\delta(t-t').\tag{7.34}$$

As to Eq. (7.31), it remains unchanged.

7.1.6.2 Normal mode analysis

To solve this problem, Rouse introduced the *normal coordinates* $\vec{X}_p$, defined by the following relations:

$$\vec{R}_n = \vec{X}_0 + 2\sum_{p=1}^{\infty}\vec{X}_p\cos\left(\frac{\pi pn}{N}\right),\tag{7.35}$$

$$\vec{X}_p = \frac{1}{N}\int_0^N\cos\left(\frac{p\pi n}{N}\right)\vec{R}_n(t)\,\mathrm{d}n \qquad (p=0,1,2,\ldots).\tag{7.36}$$

In this representation, the mode $\vec{X}_0$ represents the translation of the entire chain. As to the mode of order $p>0$, it corresponds to the vibration mode with p nodes along the chain.

Using this decomposition, the equations of motion of the chain can be rewritten as:

$$\zeta_p\frac{\partial\vec{X}_p}{\partial t} = -\kappa_p\vec{X}_p + \vec{f}_p \qquad (p=0,1,2,\ldots),\tag{7.37}$$

where we introduced the following notation:

$$\zeta_0 = N\zeta \quad\text{and}\quad \zeta_p = 2N\zeta \qquad (p=1,2,\ldots),\tag{7.38}$$

$$\kappa_p = \frac{2\pi^2\kappa p^2}{N} = \frac{6\pi^2 k_B T}{Nb^2}p^2 \qquad (p=0,1,2,\ldots).\tag{7.39}$$

As to the random forces $\vec{f}_p$, given by:

$$\vec{f}_p = \frac{\zeta_p}{\zeta}\int_0^N\frac{1}{N}\cos\left(\frac{p\pi n}{N}\right)\vec{f}_n\,\mathrm{d}n,\tag{7.40}$$

according to Eqs. (7.31) and (7.32), they obey the following relations:

$$\langle f_{p\alpha}(t)\rangle = 0,\tag{7.41}$$

$$\langle f_{p\alpha}(t)\, f_{q\beta}(t')\rangle = 2\zeta_p k_B T\delta_{pq}\delta_{\alpha\beta}\,\delta(t-t').\tag{7.42}$$

The remarkable interest of this representation is that the modes $\vec{X}_p$ are *independent of each other*, since the random forces are independent. That is why one speaks of *normal modes* (the energy of the set of these modes is the sum of their individual energies, $\langle\vec{X}_p^2\rangle = 3k_B T/\kappa_p$).

It is easy to check that the general solution to Eqs. (7.37) can be written in the following form (satisfying the causality principle):

$$\vec{X}_p(t) = \frac{1}{\zeta_p} \int_{-\infty}^{t} dt' \exp\left(-\frac{t-t'}{\tau_p^*}\right) \vec{f}_p(t'),$$ (7.43)

with

$$\tau_p^* = \frac{\zeta_p}{\kappa_p} = \frac{\tau_R}{p^2} \qquad (p > 0).$$ (7.44)

We introduced here the *terminal time* τ_R (also called *Rouse time*), which is the longest among all times τ_p^*:

$$\tau_R = \frac{\zeta N^2 b^2}{3\pi^2 k_B T}.$$ (7.45)

Its physical significance will be given below.

The translation mode must be treated separately, since $\kappa_0 = 0$. The corresponding equation is:

$$\zeta_0 \frac{\partial \vec{X}_0}{\partial t} = \vec{f}_0.$$ (7.46)

It admits as solution:

$$\vec{X}_0(t) = \frac{1}{\zeta_0} \int_0^t \vec{f}_0(t') \, dt' + \vec{X}_0(0).$$ (7.47)

7.1.6.3 Calculating the diffusion coefficient of the molecule

To obtain this parameter, let us study the displacement of the centre of mass G of the molecule, with radius vector $\vec{R}_G$. From Eqs. (7.20) and (7.35), one has:

$$\vec{R}_G = \frac{1}{N} \int_0^N \vec{R}_n \, dn = \vec{X}_0$$ (7.48)

yielding:

$$\left\langle \left[\vec{R}_G(t) - \vec{R}_G(0)\right]^2 \right\rangle = \sum_{\alpha=x,y,z} \left\langle [X_{0\alpha}(t) - X_{0\alpha}(0)]^2 \right\rangle.$$ (7.49)

Using Eqs. (7.42) and (7.47) leads to:

$$\left\langle (X_{0\alpha}(t) - X_{0\alpha}(0))^2 \right\rangle = \frac{1}{\zeta_0^2} \left\langle \int_0^t f_{0\alpha}(t') \, dt' \int_0^t f_{0\alpha}(t'') \, dt'' \right\rangle$$

$$= \frac{1}{\zeta_0^2} \int_0^t \int_0^t 2\zeta_0 k_B T \, \delta(t' - t'') \, dt' \, dt''$$

$$= 2 \frac{k_B T}{\zeta_0} t,$$ (7.50)

whence, using Eq. (7.49):

$$\left\langle \left[\vec{R}_G(t) - \vec{R}_G(0) \right]^2 \right\rangle = 6 \frac{k_B T}{\zeta_0} t. \tag{7.51}$$

Thus, the centre of mass of the molecule diffuses over time with the *diffusion coefficient*:

$$D_G = \frac{k_B T}{\zeta_0} = \frac{k_B T}{N \zeta}. \tag{7.52}$$

The Rouse model predicts that the diffusion coefficient of the molecule is inversely proportional to its molecular mass:

$$D_G \propto \frac{1}{M}. \tag{7.53}$$

7.1.6.4 Time correlation function of the end-to-end vector $\vec{R}$: rotational diffusion time

One can also calculate the time correlation function of the end-to-end vector $\vec{R}(t) = \vec{R}_N(t) - \vec{R}_0(t)$. The same type of calculation as performed above leads to the following result:

$$\left\langle \vec{R}(t) \, \vec{R}(0) \right\rangle = N b^2 \sum_{p \text{ odd}} \frac{8}{\pi^2 p^2} \exp\left(-\frac{t p^2}{\tau_R} \right). \tag{7.54}$$

This formula reduces to the average value of the end-to-end vector in the $t = 0$ limit: $\overline{R} = \sqrt{N} b$.

It also shows that, at times much longer than τ_R:

$$\left\langle \vec{R}(t) \, \vec{R}(0) \right\rangle \approx \frac{8}{\pi^2} N b^2 \exp\left(-\frac{t}{\tau_R} \right) \qquad (t \gg \tau_R). \tag{7.55}$$

Thus, τ_R can be interpreted as the time after which the molecule globally loses its initial orientation. Hence, it is the *rotational diffusion time* (hence the R subscript) of the entire molecule in the absence of entanglements. We will therefore call it the *Rouse time* of the entire chain.

Let us now consider the dynamic properties of the chain under shear.

7.1.6.5 Equation of motion for the chain under flow

One must now take into account the macroscopic flow characterised by the tensor of velocity gradients $(\vec{\nabla}\vec{v})_{ij} = \partial v_i / \partial x_j$. In this case, what matters is no longer the absolute velocity of the molecules (or, more exactly, of the beads on the Rouse chain), but rather their relative velocity with respect to the average flow. Under these conditions, the evolution equation (7.37) for mode p of the chain becomes:

$$\frac{D\vec{X}_p}{Dt} - \left(\vec{\nabla}\vec{v} \right) \vec{X}_p = -\frac{\kappa_p}{\zeta_p} \vec{X}_p + \frac{1}{\zeta_p} \vec{f}_p \qquad (p = 0, 1, 2, \ldots), \tag{7.56}$$

where D/Dt is the material derivative. Note that the left-hand side of this equation, giving the hydrodynamic friction force exerted by the medium on the molecule, vanishes as this latter is advected by the flow and undergoes affine deformation in the local velocity gradient. The next step consists of calculating the expression of the stress tensor as a function of the normal components $\vec{X}_p$.

7.1.6.6 Expression of the stress tensor as a function of the normal components

In Section 7.1.5, we established a microscopic expression for the stress tensor. This formula can be rewritten as a function of the normal coordinates $\vec{X}_p$. Indeed, it is enough to replace the $\vec{R}_n$ by their expressions (7.35) in Eq. (7.28), followed by integration over n. This calculation leads to the following simple result:

$$\sigma_{\alpha\beta} = \frac{c}{N} \sum_{p=1}^{\infty} \kappa_p \langle X_{p\alpha} X_{p\beta} \rangle, \tag{7.57}$$

where we recall that $\kappa_p = 6\pi^2 k_B T p^2 / (Nb^2)$.

We are now able to calculate the stress tensor in the presence of an arbitrary flow. This calculation (as we will see, it gives back the upper convected Maxwell equation introduced in a phenomenological way in the previous chapter) is rather tedious. For this reason, we postpone it to the end of this section (Section 7.1.6.10), beginning instead with the – easier – calculations that will provide the viscosity η_0 and the relaxation modulus $G(t)$, associated with the complex modulus $G^*(\omega)$.

7.1.6.7 Viscosity η_0 under simple shear ($\dot{\gamma} \to 0$)

The expression (7.57) shows that the stress is expressed as a function of the average quantities $\langle X_{p\alpha} X_{p\beta} \rangle$. These are obtained from the dynamic equation (7.56). To simplify the calculations as much as possible, we assume the fluid to be sheared in the (x, y) plane at a constant shear rate $\dot{\gamma}$ such that $v_x = \dot{\gamma} y$. In the limit of very low shear rates ($\dot{\gamma} \to 0$), the only finite component of the stress tensor is σ_{xy} (equal to σ_{yx}) (the normal stress differences tending towards 0 as $\dot{\gamma}^2$). One must then calculate the quantities $\langle X_{px} X_{py} \rangle$. They are obtained by combining the evolution equations for each of the components X_{px} and X_{py}. Some elementary algebra leads to the following result:

$$\frac{\partial}{\partial t} \langle X_{px} X_{py} \rangle = -2 \frac{\kappa_p}{\zeta_p} \langle X_{px} X_{py} \rangle + \dot{\gamma} \langle X_{py}^2 \rangle, \tag{7.58}$$

where we used the fact that $\langle f_{px} X_{py} \rangle = \langle f_{py} X_{px} \rangle = 0$ since the x (or y) displacements are not correlated with the y (or x, respectively) component of the random force (for more details, see Section 7.1.6.10).

In the $\dot{\gamma} \to 0$ limit, $\langle X_{py}^2 \rangle$ can be replaced by its average equilibrium value, easily obtained from Eqs. (7.42) and (7.43): $\langle X_{py}^2 \rangle = k_B T / \kappa_p$. Substituting in the expression above yields:

$$\frac{\partial}{\partial t} \langle X_{px} X_{py} \rangle = -2 \frac{\kappa_p}{\zeta_p} \langle X_{px} X_{py} \rangle + \frac{k_B T}{\kappa_p} \dot{\gamma}. \tag{7.59}$$

In the stationary régime ($\dot{\gamma} = $ Cst), the left-hand side vanishes, and so the stress can be calculated from expression (7.57):

$$\sigma_{xy} = \frac{c}{N} \sum_{p=1}^{\infty} \kappa_p \langle X_{px} X_{py} \rangle = \frac{1}{2} \frac{c}{N} \left(\sum_{p=1}^{\infty} \frac{\zeta_p}{\kappa_p} \right) k_B T \dot{\gamma}. \tag{7.60}$$

By definition, $\sigma_{xy} = \eta_0 \dot{\gamma}$, which gives, using Eq. (7.44):

$$\eta_0 = \frac{1}{2} \frac{c}{N} \left(\sum_{p=1}^{\infty} \frac{\zeta_p}{\kappa_p} \right) k_B T = \frac{1}{2} \frac{c}{N} k_B T \tau_R \sum_{p=1}^{\infty} \frac{1}{p^2}. \tag{7.61}$$

Finally, the viscosity is calculated using Eq. (7.45) and keeping in mind that $\sum_{p=1}^{\infty} 1/p^2 = \pi^2/6$:

$$\eta_0 = \frac{c}{36} \zeta N b^2. \tag{7.62}$$

We see thus that, in the Rouse model, the viscosity (in the linear regime) is proportional to N, and thus to the molecular mass M of the polymer:

$$\eta_0 \propto M. \tag{7.63}$$

This result is in agreement with the experiment for unentangled polymers.

7.1.6.8 Calculating the relaxation modulus $G(t)$

One can also calculate the relaxation function $G(t)$ starting from Eq. (7.59). Indeed, for a time-dependent $\dot{\gamma}$ this equation has the general solution:

$$\langle X_{px} X_{py} \rangle = \frac{k_B T}{\kappa_p} \int_{-\infty}^{t} dt' \exp\left[-(t - t')/\tau_p\right] \dot{\gamma}(t'), \tag{7.64}$$

where we defined:

$$\tau_p = \frac{\tau_p^*}{2} = \frac{\zeta N^2 b^2}{6\pi^2 p^2 k_B T}. \tag{7.65}$$

From the general expression (7.57) of the stress tensor, one has:

$$\sigma_{xy} = k_B T \frac{c}{N} \sum_{p=1}^{\infty} \int_{-\infty}^{t} dt' \exp\left[-(t - t')/\tau_p\right] \dot{\gamma}(t'). \tag{7.66}$$

By identification with the formula (6.20) given in the previous chapter, one obtains the following expression for the relaxation modulus $G(t)$:

$$G(t) = k_B T \frac{c}{N} \sum_{p=1}^{\infty} \exp(-t/\tau_p). \tag{7.67}$$

This formula shows that the fluid behaves as a *Maxwell fluid with a discrete distribution of relaxation times* τ_p. Note that, here, all Maxwell elements have the same elastic modulus:

$$G_p^{\text{Rouse}} = G_R = \frac{k_B T c}{N} = \frac{\rho R T}{M} \tag{7.68}$$

(where ρ is the density and R the perfect gas constant). Their viscosities, on the other hand, are all different, being given by:

$$\eta_p = G_R \tau_p = \frac{b^2 \zeta c N}{6\pi^2 p^2}. \tag{7.69}$$

In conclusion, let us describe the asymptotic behaviour of the relaxation modulus.

At long times, for $t > \tau_1 = \tau_R/2$, the longest mode $p = 1$ quickly dominates all the others. The relaxation is mono-exponential:

$$G(t > \tau_1) \approx k_B T \frac{c}{N} \exp(-t/\tau_1). \tag{7.70}$$

At short times, on the other hand, as $t < \tau_1 = \tau_R/2$, a large number of modes must be taken into account. The discrete sum in Eq. (7.67) can then be replaced by an integral, resulting in:

$$G(t < \tau_1) = k_B T \frac{c}{N} \int_0^\infty dp \exp(-2tp^2/\tau_R) = k_B T \frac{c}{2\sqrt{2N}} \left(\frac{\tau_R}{t}\right)^{1/2}. \tag{7.71}$$

To summarise, the relaxation modulus decreases slowly at the start, following a $1/\sqrt{t}$ behaviour, and then exponentially at long times.

7.1.6.9 Linear viscoelasticity functions G' and G"

Once the relaxation modulus is known, the linear viscoelasticity functions G' and G'' can be determined using the general formulas (6.70) and (6.71). It is even simpler to start from the results of the generalised Maxwell model, since the characteristics of the composing elements are known.

Thus, the elastic modulus G' is directly given by the formula (6.73), resulting here in:

$$G'(\omega) = G_R \sum_{p=1}^{\infty} \frac{\omega^2 \tau_p^2}{1 + \omega^2 \tau_p^2}. \tag{7.72}$$

As to the loss modulus G'', it is given by Eq. (6.74) as:

$$G''(\omega) = G_R \sum_{p=1}^{\infty} \frac{\omega \tau_p}{1 + \omega^2 \tau_p^2}. \tag{7.73}$$

We recall that $G_R = k_B T c/N$ and $\tau_p = \zeta N^2 b^2/(6\pi^2 k_B T p^2)$.

Note that the *terminal relaxation time* (the longest time of the Maxwell distribution) is here given by $\tau_1 = \tau_R/2$.

It is also interesting to study the limiting behaviour of these formulas at low and high frequencies.

In the first case ('low' frequencies), one has:

$$G'(\omega < 1/\tau_1) \approx G_R(\omega\tau_1)^2 \sum_{p=1}^{\infty} \frac{1}{p^4} = \frac{\pi^4}{90} G_R(\omega\tau_1)^2 \tag{7.74}$$

and

$$G''(\omega < 1/\tau_1) \approx G_R\omega\tau_1 \sum_{p=1}^{\infty} \frac{1}{p^2} = \frac{\pi^2}{6} G_R\omega\tau_1. \tag{7.75}$$

As expected, $G' \propto \omega^2$ and $G'' \propto \omega$ in the *terminal zone*, in agreement with the experiments.

In the second case ('high' frequencies), the discrete sums can be replaced by integrals, resulting in:

$$G'(\omega > 1/\tau_1) \approx G_R \int_0^{\infty} \mathrm{d}p \frac{\omega^2\tau_1^2}{p^4 + \omega^2\tau_1^2} = \frac{\pi}{4} G_R\sqrt{\omega\tau_1} \tag{7.76}$$

and

$$G''(\omega > 1/\tau_1) \approx G_R \int_0^{\infty} \mathrm{d}p \frac{p^2\omega\tau_1}{p^4 + \omega^2\tau_1^2} = \frac{\pi}{4} G_R\sqrt{\omega\tau_1}. \tag{7.77}$$

Thus, the model predicts that, *at high frequency, the two moduli are equal and vary as the square root of the frequency* ($G' = G'' \propto \sqrt{\omega}$).

This important result is well borne out by the experimental data on polymer melts. It represents one of the reasons for the success of this model.

We emphasise that the 1/2 exponent is a direct consequence of the Gaussian chain approximation. Its experimental observation confirms, once again (see Section 7.1.4.4), that the *chain exhibits ideal behaviour in melts*, a highly non-trivial result.

To conclude this section dealing with the Rouse model, let us calculate the complete expression of the stress tensor for an arbitrary flow.

7.1.6.10 General expression of the stress tensor and upper-convected Maxwell equation

We have shown that the stress tensor $\sigma_{\alpha\beta}$ can be written as a discrete sum:

$$\sigma_{\alpha\beta} = \sum_{p=1}^{\infty} \sigma_{\alpha\beta}^p, \tag{7.78}$$

where each term $\sigma_{\alpha\beta}^p$ represents the stress associated with the Maxwell element corresponding to the vibration eigenmode $\vec{X}_p$ of the molecule:

$$\sigma_{\alpha\beta}^p = \frac{c}{N}\kappa_p \langle X_{p\alpha} X_{p\beta}\rangle. \tag{7.79}$$

To calculate $\sigma_{\alpha\beta}^p$, let us use the evolution equation (7.56). According to this equation:

$$\frac{D}{Dt}X_{p\alpha} = \frac{\partial v_\alpha}{\partial x_\mu}X_{p\mu}(t) - \frac{\kappa_p}{\zeta_p}X_{p\alpha} + \frac{1}{\zeta_p}f_{p\alpha}(t), \tag{7.80}$$

$$\frac{D}{Dt}X_{p\beta} = \frac{\partial v_\beta}{\partial x_\mu}X_{p\mu}(t) - \frac{\kappa_p}{\zeta_p}X_{p\beta} + \frac{1}{\zeta_p}f_{p\beta}(t), \tag{7.81}$$

hence, by combining the two formulas:

$$\frac{D}{Dt}\langle X_{p\alpha}X_{p\beta}\rangle = \frac{\partial v_\alpha}{\partial x_\mu}\langle X_{p\mu}X_{p\beta}\rangle + \frac{\partial v_\beta}{\partial x_\mu}\langle X_{p\mu}X_{p\alpha}\rangle$$

$$- \frac{2\kappa_p}{\zeta_p}\langle X_{p\alpha}X_{p\beta}\rangle + \frac{1}{\zeta_p}\left(\langle f_{p\alpha}X_{p\beta}\rangle + \langle f_{p\beta}X_{p\alpha}\rangle\right). \tag{7.82}$$

In this equation, each term can be expressed as a function of the stress tensor, except the last two, which need further treatment. To do this, the trick consists in expanding the evolution equation (7.56) in the short time limit. Explicitly, this gives:

$$X_{p\alpha}(t) = X_{p\alpha}(t - \Delta t) + \Delta t \left(-\frac{\kappa_p}{\zeta_p}X_{p\alpha}(t - \Delta t) + \frac{\partial v_\alpha}{\partial x_\mu}(t - \Delta t)X_{p\mu}(t - \Delta t)\right)$$

$$+ \int_{t-\Delta t}^{t} dt' \frac{f_{p\alpha}(t')}{\zeta_p}. \tag{7.83}$$

It immediately follows that:

$$\langle X_{p\alpha}(t)f_{p\beta}(t)\rangle = \langle X_{p\alpha}(t - \Delta t)f_{p\beta}(t)\rangle + \Delta t(\ldots) + \int_{t-\Delta t}^{t} dt' \frac{\langle f_{p\alpha}(t')f_{p\beta}(t)\rangle}{\zeta_p}. \tag{7.84}$$

Noting that the first term on the right-hand side is zero (the time arguments are different) one has, after letting Δt tend towards 0 and using Eq. (7.42):

$$\langle X_{p\alpha}(t)f_{p\beta}(t)\rangle = \int_{t-\Delta t}^{t} 2\delta_{\alpha\beta}k_B T\delta(t - t')dt' = \delta_{\alpha\beta}k_B T. \tag{7.85}$$

Using this result and Eq. (7.79), we can finally rewrite Eq. (7.82) in the form:

$$\frac{D}{Dt}\sigma_{\alpha\beta}^p - \frac{\partial v_\alpha}{\partial x_\mu}\sigma_{\mu\beta}^p - \frac{\partial v_\beta}{\partial x_\mu}\sigma_{\mu\alpha}^p + \frac{1}{\tau_p}\sigma_{\alpha\beta}^p - \frac{1}{\tau_p}\frac{c}{N}k_B T\delta_{\alpha\beta} = 0. \tag{7.86}$$

One recognizes here the upper-convected time derivative $\overset{\triangledown}{}$ defined in the previous chapter (see Eq. (6.262)). Using this notation, the equation above can be written in the contracted form:

$$\overset{\triangledown}{\sigma}{}^P_{\alpha\beta} + \frac{1}{\tau_p}\sigma^P_{\alpha\beta} - \frac{1}{\tau_p}\frac{c}{N}k_BT\delta_{\alpha\beta} = 0. \tag{7.87}$$

Thus written, the equation seems different from the upper-convected Maxwell equation given in the previous chapter (Eq. (6.261)). To solve this little mystery, one must recall that, in an incompressible fluid, the stress tensor is defined up to a pressure term. One can thus redefine the stress tensor, using:

$$\sigma^P_{\alpha\beta} \equiv \sigma^P_{\alpha\beta} - \frac{c}{N}k_BT\delta_{\alpha\beta} = \sigma^P_{\alpha\beta} - G_R\delta_{\alpha\beta}. \tag{7.88}$$

With this new definition for the stress, it is easily checked that Eq. (7.87) can be rewritten in the usual form:

$$\tau_p\overset{\triangledown}{\underline{\sigma}}{}^P + \underline{\sigma}^P = 2\eta_p\underline{A}, \tag{7.89}$$

where $\underline{A}$ is the strain rate tensor and $\eta_p = G_R\tau_p$ is the viscosity associated with the Maxwell element corresponding to the pth mode (see Eq. (7.69)).

We can see thus that the microscopic Rouse model backs up the results announced without proof in the previous chapter, namely that the *contravariant form of the convected derivative* must indeed be used in polymer melts.

Consequently, the Rouse model yields not only the linear viscoelasticity functions G' and G'' of the polymer, but also, using Eqs. (7.89), its shear viscosity η, as well as its normal stress differences N_1 and N_2 and its elongational viscosity η_e. To this end, one only needs to use the results already demonstrated (see Section 6.4.5). It is then easily checked that the viscosity η remains equal to the viscosity η_0 calculated in the limit of vanishing shear rate, itself given by Eq. (7.62) as long as $\dot\gamma < 1/\tau_1$. As to the normal stress differences, in the framework of this model they are given by Eqs. (6.294) and (6.295), where the sum is over all contributions from the various Rouse modes. This yields:

$$N_1 = \sum_{p=1}^{\infty} 2\frac{\eta_p^2}{G_R}\dot\gamma^2 \tag{7.90}$$

and

$$N_2 = 0. \tag{7.91}$$

Using Eqs. (7.68) and (7.69), the first normal stress difference becomes, explicitly:

$$N_1 = \frac{N}{k_BTc}\frac{b^4\zeta^2c^2N^2}{18\pi^4}\left(\sum_{p=1}^{\infty}\frac{1}{p^4}\right)\dot\gamma^2 = \frac{1}{1620}\frac{b^4\zeta^2cN^3}{k_BT}\dot\gamma^2. \tag{7.92}$$

This calculation shows that, in the linear deformation regime of the polymer coils, N_1 is proportional to $\dot{\gamma}^2$ and to N^3. Since the monomer concentration per unit volume c is almost independent of N in a polymer melt, it follows that N_1 varies as M^3. This is a result already announced qualitatively in the discussion of Section 7.1.3.4, namely that the longer the polymer, the more important the normal stresses. Remember however that the formula (7.92) only applies for $M < M_e$.

In a similar way, one can calculate the elongational viscosity η_e from Eq. (6.296). Note that this viscosity diverges as the elongation rate $\dot{\varepsilon}$ tends towards $1/2\tau_1$. This problem is obviously related to the – incorrect – assumption of infinite molecular extensibility.

In conclusion, the Rouse model provides a correct description (only in the *linear regime*, however) for the behaviour of polymers with a molecular mass below the entanglement mass M_e.

On the other hand, it cannot explain their shear-thinning behaviour at high shear rates (for $\dot{\gamma} > 1/\tau_1$), nor does it predict the saturation of their elongational viscosity at high elongation rates ($\dot{\varepsilon} > 1/2\tau_1$). To this end, one must take into account the local nonlinearities. Two models of this type (the Giesekus and the FENE models), already mentioned in the previous chapter, will be presented schematically in Appendix 7.B. For simplicity's sake, these models will be derived from the Hookean dumbbell model, which is a single relaxation time (τ_1) version of the Rouse model (see Appendix 7.A).

More important at this point is that the Rouse model does not explain the presence, in linear viscoelasticity, of a rubber plateau in high molecular mass polymers (for molecular mass values above M_e). In the next section, we will show that the *entanglements* and the *reptation* of the chains must be taken into account to explain this phenomenon.

7.1.7 Doi–Edwards model for entangled polymers

The goal of this section is to explain the appearance of a rubber plateau in high molecular mass polymers ($M > M_e$). This plateau is observed over a frequency range ω_{min}, ω_{max} which is the more extended, the higher the molecular mass of the polymer. In these materials, ω_{max} is almost independent of M, while ω_{min} typically decreases as $1/M^3$. To explain these observations, one must invoke new concepts; we begin with their qualitative description in the next section.

7.1.7.1 Tubes and entanglements

When the polymers are very long, the chains entangle. The motion of each chain is then restricted by a forest of topological constraints due to its neighbours. An easy way of visualising its motion in this case consists in assuming that it is confined within a *tube* (Edwards, 1967; [31, 32]) whose walls are roughly defined by the *entanglement points* with the neighbouring chains. A schematic representation of this tube is given in Fig. 7.14.

To find the typical dimensions of the tube, let us denote by N_e the average number of monomers (of Kuhn segments, more precisely) between two entanglement points. This number (which is of the order of 100, or even more) is proportional to the entanglement mass M_e, defined in the experimental section. If the deformations are small (linear

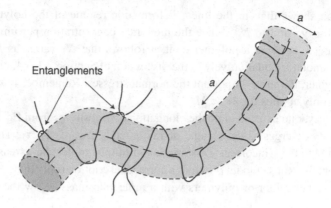

Fig. 7.14 Tube and an entangled polymer.

regime), each chain remains locally in thermodynamic equilibrium and its statistics remain Gaussian at all scales (large compared to the monomer size). In particular, this is the case for any chain segment between two entanglements, whose average spacing 'a' sets the tube diameter:

$$a = \sqrt{N_e}b. \tag{7.93}$$

As to the length L of the tube, it remains on the average constant over time, given by:

$$L = \frac{N}{N_e}a = N\frac{b^2}{a}. \tag{7.94}$$

We will start by showing that these two scales fix the limits of the rubber plateau.

To this end, let us imagine that the material is suddenly deformed by an amount γ_0 and let us determine the stress relaxation with time (at an imposed deformation). This type of experiment, as we well know, provides directly the relaxation modulus $G(t)$ that can be used to yield G' and G'' (since $\sigma(t) = \gamma_0 G(t)$, see Section 6.2.1.5 and Fig. 6.5). At the initial moment, the tube and the chain it contains are slightly deformed (linear regime). The question is, how does this (stress-transmitting) chain relax the deformation with time? The Doi–Edwards model provides an answer. It distinguishes between short- and long-time behaviour of the chain. In the following, we will describe these two regimes, and then show how they can be patched together. This procedure will allow us to define the limits of the rubber plateau.

7.1.7.2 Short-time relaxation modulus $G(t)$: application to the Rouse model

The first relaxation modes that come into play after the sudden application of the deformation γ_0 are the fastest Rouse modes, characterised by short wavelengths and by high ranks p. These modes are not sensitive to the entanglements as long as their wavelength Nb/p is smaller than the distance $N_e b$ between entanglements along the chain. This condition fixes the order of the longest Rouse mode contributing to stress relaxation in this regime

$$p_{\min} = \frac{N}{N_e}. \tag{7.95}$$

Associated with this order is the longest Rouse relaxation time of a chain segment of length $N_e b$. From Eq. (7.65), this time is

$$\tau_e = \frac{\zeta N_e^2 b^2}{6\pi^2 k_B T} = \left(\frac{N_e}{N}\right)^2 \frac{\tau_R}{2}. \tag{7.96}$$

This time, proportional to M_e^2 and independent of the molecular mass, determines the duration of the initial transient regime in a stress relaxation experiment. We can thus write that

$$G(t < \tau_e) = G_{\text{Rouse}}(t). \tag{7.97}$$

For a polymer of high molecular weight $(M \gg M_e)$, $\tau_e \ll \tau_R$, such that the asymptotic formula (7.71) yields

$$G(t < \tau_e) = k_B T \frac{c}{2\sqrt{2}N} \left(\frac{\tau_R}{t}\right)^{1/2}. \tag{7.98}$$

At long times $(t > \tau_e)$, *this model no longer applies, due to the entanglements.* Another mechanism comes into play, and we will discuss it below.

7.1.7.3 Long-time relaxation modulus $G(t)$: reptation of the primitive chain

To explain the stress relaxation at long times, the theorists imagined an additional mechanism: that of *reptation of the primitive chain.*

To understand the concepts involved, let us start by defining the term *primitive chain* (Edwards, 1977; [31, 32]). As we have already seen, the chain is confined within the tube formed by its neighbours. By definition, the *primitive chain corresponds at each moment to the tube axis.* Its length is therefore constant in time, and equal to the tube length L given by Eq. (7.94). One can also define the primitive chain as the real chain 'smoothed' over the distance 'a' between entanglements.

The second concept that becomes relevant at long times is that of *reptation,* a term coined by de Gennes in 1971 (de Gennes, 1971; [31, 32]). The main idea is that, beyond the time τ_e, the chain will have relaxed the stress over the distance 'a' between entanglements, so that it can be replaced by its primitive chain. On the other hand, the primitive chain still transmits the stress related to the deformation of the tube that contained it initially. In the linear regime, this tube (the initial tube, as we will call it) is only weakly deformed, so that its geometrical characteristics are essentially the same immediately before and after the deformation (namely, at the moments $t = 0^-$ and $t = 0^+$).

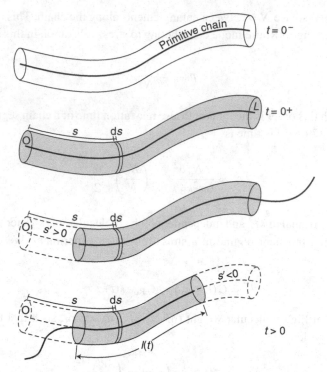

Fig. 7.15 Reptation of the primitive chain with time. At the moment $t = 0^-$, the tube and the primitive chain it contains are not deformed; at time $t = 0^+$, the tube and the primitive chain are slightly deformed; the macroscopic deformation is maintained at subsequent times. The primitive chain returns to an equilibrium configuration by sliding back and forth within its tube. The last two sketches show that, each time one of the ends of the primitive chain moves out of the initial tube (shown in grey), the tube shortens (here, the part of the initial tube drawn in dotted lines represents the fraction having disappeared at an instant t). After a time t, only the remaining fraction of the initial tube, with length $l(t)$, contributes to the stress.

The main feature of the reptation model is that the primitive chain 'slithers' like a snake along its tube due to the Brownian motion of its centre of mass (Fig. 7.15). Due to this motion, the extremities of the primitive chain progressively leave the initial tube and enter new tubes, which are isotropically distributed and, therefore, do not contribute to the stress. Thus, the length of the initial tube (the only one to undergo deformation) decreases progressively and ends by vanishing over very long times: the stress will have completely relaxed.

Before writing down the equation for this back and forth Brownian motion of the chain along its initial tube, let us estimate the typical time that the primitive chain takes to leave it completely. Since we are talking about one-dimensional diffusion over a length L, this time is given by:

$$\tau_{\text{rep}} = \frac{L^2}{\pi^2 D_{\text{c}}},$$

(7.99)

where D_c is the *curvilinear diffusion coefficient* of the chain along the tube and π^2 a numerical factor introduced here without proof, which will be given in the next section. The diffusion coefficient is given by the Einstein relation, using $N\zeta$ for the friction coefficient of the complete chain, where ζ is the friction coefficient for a monomer:

$$D_c = \frac{k_B T}{N\zeta}. \tag{7.100}$$

Using this equation and Eqs. (7.94) and (7.99) yields:

$$\tau_{rep} = \frac{1}{\pi^2} \frac{N^2 b^4}{a^2} \frac{N\zeta}{k_B T} = \frac{1}{\pi^2 N_e} \frac{b^2 \zeta}{k_B T} N^3. \tag{7.101}$$

This time can be much longer than τ_e for 'high' polymers, since it increases as M^3, while τ_e is independent of M. It is precisely in this kind of materials that one encounters a rubber plateau. Before discussing this point, let us analyse in more detail the reptation motion and calculate the associated relaxation spectrum.

7.1.7.4 Relaxation spectrum associated with reptation: calculating $G(t)$

We will follow step-by-step the demonstration given by Doi and Edwards [32]. The first step consists of calculating the average length $\langle l(t) \rangle$ of the remaining fraction of the initial tube at time t. To do this, consider a portion of this tube, of curvilinear abscissa s measured from one end of the tube and of length ds (Fig. 7.15). This tube segment 'disappears' when touched by either one of the two extremities of the primitive chain. Let $\psi(s, t)$ be the probability that this tube segment still exists at time t. One then has:

$$\langle l(t) \rangle = \int_0^L ds\, \psi(s, t). \tag{7.102}$$

To calculate $\psi(s, t)$, let us introduce the probability $\Psi(s', t, s)$ that the primitive chain has moved by s' at time t, knowing that at this moment neither of its ends has yet reached the segment of abscissa s. This probability is a solution of the diffusion equation:

$$\frac{\partial \Psi}{\partial t} = D_c \frac{\partial^2 \Psi}{\partial s'^2}, \tag{7.103}$$

where D_c is the curvilinear diffusion coefficient defined above (Eq. (7.100)). The solution must obey the boundary condition:

$$\Psi(s', 0, s) = \delta(s'), \tag{7.104}$$

which states simply that at the instant $t = 0$ the chain has not yet moved. On the other hand, we know that the tube segment with abscissa s disappears when reached by either end of the primitive chain. This occurs for $s' = s > 0$ or $s' = s - L < 0$ (Fig. 7.15), which amounts mathematically to the following additional boundary conditions:

$$\Psi(s', t, s) = 0 \quad \text{for} \quad s' = s \quad \text{or} \quad s' = s - L. \tag{7.105}$$

It is easily checked that the solution to this problem can be expressed in the form:

$$\Psi(s', t, s) = \sum_{p=1}^{\infty} \frac{2}{L} \sin\left(\frac{p\pi s}{L}\right) \sin\left(\frac{p\pi(s - s')}{L}\right) \exp\left(-\frac{p^2 t}{\tau_{\text{rep}}}\right), \tag{7.106}$$

where τ_{rep} is defined by Eq. (7.101). This formula then yields the probability $\psi(s, t)$:

$$\psi(s, t) = \int_{s-L}^{s} ds'\, \Psi(s', t, s) = \sum_{p \text{ odd}} \frac{4}{p\pi} \sin\left(\frac{p\pi s}{L}\right) \exp\left(-\frac{p^2 t}{\tau_{\text{rep}}}\right) \tag{7.107}$$

and then the remaining length of the initial tube at time t:

$$\langle l(t) \rangle = \int_{0}^{L} ds\, \psi(s, t) = L \sum_{p \text{ odd}} \frac{8}{\pi^2 p^2} \exp\left(-\frac{p^2 t}{\tau_{\text{rep}}}\right). \tag{7.108}$$

The second step of the calculation consists of assuming that the remaining stress, and hence the relaxation modulus $G(t)$, is proportional to $\langle l(t) \rangle$. Setting

$$\psi(t) = \sum_{p \text{ odd}} \frac{8}{\pi^2 p^2} \exp\left(-\frac{p^2 t}{\tau_{\text{rep}}}\right) \tag{7.109}$$

gives:

$$G(t) = G_N^0 \psi(t) = G_N^0 \sum_{p \text{ odd}} \frac{8}{\pi^2 p^2} \exp\left(-\frac{p^2 t}{\tau_{\text{rep}}}\right), \tag{7.110}$$

where G_N^0 is a constant (as yet unknown).

This calculation shows that, over times much longer than τ_e, the material behaves as a generalised Maxwell fluid. Its relaxation time spectrum is still discrete, given this time by:

$$\tau_p^{\text{rep}} = \frac{\tau_{\text{rep}}}{p^2} = \frac{1}{N_e} \frac{b^2 \zeta}{k_B T p^2} N^3 \qquad (p \text{ odd}). \tag{7.111}$$

Each Maxwell element has an associated elastic modulus:

$$G_p^{\text{rep}} = \frac{8}{\pi^2 p^2} G_N^0 \qquad (p \text{ odd}). \tag{7.112}$$

The elastic modulus G_N^0 remains to be calculated. To this end, we will patch together the Rouse and reptation regimes.

7.1.7.5 Crossover between the two regimes and width of the rubber plateau as a function of the molecular mass

The two regimes merge at times $t = \tau_e$. For shorter times, the relaxation modulus is given by the asymptotic formula (7.71), since $\tau_e \ll \tau_R$:

$$G(t < \tau_e) = k_B T \frac{c}{2\sqrt{2}N} \left(\frac{\tau_R}{t}\right)^{1/2}. \qquad (7.113)$$

At longer times:

$$G(t > \tau_e) = G_N^0 \psi(t), \qquad (7.114)$$

where the $\psi(t)$ function is given by Eq. (7.109). On the other hand, $\tau_e \ll \tau_{rep}$ for 'high' polymers. As a consequence, $\psi(\tau_e) \approx \psi(0) = 1$. It follows that:

$$G_N^0 = k_B T \frac{c}{2\sqrt{2}N} \left(\frac{\tau_R}{\tau_e}\right)^{1/2} \qquad (7.115)$$

and, from Eq. (7.96):

$$G_N^0 = \frac{k_B T}{2} \frac{c}{N_e}. \qquad (7.116)$$

This formula can be rewritten in terms of the *number of entanglements per unit volume* $\mathcal{N}_e$. Since it takes two molecules to form an entanglement, $\mathcal{N}_e = c/(2N_e)$, and we retrieve a formula already given in Chapter 1 (Eq. (1.25)):

$$G_N^0 = \mathcal{N}_e k_B T. \qquad (7.117)$$

To summarise this discussion, we plot in Fig. 7.16 the shape of the relaxation modulus $G(t)$ in entangled polymers (in logarithmic scale), when the characteristic times τ_e and τ_{rep} are clearly separated. We introduced the following dimensionless parameters:

$$\tilde{G} = \frac{G}{G_N^0}, \qquad \tilde{t} = \frac{t}{\tau_e}, \qquad \tilde{r} = \frac{\tau_{rep}}{\tau_e} = 6 \left(\frac{M}{M_e}\right)^3. \qquad (7.118)$$

With these notations, in the Rouse regime:

$$\tilde{G}\left(\tilde{t} \le 1\right) = \tilde{t}^{-1/2}, \qquad (7.119)$$

while in the reptation-dominated regime:

$$\tilde{G}\left(\tilde{t} \ge 1\right) = \sum_{p \text{ odd}} \frac{8}{\pi^2 p^2} \exp\left(-p^2 \tilde{t}/\tilde{r}\right). \qquad (7.120)$$

These curves are in good agreement with the experimental results. In practice, it is easier to measure the G' and G'' moduli. These can be immediately calculated from the function $G(t)$. We give in the next section the expression of G', as well as the theoretical value of η_0.

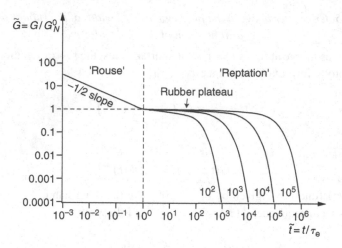

Fig. 7.16 Shape of the relaxation modulus in entangled polymers of high molecular mass. The width of the rubber plateau increases as the cube of the molecular mass. On the other hand, the plateau modulus is independent of the molecular mass. We indicated alongside each curve the value of the ratio τ_{rep}/τ_e (corresponding to $M/M_e = 2.5, 5.5, 12,$ and 26).

7.1.7.6 Viscosity η_0 and elastic modulus G'

The viscosity under continuous shear is given by the formula (6.49):

$$\eta_0 = \int_0^\infty G(t)\, dt. \tag{7.121}$$

This formula applies as long as the system remains in the linear regime, which implies that $\dot{\gamma} < 1/\tau_{rep}$. In this case, the contribution of the Rouse modes is negligible, so that:

$$\eta_0 = G_N^0 \int_0^\infty \psi(t)\, dt = \frac{\pi}{12} G_N^0 \tau_{rep} = \frac{1}{24} cb^2 \zeta \frac{N^3}{N_e^2}. \tag{7.122}$$

This formula predicts that the viscosity increases as M^3. This exponent is somewhat lower than that observed experimentally (namely 3.3). Refinements of the model (which are beyond the scope of this book), have reduced this discrepancy slightly [32].

In the same way, one can calculate the elastic modulus G' starting from Eq. (6.70). Using the expressions (7.119) and (7.120) for $G(t)$ yields:

$$G'(\omega\tau_e \geq 1) = G_N^0 \left(\frac{\pi}{2}\omega\tau_e\right)^{1/2}, \tag{7.123a}$$

$$G'(\omega\tau_e \leq 1) = G_N^0 \sum_{p \text{ odd}} \frac{8}{\pi^2 p^2} \frac{(\omega\tau_{rep}/p^2)^2}{1 + (\omega\tau_{rep}/p^2)^2}. \tag{7.123b}$$

This function is plotted in Fig. 7.17 (where we set $\tilde{G}' = G'/G_N^0$ and $\tilde{\omega} = \omega\tau_e$) for the same values of the ratio $\tilde{r} = \tau_{rep}/\tau_e$ as in Fig. 7.16. The resulting curves have the same shape as the experimental data (see Fig. 7.7), validating the proposed theory.

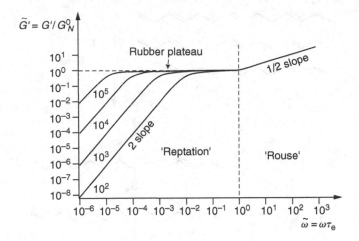

Fig. 7.17 Shape of the elastic modulus $G'(\omega)$ for different values of the ratio τ_{rep}/τ_e (indicated alongside each curve).

In the next section, we will consider the topic of elastomers.

7.2 Example of an elastomer from the family of viscoelastic solids: vulcanised rubber

We have already spoken of natural rubber in Chapter 1. This material is obtained from latex, which is the sap of the hevea tree. Latex is composed of polyisoprene, a linear polymer with the chemical formula $\{CH_3-CH=CH-CH_3\}$. After reacting with sulphur, it *reticulates* (Fig. 7.18). All the macromolecules bond to each other, finally forming a single huge molecule one can hold in the hand. This material belongs to the family of *elastomers*, a concept to be discussed in the next section.

7.2.1 Definition of an elastomer

An elastomer is, by definition, a soft solid that can deform reversibly (i.e. without creeping) by at least 10%. In practice, the deformation can even reach several hundred per cent. This is the case for rubber bands that postmen use every day to hold together stacks of envelopes.

Elastomers are obtained by linking flexible polymer chains, with a large number of conformations, which implies a short persistence length. In the case of natural rubber, the reticulation points are *chemical in nature* (sulphur bridges, more specifically). However, this is not the only possibility: in silicones, for instance, the chains are attached to each other by absorbtion onto small beads of amorphous silica. The same principle applies to hydrocarbon chains, where carbon black particles are used. Block copolymers can also be used as, for instance, in certain *thermoplastic elastomers*, where the chain is composed of at least two blocks, one of them a crystallisable homopolymer, with a melting temperature

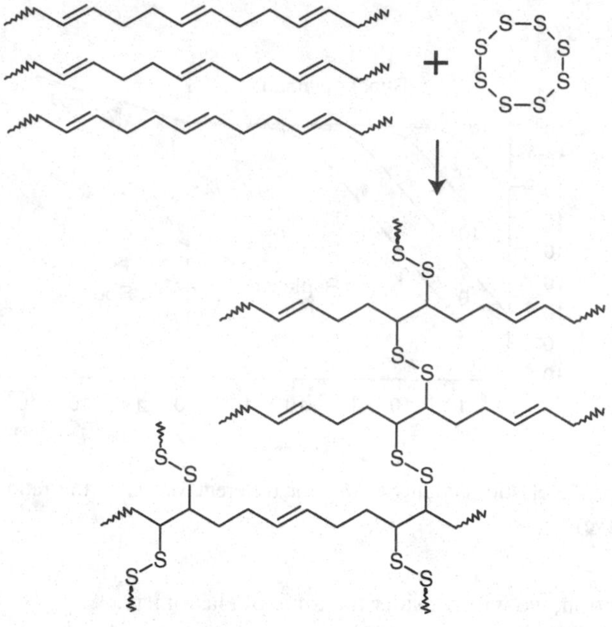

Fig. 7.18 Latex reticulation by sulphur: natural rubber is obtained.

T_m above the operating temperature, and the other a flexible chain of a different homopolymer, with a glass transition temperature T_g below the operating temperature. In this case, the crystallised zones connect the chains between them. The great advantage of these materials over classical elastomers such as natural rubber is that they can be processed at high temperature, since they melt above T_m (hence the term thermoplastic elastomer). In this case, as for silicone or hydrocarbon chains, the reticulation points are of a *physical nature*.

The specific nature of the reticulation points is of little consequence: what matters is that they are resilient enough to withstand prolonged elongation (or shear). If this condition is fulfilled, the elastomer behaves as a *viscoelastic solid*. Its main feature is then the ability to withstand huge deformations *elastically*, meaning that the deformation disappears completely after stress removal. Such exceptional elasticity is of *entropic* nature, as we have already emphasised in Chapter 1. We develop this point in the next section.

7.2.2 *The Young and shear moduli of an elastomer*

To calculate the Young modulus E of an elastomer, the material should be subjected to an elongational deformation and the variation in its free energy F determined.[7]

---

[7] We emphasise that, rigorously speaking, the theory developed in this section only applies to 'hard' – strongly reticulated – elastomers, where the topological entanglements of the chains between two reticulation points can be neglected. The case of 'soft' elastomers (making up the most common elastic materials) is more complex, since the chains are very entangled between two reticulation points, which contributes to their elastic response, even at zero frequency. This opposite limit was first studied by S. F. Edwards (Edwards, 1977).

Before explicitly performing this calculation, one should realise that the entropy of the chains decreases under elongation, as the number of accessible conformations decreases. On the other hand, their internal energy changes very little if they are long and flexible enough, since their length can be changed easily while preserving the length of the bonds between monomers. In these conditions, the variation in free energy is proportional to the variation in entropy since:

$$\Delta F = \Delta(U - TS) \cong -T\,\Delta S. \tag{7.124}$$

To calculate ΔS, we consider the set of end-to-end vectors $\vec{R}$ connecting two successive reticulation points on the same chain.

At rest, their distribution $\{\vec{R}_0\}$ is isotropic ($\langle \vec{R}_0 \rangle = 0$) with, from Eq. (7.11):

$$\overline{R}_0 = \sqrt{N_c} b, \tag{7.125}$$

where N_c is the average number of monomers (rigorously speaking, of statistical Kuhn segments) between two successive knots along a chain.

After elongation, each vector $\vec{R}_0$ is transformed into a vector $\vec{R}$ given by Eq. (6.241) as:

$$\vec{R} = \begin{pmatrix} \lambda_1 & 0 & 0 \\ 0 & \lambda_2 & 0 \\ 0 & 0 & \lambda_3 \end{pmatrix} \vec{R}_0, \tag{7.126}$$

where λ_i is the elongation in the $\vec{x}_i$ direction. We infer that:

$$\overline{R}^2 = \left(\lambda_1 \overline{R}_{01}\right)^2 + \left(\lambda_2 \overline{R}_{02}\right)^2 + \left(\lambda_3 \overline{R}_{03}\right)^2 = \frac{\overline{R}_0^2}{3}\left(\lambda_1^2 + \lambda_2^2 + \lambda_3^2\right) \tag{7.127}$$

since at rest:

$$\left(\overline{R}_{01}\right)^2 = \left(\overline{R}_{02}\right)^2 = \left(\overline{R}_{03}\right)^2 = \frac{\overline{R}_0^2}{3}. \tag{7.128}$$

The probability that the end-to-end vector between two knots takes the value $\vec{R}$ is given (in the Gaussian approximation) by the distribution function (7.15):

$$\Phi\left(\vec{R}, N_c\right) = \Phi_0 \exp\left(\frac{-3R^2}{2\overline{R}_0^2}\right). \tag{7.129}$$

Since the Boltzmann entropy of a chain is:

$$S_{\text{chain}} = k_B \ln(\Phi), \tag{7.130}$$

it follows that the variation in elastomer entropy after deformation is:

$$\Delta S = \frac{-3k_B}{2} \frac{N}{N_c} \frac{\overline{R}^2 - \overline{R}_0^2}{\overline{R}_0^2}, \tag{7.131}$$

where N is the number of monomers in the sample. Note that, by multiplying here with N/N_c, we assumed implicitly that all chains are reticulated. Using Eqs. (7.124), (7.127) and (7.131), one finally obtains the free energy variation per unit volume:

$$\Delta F = \frac{k_B T}{2} \frac{N}{N_c} \left(\lambda_1^2 + \lambda_2^2 + \lambda_3^2 - 3 \right). \tag{7.132}$$

This formula simplifies for simple elongation, along direction 3, for instance. In this case $\lambda_3 = \lambda$ and $\lambda_1 = \lambda_2 = \lambda^{-1/2}$, assuming that the material is incompressible ($\lambda_1 \lambda_2 \lambda_3 = 1$). Substituting in the equation above yields:

$$\Delta F = \frac{k_B T}{2} \frac{N}{N_c} \left(\lambda^2 + 2\lambda^{-1} - 3 \right). \tag{7.133}$$

This formula gives access to the stress that must be applied to the sample in order to deform it. Indeed, let us denote by L_0 the initial length of the sample along direction 3 and by S_0 its initial cross-section in the perpendicular plane. During isothermal elongation, the free energy variation for the sample equals the work of the force f applied at its free end (the other end is assumed fixed):

$$\Delta F = \int_{L_0}^{\lambda L_0} f(x) \, dx. \tag{7.134}$$

It follows that the force f one must apply to the sample along the direction 3 in order to stretch it by $(\lambda - 1)L_0$ is:

$$f = \left(\frac{\partial \Delta F}{\partial \lambda L_0} \right)_T = \frac{k_B T}{L_0} \frac{N}{N_c} \left(\lambda - \lambda^{-2} \right). \tag{7.135}$$

The stress is obtained by dividing this force by the cross-section area after deformation S_0/λ:

$$\sigma = \frac{k_B T}{S_0 L_0} \frac{N}{N_c} \left(\lambda^2 - \lambda^{-1} \right). \tag{7.136}$$

Denoting by $c = N/(S_0 L_0)$ the monomer concentration per unit volume, this formula becomes:

$$\sigma = \frac{c k_B T}{N_c} \left(\lambda^2 - \lambda^{-1} \right). \tag{7.137}$$

It can also be expressed as a function of the density ρ and the average molecular mass M_c between reticulation points:

$$\sigma = \frac{\rho R T}{M_c} \left(\lambda^2 - \lambda^{-1} \right), \tag{7.138}$$

with R the perfect gas constant.

Note that this formula is only valid if the chains exhibit ideal Gaussian behaviour, so it only applies to moderate deformations. In particular, in the very small deformation limit

ε ($|\varepsilon| = |\lambda - 1| \ll 1$), it becomes $\sigma = E\varepsilon$, where E is the Young modulus. Comparison with Eq. (7.138) yields:

$$E = \frac{3RT\rho}{M_c}.\tag{7.139}$$

Using the Trouton rule (see Eq. (6.110)), one can also obtain the shear modulus of the elastomer (at low deformation):

$$G = \frac{RT\rho}{M_c}.\tag{7.140}$$

In conclusion, elastomers are soft solids, since their elasticity is of purely entropic origin. Their elastic modulus decreases as the reciprocal of the average distance between reticulation points. Thus, the more reticulated the elastomer, the stiffer it is.

As an order of magnitude, the elastic modulus of an elastomer is 1000 to 10 million times lower than that of a metal.

We emphasise that these materials are also *viscoelastic*, since under deformation the Rouse modes of the chain segments between reticulation points become excited. The longest relaxation time τ_R associated with these modes is given by formula (7.45), where N is replaced by N_c. Thus, the elastomer dissipates little energy as long as it is stressed at frequencies below $1/\tau_R$. In the opposite case, the dissipative effects start to become important.

In the next section, we return to the case of viscoelastic liquids and discuss in detail the case of polymers in semi-dilute solution.

7.3 Polymers in semi-dilute solution

When checking the ingredients of a shampoo or dish detergent, a polymer is almost always included. Most often, it is added at a very low concentration, in order to render the liquid more viscous: it therefore plays the role of *thickener*. Polymers are similarly encountered in food products such as ice cream: in this case, it is however better to use a bio-polymer, to facilitate its digestion.

In this section, we will try to highlight a few fundamental concepts dealing with the structure and rheology of polymer solutions. We will mostly be concerned with semi-dilute solutions and their universal features, referring the reader to Section 7.5.3 for a description of the rheological properties of dilute solutions.

7.3.1 General points on polymers in solution

We will begin by describing the structure of polymer coils in the dilute regime, when they do not interact, and then in the semi-dilute regime, when they are strongly entangled.

7.3.1.1 The concept of good and bad solvents

The physical properties of a polymer solution depend on three parameters: the solvent, the temperature and the polymer concentration.

For each polymer, there are two types of solvents [31–33]:

- the *good solvents*, presenting strongly attracting interaction energy with the polymer. In this case, the polymer segments tend to repel each other, since they prefer contact with the solvent molecules rather than among themselves. Good solvents can dissolve the polymer over a wide concentration and temperature range;
- the *bad solvents*, for which the interactions with the polymer are, on the contrary, repulsive. In this case, the polymer segments attract, since they prefer their own contact to that of the solvent molecules. The polymer coils then tend to collapse on themselves, or even adhere to each other: as the concentration increases, the polymer precipitates. Thus, bad solvents only dissolve small amounts of polymer, especially at low temperature.

To quantify mathematically the quality of the polymer–solvent interaction, the theorists introduced the concept of *excluded volume* v (Kuhn, 1934; Flory, 1949; [31–33]) This parameter, with volume units, is positive for a good solvent and negative for a bad solvent. In the next section we will give its definition, as well as that of Θ solvent, where v becomes zero.

7.3.1.2 Excluded volume and Θ solvent

In polymer melts, we assumed that there was no interaction between the Kuhn segments composing the chains. This assumption helped us show that the chains have Gaussian statistics at all scales, in agreement with experiment.

As the polymer is dissolved in a good solvent, the monomers (or, more precisely, the Kuhn segments) start to experience repulsive interactions, of the steric type. Let us denote by $\vec{R}_n$ and $\vec{R}_m$ the radius vectors of two segments with indices n and m along the chain. One usually writes the steric interaction energy between these two segments in the form $v k_B T \delta(\vec{R}_n - \vec{R}_m)$, where v is the *excluded volume parameter*. This parameter has dimensions of a volume and translates the competition between the attractive and repulsive interactions between the two monomers.

In polymer theory, it can be shown that, if the interaction potential u between two segments only depends on the distance separating them (Fig. 7.19), then:

$$v = \int d\vec{r} \left[1 - \exp\left(-\frac{u(r)}{k_B T}\right) \right]. \tag{7.141}$$

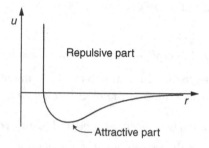

Fig. 7.19 Shape of the potential $u(r)$.

Under these conditions, there is a particular temperature, denoted by Θ, for which v vanishes. One can thus write the excluded volume in the form [33]:

$$v = v_0\left(1 - \frac{\Theta}{T}\right), \tag{7.142}$$

where $v_0 > 0$.

For $T > \Theta$, $v > 0$: one then speaks of a *good solvent*.

For $T < \Theta$, $v < 0$: one is then in the presence of a *bad solvent*. In this case, the polymer is weakly soluble and can precipitate.[8]

In the remainder of this chapter, we will always assume that the temperature is greater than or equal to Θ.

7.3.1.3 Non-Gaussian statistics of a chain in a good solvent

In sufficiently dilute solution, the chains are separated from each other and do not entangle. They form independent coils (swollen by the solvent) whose statistics in good solvent can be studied. One must then take into account the interaction energy between all the segments, with the general expression:

$$U = \frac{1}{2}v k_B T \int_0^N dn \int_0^N dm \, \delta\left(\vec{R}_n - \vec{R}_m\right), \tag{7.143}$$

where N is the total number of Kuhn segments along the chain. It follows that the distribution function of the $\vec{R}_n$ has the following form (see Eq. (7.17)):

$$\Psi\left(\{\vec{R}_n\}\right) = Cst \exp\left[-\frac{3}{2b^2}\int_0^N dn \left(\frac{\partial \vec{R}_n}{\partial n}\right)^2 \right.$$
$$\left. -\frac{1}{2}v \int_0^N dn \int_0^N dm \, \delta\left(\vec{R}_n - \vec{R}_m\right)\right]. \tag{7.144}$$

Unfortunately, this expression is too complicated to permit an exact calculation of the average radius $\overline{R}$ of the coil.

For this reason, we will employ an approximate theory, of the *mean field* type, due to P. J. Flory [33].

Let us begin with the formal definition of the local concentration of Kuhn segments:

$$c\left(\vec{r}\right) = \int_0^N dn \, \delta\left(\vec{r} - \vec{R}_n\right). \tag{7.145}$$

Using this definition, it follows that:

$$U = \int d\vec{r} \, \frac{1}{2}v k_B T c\left(\vec{r}\right)^2. \tag{7.146}$$

---

[8] Consequently, the same solvent can be 'good' or 'bad', depending on the temperature.

Let us now consider a coil with a fixed end-to-end vector, equal to $\vec{R}$. Its free energy contains two terms:

- an excluded volume term, given in the first approximation by:

$$U \approx v k_B T \tilde{c}^2 R^3. \tag{7.147}$$

This expression is derived from Eq. (7.146), where $c(\vec{r})$ was replaced by its average value inside the coil $\tilde{c} \approx N/R^3$;

- an entropic term, equal to $-k_B T \ln \Phi(\vec{R}, N)$, given by Eq. (7.15):

$$-TS \approx k_B T \frac{3R^2}{2Nb^2} + \text{(term independent of R)}. \tag{7.148}$$

Minimising the free energy $F = U - TS$ as a function of R leads to the following expression for the mean square value of the end-to-end vector $\vec{R}$ (Flory radius):

$$\overline{R}_d \approx v^{1/5} b^{2/5} N^{3/5}, \tag{7.149}$$

where subscript 'd' indicates a calculation performed in the dilute regime. One can thus see that *the chain statistics are no longer Gaussian in a good solvent*, the coil size varying as N^ν, with $\nu = 3/5$ instead of $\nu = 1/2$ for the ideal case. We should however note that, at the Θ temperature, the behaviour of the chain is again Gaussian since $\nu = 0$ at this temperature, by definition.

These results were verified experimentally by small-angle static light scattering (Cotton, 1980).

In the next section, we will describe what happens as the polymer concentration increases.

7.3.1.4 Crossover from the dilute to the semi-dilute regime

Let ρ be the polymer concentration, i.e. the mass of dissolved polymer per unit volume. Experiment shows that the rheological properties of the solution change rather sharply when the concentration exceeds a certain critical value ρ^*, depending on the size of the polymer. This spectacular effect will be exemplified later for the case of giant micelle solutions, which behave in exactly the same way (see Section 7.4.2). From a general point of view, the concentration ρ^* defines the crossover between the dilute regime, where the coils are independent (Fig. 7.20a), to the semi-dilute regime, where the coils entangle (see next section).

By definition, at the concentration ρ^* the coils just touch (Fig. 7.20b). In this case, the local polymer density is the same everywhere (on average), so that in the first approximation:

$$\frac{\rho^* N_A}{M} \overline{R}_d^3 \approx 1. \tag{7.150}$$

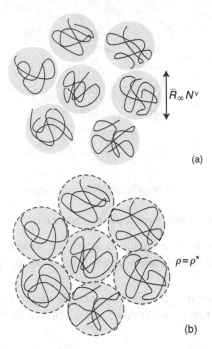

(a)

(b)

Fig. 7.20 (a) Polymer in the dilute regime, with $\nu = 1/2$ in Θ solvent and $\nu = 3/5$ in a good solvent; (b) polymer at the critical concentration ρ^*.

Since the average radius $\overline{R}_d$ varies as N^ν, and hence as M^ν, it follows that the overlap concentration ρ^* follows the scaling law:

$$\rho^* \propto M^{1-3\nu} = M^{-4/5} \qquad \text{(for } \nu = 3/5\text{)}. \tag{7.151}$$

This formula shows that ρ^* can take very small values if the molecular mass of the polymer is high. Thus, for $M \approx 10^6$ g/mol, $\rho^* \approx 0.005$ g/cm^3, which is about 0.5 wt% of polymer in the solution.

In the following, we will only be concerned with the rheological behaviour of semi-dilute solutions ($\rho > \rho^*$) (for the case of dilute solutions, see Section 7.5.3 at the end of the chapter). In this case, the chains interact strongly with each other and tend to entangle. Clearly, the rheology of semi-dilute solutions falls within the range of the Doi–Edwards reptation model. Before applying this model, it is however necessary to specify the structure of the polymer over the different scales. This is the goal of the next two sections.

7.3.1.5 *The correlation length* ξ

To define this fundamental length, let us follow the graphic reasoning of P.-G. de Gennes [31]. Imagine taking a snapshot of the polymer at a given time: the system will resemble a disordered network with a typical mesh size ξ (Fig. 7.21). This length sets the average distance (along the chain) between two contact points with two other chains. We will call it the *correlation length*.

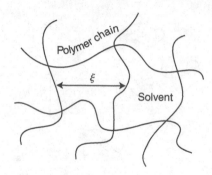

Fig. 7.21 Correlation length ξ in a semi-dilute polymer.

To calculate it, de Gennes used two simple arguments:

- the first one is that, for a concentration $\rho > \rho^*$, the structure of the network at the ξ scale must only depend on the concentration, and not on the degree of polymerisation N of the polymer;
- the second one is that, for $\rho \approx \rho^*$, the polymer coils are just touching each other, setting the value of the correlation length at this concentration ($\xi = \overline{R}_d$ at $\rho \approx \rho^*$, where $\overline{R}_d$ is given by Eq. (7.149)).

These two requirements lead to the following relation:

$$\xi(\rho) \approx \overline{R}_d \left(\frac{\rho^*}{\rho} \right)^{\nu_\xi}, \tag{7.152}$$

where the exponent ν_ξ remains to be determined. This is very easily done, since we know that $\overline{R}_d \propto N^{3/5}$ and $\rho^* \propto N^{-4/5}$ from Eqs. (7.149) and (7.151). In these conditions, ξ is independent of N for $\nu_\xi = 3/4$, finally yielding (in good solvent conditions):

$$\xi \approx v^{-1/4} b^{-1/2} \left(\frac{\rho}{m} \right)^{-3/4} \qquad (\rho^* \ll \rho \ll \rho^{**}), \tag{7.153}$$

where m stands for the mass of the Kuhn segment and ρ^{**} for the density of dissolved polymer above which the mixture behaves as a melt ($\rho^{**} \approx vm/N_A b^6$ according to Doi and Edwards [32]). This formula shows that, in a good solvent, the correlation length follows the scaling law:

$$\xi \propto \rho^{-3/4}. \tag{7.154}$$

7.3.1.6 The concept of 'blob'

We will now consider the structure of a chain in semi-dilute solution.

Following de Gennes, the chain can be seen as a sequence of elementary units of size ξ, which he named *'blobs'* [31]. Within each blob, the chain is in contact only with itself (from the very definition of the correlation length ξ). As a consequence, the chain statistics within the blob are those of a self-avoiding random walk, dominated by excluded volume

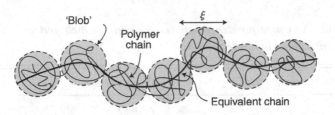

Fig. 7.22 Decomposition of a polymer chain into 'blobs' of size ξ. The equivalent chain (of curvilinear length L_ξ) is drawn as a thick line.

effects ($v \neq 0$). If g represents the number of Kuhn segments within a blob, one has from relation (7.149):

$$\xi = v^{1/5} b^{2/5} g^{3/5}, \tag{7.155}$$

yielding, after use of Eq. (7.153):

$$g = v^{-3/4} b^{-3/2} \left(\frac{\rho}{m}\right)^{-5/4}. \tag{7.156}$$

Hence, the number of Kuhn segments within a blob follows the scaling law:

$$g \propto \rho^{-5/4}. \tag{7.157}$$

At length scales larger than ζ, on the other hand, the chains have numerous contacts with each other. In these conditions, they behave as *effective chains*, formed by a succession of size ξ blobs that replace the Kuhn segments of the real chain (Fig. 7.22). Since the effective chains can be treated as ideal chains in a melt, their statistics are Gaussian. This result yields the average radius of each polymer coil in the semi-dilute regime (Flory radius):

$$\overline{R}_{sd}(\rho) = \sqrt{\frac{N}{g}}\xi, \tag{7.158}$$

so that, from Eqs. (7.153) and (7.156):

$$\overline{R}_{sd}(\rho) = N^{1/2} v^{1/8} b^{1/4} m^{1/8} \rho^{-1/8}. \tag{7.159}$$

In conclusion, one can see that the size of the coils does not follow the same scaling law in the dilute and the semi-dilute regimes.

In the dilute regime, $\overline{R}$ is independent of ρ and varies as:

$$\overline{R}_d(\rho < \rho^*) \propto N^{3/5}, \tag{7.160}$$

while in the semi-dilute regime:

$$\overline{R}_{sd}(\rho > \rho^*) \propto N^{1/2} \rho^{-1/8}. \tag{7.161}$$

Table 7.1. *Correspondence between a chain in a melt and the effective chain in dilute solution*

Chain in a melt	Effective chain
c	$\rho/(mg)$
(number of Kuhn segments per unit volume)	(number of blobs per unit volume)
N	N/g
(number of Kuhn segments along a chain)	(number of blobs along a chain)
b	ξ
(length of the Kuhn segment)	(blob size or correlation length)
$\zeta \approx 6\pi\eta_s b$	$\zeta_{\text{blob}} \approx 6\pi\eta_s\xi$
(friction coefficient for the monomer)	(friction coefficient for the blob)

Having established these results, we can now predict the rheological behaviour of a semi-dilute polymer solution.

7.3.1.7 *Application of the Doi–Edwards model*

Since, at length scales much larger than ξ, the real chains can be replaced by effective chains exhibiting ideal behaviour, it is enough to apply the Doi–Edwards model to the latter in order to predict the rheological behaviour of the solution in the semi-dilute regime. In particular, this model will be well adapted to describing the low-frequency behaviour of the solution, in the terminal regime and on the rubber plateau.

In Table 7.1, we present the equivalence between the ideal chains in a polymer melt and the effective chains in a semi-dilute solution. This correspondence immediately yields the curvilinear diffusion coefficient of an effective chain along its tube. Using Eq. (7.100), and then Eqs. (7.153) and (7.156), leads to:

$$D_c \approx \frac{k_B T}{6\pi\eta_s\xi} \frac{g}{N} = \frac{k_B T}{6\pi\eta_s} N^{-1} v^{-1/2} b^{-1} \left(\frac{\rho}{m}\right)^{-1/2}. \tag{7.162}$$

The plateau modulus and the reptation time can be calculated in a similar way, using Eqs. (7.116) and (7.101).

For the first parameter, we obtain:

$$G_N^0 = \frac{k_B T}{2} \frac{\rho}{mgN_e}, \tag{7.163}$$

yielding, by using Eqs. (7.153) and (7.156):

$$G_N^0 = \frac{k_B T}{2N_e\xi^3} = \frac{k_B T}{2N_e} v^{3/4} b^{3/2} \left(\frac{\rho}{m}\right)^{9/4}. \tag{7.164}$$

As to the reptation time, it is given by the following expression:

$$\tau_{\text{rep}} = \frac{1}{\pi^2 N_e} \frac{\xi^2 \zeta_{\text{blob}}}{k_B T} \left(\frac{N}{g}\right)^3,$$ (7.165)

yielding, from Eqs. (7.153) and (7.156) and Table 7.1:

$$\tau_{\text{rep}} \approx \frac{6\eta_s}{\pi N_e} \frac{v^{3/2} b^3}{k_B T} N^3 \left(\frac{\rho}{m}\right)^{3/2}.$$ (7.166)

Note that in these formulas N_e is a unitless parameter characterising the distance between entanglements. This number, which depends on the polymer–solvent system, is of the order of 50 (Fetters *et al.*, 1999).

To summarise, from Eqs. (7.164) and (7.166) we obtain the following scaling laws:

$$G_N^0 \propto \rho^{9/4},$$ (7.167)

$$\tau_{\text{rep}} \propto M^3 \rho^{3/2}.$$ (7.168)

These formulas can then be used to calculate the characteristics of the Maxwell elements associated with the polymer in semi-dilute solution. These calculations are done using Eqs. (7.111) and (7.112), which we recall here:

$$\tau_p^{\text{rep}} = \frac{\tau_{\text{rep}}}{p^2} \quad \text{and} \quad G_p^{\text{rep}} = \frac{8}{\pi^2 p^2} G_N^0 \quad (p \text{ odd}).$$ (7.169)

Finally, we can also calculate the low-shear rate viscosity η_0. It is given by Eq. (7.122), so with the necessary substitutions (see Table 7.1):

$$\eta_0 = \frac{\pi \eta_s}{4} \frac{\rho}{mg^4} \xi^3 \frac{N^3}{N_e^2}.$$ (7.170)

Using Eqs. (7.153) and (7.156) we have finally:

$$\eta_0 \approx \frac{\pi \eta_s}{4 N_e^2} v^{9/4} b^{9/2} N^3 \left(\frac{\rho}{m}\right)^{15/4},$$ (7.171)

leading to the following scaling law:

$$\eta_0 \propto M^3 \rho^{15/4}.$$ (7.172)

One can see that, as for melts, the viscosity increases as M^3. This universal law is due to chain entanglement, which increases rapidly with the molecular mass. The theory also predicts a strong increase of the viscosity with the polymer concentration, more or less as ρ^4. For this reason, the viscosity can very quickly exceed that of the solvent, so that the polymer is a very good thickener.

This behaviour is compatible with the experimental data, which give however somewhat larger exponents:

$$\eta_0(\exp) \propto M^{3-4}\rho^{4-5}. \tag{7.173}$$

To explain these deviations, the theory, and in particular the reptation model, must be improved. Possible refinements are presented in the book of Doi and Edwards [32]. They lead to slightly higher exponents, in better agreement with the experiments.

In the next section, we will analyse the special case of giant micelle solutions; notwithstanding their obvious similarity with semi-dilute polymer solutions, their behaviour is very different.

7.4 'Living polymers': the example of giant micelles formed in surfactant solutions

Micelles are obtained by mixing water with an adequate amount of surfactant. Most times, these micelles have a spherical shape. There are however cases when the micelles form very elongated cylinders. One then speaks of *giant micelles*. They are very similar to the long chains of linear polymers in solution that we described in the previous section. There is however a crucial difference between the two systems, namely that the micelles can *break* and *recombine* between them (hence the term *'living' polymers*). In this section, we will show how this mechanism imparts to giant micelle solutions completely novel rheological properties. Before that, let us begin with a few details on the chemistry and thermodynamics of these systems.

7.4.1 How to obtain giant micelles

They are generally obtained by dissolving in water *amphiphilic* (or *surfactant*, as they act directly on the surface tension of water) molecules. These molecules (among them detergents, non-ionic surfactants as well as numerous biological molecules such as lecithin, which is a phospholipid) consist of two moieties with very different affinities: a *polar head* that likes water, and a *hydrophobic tail* (often an alkyl chain) that avoids water by all means possible.

When such molecules are dissolved in water, they rapidly form *aggregates* such as *micelles* or *vesicles*, and sometimes more complicated objects with labyrinthine shape that form *sponge phases*. The minimal concentration above which these objects appear is the *CMC* (for Critical Micellar Concentration). In an aggregate, the amphiphilic molecules form a *monolayer*, with the polar heads to one side and the hydrophobic tails to the other. The topology of the aggregate varies with the type of surfactant employed and the *spontaneous curvature* c_s of the monolayer (Fig. 7.23). Thus, for high spontaneous curvature, the aggregate has the shape of a spherical micelle. For intermediate values of the spontaneous curvature, the micelle takes the shape of a cylinder terminated by two hemispheres. In the

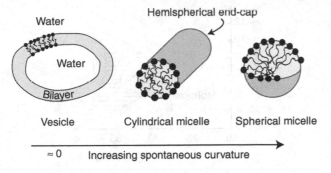

Fig. 7.23 Shape of the micelles as a function of the spontaneous curvature of the monolayer.

CPCl   NaSal

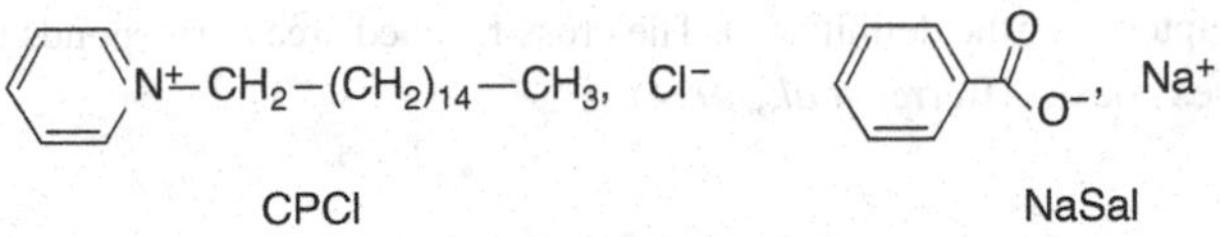

Fig. 7.24 An example of molecules exhibiting giant micelle phases.

specific case we will consider in the following, these cylinders are very long, reaching several micrometres, or even much more. As the spontaneous curvature tends towards 0, two monolayers associate and form a *bilayer*, which can close on itself: the aggregate then takes the shape of a *vesicle* (as in the case of lecithin). In some situations, the bilayer forms more complex structures, of the *sponge phase* type: the local topology of the bilayer is then close to that of a cubic bicontinuous phase.

Clearly, by tuning the value of the spontaneous curvature of the monolayer and by increasing the energy of the hemispheric end-caps, giant cylindrical micelles can be obtained. An example is given in the inset below.

Several systems are known to form giant micelles. One of them (among others) is obtained by mixing a *cationic detergent*, cetylpyrimidium chloride (CPCl) with salty water (brine). To favour the formation of giant micelles, one adds to this mixture a salt, sodium salicylate (NaSal), which is a small molecule known to induce a dramatic elongation of the micelles, even at low CPCl concentration (Fig. 7.24).

This molecule inserts itself into the monolayer and acts on its spontaneous curvature, while at the same time reducing the global electrical charge of the micelle. The phase diagram of a CPCl/Sal/brine mixture is presented in Fig. 7.25. One finds a large existence domain for an isotropic solution of cylindrical micelles, in the dilute and semi-dilute regimes (where, as we will see, the micelles entangle). We will be more particularly concerned with the rheological properties in the latter regime.

In the next section we will discuss the rheological behaviour of giant micelle solutions.

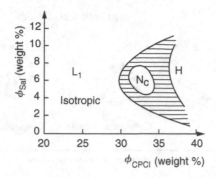

Fig. 7.25 Phase diagram at $30\,^{\circ}$C of the CPCl/Sal/brine system (with 0.5 mol/l NaCl). The salycilate mass fraction is plotted as a function of the mass fraction of surfactant. The isotropic giant micelle phase is denoted by L_1. The N_c phase corresponds to a nematic phase and H corresponds to a hexagonal phase (see Chapter 1 for the definitions). The cross-hatched area corresponds to a demixed zone, where several phases coexist (Berret *et al.*, 1994).

7.4.2 Rheological behaviour

Experimentally, the viscosity η_0 was measured as a function of the *volume fraction of micelles* ϕ which is, by definition:

$$\phi = \frac{X v_{\text{surf}}}{(1-X)v_s + X v_{\text{surf}}} \approx X \frac{v_{\text{surf}}}{v_s} \qquad (X \ll 1). \tag{7.174}$$

In this expression, v_{surf} is the volume of a surfactant molecule in an aggregate and v_s the volume of a solvent molecule. X is the total number concentration of surfactant molecules ($X = n_{\text{surf}}/(n_{\text{surf}} + n_s) \approx n_{\text{surf}}/n_s$). One usually identifies ϕ with the mass fraction of micelles:

$$\phi = \frac{m_{\text{surf}}}{m_{\text{surf}} + m_s}, \tag{7.175}$$

since the density of the micelles is close to that of the solvent. Note that, for the CPCl/Sal/brine mixture, $n_{\text{surf}} = n_{\text{CPCl}} + n_{\text{NaSal}}$ and $m_{\text{surf}} = m_{\text{CPCl}} + m_{\text{NaSal}}$.

The plot in Fig. 7.26 shows the variation of the viscosity η_0 as a function of ϕ in this mixture. One can see that two regimes must be distinguished:

- a low-concentration regime ($\phi < \phi^*$), where the viscosity varies linearly with ϕ (the solid line is the best fit with an Einstein law $\eta_0 = \eta_s(1 + C\phi)$, see Section 7.5.1.5);
- a high-concentration regime ($\phi > \phi^*$), where the viscosity follows a power law (the dotted line corresponding here to the best fit with a law of the type $\eta = \text{Cst}\,\phi^\nu$). The exponent ν is here 3.5 ± 0.15.

The crossover from one regime to the other occurs at a well-defined concentration ϕ^* (of the order of 0.3% in our case).

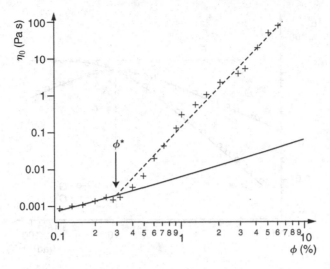

Fig. 7.26 Viscosity in the linear regime of the lyotropic mixture CPCl/Sal/brine as a function of the volume fraction of micelles. The molar ratio [Sal]/[CPCl] is fixed at 0.5 (from Massiera, 2002). The solid line is the best fit of the points below ϕ^* with a linear law. The dotted line is the best fit of the points above ϕ^* with a power law, giving an exponent 3.5 ± 0.15.

These results are very similar to those presented for polymers in solution; they suggest that below ϕ^* the micelles are independent, while above this concentration they start to entangle.

The linear viscoelasticity functions have also been measured. These experiments were performed in the semi-dilute regime, where the viscoelastic properties of the mixture are much more readily apparent. Figure 7.27 shows the behaviour of the moduli G' and G'' in the semi-dilute regime, for a micelle fraction $\phi = 12\%$. The solid lines in the plot are the best fits of the experimental data with a Maxwell model with a single relaxation time over the entire frequency range investigated:

$$G' = \frac{G\tau^2\omega^2}{1 + \tau^2\omega^2}, \quad G'' = \frac{G\tau\omega}{1 + \tau^2\omega^2}. \tag{7.176}$$

The agreement is very good, yielding $G = 235$ Pa and $\tau = 0.11$ s.

This purely Maxwellian behaviour is very surprising, since the giant micelles have an extremely wide size distribution, as we will show in the next section. We know however that in polymers the polydispersity strongly tends to widen the relaxation spectrum, in contrast with the present observations.

This means that in giant micelle systems another – completely new – mechanism is at work, which hides the effects of polydispersity, and which does not come into play in classical polymers.

Before discussing it, let us recall the basics of the thermodynamics of aggregation in giant micelle systems, justifying their surname of 'living polymers'.

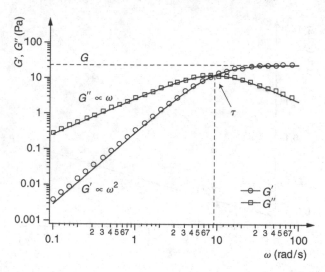

Fig. 7.27 Elastic and loss moduli measured at $30\,°C$ as a function of the frequency in the CPCl/Sal/brine mixture (with a molar ratio [NaSal]/[CPCl] $= 0.5$ and a micelle fraction of 12%) (Massiera, 2002).

7.4.3 Thermodynamics of aggregation

For a giant cylindrical micelle, the monolayer curvature is optimal everywhere except in the two hemispherical end-caps. Hence, the standard chemical potential $\mu^0_{\mathrm{cyl.}}$ of a molecule belonging to the cylindrical body is different from that $\mu^0_{\mathrm{sph.}}$ of a molecule belonging to one of the end-caps. Consider an aggregate containing N molecules, of which n are in the cylindrical body and n_0 in the two end-caps ($N = n + n_0$). The micellisation free energy in this case is:

$$F^N_{\mathrm{mic.}} = (N - n_0)\mu^0_{\mathrm{cyl.}} + n_0\mu^0_{\mathrm{sph.}} = N\mu^0_{\mathrm{cyl.}} + E, \qquad (7.177)$$

where

$$E = n_0\left(\mu^0_{\mathrm{sph.}} - \mu^0_{\mathrm{cyl.}}\right) \qquad (7.178)$$

represents the energy to be expended in order to create two end-caps. In the following, we will refer to it as the *'scission energy'*.

From the thermodynamical point of view, the size distribution of the micelles results from the balance between the mixing entropy (tending to shorten them) and the energy expended in end-cap formation (tending to elongate them). That is why one speaks of 'equilibrium polymers' or, more suggestively, of 'living polymers' (certain protein filaments having similar properties).

To find the size distribution, let us apply the *mass action law*. According to this law, all chemical potentials μ_N of molecules belonging to an aggregate of any size N are equal:

$\mu_1 = \mu_N$ ($\forall N$). For an ideal solution of aggregates:

$$\mu_N = \mu_N^0 + \frac{k_B T}{N} \ln\left(\frac{X_N}{N}\right), \tag{7.179}$$

where X_N is the (number) concentration of surfactant molecules belonging to aggregates of size N. Imposing the equality of all chemical potentials yields:

$$X_N = N \left\{ X_1 \exp\left[\left(\mu_1^0 - \mu_N^0\right)/k_B T\right]\right\}^N, \tag{7.180}$$

where the concentration of free molecules X_1 is given by the closure relation: $X = \sum X_N$. One still needs to express the difference in standard average chemical potential between the surfactant molecules that are free in solution (μ_1^0) and those belonging to a micelle of size N (μ_N^0). From Eq. (7.177), this difference is:

$$\mu_1^0 - \mu_N^0 = \mu_1^0 - \mu_{\text{cyl.}}^0 - \frac{E}{N} = E_\infty - \frac{E}{N}, \tag{7.181}$$

where $E_\infty = \mu_1^0 - \mu_{\text{cyl.}}^0$ is the free energy gained as a free molecule joins an infinite micelle. Replacing $\mu_1^0 - \mu_N^0$ by its expression in the mass action law (Eq. (7.180)) yields the size distribution for the micelles. J. N. Israelachvili showed that, at concentrations much higher than the CMC ($X \gg e^{-E/k_B T}$), this distribution can be written in the following simplified form (Israelachvili, 1992):

$$X_N = N \left[1 - \left(Xe^{E/k_B T}\right)^{-1/2}\right]^N e^{-E/k_B T}$$

$$\approx Ne^{-N/\sqrt{Xe^{E/k_B T}}} \qquad \text{for } N \text{ large.} \tag{7.182}$$

Hence, the average number of molecules in a micelle (or *mean aggregation number*) is:

$$\langle N \rangle = \frac{1}{X} \int_0^\infty N X_N \, dN = 2X^{1/2} e^{E/2k_B T}, \tag{7.183}$$

so that the distribution X_N can be rewritten as:

$$X_N = Ne^{-2N/\langle N \rangle}. \tag{7.184}$$

This formula shows that *the concentration X_N/N of micelles of size N decreases exponentially with N*. The size distribution of the giant micelles is thus extremely large.

In conclusion, we will provide the expression of the volume fraction $\phi(N)$ of micelles with size N (which is an experimentally relevant parameter). By definition, $\phi(N) = X_N v_{\text{surf}}/v_s$, yielding from Eq. (7.184):

$$\phi(N) \approx N \frac{v_{\text{surf}}}{v_s} \exp\left(-\frac{2N}{\langle N \rangle}\right), \tag{7.185}$$

where, according to Eqs. (7.174) and (7.183):

$$\langle N \rangle = 2 \left(\frac{v_{\mathrm{s}}}{v_{\mathrm{surf}}} \right)^{1/2} \phi^{1/2} \exp\left(\frac{E}{2k_{\mathrm{B}}T} \right). \tag{7.186}$$

We emphasise that these calculations are only valid in the absence of electrostatic interactions. This assumes that the micelles are either neutral, or – if charged – in a solvent containing large amounts of salt in order to screen the electrostatic interactions (the case of the system described in the above section).

Now that the thermodynamic properties of these systems are better known, let us give the expression of the concentration ϕ^* defining the crossover between the dilute and the semi-dilute regimes.

7.4.4 Crossover between the dilute and the semi-dilute regimes

In practice, the Maxwellian behaviour of giant micelle solutions is observed in the *semi-dilute regime*. This regime clearly corresponds to the moment when the micelles start to entangle with each other, like polymer chains in semi-dilute solution. This phenomenon occurs above a concentration ϕ^* that we will try to specify.

To this end, let us start by calculating the average Flory radius $\overline{R}_{\mathrm{d}}$ of a micelle with aggregation number N in the dilute regime. We denote its cross-section by s and the length of a Kuhn segment (equal to twice the persistence length) by b. These quantities can be measured by neutron scattering. In the CPCl/Sal/brine system, the measurements yield $s \approx 2800 \,\text{Å}^2$, corresponding to a micelle radius close to 30 Å, and $b \approx 360$ Å (Massiera, 2002). These values are typical for all giant micelle systems studied to date. Since the micelles have a strong affinity with their solvent, they behave as a polymer chain in a good solvent. Their statistics are thus similar to those of a self-avoiding random walk giving, from Eq. (7.149):

$$\overline{R}_{\mathrm{d}} \approx v^{1/5} b^{2/5} \widetilde{N}^{3/5}, \tag{7.187}$$

where v is again the excluded volume and $\widetilde{N} = N v_{\mathrm{surf}}/bs$ the number of Kuhn segments along the micelle (not to be mistaken for the aggregation number N).

As for polymer solutions, the crossover between the dilute and the semi-dilute regimes occurs as the coils formed by the micelles just touch (Fig. 7.20b). In these conditions, the volume fraction of surfactant molecules within a micelle is equal to their global volume fraction, yielding immediately:

$$\phi^* \approx \frac{sb \langle \widetilde{N} \rangle}{\langle \overline{R}_{\mathrm{d}} \rangle^3}. \tag{7.188}$$

Note that, in this expression, we took the average of $\widetilde{N}$ and $\overline{R}_{\mathrm{d}}$ over the entire micelle size distribution:

$$\langle \widetilde{N} \rangle = \frac{\langle N \rangle v_{\mathrm{surf}}}{bs} \quad \text{and} \quad \langle \overline{R}_{\mathrm{d}} \rangle = v^{1/5} b^{2/5} \langle \widetilde{N} \rangle^{3/5}, \tag{7.189}$$

where $\langle N \rangle$ is given by Eq. (7.186). These formulas finally give:

$$\phi^* = (4v_s v_{surf})^{-2/7} (b/v)^{3/7} s^{9/7} \exp\left(-\frac{2E}{7k_B T}\right). \tag{7.190}$$

One can see thus that a way of estimating the values of the excluded volume and the scission energy is to measure ϕ^* as a function of temperature. Another technique consists of measuring by light scattering the hydrodynamic radius of the micelles R_h in dilute solution as a function of temperature at several concentrations (this radius will be defined in Section 7.5.3.2, where we will show that it is proportional to $\overline{R_d}$). In the CPCl/Sal/brine system, this type of measurement leads to a scission energy of the order of $26 k_B T$ and an excluded volume value of the order of 8.2×10^7 Å^3 (Massiera, 2002).

7.4.5 The semi-dilute regime

For $\phi > \phi^*$, the micelles entangle. In these conditions one can define, as for polymers in semi-dilute solution, the correlation length ξ. With this length is associated the blob, which represents the largest portion of an individual micelle for which the Kuhn segments (in a number g) display non-Gaussian statistics, of the type of a self-avoiding random walk. It is noteworthy that these parameters do not depend on the length of the micelles (as they did not depend on the length of the chains in polymer solutions). One can thus write:

$$\phi = \frac{sbg}{\xi^3} \tag{7.191}$$

and

$$\xi = v^{1/5} b^{2/5} g^{3/5}. \tag{7.192}$$

The first relation states that the overall density of the system is the same as within the blobs. The second one implies that the statistics of the micelle are those of a self-avoiding random walk within each blob.

From these two relations, one can determine ξ and g:

$$\xi = s^{3/4} b^{1/4} v^{-1/4} \phi^{-3/4}, \tag{7.193}$$
$$g = s^{5/4} b^{-1/4} v^{-3/4} \phi^{-5/4}. \tag{7.194}$$

As in classical polymer solutions, they follow the scaling laws:

$$\xi \propto \phi^{-3/4}, \tag{7.195}$$
$$g \propto \phi^{-5/4}. \tag{7.196}$$

The scaling law for ξ can be tested by using static light scattering and by measuring the elastic modulus on the rubber plateau (see next section).

For instance, in the CPCl/Sal/brine system, static light scattering gives $\xi \propto \phi^{-0.65}$, while the elastic modulus measurements yield $\xi \propto \phi^{-0.72}$. These results are thus in fairly good agreement with the theory, which predicts an exponent of -0.75 (Berret, 1993).

In the next section, we discuss the rheology of these phases.

7.4.6 *Predictions for the linear viscoelasticity functions G' and G'' and for the viscosity η_0*

The crucial experimental result is that, in the semi-dilute regime, the system behaves over a large frequency range as a *Maxwell fluid* (with a single relaxation time). Clearly, at the low frequencies investigated (below 100 Hz), the Rouse modes can safely be neglected. On the other hand, the viscoelastic character of the mixture at these frequencies is (as in the case of classical polymers) intimately connected to the entanglement and reptation of the micelles. With this process is associated a typical mean time τ_{rep}, which we will start by estimating. We will then see that possible breaking and recombination of the micelles must be taken into account. With this process is associated a second typical mean time τ_{life}, which gives the lifetime for a micelle of average size. The theory of M. E. Cates, which gives the linear viscoelasticity functions and the viscosity (Cates, 1988; Cates and Candau, 1990; Granek and Cates, 1992) is built on these fundamental parameters.

7.4.6.1 *The mean reptation time*

We have already calculated this time in the case of classical polymers in semi-dilute solution. The physics being exactly the same in the case of giant micelles, the formula (7.166) still applies, provided that N is replaced by $\langle \widetilde{N} \rangle$, the number of Kuhn segments along a micelle with the average aggregation number $\langle N \rangle$. Knowing that $\rho/m = g/\xi^3$, and using Eq. (7.192), one finds the mean reptation time:

$$\tau_{rep} = \frac{6}{\pi} \frac{\eta_s}{N_e} \frac{vb^2}{k_B T} \frac{\langle \widetilde{N} \rangle^3}{\xi^2}. \tag{7.197}$$

This time can be calculated using Eqs. (7.186), (7.189) and (7.193), yielding explicitly:

$$\tau_{rep} \approx \frac{4}{\pi} \frac{8\eta_s}{N_e k_B T} \left(\frac{v v_s v_{surf}}{b s^3} \right)^{3/2} \phi^3 \exp\left(\frac{3E}{2k_B T} \right). \tag{7.198}$$

This formula shows that the mean reptation time follows the scaling law:

$$\tau_{rep} \propto \phi^3. \tag{7.199}$$

7.4.6.2 *The breaking–recombination time*

Since we are dealing with 'living polymers', we must account for the fact that the chains are incessantly breaking and recombining. At equilibrium, the breaking time of a chain is equal to the time of recombination between two chains. This breaking time also defines the *lifetime* of a chain. In the Cates model, this time is taken as inversely proportional to the total length of the micelle, so that for a micelle of length L:

$$\tau_{\text{life}}(L) = \frac{1}{c_1 L},$$ (7.200)

where c_1 is the breaking rate per unit length, taken as constant.

In the following, we will denote by τ_{life} the lifetime of a micelle of mean length $\langle L \rangle = \langle \tilde{N} \rangle b$:

$$\tau_{\text{life}} = \frac{1}{c_1 \langle L \rangle}.$$ (7.201a)

Once again, this time can be calculated using Eqs. (7.186) and (7.189), yielding:

$$\tau_{\text{life}} = \frac{s}{2c_1 (v_s v_{\text{surf}})^{1/2}} \phi^{-1/2} \exp\left(-\frac{E}{2k_B T}\right).$$ (7.201b)

We see that this time follows the scaling law

$$\tau_{\text{life}} \propto \phi^{-1/2}.$$ (7.202)

In the next section, we will show how this breaking–recombination process couples to reptation and results in accelerated stress relaxation.

7.4.6.3 The Cates model

At the risk of repeating ourselves, we recall that, in the framework of the reptation model, each micelle behaves as a chain in a polymer melt, formed by a sequence of blobs, with a size ξ and an effective length $L_\xi = \tilde{N}\xi/g$ (see Fig. 7.22). It is then convenient to reason in terms of the effective chain when discussing reptation, while it is the complete length of the micelle $L = \tilde{N}b$ that comes into play when calculating its lifetime. We applied this distinction when calculating the two characteristic times τ_{life} and τ_{rep}, both defined for a micelle with the mean aggregation number $\langle N \rangle$.

Assume now that $\tau_{\text{life}} \ll \tau_{\text{rep}}$ (which is indeed the case in the experimental system we described). In this case, new ends will appear along a micelle before it has time to escape completely from its initial tube by reptation. Then, each newly formed end will diffuse over a certain distance λ_ξ before recombining with the end of a neighbouring micelle (the subscript ξ reminds us that the reasoning is in terms of the effective chain). At this moment, the initial conformation of the micelle is forgotten over the distance λ_ξ. We will see that this mechanism leads to a considerable acceleration in stress relaxation.

To show this, let us estimate the time τ after which the stress will have relaxed completely.

Since the curvilinear distance λ_ξ covered by the effective chain between two consecutive breaking (or recombination) events is also the distance over which it travels by diffusion during the mean lifetime of a micelle, we have:

$$\lambda_\xi = \sqrt{D_c \tau_{\text{life}}},$$ (7.203)

where D_c is the mean curvilinear diffusion coefficient of an effective chain with the mean length $\langle L_\xi \rangle$ along its tube.

For the stress to relax completely, the initial orientation of the chain must be lost over its entire length. This is the case if, after a time τ, a breaking event occurs at a distance smaller than λ_ξ from any point of the effective chain (with mean length $\langle L_\xi \rangle$), since by then all micelle segments with length λ_ξ will have lost their initial orientation by reptation. In other words, the relaxation time τ is the time after which any segment of the equivalent chain with length λ_ξ will experience a breaking event. This time is therefore simply the lifetime of a segment with length λ_ξ. Since this lifetime is inversely proportional to the length of the segment, it is simply $\tau_{\text{life}} \langle L_\xi \rangle / \lambda_\xi$; finally, using Eq. (7.203):

$$\tau = \sqrt{\tau_{\text{life}}} \sqrt{\frac{\langle L_\xi \rangle^2}{D_c}}. \tag{7.204a}$$

As $\tau_{\text{rep}} = \langle L_\xi \rangle^2 / \pi^2 D_c$ from Eq. (7.99), we obtain the following result:

$$\tau \approx \pi \sqrt{\tau_{\text{life}} \tau_{\text{rep}}}. \tag{7.204b}$$

We have thus demonstrated that the relaxation time of the stress is given (up to a factor of π) by the geometrical average of the mean lifetime of the micelles and the reptation time of an unbreakable micelle with mean length $\langle L \rangle$.

It is noteworthy here that this time is unique in the $\tau_{\text{life}} \ll \tau_{\text{rep}}$ limit, since – during relaxation – each segment of the initial micelle will find itself at different positions on several micelles with various lengths.

The existence of a unique relaxation time is thus a result of this micellar reshuffling with time. That is why during the calculation we only considered the average quantities, defined for a micelle with the mean aggregation number $\langle N \rangle$.

This result is in agreement with the purely Maxwellian behaviour observed in experiments performed on concentrated systems of giant micelles (see Fig. 7.27).

This theory also predicts the scaling law of the relaxation time τ as a function of ϕ, since we know that $\tau_{\text{rep}} \propto \phi^3$ and that $\tau_{\text{life}} \propto \phi^{-1/2}$. It follows that:

$$\tau \propto \phi^{5/4}. \tag{7.205}$$

Furthermore, it yields the value of the plateau modulus. Given by relation (7.164), it is obtained from Eq. (7.193) as:

$$G_N^0 = \frac{k_B T}{2 N_e \xi^3} = \frac{k_B T}{2 N_e} \left(\frac{v}{b s^3} \right)^{3/4} \phi^{9/4}. \tag{7.206}$$

Thus, the plateau modulus follows the scaling law:

$$G_N^0 \propto \phi^{9/4}. \tag{7.207}$$

The solution exhibiting Maxwellian behaviour, the expression of the viscosity in the linear regime follows immediately:

$$\eta_0 = \tau G_N^0 \tag{7.208}$$

and, from Eqs. (7.205) and (7.207) then follows the scaling law:

$$\eta_0 \propto \phi^{7/2}. \tag{7.209}$$

The rheology experiments performed in the CPCl/Sal/brine system give for τ an exponent of 1.13, slightly below the theoretical value of 5/4, and for G_N^0 an exponent of 3.1, this time slightly above the theoretical value of 9/4. On the other hand, the experimentally measured exponent for η_0, equal to 3.5 (see Fig. 7.26), is in excellent agreement with the theoretical value of 7/2 (Berret *et al.*, 1993; Massiera, 2002).

These measurements, as well as those performed on different systems, confirm the main lines of the theory presented here. Some refinements are however needed to better account for the measured exponents: one of them consists of accounting for the possible side-branching of the micelles. These branching points are very mobile, in contrast with the entanglements. That is why they tend to decrease the viscosity of the solution. Although their existence was clearly demonstrated in certain systems of non-ionic surfactants (Constantin *et al.*, 2003), it remains nonetheless a subject of controversy in the giant micellar systems presented in this section.

7.5 Dispersions

Dispersions are obtained by dispersing in a liquid (the solvent), either solid particles, or droplets of another (immiscible) liquid: one speaks of *suspensions* in the first case, and of *emulsions* in the second.

From the rheological point of view, the particles must be small enough that the solvent flow remains laminar everywhere (requiring that the Reynolds number associated with the flow around each particle be much smaller than 1: $Re = \dot{\gamma} R^2 / v_s \ll 1$ where R is the typical radius of the particles and v_s the kinematic viscosity of the solvent), but large enough compared to the molecules of the solvent so that the latter can be treated as a continuous medium. With this definition, dilute solutions of macromolecules are related to dispersions and do, in fact, exhibit very similar properties.

In this section, we will discuss some rheological properties of these three types of dispersions.

7.5.1 Suspensions of solid particles

A suspension is obtained by dispersing solid particles in a viscous liquid (assumed Newtonian in the following).

Two kinds of suspensions must be distinguished, according to the size of the particles employed: *colloidal suspensions*, where the *Brownian motion* of the particles is important, and the others, where this effect can be neglected.

This will bring us to defining the *Péclet number*, which, as we will see later, plays an essential role in the rheology of suspensions.

7.5.1.1 Péclet number of a suspension

Consider a solid particle immersed in a liquid. This particle experiences hydrodynamic forces due to the liquid flow, and thermal forces due to the incessant collisions with the molecules in the liquid. We recall that these collisions are responsible for the *Brownian motion* of the particle and for its *diffusion* over time. The *Péclet number* quantifies the relative importance of these two effects. Its general definition is:

$$Pe = \frac{\text{hydrodynamic effects}}{\text{thermal effects}} = \frac{UL}{D}, \tag{7.210}$$

where U is a typical flow velocity, L the size of the particle and D its diffusion coefficient.

To comprehend this definition, let us consider a spherical particle with a radius R, immersed in a simple shear flow, with shear rate $\dot{\gamma}$. In this case, the relevant hydrodynamic quantity is not the drift velocity of the bead, but the amount of shear it feels, i.e. by how much the liquid velocity varies over the distance R: thus, one should take $U = \dot{\gamma} R$. On the other hand, we know that the diffusion coefficient of the particle is given by the Stokes–Einstein relation:

$$D = \frac{k_B T}{6\pi \eta_s R}. \tag{7.211}$$

This leads to a new expression for the Péclet number:

$$Pe = \frac{6\pi \eta_s \dot{\gamma} R^3}{k_B T}. \tag{7.212}$$

This formula shows that the crossover from Brownian ($Pe \ll 1$) to non-Brownian ($Pe \gg 1$) behaviour is strongly dependent on the size of the particle and, to a lesser extent, on the shear rate. To fix the ideas, consider the case of water ($\eta_s = 10^{-3}$ Pa s) and a typical shear rate of 1 s^{-1}. It follows that, at room temperature, $Pe \approx 1$ for a particle diameter $2R$ of the order of 1 μm. In this case, $D \approx 1$ μm^2/s and the particle takes typically 1 s to diffuse over the distance $2R$.

This order of magnitude calculation fixes the frontier between Brownian and non-Brownian particles at a size of the order of 1 μm.

We emphasise that, in the colloidal domain, other forces (of the van der Waals type, or electrostatic for charged particles in polar solvents) can come into play when the distances between particles are lower than about a thousand angstroms. We will briefly discuss their effect on the rheological properties of suspensions in Section 7.5.1.8. Let us however begin with a few basic notions on colloidal suspensions.

7.5.1.2 Obtaining a colloidal suspension

In 1856, M. Faraday prepared the first suspension of colloidal gold. To obtain it, he mixed 100 cm^3 of distilled water with 1 cm^3 of a 1% solution of gold chloride (HAuCl$_4$.3H$_2$O).

He brought the solution to boiling point and added 2.5 cm^3 of a 1% solution of sodium citrate. After a few minutes at this temperature, he noticed the appearance of a nice blue colouration, which rapidly turned to vermilion red (the colour of rubies). He had just obtained the first dispersion of colloidal gold. Faraday understood this after noticing that the colouration could be removed by the addition of reactants that dissolve metallic gold. He concluded that the gold was not in ionic form in the solution, but rather as small solid particles suspended in the liquid. He also noticed that the solution scattered light very strongly and that, after adding a small amount of various salts, it turned blue, and then sedimented, forming a pasty precipitate at the bottom of the container. He had discovered two other essential properties of suspensions, namely that they strongly scatter light and that they coagulate on adding salt (which induces the precipitation of the particles).

Other recipes for suspensions of solid particles (containing sulphur, silver bromide, or iron hydroxide) are given in the book by D. H. Everett on colloidal science [37]. One can also produce suspensions of perfectly spherical and monodisperse latex particles (made of polystyrene), already discussed at the end of Chapter 4, in the section dealing with colloidal crystals. Their chemistry is however more complex, and they cannot be produced in one's kitchen. We simply note here that the electrostatic interactions between beads must be screened (by salt addition, for instance) to avoid crystallisation (see Section 4.3.1.2).

Let us now say a few words on the sedimentation of disordered suspensions.

7.5.1.3 On the sedimentation of disordered suspensions

If the particles are too large (and their density higher than that of the solvent), they will tend to sediment at the bottom of the container. We have already studied this effect for colloidal crystals (see Section 4.3.2.1). Clearly, it also occurs in disordered suspensions such as those of Faraday, as they turn blue after the addition of a small amount of salt. On the other hand, in the vermilion red suspension he obtained at the beginning, the particles are so small that their Brownian motion keeps them in suspension: the sedimentation is then negligible.

To quantify this effect, let us construct the relevant Péclet number. In this case, one must take for U the sedimentation velocity of a particle under the action of its own weight. For a spherical particle with radius R, this velocity is obtained by balancing the Stokes force and the gravity:

$$6\pi \eta_s R U = \frac{4}{3}\pi R^3 (\rho - \rho_s)g. \tag{7.213}$$

With $\Delta\rho = \rho - \rho_s$, this yields:

$$U = \frac{2}{9}R^2 \frac{\Delta\rho g}{\eta_s}. \tag{7.214}$$

Replacing U by its expression in Eq. (7.210) and taking as typical size the diameter of the particle ($L = 2R$), one obtains from Eq. (7.211):

$$Pe = \frac{8}{3}\frac{\Delta\rho g R^4}{k_B T}. \tag{7.215}$$

For gold, this number is 1 when the diameter of the particles is 0.6 μm. This is consistent with the order of magnitude already announced as the limit between the colloidal and the hydrodynamic regimes.

It follows that the suspension will tend to sediment and form a pasty precipitate at the bottom of the container if the particle size exceeds the micrometre.

On the other hand, smaller particles tend to remain in suspension due to their Brownian motion. In this case, one can easily calculate their concentration profile in the gravity field. Let c be the concentration at a height h with respect to the bottom of the container and set $c(h = 0) = c_0$. The direct application of the Boltzmann distribution law gives:

$$c = c_0 \exp(-\Delta G/k_B T), \tag{7.216}$$

where ΔG is simply the potential weight energy of a particle. If v is the volume of the particle, ρ its density, ρ_s that of the solvent, then $\Delta G = v(\rho - \rho_s)gh$ with g the gravity acceleration. Replacing in the equation above gives the density profile:

$$c(h) = c_0 \exp\left[\frac{-v(\rho - \rho_s)gh}{k_B T}\right]. \tag{7.217}$$

The concentration is obviously uniform if the densities are equal, a condition almost exactly fulfilled for latex suspensions, but not for metals in water. In the case of gold (a metal denser than lead) $\rho = 19.3$ g/cm^3. It can be calculated that the concentration varies by less than 10% over a height of 10 cm if the particle volume is below 2.3×10^{-20} cm^3, corresponding (in the case of spheres) to diameters below 3.4×10^{-7} cm, or 34 Å. The metal particles must therefore be really small for their concentration to be considered as constant. We will neglect sedimentation effects in the following, which amounts to assuming that the measurements are performed over times much shorter than the sedimentation time.

After this preliminary discussion, let us now consider the problems of suspension rheology proper. To simplify, we will start with the case of spherical particles.

7.5.1.4 Viscosity under continuous shear for model suspensions of spherical particles

In this section, we will discuss results obtained for suspensions of spheres with a monodisperse size. The spheres are made of polymer or silica and they are dispersed in various solvents. The systems are model suspensions, insofar as the only interactions between spheres are of the excluded volume type. In the following, we will denote by ϕ the volume fraction of spheres.

Figure 7.28 shows the variation of the viscosity under continuous shear for three different suspensions as a function of shear rate, for a volume fraction $\phi = 0.5$. The particles are made of polystyrene. Their size varies between 70 and 200 nm, typically, and they are dispersed in solvents with different viscosities (water, benzylic alcohol and meta-cresol). This plot shows that all curves are superposed if the viscosity of the suspension is scaled by that of the solvent η_s, and if the shear rate is replaced by the Péclet number (defined by Eq. (7.212)). One can notice the existence of two asymptotic regimes, where the viscosities saturate:

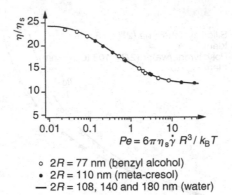

o $2R = 77$ nm (benzyl alcohol)
• $2R = 110$ nm (meta-cresol)
— $2R = 108, 140$ and 180 nm (water)

Fig. 7.28 Viscosity as a function of the shear rate for three different dispersions (with $\phi = 0.5$). The curves are superimposed if the viscosity is scaled by that of the solvent and if the shear rate is replaced by the Péclet number (Krieger, 1972).

• a regime at low Péclet numbers ($Pe < 0.1$), where the viscosity (denoted by η_0) is constant. This is the linear regime, where the statistical distribution of the beads (governed by the Brownian motion) is hardly affected by the flow;
• a regime at high Péclet numbers ($Pe > 1$), where the viscosity (denoted by η_∞) is also constant, but lower than η_0. In this strongly nonlinear regime, the hydrodynamic effects dominate over the Brownian motion of the particles. In this case, the statistical distribution of the particles is no longer the equilibrium one.

Between the two regimes, the viscosity exhibits a monotonous decrease. This shear-thinning behaviour indicates a structuring of the particles under shear.

It is also very interesting to study the variation of the two asymptotic viscosities η_0 and η_∞ as a function of the volume fraction of particles (Fig. 7.29). These curves were obtained for polystyrene beads in water and for silica beads in cyclohexane.

The experimental points are almost superposed for volume fractions below $0.25-0.3$. In this regime, the best fit to the experimental data with a third order polynomial in ϕ gives, for each viscosity (Kruif et al., 1985):

$$\eta_0/\eta_s = 1 + 2.5\,\phi + (4 \pm 2)\phi^2 + (42 \pm 10)\phi^3 + \cdots , \qquad (7.218)$$

$$\eta_\infty/\eta_s = 1 + 2.5\,\phi + (4 \pm 2)\phi^2 + (27 \pm 7)\phi^3 + \cdots . \qquad (7.219)$$

At higher volume fraction, the curves are well described by laws of the type:

$$\frac{\eta_0}{\eta_s} = \left(1 - \frac{\phi}{0.63}\right)^{-2} , \qquad (7.220)$$

$$\frac{\eta_\infty}{\eta_s} = \left(1 - \frac{\phi}{0.71}\right)^{-2} . \qquad (7.221)$$

These curves are plotted as solid lines in Fig. 7.29.

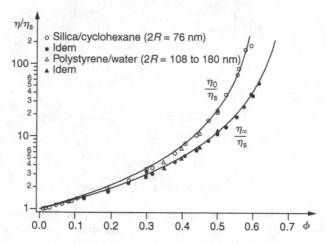

Fig. 7.29 Limiting viscosities as a function of the volume fraction of beads. The solid lines were calculated using Eqs. (7.220) and (7.221) (Kruif *et al.*, 1985; Krieger, 1972).

The first expression shows that the viscosity diverges for a volume fraction of 0.63, which is very close to the value corresponding to a random close packing ($\phi_m = 0.637$). This result is perfectly compatible with the idea that the flow does not modify the distribution function of the particles in the linear regime, at very low shear rates.

The second formula shows that the suspension can still flow at high shear rates for volume fractions above ϕ_m, since its viscosity η_∞ only diverges as its volume fraction tends towards 0.71 (a value above ϕ_m, but below the value $\phi_M = \pi/(3\sqrt{2}) = 0.74$ corresponding to an f.c.c. or h.c.p. close packing of spheres). This result suggests that the suspension organises locally in ordered microdomains.

In the next section, we will consider the dilute regime ($\phi < 0.05$). This is the limit where one can use the famous formula first demonstrated by Einstein in 1906.

7.5.1.5 The Einstein formula

Fitting the experimental data in Fig. 7.29 shows that the very dilute hard-sphere suspensions exhibit Newtonian behaviour over the entire shear rate range investigated. In this regime, their average viscosity increases linearly with the volume fraction of spheres according to the law:

$$\eta = \eta_s(1 + 2.5\phi).\tag{7.222}$$

This formula (only valid for spheres) was first obtained theoretically in 1906 by A. Einstein (Einstein, 1906; 1911). It is demonstrated in Appendix 7.C.

It is noteworthy that this law, bearing Einstein's name, is valid for *spherical* particles, and provided that their interactions can be neglected. This restricts its application to *very low concentrations*, since hydrodynamic interactions are *long-ranged*. Note that this formula applies for any type of suspension (colloidal or not). The only limitation is that the particles must be large enough that solvent can be treated as a continuum.

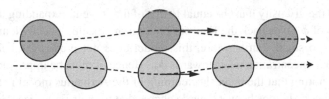

Fig. 7.30 As two particles collide, their trajectories deviate, leading to an increase in dissipation.

In the next section, we will attempt a qualitative analysis of the effects of increasing particle concentration.

7.5.1.6 *Concentration effect: phenomenological models of Arrhenius and of Ball and Richmond*

When the bead concentration exceeds a small percentage, experiment shows that the viscosity increase as a function of ϕ is faster than predicted by the Einstein law. This effect is due to particle collisions. Indeed, consider two particles such that (as long as they are far apart) their trajectories are separated by a distance below $2R$. When the particles come closer, they eventually collide. As a result of this collision, their trajectories are deformed as shown in Fig. 7.30. This distortion leads to an additional dissipation, increasing the effective viscosity of the suspension.

Since the probability of binary collisions is proportional to the square of the particle volume fraction, a ϕ^2 correction must be added to the Einstein equation. In the same way, three-body collisions will introduce ϕ^3 corrections, etc. Thus, the viscosity can be expanded in the following series:

$$\eta = \eta_s(1 + 2.5\phi + k_2\phi^2 + k_3\phi^3 + \cdots).\qquad(7.223)$$

The k_2 coefficient was calculated exactly by G. K. Batchelor in 1977: its value is 6.2 (Batchelor, 1977). Considering the experimental uncertainties, this result is in good agreement with the experiments (see Eq. (7.218)). As to the other coefficients, they have not yet been obtained, the calculations being exceedingly complex.

As illustrated by the experiments, this type of expansion applies as long as the volume fraction of beads is lower than 0.3. Above this value, a different approach must be used.

One possible model is that of Arrhenius, based on the *effective medium* hypothesis (Arrhenius, 1917). The idea behind this model is extremely simple. It consists in treating the suspension with volume fraction ϕ as an effective medium with a viscosity $\eta(\phi)$. If the concentration is increased by $d\phi$, then, from the Einstein law, the viscosity of the suspension varies by

$$d\eta = 2.5\eta(\phi)\,d\phi.\qquad(7.224)$$

This equation is immediately integrated, yielding:

$$\eta = \eta_s \exp(2.5\phi),\qquad(7.225)$$

since at $\phi = 0$ the viscosity must be equal to that of the solvent. Expanding this equation as a power series of ϕ gives back the Einstein law, but the resulting coefficient $k_2 = 25/8 = 3.1$ is clearly too small. Furthermore, this model does not predict the behaviour of the viscosity as the concentration tends towards ϕ_m. Hence, it must be improved.

We start by noting that the main shortcoming of the Arrhenius model is that it neglects completely the correlations between hard spheres due to their finite size. Hence, as a particle is added to a concentrated solution, this requires more space than simply $d\phi$, due to stacking constraints. For this reason, Ball and Richmond proposed in 1980 to replace $d\phi$ by $d\phi/(1 - K\phi)$, where K is a constant to be determined (Ball and Richmond, 1980). In this case, the Arrhenius equation (7.224) becomes:

$$d\eta = 2.5\eta \frac{d\phi}{1 - K\phi}. \tag{7.226}$$

Integrating it gives the rheological law:

$$\eta = \eta_s (1 - K\phi)^{-2.5/K}. \tag{7.227}$$

With this formula, the viscosity diverges for $\phi = 1/K$. Intuitively, at low shear rates this must occur for $\phi = \phi_m$, where $\phi_m \approx 0.637$ is the maximum packing fraction for random stacking. This choice yields $K = 1/\phi_m = 1.57$. The resulting expression is fairly close to Eq. (7.220) used for fitting the experimental data in Fig. 7.29. It can also be expanded as a power series of ϕ; one thus retrieves the Einstein equation, as well as a coefficient $k_2 \approx 5.1$ fairly close to that calculated exactly by Batchelor.

In the next section, we will briefly discuss the viscoelastic properties of these suspensions.

7.5.1.7 Elastic modulus G' and viscosity η' under oscillating shear

The linear viscoelasticity functions were also measured in model sphere suspensions. Some data are shown in Fig. 7.31. As for continuous shear, these rheological data are plotted as a function of the Péclet number for the dispersion constructed using the angular frequency ω:

$$Pe = \frac{R^2 \omega}{D} = \frac{6\pi \eta_s R^3 \omega}{k_B T}. \tag{7.228}$$

It is apparent that, at a fixed volume fraction, the curves for the viscosity η', like those for the elastic modulus G', superpose if the viscosity is rescaled by that of the solvent η_s, and the elastic modulus by the ratio $k_B T / R^3$.

These curves also show that, at low Péclet number, the viscosity η' is equal to the viscosity η_0 measured under continuous shear at low shear rate. On the other hand, the elastic modulus G' is very low in this regime. It can be inferred that the flow has hardly any effect on the statistical distribution of the particles, which remains very close to that at equilibrium, due to the Brownian motion of the particles.

At high Péclet number, on the other hand, the distribution of the particles is affected by the flow: the viscosity η' decreases (once again leading to shear-thinning behaviour), and

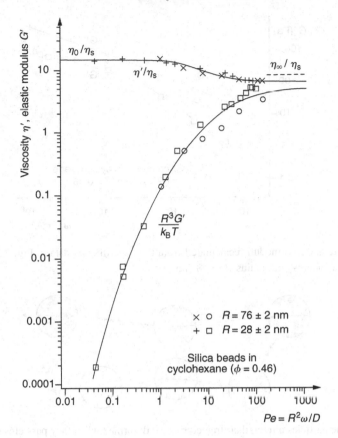

Fig. 7.31 Viscosity η' and elastic modulus G' as a function of the Péclet number (Mellema, 1987).

finally saturates at very high Péclet number (which corresponds to high frequency), as it did under continuous shear at very high shear rate. Nevertheless, experiment shows that the viscosity reached in this regime is lower under oscillating shear (in the linear regime) than under continuous shear (in the strongly nonlinear regime). This effect is certainly related to the frequency of particle collisions, which is presumably much higher under continuous shear (thus increasing the viscosity, see Fig. 7.30) than under oscillating shear at very low amplitude.

Note that, in suspensions, the elastic effects are not due to a deformation of the particles, as in the case of polymers, but rather to a change in the statistical distribution function of their centres of mass under shear. Intuitively, one would expect these elastic effects to be less important in suspensions than in polymer materials. Since the data in Fig. 7.31 do not illustrate this feature clearly enough, we plot in Fig. 7.32 the 'raw' data $G'(\omega)$ and $G''(\omega)$ corresponding to the suspension of spheres with radius $R = 28$ nm. Clearly, the elastic modulus G' is much smaller than the loss modulus G'' at all investigated frequencies, which confirms our intuition.

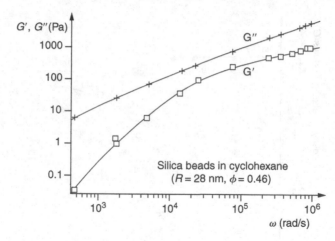

Fig. 7.32 Elastic and loss moduli recalculated from the experimental data in Fig. 7.31 for the suspension of silica spheres with radius $R = 28$ nm.

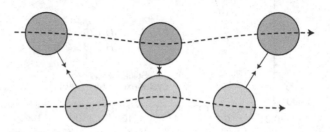

Fig. 7.33 As the particles attract, their trajectories are deformed when they pass close to each other. This leads to a viscosity increase.

7.5.1.8 Role of the interactions between particles

So far, we have always neglected the interactions between particles, with the exception of the excluded volume, which become very important at high values of the concentration ϕ (above 30%, typically) as we have seen.

However, in most colloidal suspensions, the particles also interact by van der Waals, electrostatic and other forces. These interactions, of attractive or repulsive nature, become important when the particles are close enough to each other, typically at distances below 1000 Å (which can be very large with respect to the particle size).

Under these conditions, the trajectories of the particles deform more than they would under the exclusive influence of hydrodynamic interactions. Hence, this mechanism leads to an additional dissipation, and thus to an increase in the apparent viscosity of the suspension (Fig. 7.33).

An example of this type was studied in detail by C. Allain and coworkers (Allain *et al.*, 1994). They used suspensions of silica beads (with a diameter $2R = 18$ nm) in brine. For NaCl concentrations between 10^{-3} and 10^{-1} mol/litre, the pH is almost constant, with a value close to 9. At this pH value, the beads are known to bear a high negative charge (about 0.5 OH$^-$ per nm^2). The interactions between beads are thus of electrostatic nature, and they are partially screened by the Debye layers formed around the beads (see Section 4.3.1.2). These authors showed that the low-shear-rate viscosity of these suspensions follows a law of the type

$$\frac{\eta_0}{\eta_s} = \left(1 - \frac{\phi_{\text{eff}}}{0.55}\right)^{-2}, \tag{7.229}$$

where ϕ_{eff} is an effective concentration given by

$$\phi_{\text{eff}} = \phi(1 + \alpha\lambda_D/R)^3, \tag{7.230}$$

with ϕ the concentration of beads, λ_D the thickness of the Debye layer around each bead (given by Eq. (4.250)) and α a numerical coefficient, with an experimental value close to 0.5. As expected, this formula leads to a higher viscosity than predicted by Eq. (7.220). It also shows that the increase in viscosity with respect to a model suspension is more important as ϕ increases, since the viscosity diverges for a critical value of ϕ_{eff} of the order of 0.55, noticeably lower than ϕ_m. This experimental study shows the crucial importance of the interactions between particles in the rheology of suspensions (for more details, see [27]).

We end this section by noting that interactions other than those of the excluded volume type should not change the Einstein term in the viscosity expansion as a power series of ϕ. Indeed, this additional dissipation only occurs when two (or more) particles meet. Consequently, only the ϕ^n terms of order n at least equal to two are affected, since they account for binary (ternary, etc.) collisions.

In some cases, the particles aggregate as they come sufficiently close to each other. Two situations can then be distinguished:

- either the aggregates are transient, and break up after a certain time. Under these conditions, the results above remain applicable (in particular, the Einstein term remains unchanged);
- or the aggregates are permanent. In this case, the value of the Einstein coefficient, also known as the *intrinsic viscosity*, and given by:

$$[\eta] = \lim_{\phi \to 0} \frac{\eta - \eta_s}{\phi\eta_s}, \tag{7.231}$$

must change, since the aggregates have no particular reason to be spherical. This brings us to discussing (briefly) the case of suspensions of non-spherical particles.

7.5.1.9 Suspensions of non-spherical particles

When the particles are non-spherical, the situation becomes extremely complicated, even in the dilute regime, since the excess dissipation due to the flow around each particle, which is responsible for the increase in the viscosity of the suspension, depends on the particle orientation with respect to the shear plane, defined by the velocity in the mean flow and its gradient. To fix the ideas, let us assume that all solid particles have the shape of a revolution ellipsoid with half-axes a, b and b, with the aspect ratio $r = a/b$. For $r \gg 1$, the ellipsoid has a rod-like shape; for $r \ll 1$, it resembles a flat disc.

These two types of particles do exist. For instance, certain viruses have the shape of an elongated rigid rod, with a perfectly calibrated size. The most famous is the tobacco mosaic virus, discovered in 1892 by the Russian researcher Dimitry Ivanovssky, and then isolated in 1935 by the American biochemist Wendell Meredith Stanley. This virus has the shape of a hollow cylinder, 300 nm long and 18 nm in diameter, with a central channel 4 nm in diameter. One can also obtain rod suspensions using glass fibres, as well as certain high-molecular mass polymers, such as poly(γ-benzyl-L-glutamate) (or PBLG) which forms very stiff rods. On the other hand, clays are composed of small flat particles (themselves made up of sheets), whose shapes and sizes are seldom regular (in the case of natural clays). There are however exceptions, such as Laponite, a clay obtained by synthesis, where the disc-like particles have well-defined diameter and thickness (300 and 10 Å, respectively).

Using this type of material, one can obtain reasonably monodisperse colloidal suspensions of flat discs or elongated cylinders. In the following, we will only discuss the case of cylinders (modelled as very elongated ellipsoids).

Let us now analyse the behaviour of such a suspension under shear. If the Brownian motion is negligible (the range of validity of this assumption will be discussed later), the answer is graphically presented in Fig. 7.34.

This figure shows the trajectory of an isolated rod under shear (also known as the '*Jeffery orbit*'). Since the fluid exerts a torque on the particle, the latter starts to spin. However, since the particle is not spherical, this torque depends on the orientation of its major axis with respect to the velocity. It is easy to see that this torque reaches a minimum (or a maximum) when this axis is parallel (or perpendicular, respectively) to the velocity. Hence, in the hydrodynamic regime, the trajectory of each particle can be decomposed into a sequence of rather fast spins, separated by fairly long intervals during which the particle remains aligned along the velocity (hence, the dissipation is minimal). Since most of the time the particles are parallel to the velocity, the medium becomes anisotropic. This effect manifests itself optically by the appearance of flow birefringence, which can be detected experimentally.

We emphasise that the Jeffery orbit can be calculated for an ellipsoid with a generic aspect ratio r (above or below 1) (Jeffery, 1922). Denoting by θ the angle between the

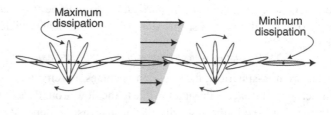

Fig. 7.34 Trajectory of an ellipsoidal particle under shear flow.

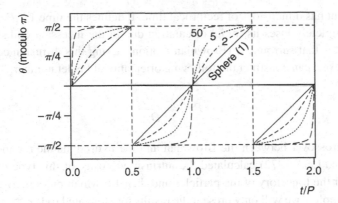

Fig. 7.35 Variation of the rotation angle as a function of time. The higher the aspect ratio r (indicated for each curve), the longer the rod remains parallel to the velocity ($\theta = \pi/2 \pmod{\pi}$).

revolution axis of the ellipsoid and the velocity gradient, Jeffery obtained:

$$\tan \theta = r \tan\left(\frac{\dot{\gamma} t}{r + 1/r}\right),\tag{7.232}$$

knowing that at the initial time $\theta = 0$ (vertical rod in Fig. 7.34). The period of this motion is:

$$P = \frac{\pi}{\dot{\gamma}}\left(r + \frac{1}{r}\right).\tag{7.233}$$

The function $\theta(t/P)$ is plotted in Fig. 7.35. This graph shows that, in the case of rods, the alignment effect is more important as the particles are more elongated (r is higher).

The situation is clearly very different if the Brownian motion of the molecules dominates over the rotational effects associated with the hydrodynamic flow. In this case, the particle spins in a random fashion, as a result of the thermal fluctuations. From the hydrodynamic point of view, it behaves as a sphere with an effective volume larger than the real volume of the particle. It follows that, in this limit, the intrinsic viscosity of the suspension (defined by Eq. (7.231)) must be larger than the value 2.5 obtained for spheres.

This discussion shows that two regimes must be distinguished in dilute suspensions of non-spherical particles (obviously, this does not concern the spheres, where the orientation is not relevant).

The crossover between these two regimes is defined by the competition between the rotation Brownian motion of the particles, described by the rotational diffusion coefficient of the particle D_{rot}, and the shear rate $\dot{\gamma}$. We recall that, for an elongated ellipsoid (a rod), one has [27]:

$$D_{rot} = \frac{k_B T}{\dfrac{8\pi}{3\ln(2r) - 0.5}\eta_s a^3}.\tag{7.234}$$

This coefficient has dimensions of reciprocal time. It defines the time $1/6D_{rot}$ after which a particle completely loses its initial orientation due to the incessant collisions with the solvent particles that surround it (Brownian motion). Using this diffusion coefficient and the shear rate, one can construct the following orientational Péclet number:

$$Pe^* = \frac{\dot{\gamma}}{D_{rot}} \tag{7.235}$$

defining the crossover between the Brownian and the hydrodynamic regimes. The theorists (Hinch and Leal, 1972) calculated the intrinsic viscosity of this type of suspension, accounting for the trajectory of the particles and their Brownian motion. The calculations being very complex, we will only present the results for elongated rods $r \gg 1$. In this case, the intrinsic viscosity defined by the relation (7.231) is given by:

$$[\eta] = \frac{4}{15} \frac{r^2}{\ln r} \qquad Pe^* \ll 1, \tag{7.236}$$

$$[\eta] = 0.32 \frac{r}{\ln r} \qquad Pe^* \gg 1 \tag{7.237}$$

and, in the intermediate regime:

$$[\eta] = 0.5 \frac{r^2}{\ln r} \left(\frac{D_{rot}}{\dot{\gamma}} \right)^{1/3}. \tag{7.238}$$

The shape of the intrinsic viscosity curve is shown in Fig. 7.36. To fix the ideas, we took $r = 17$, which is the aspect ratio of the tobacco mosaic virus. This plot shows that the crossover from the Brownian regime, where the suspension behaves as a Newtonian fluid, to an intermediate regime, where it is strongly shear-thinning, occurs at a shear rate $\dot{\gamma}_1 \approx 6.6 D_{rot}$ (independent of r) corresponding to a Péclet number $Pe_1^* = 6.6$. The shear-thinning behaviour of the suspension then ceases, for shear rates above $\dot{\gamma}_2 = 3.8 r^3 D_{rot}$, corresponding to a Péclet number $Pe_2^* = 3.8 r^3$: in this regime, the hydrodynamics dominates over the Brownian motion, and the suspension behaves again as a Newtonian fluid.

In the case of the tobacco mosaic virus in water, $D_{rot} \approx 3 \times 10^5$ s^{-1}, yielding $\dot{\gamma}_1 \approx 10^7$ s^{-1} and $\dot{\gamma}_2 \approx 5 \times 10^9$ s^{-1}. These shear rates are very hard to reach in experiments so that, in practice, with objects as small as the viruses, the Brownian motion dominates: the intrinsic viscosity is then given by the formula (7.236).

To test these theoretical predictions one must use longer particles, of micrometre size, typically. Indeed, for $a = 1$ μm, $D_{rot} \approx 8$ s^{-1}, giving $\dot{\gamma}_1 \approx 50$ s^{-1} and $\dot{\gamma}_2 \approx 1.5 \times 10^5$ s^{-1} (for the same value of $r = 17$). These shear rates are attainable experimentally.

Such an example is shown in Fig. 7.37, for a dilute solution of poly(γ-benzyl-L-glutamate) (PBLG) in meta-cresol. Here, the intrinsic viscosity experiences a strong decrease as the shear rate exceeds a few hundred s^{-1}, signalling an alignment of the PBLG molecules along the velocity. Note that this curve provides an estimate for the rotational diffusion coefficient of PBLG: $D_{rot} \approx 100$ s^{-1}.

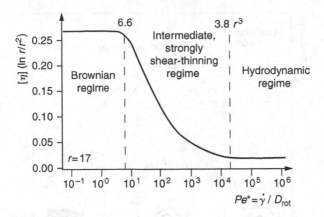

Fig. 7.36 Intrinsic viscosity curve as a function of the shear rate for a dilute suspension of rods (from the lecture notes of J. Hinch, 'École de physique des Houches,' winter 2003).

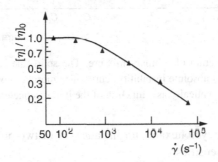

Fig. 7.37 Scaled intrinsic viscosity of a dilute PBLG solution (with a molecular mass of 208 000) in meta-cresol (Yang, 1958).

7.5.2 Emulsions

An emulsion is a suspension of droplets of a liquid B within an (immiscible) liquid A. We have already mentioned them in the first chapter, when we defined the concept of Jeffrey fluid. In this case, the origin of the elasticity is more readily apparent than for suspensions, since the dominant effect – no longer related to the changes in the statistical distribution of the particles – is a direct result of their deformation under shear, leading to an increase in the interfacial energy stored in the system. In this section, we will complete the discussion started in the first chapter with a more detailed account of their behaviour under continuous or oscillating shear. Let us however start by explaining how to prepare an emulsion.

7.5.2.1 Preparing an emulsion

Several methods can be employed.

One of them is based on the principle of *phase separation*. Figure 7.38 displays the typical phase diagram of a typical binary mixture of two liquids, composed either of small

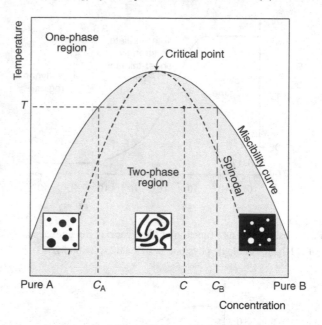

Fig. 7.38 Typical phase diagram of a binary mixture. The solid line is the miscibility curve; the dotted line is the spinodal (or absolute instability) curve. The morphology of the emulsions obtained by the quench is shown schematically as a function of the initial concentration of the mixture.

molecules (methanol and cyclohexane, for instance) or of two polymers with comparable molecular masses and viscosities.

Under these conditions, the temperature–concentration diagram is more or less symmetric, and divided into two regions separated by a bell-shaped curve shown as a solid line. Above this (*miscibility*) curve, the liquids are miscible in all proportions and form a continuous phase (one-phase region). Below it, they separate into two phases with different compositions, one of them rich in A, and the other in B (two-phase region). As shown in Fig. 7.38, the respective concentrations C_A and C_B of the two phases are given, at a temperature T, by the miscibility curve. Finally, the volume fraction ϕ of the A-rich phase in the B-rich phase can be calculated at thermodynamic equilibrium, as long as the average concentration C of the mixture is known. One is however left with the fundamental question of the morphology of the emulsion after the system is thermally quenched from a temperature T_i in the one-phase region to a temperature T_f in the two-phase region. Experiment shows that the morphology depends on whether T_f is above or below the *spinodal limit* of the monophase, which is by definition its absolute instability limit (this curve is plotted in the diagram as a dotted line).

Thus, the emulsion is formed by droplets dispersed in a continuous matrix if the system is quenched above the spinodal limit, in which case the separation proceeds by a nucleation–growth mechanism. On the contrary, the two phases will tend to form imbricated domains if the system is quenched below the spinodal limit (this can even lead to a 'bicontinuous' structure if the quench takes place at the critical concentration). In the following, we will not discuss the rheological properties of this particular type of suspension. We emphasise that, in the usual case, the faster and 'deeper' the quench (i.e. the closer T_f is to the spinodal limit) the smaller and more numerous the resulting droplets.

This 'chemical' method does not result in monodisperse suspensions. They can however be achieved using mechanical methods.

One such method consists of injecting the liquid L_d (where subscript 'd' stands for the dispersed liquid) into the liquid L_s (with 's' designating the 'solvent') using a pipette. Thus, the droplets are produced one by one. This (very tedious) method yields emulsions with a perfectly monodisperse size, which is very useful for testing the theoretical models.

Another solution consists of mixing the two liquids without taking any precautions, and then shearing them. This is what happens when we prepare a vinaigrette dressing by adding oil to vinegar. If the mixture is stirred long enough, the final result is an emulsion of oil droplets in water. When the stirring ceases, the vinaigrette 'breaks', the oil droplets tending to *coalesce* and form an oily layer floating at the water surface. In this way, the mixture reduces its interfacial energy. However, experiment shows that the coalescence process is slow – we have time enough to eat before the vinaigrette coalesces.

But if the droplets tend to coalesce, why do they form under shear?

The answer to this question is not that simple; moreover, it depends on the type of flow employed.

Let us illustrate this by analysing the behaviour of a droplet in elongational flow. The experimental geometry is depicted in Fig. 7.39a. The droplet of the L_d liquid that we intend to break is placed at the centre of an elongational flow produced by four rotating cylinders. The sequence of images in Fig. 7.39b shows the result of a real experiment, performed using two oils of identical viscosity (0.7 Pa s). The droplet of L_d (castor oil) is spherical at the start, with a diameter of 1.5 mm. As the liquid L_s (silicone oil) is sheared, the droplet starts to deform, stretching along the stream lines. Above a critical threshold, it takes the shape of an elongated cylinder, which starts to pinch at its centre. This event marks the beginning of the instability. If the shear is stopped at this point, the droplet of L_d keeps on pinching, transiently forming two droplets connected by a thin filament. Finally, the two droplets separate, while the filament gives a minuscule droplet, clearly visible in the last image. It should be noted here that, though the flow is responsible for the stretching of the droplet, it is not the origin of the break-up, which is due instead to the decrease in surface energy as the cylinder is pinched (a phenomenon discovered by Rayleigh, similar in nature to the instability of cylindrical jets).

In summary, this experiment shows that the break-up of a droplet is due to the competition between the stretching due to the viscous stress exerted by the liquid L_s at the surface of the L_d drop and the surface tension which, at least in the beginning, opposes the deformation of the droplet. As an order of magnitude, the viscous stresses (σ_{11} and σ_{22} in the experiment described above) are given by $2\eta_s\dot{\varepsilon}$, where $\dot{\varepsilon}$ is the local stretching rate. As to the interfacial tension forces that maintain the cohesion of the droplet, they are quantified by the capillary pressure difference Γ/R, where Γ is the surface tension between the two

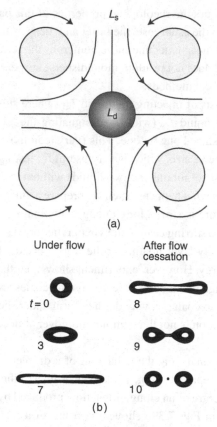

(a)

Under flow   After flow cessation

$t = 0$ 8

3   9

7   10

(b)

Fig. 7.39 (a) Sketch of the experimental setup; (b) image sequences showing the break-up of a droplet under the action of elongational flow. Images in the right column were taken after flow cessation. The time is given in seconds. Liquid A is a silicone oil and liquid B is castor oil (Janssen, 1993).

liquids. With these two quantities, one can build the following unitless number:

$$Ca = \frac{2\eta_s \dot{\varepsilon} R}{\Gamma} \qquad \text{(elongational flow).} \qquad (7.239)$$

This is the *capillary number*, representing the ratio between the (destabilising) viscous forces and the (stabilising) surface tension forces. One can thus predict that, for each pair of liquids L_s and L_d, there exists a *critical capillary number* Ca_{crit} above which the droplets of L_d become unstable. It turns out that this number depends on the ratio of the viscosity coefficients between the two liquids. The variation of Ca_{crit} as a function of the viscosity ratio $M = \eta_d/\eta_s$ is shown in Fig. 7.40 for elongational flow.

In this graph, we also plotted the critical capillary number for simple shear flow. In this case, the capillary number is defined as:

$$Ca = \frac{\eta_s \dot{\gamma} R}{\Gamma} \qquad \text{(simple shear flow),} \qquad (7.240)$$

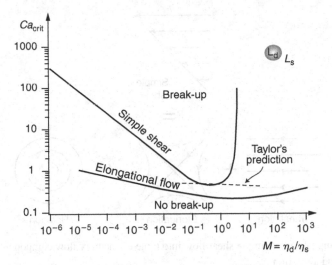

Fig. 7.40 Critical capillary number as a function of the ratio between the viscosity of the dispersed liquid (droplets) and that of the continuum (solvent) (Grace, 1982).

where $\dot{\gamma}$ is the usual shear rate.

The latter graph is above the former, showing that simple shear is less effective than elongation in breaking up the droplet. Another surprising observation is that, above a viscosity ratio of the order of 4, the droplet can no longer be broken by simple shear, irrespective of $\dot{\gamma}$.

A qualitative explanation for these differences can be found by first recalling that a simple shear is decomposed into three fundamental flow components (Fig. 7.41):

• a translation, which has only the effect of carrying the droplet along at the average flow velocity at its current position;
• an elongational flow, corresponding to the irrotational part of the velocity field and tending to stretch the droplet along a direction making a 45° angle with the direction of the velocity;
• a rotation flow, tending to spin the droplet about its axis; we have already discussed the importance of this effect on the rheology of rod suspensions.

Thus, simple shear tends to stretch the droplet at a 45° angle with respect to the flow direction, and also to spin it, bringing its stretching axis closer to the velocity direction. This second effect decreases the effectiveness of simple shear with respect to a purely elongational flow.

To illustrate this competition, we show in Fig. 7.42 the experimental record of a droplet breaking up under simple shear flow. This sequence of images shows indeed that the stretching axis of the droplet makes an angle smaller than 45° with the velocity direction. In the last image, one can clearly see the string of minute droplets formed immediately after break-up.

This discussion shows that, for each liquid mixture of L_s (the continuous medium, or solvent) and L_d (the dispersed medium), there exists a critical capillary number, whose value (calculated numerically) is plotted in Fig. 7.40.

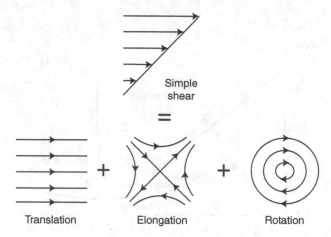

Fig. 7.41 Decomposition of simple shear flow into three elementary flow components: a translation, an elongation and a rotation.

Fig. 7.42 Silhouettes of a droplet (obtained from photographs) showing its break-up under simple shear. Once again, this is a castor oil droplet sheared in silicone oil with the same viscosity (Rumscheidt and Mason, 1961).

We should however say that G. I. Taylor was the first to provide an estimate for this number in 1934. He obtained it by calculating the elongation of the droplet under shear. Assuming that it takes the shape of an ellipsoid with major axis a and minor axis b, he calculated its deformation D:

$$D = \frac{a - b}{a + c} = Ca\, f(M), \tag{7.241}$$

where $f(M)$ is a function (of unit order) of the viscosity ratio $M = \eta_d/\eta_s$, given by:

$$f(M) = \frac{19M + 16}{16(M + 1)}. \tag{7.242}$$

Taylor then assumed that the droplet breaks as its deformation is of the order of 0.5. He obtained:

$$Ca_{\text{crit}}(\text{Taylor}) = \frac{0.5}{f(M)}. \tag{7.243}$$

Taylor's prediction is shown as a dotted line in Fig. 7.40. It is not too bad as long as M remains in the range 0.1 to 1.

Taylor went on to discuss the mechanism of emulsion formation. His reasoning is extremely simple, founded on the idea that, as the mixture is sheared, the L_d droplets keep on breaking until they are too small to break. This reasoning allowed him to predict the final size of the droplets under shear:

$$R = \frac{\Gamma Ca_{crit}}{\eta_s \dot{\gamma}}. \tag{7.244}$$

This famous formula shows, in particular, that the *radius of the droplets is inversely proportional to the shear rate*. This result is in good agreement with the experiments.

Note that we neglected here the possible droplet coalescence, leading to their growth. This assumption is legitimate insofar as coalescence is a rare event, experimentally.

Finally, we emphasise that this technique for emulsion preparation only works if the viscosity of the liquid to be dispersed is not too large (below about four times that of the solvent), a detail that Taylor did not predict.

In the next section, we describe the behaviour of an emulsion under continuous shear.

7.5.2.2 Viscosity η_0: the Taylor and Pal formulas

In 1932, Taylor also generalised the Einstein formula by taking into account the fluid motion within each droplet (Taylor, 1932). To simplify the calculations, he assumed that the capillary number, defined by the relation (7.240), is much smaller than 1, which amounts to saying that the shear rate is not too high, or that the droplets are very small.[9] Under these conditions, the droplets deform very little (linear regime) and can be considered spherical. Taylor then calculated the viscosity of the suspension:

$$\eta_0 = \eta_s \left(1 + \frac{1 + 2.5M}{1 + M}\phi\right). \tag{7.245}$$

As before, η_s is the viscosity of the solvent, M the viscosity ratio η_d/η_s and ϕ the volume fraction of L_d in L_s.

This formula reduces to the Einstein result for $M \to \infty$, but predicts lower viscosities in all other situations. This result is understandable insofar as the liquid within each sphere can flow, reducing the perturbation of the outer flow. In particular, the Taylor formula leads to $\eta_0 = \eta_s(1+\phi)$ for $M \to 0$ (this limit is attainable experimentally by using gas bubbles). This formula was verified experimentally in 1958 by M. A. Nawab and S. G. Mason, for M between 0.5 and 5 (Nawab and Mason, 1958).

Semi-empirical formulas, similar to that of Ball and Richmond for hard sphere, suspensions, were also proposed for concentrated emulsions in the linear regime at low shear rate. One of them was proposed by F. Pal, in 1992 (Pal, 1992). It reads:

$$\left(\frac{\eta_0}{\eta_s}\right)^{1/K} = \exp\left(\frac{2.5\phi}{1 - \phi/\phi_m}\right), \tag{7.246}$$

---

[9] This assumption implicitly requires that the droplets do not coalesce during the experiment. Indeed, this phenomenon might occur after a certain time, as the shear rate is presumed small compared to the value that fixes the droplet size (given by formula (7.244)). One way of avoiding this problem is to treat the surface of the droplets (using a surfactant, for instance) so that they repel at short range, thus stabilising the emulsion.

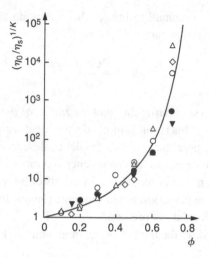

Fig. 7.43 Viscosity of concentrated emulsions. The solid line is the best fit with the Pal law (Eq. (7.246)) (Pal, 1992). The fit yields $\phi_m = 0.91$.

where ϕ_m is a sort of volume fraction of maximal stacking and K a constant given by:

$$K = \frac{0.4 + M}{1 + M}. \tag{7.247}$$

This formula reduces to the Taylor formula in the $\phi \to 0$ limit and predicts a divergence of the viscosity at ϕ_m. It accounts very well for the experimental results, as shown in Fig. 7.43.

Note that this experiment was performed using concentrated emulsions of vaseline oil in aqueous solutions of Triton X-100. The addition of surfactant is important since, by forming a monolayer at the surface of the droplets, it creates a repulsive barrier between them and prevents their coalescence.

We emphasise that the Pal formula also accounts for the experimental results obtained in the nonlinear regime. Fitting the experimental data then yields ϕ_m values above 0.91, increasing with the shear rate.

Finally, note that the concentrated suspensions generate considerable normal stresses in the nonlinear regime due to the strong deformation of the droplets under shear. Theoretical expressions for the normal stresses were proposed by J. C. Choi and W. R. Schowalter in 1975, but they have not yet been tested experimentally (Choi and Schowalter, 1975).

7.5.2.3 Linear viscoelasticity: the expression of G' and G''

In this section, we only consider the case of dilute emulsions, still under the assumptions that both liquids are Newtonian. We also assume that the surface tension Γ remains constant as the droplets deform. For simplicity's sake, we will assume that the two liquids have the same viscosity $\eta = \eta_s = \eta_d$. In this case (and only in this case), the emulsion behaves as a Jeffrey liquid, for which we can simply calculate the complex modulus G^*.

Before proceeding with this calculation, let us recall the expression for the complex modulus of a Jeffrey liquid. As in the previous chapter, we denote by G_p the elastic modulus associated with the particles (here, the droplets of the L_d liquid, with radius R, at rest) and by η_P the corresponding viscosity. In this case, the complex modulus is given by Eqs. (6.76) and (6.77):

$$G^* = i\omega \left(\eta + \frac{\eta_P}{1 + i\omega\tau_P} \right), \tag{7.248}$$

where $\tau_P = \eta_P/G_P$ is the viscoelastic relaxation time associated with the droplets. The first term accounts for the effect of the solvent and the second one for that of the droplets.

Let us now calculate η_P and G_P.

To find the expression for η_P, one need only remember that, at zero shear rate, the viscosity η_0 of the suspension is the sum of the viscosities associated with the two Maxwell elements representing the emulsion:

$$\eta_0 = \eta + \eta_P. \tag{7.249}$$

Using the Taylor formula (7.245) immediately yields the expression for the viscosity η_P (with $M = 1$):

$$\eta_P = \frac{7}{4}\eta\phi. \tag{7.250}$$

To calculate the elastic modulus G_P, let us go to the high-frequency limit ($\omega \to \infty$). Now, Eq. (7.248) becomes $G^* = i\omega\eta + G_P \approx i\omega\eta$. This expression shows that the suspension behaves as the pure solvent. However, beware that *this is only true if the viscosities of the solvent and of the dispersed phase are equal.* Indeed, although the surface tension effects can always be neglected at high frequency (since the associated pressure terms are independent of ω, whereas the viscous stress increases as ω), the equality of the viscosities is required in order for the velocity profile to remain linear throughout the system, as if the droplets did not exist. This is a necessary condition for the high-frequency dynamic viscosity η' of the suspension to equal that of the solvent. An immediate consequence is that *the deformation of the droplets under shear is affine*, allowing a simple determination of the elastic modulus. Indeed, assume that the suspension is 'suddenly' sheared by an amount γ. In this case, each droplet of radius R takes an ellipsoidal shape with a conserved volume, but with a surface area that is larger by an amount:

$$\Delta S = \gamma^2 \int_{-R}^{+R} \int_{-\sqrt{R^2-x^2}}^{+\sqrt{R^2-x^2}} \sqrt{1 - \frac{x^2 + y^2}{R^2}} \left(1 - \frac{x^2}{R^2} \right) = \frac{8\pi}{15} R^2 \gamma^2. \tag{7.251}$$

Note that this ellipsoid has no revolution symmetry, in contrast with the low shear rate situation. It follows that the energy stored in the medium is given by:

$$E_{\text{elast}} = \frac{\phi}{\frac{4}{3}\pi R^3} \Gamma \Delta S. \tag{7.252}$$

On the other hand, we have shown in the previous chapter that this quantity was related to the elastic modulus by the relation:

$$E_{\text{elast}} = G_P \frac{\gamma^2}{2},\tag{7.253}$$

where we set $G' = G_P$ in Eq. (6.90), since the deformation is considered instantaneous. Comparing this expression with the one above and using Eq. (7.251) yield the elastic modulus:

$$G_P = \frac{4}{5}\frac{\Gamma}{R}\phi.\tag{7.254}$$

Finally, the complex modulus of the suspension is obtained by replacing η_P and G_P by their expressions (7.250) and (7.254) in Eq. (7.248):

$$G^* = i\omega\eta\left(1 + \frac{5}{2}\phi\frac{56(\Gamma/R)}{80(\Gamma/R) + 175i\omega\eta}\right).\tag{7.255}$$

This expression was generalised by J. G. Oldroyd in the 1950s, first to the case where the two liquids have different viscosities, by using an *effective medium* theory (Oldroyd, 1953), and then to account for the complex character of the surface tension between the two fluids (including, in particular, a surface viscosity, which is not necessarily negligible when the droplets are covered by a layer of surfactant or polymer) (Oldroyd, 1955). More recently, J.-F. Palierne extended the Oldroyd theory to the situation when the two fluids themselves can be viscoelastic (Palierne, 1990; [27]).

These theories, with that of Palierne being the most complete, were successfully tested in experiments (Kitade *et al.*, 1997; [27])).

In the next section, we will deal with the case of dilute polymer solutions. We will see that they are also viscoelastic and that they behave, under simple shear and in the $\dot{\gamma} \to 0$ limit, as suspensions of spherical particles, with a radius that is proportional to the gyration radius of the polymer.

7.5.3 Dilute polymer solutions

We showed in Section 7.3.1 that a polymer can be dissolved in a solvent, provided they are compatible. If the polymer concentration is very low, the chains are independent from one another. One might then think that the Rouse theory, described in the section on polymer melts, is directly applicable to these systems. Experiment shows that this assumption is unwarranted, and that new effects (neglected in the Rouse theory) must be taken into account. The most important among them was already touched upon briefly in the section dealing with polymers in semi-dilute solution: it has to do with the *hydrodynamic interactions* between monomers. We will see that the *Zimm model* can take this into account, in an approximate manner. Another effect concerns the *excluded volume interactions*. We have already described them when discussing polymer solutions, showing that they are responsible for the non-Gaussian behaviour of the chain. These interactions are manifest in *'good' solvents*, where one must take $v \neq 0$. In this case, the chain behaviour is no longer ideal, and its gyration radius goes as $M^{3/5}$ (instead of $M^{1/2}$ in the Gaussian case). This additional difficulty can however be avoided by using a Θ solvent where $v = 0$ by definition.

In the following, to simplify the calculations, we will restrict ourselves to this very specific (and nonetheless perfectly attainable in practice) condition. We will present some experimental results on the linear regime viscoelasticity of a dilute polymer solution in Θ solvent. We will show that these results are incompatible with the Rouse theory, the Gaussian character of the chain notwithstanding. We will then describe the Zimm model, which is a generalisation of the Rouse model to account for the hydrodynamic interactions between monomers. We will see that this model is in good agreement with the experimental results in Θ solvent (as well as in good solvent, a case for which we will only present some classical results). We will conclude by giving the expression for the viscosity η_0 of a dilute solution under simple shear. At the same time, we will discuss the definition of the *hydrodynamic radius* of a polymer coil in solution.

Let us begin by presenting some experimental results on the viscoelastic behaviour of a dilute polymer solution in Θ solvent.

7.5.3.1 Some linear viscoelasticity results

The elastic and loss moduli of dilute polyethylene solutions in Θ solvents were measured, and the data are shown in Fig. 7.44.

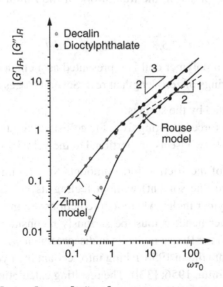

Fig. 7.44 Intrinsic moduli $\left[G'(\omega)\right]_R$ and $\left[G''(\omega)\right]_R$ as a function of the scaled frequency $\omega\tau_0$ for two PE solutions with molecular mass 860 000 in Θ solvent. The solid lines correspond to the predictions of the Zimm model, to be presented in the next section (Johnson *et al.*, 1970).

Along the y-axis are plotted the *intrinsic moduli*, defined as:

$$[G'(\omega)]_R = \lim_{\rho \to 0} \frac{M}{\rho RT} G'(\omega), \tag{7.256}$$

$$[G''(\omega)]_R = \lim_{\rho \to 0} \frac{M}{\rho RT} \left[G''(\omega) - \omega \eta_s\right], \tag{7.257}$$

where ρ is the polymer mass dissolved per unit volume of solution and η_s the viscosity of the solvent. Note that, in the case of G'', the contribution of the pure solvent was subtracted. The parameters thus obtained are therefore due only to the polymer (and its interactions with the solvent).

As to the frequency, it is rescaled by multiplication with the typical time τ_0, defined as:

$$\tau_0 = \lim_{\rho \to 0} (\eta_0 - \eta_s) \frac{M}{\rho RT}, \tag{7.258}$$

where η_0 is the viscosity of the solution measured in the linear regime. We can already note that, in the Zimm model, this time is related to the terminal relaxation time τ_1 of the equivalent Maxwell distribution by the relation $\tau_0 = 1.64\tau_1$.

These curves show that, at high frequencies ($\omega\tau_0 \gg 1$), the two intrinsic moduli are not equal, and they vary as $\omega^{2/3}$. This result is in disagreement with the Rouse model, which predicts an $\omega^{1/2}$ dependence and moduli that are equal in this regime.

At low frequencies ($\omega\tau_0 \ll 1$), on the other hand, the elastic modulus varies as ω^2 and the loss modulus as ω, which is the usual dependence in the terminal regime.

These data can be interpreted in the framework of the Zimm model that we will now present.

7.5.3.2 The Zimm model

In the Rouse model, each polymer coil is represented as a chain of beads, each one connected to the next by springs of zero length at rest. Several forces act on the beads:

- the elastic forces exerted by the springs;
- a stochastic Langevin force, resulting from the collision with the solvent particles;
- a friction force, taken by Rouse as proportional to the velocity of the bead.

This simplified form of the friction force amounts to assuming that each bead moves across the medium (here, the solvent) without feeling its neighbours. This assumption seems reasonable in polymer melts, where its predictions are in good agreement with the experiments. On the other hand, it must be seriously reconsidered in the case of dilute polymer solutions, where the Rouse model is in disagreement with the experiments.

This is what B. H. Zimm did in 1956, taking into account the hydrodynamic interactions between all the beads (Zimm, 1956; [32]). The resulting calculations are much more complicated, since each bead interacts with all the others, as the hydrodynamic interactions are well known to be slowly decreasing with distance. For this reason, we will simply sketch the main lines of the Zimm model, without going into the calculation details.

To state the problem in the most general terms possible, let us show how the forces $\{\vec{F}\} = \{\vec{F}_1, \vec{F}_2, \ldots, \vec{F}_N\}$ acting on the N beads are connected to their positions $\{\vec{R}\} = \{\vec{R}_1, \vec{R}_2, \ldots, \vec{R}_N\}$. In the Rouse model, one simply writes for each particle that $\vec{F}_i = \zeta \, d\vec{R}_i/dt$, where ζ is a phenomenological friction coefficient. In fact, the velocity of each particle depends on the forces acting on all the others, due to the hydrodynamic interactions between them. One must then write:

$$\frac{d\vec{R}_n}{dt} = \sum_m H_{nm} \vec{F}_m, \tag{7.259}$$

where H_{nm} is a matrix to be obtained by solving the hydrodynamic velocity field around the particles. To understand how the reasoning goes, let us write the equations of motion for the fluid. These include the incompressibility equation:

$$\operatorname{div} \vec{v} = 0 \tag{7.260}$$

and the Navier–Stokes equation (written in the $Re \ll 1$ limit):

$$\eta_s \, \Delta \vec{v} + \overrightarrow{\operatorname{grad}} P + \sum_n \vec{F}_n \, \delta\left(\vec{r} - \vec{R}_n\right) = 0. \tag{7.261}$$

This equation contains the forces acting on the beads (and thus on the fluid, due to the no-slip condition on the beads, taken here as point-like). Oseen showed that the solution to this problem is (for a demonstration see [32], p. 88):

$$\vec{v}(\vec{r}) = \sum_n \underline{H}\left(\vec{r} - \vec{R}_n\right) \vec{F}_n, \tag{7.262}$$

where:

$$\underline{H}(\vec{r}) = (\underline{I} + \vec{u}\vec{u}) \frac{1}{8\pi \eta_s r} \tag{7.263}$$

with $\vec{u} = \vec{r}/r$ and $(\vec{u}\vec{u})_{ij} = u_i u_j$.[10] As, on the other hand:

$$\vec{v}(\vec{R}_n) = \frac{d\vec{R}_n}{dt} \tag{7.264}$$

due to the no-slip condition on the particles, one finally has:

$$\underline{H}_{nm} = \underline{H}\left(\vec{R}_n - \vec{R}_m\right) \quad (n \neq m). \tag{7.265}$$

---

[10] In general, we will simply denote by $\vec{u}\vec{v}$ the *dyadic product* of two vectors $\vec{u}$ and $\vec{v}$. This is the second-rank tensor with components $u_i v_j$ in Cartesian coordinates. The mathematicians often write this product as $\vec{u} \otimes \vec{v}$.

It should be noted that, due to the assumption of point-like particles, $\underline{H}_{nn} = \underline{H}(0)$ diverges. This divergence can be removed by assuming, as in the Rouse model, that the beads have a finite radius r. Under these conditions:

$$\underline{H}_{nn} = \frac{\underline{I}}{\zeta} \tag{7.266}$$

with $\zeta = 6\pi \eta_s r$.

This purely hydrodynamical problem having been solved, we can now write the general equation for the motion of beads in a Θ solvent (where $v = 0$):

$$\frac{\partial \vec{R}_n}{\partial t} = \sum_m \underline{H}_{nm} \left(\kappa \frac{\partial^2 \vec{R}_m}{\partial m^2} + \vec{f}_m(t) \right), \tag{7.267}$$

where κ is the stiffness of the springs and $\vec{f}_m$ the Langevin force acting on the bead m. This equation extends the Rouse equation (7.33). It is however much more complicated, since the matrix elements $\underline{H}_{nm}$ are nonlinear functions of $\vec{R}_n - \vec{R}_m$.

This problem being intractable, Zimm proposed to simplify it by replacing the matrices $\underline{H}_{nm}$ by their equilibrium statistical average $\langle \underline{H}_{nm} \rangle_{\text{eq}}$. This procedure yields:

$$\langle \underline{H}_{nm} \rangle_{\text{eq}} = \frac{\underline{I}}{(6\pi^3 |n - m|)^{1/2} \eta_s b} \equiv h(n - m)\underline{I} \quad (n \neq m), \tag{7.268}$$

$$\langle \underline{H}_{nn} \rangle_{\text{eq}} = \frac{\underline{I}}{\zeta} \equiv h(0)\underline{I}. \tag{7.269}$$

As a result, Eq. (7.267) can be rewritten under the following (approximate) form:

$$\frac{\partial \vec{R}_n}{\partial t} = \sum_m h(n - m) \left(\kappa \frac{\partial^2 \vec{R}_m}{\partial m^2} + \vec{f}_m(t) \right). \tag{7.270}$$

This equation shows that the hydrodynamic interactions between beads decrease very slowly, as $|n - m|^{-1/2}$. For this reason, one can expect qualitatively important behaviour differences with respect to the Rouse model, where these interactions were completely neglected.

This can be shown by solving the problem in terms of *normal modes*. This procedure (in all points identical to that employed by Rouse) leads after several (hard to justify) approximations, to the following simplified equations:

$$\zeta_p \frac{\partial}{\partial t} \vec{X}_p(t) = -\kappa_p \vec{X}_p + \vec{f}_p(t) \tag{7.271}$$

with

$$\zeta_p = (12\pi^3)^{1/2} \eta_s (Nb^2 p)^{1/2} \qquad p = 1, 2, \ldots, \tag{7.272}$$

$$\zeta_0 = \frac{3}{8} \left(6\pi^3\right)^{1/2} \eta_s \left(Nb^2\right)^{1/2}, \tag{7.273}$$

$$\kappa_p = \frac{6\pi^2 k_B T}{Nb^2} p^2 \qquad p = 0, 1, 2, \ldots \tag{7.274}$$

These formulas (where the terms due to the phenomenological friction coefficient ζ were neglected) yield the diffusion coefficient of the molecular centre of mass. Using Eq. (7.52) gives directly:

$$D_G = \frac{k_B T}{\zeta_0} = 0.196 \frac{k_B T}{\eta_s \overline{R}}, \tag{7.275}$$

where $\overline{R}$ designates the mean square value of the end-to-end vector $\vec{R}$. Since the chain exhibits ideal behaviour in Θ solvent, $\overline{R} = \sqrt{N} b$ and $D_G \propto M^{-1/2}$ (in contrast with the Rouse model, where $D_G \propto M^{-1}$ from Eq. (7.53)).

From Eqs. (7.44) and (7.65), one can also calculate the spectrum of relaxation time of the associated Maxwell model:

$$\tau_p = \frac{\tau_R}{2p^2} \qquad p = 1, 2, \ldots, \tag{7.276}$$

where τ_R is the rotational diffusion time of the molecule, with the expression:

$$\tau_R = 0.325 \frac{\eta_s \overline{R}^3}{k_B T}. \tag{7.277}$$

We can see that the longest (or terminal) relaxation time, given by $\tau_1 = \tau_R/2$, varies as $M^{3/2}$. This dependence is not the same as for the Rouse model, where τ_1 goes as M^2 (see Eq. (7.65)).

As to the elastic modulus associated with each Maxwell element, it is constant and, from Eq. (7.68), its value is:

$$G_p^{\text{Zimm}} = G_R = \frac{\rho R T}{M}, \tag{7.278}$$

where, this time, ρ represents the mass of polymer dissolved per unit volume of solution.

These formulas yield the linear viscoelasticity functions G' and G'' (from Eqs. (7.72) and (7.73)). In intrinsic form (see Eqs. (7.256) and (7.257)), these functions read:

- at low frequency ($\omega \tau_1 < 1$):

$$[G'(\omega)]_R = (\omega \tau_1)^2 \sum_{p=1}^{\infty} \frac{1}{p^3} \propto \omega^2, \tag{7.279}$$

$$[G''(\omega)]_R = \omega \tau_1 \sum_{p=1}^{\infty} \frac{1}{p^{3/2}} \propto \omega; \tag{7.280}$$

• at high frequency ($\omega\tau_1 > 1$):

$$[G'(\omega)]_R = \int_0^\infty \frac{(\omega\tau_1)^2 p^{-3}}{1 + (\omega\tau_1)^2 p^{-3}} \, dp = 1.21(\omega\tau_1)^{2/3} \propto \omega^{2/3}, \qquad (7.281)$$

$$[G''(\omega)]_R = \int_0^\infty \frac{(\omega\tau_1) p^{-3/2}}{1 + (\omega\tau_1)^2 p^{-3}} \, dp = 2.09(\omega\tau_1)^{2/3} \propto \omega^{2/3}. \qquad (7.282)$$

These functions are plotted in Fig. 7.44. Clearly, the agreement with the experimental data is excellent, especially since there is no adjustable parameter. One must however remain cautious, in view of the numerous theoretically uncontrolled approximations in Zimm's calculations. Such an excellent quantitative agreement with the experimental data is certainly accidental, undoubtedly resulting from the compensation of several approximations. It is nevertheless certain that the Zimm model, which accounts for the hydrodynamic interactions between particles, does capture the essential physics of dilute solutions.

In conclusion, let us give the expression for the solution viscosity η_0 or, equivalently, its intrinsic viscosity in the linear regime, defined as:

$$[\eta] = \lim_{\rho \to 0} \frac{\eta_0 - \eta_s}{\rho \eta_s}. \qquad (7.283)$$

In the Maxwell model, the viscosity η_0 is the sum of all the viscosities of the constitutive Maxwell elements. Using Eqs. (7.276)–(7.278), one has:

$$[\eta] = 0.267 \frac{N_A}{M} \overline{R}^3, \qquad (7.284)$$

where N_A is the Avogadro number.

This formula shows that, in a Θ solvent (where $\overline{R}$ varies as $M^{1/2}$), the intrinsic viscosity of the solution increases as $M^{1/2}$. This dependence is very different from that found in the framework of the Rouse model, where we saw that the viscosity η_0 varies as M.

The formula also shows that the solution can be treated in the linear regime as a suspension of hard spheres with an equivalent radius R_h. Indeed, for such a suspension, one can calculate from the Einstein formula, knowing that $\phi = (4/3)\pi R_h^3 (\rho/M) N_A$:

$$[\eta] = \frac{10\pi}{3} \frac{N_A}{M} R_h^3. \qquad (7.285)$$

Simple identification yields the *hydrodynamic radius* of the polymer coil in Θ solvent:

$$R_h = 0.294 \, \overline{R}. \qquad (7.286)$$

In practice, it is more convenient to express R_h as a function of the gyration radius $R_g = \overline{R}/\sqrt{6}$ (measured directly in static light scattering) yielding:

$$R_h = 0.72 \, R_g. \qquad (7.287)$$

Hence, this formula shows that each polymer coil behaves as a hard sphere, of radius slightly smaller than its gyration radius.

This prediction of the Zimm model was verified experimentally with good accuracy by S. J. Candau *et al.*, who found that $R_h/R_g \approx 0.6$ (Candau *et al.*, 1984).

To conclude this section, we emphasise that the Zimm model can be extended to the case of good solvents ($v \neq 0$) [32]. The scaling laws are different in this case. For instance, one finds that $\overline{R}$ (and hence R_g) varies as $M^{3/5}$, that the intrinsic viscosity η varies as $M^{4/5}$, and that the intrinsic linear viscoelasticity functions are given at high frequency ($\omega\tau_1 \gg 1$) by formulas of the type:

$$[G'(\omega)]_R \approx 1.14(\omega\tau_1)^{5/9}, \tag{7.288}$$

$$[G''(\omega)]_R \approx 1.38(\omega\tau_1)^{5/9}. \tag{7.289}$$

Here, the 5/9 exponent is lower than that found in a Θ solvent, namely 2/3. Thus, the Zimm model predicts that the exponent of the intrinsic moduli decreases as the solubility of the polymer increases. This effect was indeed demonstrated experimentally.

Appendix 7.A Hookean dumbbell model

This model is a simplified version of the Rouse model. It considers only the longest relaxation time τ_1 (denoted by τ in the following) of the polymer coil. Since only one vibration mode is preserved, the polymer coil can be simply depicted as a dumbbell, i.e. as two beads connected by a spring (Fig. 7.A.1). In the Hookean version of this model, the extensibility of the spring is infinite.

To obtain the stress equation, it is no longer necessary to use a normal mode representation. Indeed, let us denote by $\vec{R} = \vec{R}_2 - \vec{R}_1$ the end-to-end vector, by $\overline{\kappa}$ the stiffness of the spring (considered constant) and by v (equal to c/N) the number of dumbbells (or molecules) per unit volume. From Eq. (7.27), the stress tensor can be expressed directly as a function of $\vec{R}$ as:

$$\underline{\sigma} = v\overline{\kappa}\left\langle \vec{R}\vec{R}\right\rangle, \tag{7A.1}$$

where $\underline{S} = \left\langle \vec{R}\vec{R}\right\rangle$ is the *configuration tensor* with components $S_{\alpha\beta} = \left\langle R_\alpha R_\beta\right\rangle$.

To find the stress equation, let us start by writing the equations of motion for each of the beads in the quiescent fluid:

$$\overline{\zeta}\,\frac{d\vec{R}_1}{dt} = -\overline{\kappa}(\vec{R}_1 - \vec{R}_2) + \vec{f}_1, \tag{7A.2a}$$

$$\overline{\zeta}\,\frac{d\vec{R}_2}{dt} = -\overline{\kappa}(\vec{R}_2 - \vec{R}_1) + \vec{f}_2. \tag{7A.2b}$$

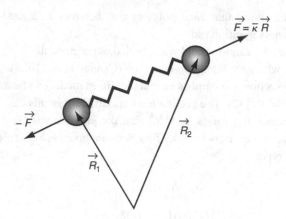

Fig. 7.A.1 Dumbbell representing a polymer coil.

In these equations, $\bar{\zeta}$ is a phenomenological friction coefficient, while $\vec{f}_1$ and $\vec{f}_2$ are random forces, with a vanishing time average, such that:

$$\langle f_{n\alpha}(t) f_{m\beta}(t') \rangle = 2\bar{\zeta} k_{\mathrm{B}} T \delta_{nm} \delta_{\alpha\beta} \delta(t - t') \qquad n, m = 1, 2. \tag{7A.3}$$

Subtracting the two equations (7A.2) leads to the evolution equation for the end-to-end vector in the quiescent fluid:

$$\bar{\zeta} \frac{d\vec{R}}{dt} = -2\bar{\kappa}\vec{R} + \vec{f}_2 - \vec{f}_1. \tag{7A.4}$$

If the fluid flows, only the relative motion of the dumbbells with respect to the average flow counts. Let $\vec{v}$ be the velocity. Then, the end-to-end vector obeys the following differential equation:

$$\frac{\mathrm{D}\vec{R}}{\mathrm{D}t} - (\vec{\nabla}\vec{v})\vec{R} = -2\frac{\bar{\kappa}}{\bar{\zeta}}\vec{R} + \frac{1}{\bar{\zeta}}(\vec{f}_2 - \vec{f}_1). \tag{7A.5}$$

Starting from this equation, one can obtain the configuration tensor $\langle \vec{R}\vec{R} \rangle$ by exactly the same procedure as in Section 7.1.6.10. The stress equation is then deduced as:

$$\overset{\triangledown}{\underline{\sigma}} + \frac{1}{\tau}(\underline{\sigma} - G\underline{I}) = 0, \tag{7A.6}$$

where τ is the *viscoelastic relaxation time*:

$$\tau = \frac{\bar{\zeta}}{4\bar{\kappa}} \tag{7A.7}$$

and G the *elastic modulus*:

$$G = \nu k_{\mathrm{B}} T. \tag{7A.8}$$

This equation yields back the upper-convected (or contravariant) Maxwell equation found in the previous chapter:

$$\tau \overset{\triangledown}{\underline{\sigma}} + \underline{\sigma} = 2\eta \underline{A} \tag{7A.9}$$

if the stress tensor is redefined as in Section 7.1.6.10: $\underline{\sigma} \equiv \underline{\sigma} - G\underline{I}$. Note that, in this model, the viscosity η is:

$$\eta = \tau G = \frac{\overline{\zeta} \nu k_B T}{4\overline{\kappa}}. \tag{7A.10}$$

In conclusion, let us discuss the relation between the phenomenological coefficients $\overline{\kappa}$ and $\overline{\zeta}$ of the dumbbell model and those (κ and ζ) of the Rouse model.

Concerning the stiffness, the spring of the dumbbell is equivalent to the N springs connected in series of the Rouse chain. One can therefore write $1/\overline{\kappa} = N/\kappa$, yielding:

$$\overline{\kappa} = \frac{\kappa}{N} = \frac{3k_B T}{Nb^2}. \tag{7A.11}$$

As to the friction coefficient $\overline{\zeta}$, it must be proportional to the number of Kuhn segments of the chain: $\overline{\zeta} = N\zeta$. This can be checked by identifying Eq. (7A.9) to that found for the stress σ_1 associated with the slowest mode of the Rouse model (see Eq. (7.89)). This comparison gives:

$$\overline{\zeta} = \frac{2}{\pi^2} N\zeta. \tag{7A.12}$$

In the following appendix, we will discuss two possible generalisations (among the many possible) of the dumbbell model to the nonlinear regime.

Appendix 7.B The Giesekus and FENE models

The model of infinitely extensible (or Hookean) dumbbells leads (like the Rouse model) to the upper-convected Maxwell model. Thus, it does not explain the shear-thinning behaviour of polymers at high shear rates (for $\dot{\gamma} > 1/\tau$), or the saturation of their elongational viscosity at high elongation rate (predicting instead a divergence for $\dot{\varepsilon} \to 1/(2\tau)$).

To improve this model, the local nonlinearities of the material must be taken into account. In this appendix, we will discuss two possible extensions (among many others) of the dumbbell model to the nonlinear regime. The first one introduces the anisotropy of the friction coefficient under flow. It leads to the Giesekus model, already presented without demonstration in Section 6.4.6.2. We will then discuss the FENE model (also mentioned in Section 6.4.6.3), which accounts for the finite extensibility of the dumbbells.

7.B.1 The Giesekus model

In 1966, Giesekus proposed an improvement to the dumbbell model, by accounting for the anisotropy of the friction coefficient $\overline{\zeta}$ (Giesekus, 1966). The idea is simple: as the polymer is sheared, the molecules are stretched and the medium becomes anisotropic (a point one can check experimentally by measuring the flow-induced birefringence[11]). In such a medium, there is no longer any reason to assume that the friction force acting on each bead is isotropic. To account for this local anisotropy, Giesekus proposed replacing in the dumbbell equation (7A.6) the mobility $1/\tau = 4\bar{\kappa}/\zeta$ by a mobility tensor $\underline{M}/\tau$. Under these conditions, the dumbbell equation becomes:

$$\overset{\triangledown}{\underline{\sigma}} + \frac{\underline{M}}{\tau}(\underline{\sigma} - G\underline{I}) = 0, \qquad (7\text{B}.1)$$

where $\underline{M}$ is a tensor depending on the stress state of the material. Since, at equilibrium, the stress tensor is isotropic, the tensor $\underline{M}$ shares this property, since $\underline{\sigma} = G\underline{I}$. On the other hand, when $\underline{\sigma}$ becomes anisotropic, so does $\underline{M}$. The simplest way of accounting for this behaviour, and this is what Giesekus did, is to assume that the anisotropy of $\underline{M}$ is proportional to that of $\underline{\sigma}$:

$$\underline{M} - \underline{I} = \alpha(\underline{\sigma}/G - \underline{I}), \qquad (7\text{B}.2)$$

where α is an empirical prefactor. Replacing $\underline{M}$ by its expression (7B.2) in Eq. (7B.1) and redefining the stress tensor as before: $\underline{\sigma} \equiv \underline{\sigma} - G\underline{I}$, leads to the Giesekus equation, already presented in the previous chapter (see Eq. 6.305):

$$\tau\overset{\triangledown}{\underline{\sigma}} + \underline{\sigma} + \alpha\frac{\tau}{\eta}\left(\underline{\sigma}\,\underline{\sigma}\right) = 2\eta\underline{A}. \qquad (7\text{B}.3)$$

This semi-phenomenological equation does indeed predict the shear-thinning behaviour of the polymers at high shear rate, and also the saturation of the elongational viscosity at high elongation rate (provided α is chosen appropriately, see Section 6.4.6.2).

7.B.2 The FENE model

Another (certainly more physical) improvement to the dumbbell model consists in accounting for their *finite extensibility L*. As a dumbbell cannot be stretched beyond L, its stiffness $\underline{\kappa}$ must become infinite for $R = L$. This behaviour is well reproduced by Warner's spring law (Warner, 1972), also known as FENE springs (for 'Finitely Extensible Nonlinear-Elastic springs'):

$$\vec{F} = \frac{\kappa}{1 - (R/L)^2}\vec{R} = H(R^2)\vec{R}. \qquad (7\text{B}.4)$$

In this expression, $\underline{\kappa}$ designates once again the Brownian tension of the polymer chain, given by Eq. (7A.11).

---

[11] It can be shown that the anisotropic part of the refractive index tensor (or birefringence tensor) $\underline{n}_P$ is proportional to the $\underline{\sigma}_P$ component (associated with the particles) of the stress tensor: $\underline{n}_P = C\underline{\sigma}_P$ (stress-optic law). The proportionality prefactor C is known as the stress-optic coefficient. It is proportional to the difference in polarisability for a Kuhn segment between the directions parallel and perpendicular to its axis (Janeschitz–Kriegl, 1983). It is noteworthy that the stress-optic law only applies in the linear regime, like expression (7A.1) of the stress tensor.

Since the force law is nonlinear, there is no longer a linear relation between the stress tensor and the configuration tensor $\langle \vec{R}\vec{R} \rangle$ (see Eq. (7A.1)). However, the demonstration given in Section 7.1.5 shows that, very generally:

$$\underline{\sigma} = \nu \langle \vec{R}\vec{F} \rangle, \tag{7B.5}$$

where ν is once again the number of dumbbells per unit volume and $\vec{F}$ the elastic (non-hydrodynamic) force acting on each bead.

The difficulty here is that, due to the nonlinearity of the $H(R)$ function, one can no longer find an analytical solution for the constitutive $\underline{\sigma}$ equation.

An approximate way of sidestepping this problem consists in replacing $H(R^2)$ by the pre-averaged quantity $H(\langle R^2 \rangle)$. This is the Peterlin approximation (Peterlin, 1961), leading to the P-FENE model. Since in this model the function H is assumed to depend only on $\langle R^2 \rangle$ (and not on R^2), Eqs. (7B.4) and (7B.5) immediately yield:

$$\underline{\sigma} = \nu \overline{\kappa} \left(1 - \frac{\text{tr}(\underline{S})}{L^2} \right)^{-1} \underline{S}, \tag{7B.6}$$

where $\underline{S} = \langle \vec{R}\vec{R} \rangle$ is once again the configuration tensor (keeping in mind that $\text{tr}(\underline{S}) = \langle R^2 \rangle$).

As to the configuration tensor, it can be shown to verify the following equation:

$$\overset{\triangledown}{\underline{S}} + \frac{1}{\tau} \left(\frac{\underline{S}}{1 - \text{tr}(\underline{S})/L^2} - \frac{k_B T}{\kappa} \underline{I} \right) = 0. \tag{7B.7}$$

The two coupled equations (7B.6) and (7B.7) represent the core of the P-FENE model. Solving it is outside the scope of this book (for the details, we refer the reader to [28], volume 2), but it predicts that at high shear rates the viscosity and the first normal stress coefficient decrease as power laws:[12]

$$\frac{\eta(\dot{\gamma})}{\eta} = \left(\frac{3L^2}{2Nb^2} \right)^{1/3} (\tau\dot{\gamma})^{-2/3} \qquad (\dot{\gamma} \gg 1/\tau), \tag{7B.8}$$

$$\frac{\Psi_1(\dot{\gamma})}{\eta\tau} = \left(\frac{3L^2}{2Nb^2} \right)^{2/3} (\tau\dot{\gamma})^{-4/3} \qquad (\dot{\gamma} \gg 1/\tau), \tag{7B.9}$$

confirming the shear-thinning behaviour of polymers. As to the second normal stress difference, this model predicts it to be zero for all shear rate values.

Finally, the elongational viscosity saturates in the limit of high elongation rates at a value:

$$\frac{\eta_e}{3\eta} = \frac{2\overline{\kappa}L^2}{k_B T} = \frac{6L^2}{Nb^2} \qquad (\dot{\varepsilon} \gg 1/2\tau). \tag{7B.10}$$

---

[12] The asymptotic behaviour is obtained by introducing the small quantity $\varepsilon = 1 - \text{tr}(\underline{S})/L^2$, and by solving the equations to lowest order in ε.

As an order of magnitude, $L \approx 0.8Nb$ in a flexible organic polymer such as silicone. Thus, the P-FENE model predicts that $\eta_e/3\eta \approx 5N$ at saturation. This ratio can then become very large for high molecular mass polymers, in agreement with the experiment.

Note that, in the entire discussion above, the contribution of the solvent was neglected. For polymers in dilute solution, one must obviously take it into account. In this case, the three preceding equations remain applicable, provided τ is replaced by τ_P, η by the viscosity at zero shear rate $\eta_0 = \eta_s + \eta_P$ and η_e by $\eta_e - 3\eta_s$. We recall that η_s is the viscosity of the solvent, η_P the contribution of the polymer to the total viscosity (given by Eq. (7A.10)) and τ_P the viscoelastic relaxation time associated with the polymer (given by Eq. (7A.7)).

Appendix 7.C Demonstration of the Einstein Formula

Einstein's proof being rather complicated, we will follow a simpler approach, inspired by the book by W. B. Russel *et al.* on colloidal suspensions [38] and a lecture by J. Hinch on the rheology of complex fluids, given at the 'École de physique des Houches' during the winter of 2003.

We begin by writing the stress field in the following general form:

$$\underline{\sigma} = -P\underline{I} + 2\eta_s\,\underline{A} + \underline{\sigma}^P. \tag{7C.1}$$

The first two terms correspond to the stress tensor in the fluid, P being the hydrostatic pressure and $\underline{A}$ the strain rate tensor $A_{ij} = (1/2)(\partial v_i/\partial x_j + \partial v_j/\partial x_i)$. The third term gives the stress field within the particles.

Since, in practice, one measures an average viscosity, this expression must be averaged over a volume containing a large number of particles. This yields:

$$\langle\underline{\sigma}\rangle = -\langle P\rangle\underline{I} + 2\eta_s\langle\underline{A}\rangle + \langle\underline{\sigma}^P\rangle \tag{7C.2}$$

with

$$\langle\underline{\sigma}^P\rangle = n\int_{\text{particle}}\underline{\sigma}^P\,dV, \tag{7C.3}$$

where n is the number of particles per unit volume.

One can obtain an alternative expression of this integral (and therein resides the subtlety of the approach), involving this time the stress field in the fluid, by rewriting the stress within each particle in the following form:

$$\sigma_{ij}^P = \partial_k(\sigma_{ik}^P x_j) - x_j\partial_k\sigma_{ik}^P = \partial_k(\sigma_{ik}^P x_j). \tag{7C.4}$$

We used here the fact that $\partial_k\sigma_{ik}^P = 0$ inside each particle (neglecting the gravity). This relation allows us to transform the volume integral (7C.3) into a surface integral:

$$\int_{\text{particle}}\underline{\sigma}^P\,dV = \int_{\text{particle}}\left(\underline{\sigma}^P\vec{v}\right)\vec{x}\,dS \tag{7C.5}$$

with $\vec{v}$ the unit vector perpendicular to the surface of the particle and pointing towards the fluid. Finally, the balance of the stress forces at the surface of the particle ($\sigma_{ij}^P v_j = \sigma_{ij} v_j$) leads to the following result:

$$\int_{particle} \underline{\sigma}^P \, dV = \int_{particle} (\underline{\sigma}\vec{v}) \, \vec{x} \, dS. \tag{7C.6}$$

This new expression shows that *it suffices to know the stress field inside the fluid* in order to calculate this integral.

Note that no assumption has yet been made as to the shape and density of the particles.

We must now solve the flow around the particles. This calculation is obviously very complex when the particles are not spherical and interact strongly with each other.

On the other hand, it becomes much simpler when the particles are *spherical* and they can also be considered as *independent* of each other, which amounts to neglecting their collisions (*dilute regime*).

Under these conditions, one has to solve the incompressibility equation:

$$\operatorname{div} \vec{v} = 0 \qquad (r > R) \tag{7C.7}$$

and the Navier–Stokes equation (where the advection terms are neglected since the Reynolds number of the particles is very small, in view of their size):

$$0 = -\overrightarrow{\operatorname{grad}} P + \eta_s \, \Delta \vec{v} \qquad (r > R). \tag{7C.8}$$

One must add to these two volume equations two boundary conditions. The first one is that, at the surface of the sphere:

$$\vec{v} = \vec{U} + \vec{\omega} \times \vec{x} \qquad (r = R), \tag{7C.9}$$

where $\vec{U}$ is the velocity of the centre of mass of the sphere and $\vec{\omega}$ its rotation vector. The second one is that, far away from the sphere, one must have (in the dilute regime):

$$\vec{v} = \langle \vec{v} \rangle + \langle \vec{\nabla}\vec{v} \rangle \vec{v} \qquad (r \to \infty), \tag{7C.10}$$

where $\langle \vec{v} \rangle$ is the average flow velocity and $\langle \vec{\nabla}\vec{v} \rangle$ its average velocity gradient (with $(\vec{\nabla}\vec{v})_{ij} = \partial v_i / \partial x_j$, by convention).

Since the bead experiences no external forces or torques, and its inertia is negligible, one must have:

$$\vec{F} = \int_{r=R} \underline{\sigma}\vec{v} \, dS = 0 \quad \text{and} \quad \vec{\Gamma} = \int_{r=R} \vec{x} \times \underline{\sigma}\vec{v} \, dS = 0. \tag{7C.11}$$

The first condition states that the particle is carried away at the same velocity as the average flow:

$$\vec{U} = \langle \vec{v} \rangle. \tag{7C.12}$$

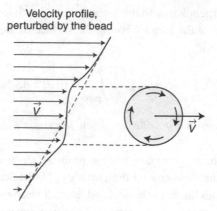

Velocity profile,
perturbed by the bead

Fig. 7.C.1 Motion of a bead in a shear flow. The bead is carried along at the average velocity of the flow and it rotates under the action of local shear. This leads to a distortion of the velocity profile within the fluid. The distortion engenders an additional dissipation, explaining the increase in the apparent viscosity.

The second one shows that the particle must rotate at the same rate as the fluid. More precisely, its rotation vector must be equal to the local rotation rate of the average flow (Fig. 7.C.1):

$$\vec{\omega} = \langle \vec{\Omega} \rangle, \tag{7C.13}$$

where $\langle \vec{\Omega} \rangle$ is connected to the vorticity by the equation $\langle \vec{\Omega} \rangle = \frac{1}{2} \overrightarrow{\mathrm{curl}} \langle \vec{v} \rangle$ (see Section 3.2.7). Thus, the boundary condition (7C.9) becomes explicitly:

$$\vec{v} = \langle \vec{v} \rangle + \langle \vec{\Omega} \rangle \times \vec{x} \qquad (r = R). \tag{7C.14}$$

As to the other boundary condition (7C.10), it can be rewritten as:

$$\vec{v} = \langle \vec{v} \rangle + \langle \underline{A} \rangle \vec{x} + \langle \vec{\Omega} \rangle \times \vec{x} \qquad (r \to \infty). \tag{7C.15}$$

The solution of Eqs. (7C.7) and (7C.8) subject to the boundary conditions (7C.14) and (7C.15) is a classical problem in hydrodynamics. We will only give the solution:

$$\vec{v} = \langle \vec{v} \rangle + \langle \underline{A} \rangle \vec{x} + \langle \vec{\Omega} \rangle \times \vec{x} - \langle \underline{A} \rangle \vec{x} \frac{R^5}{r^5} - \vec{x}(\vec{x}.\langle \underline{A} \rangle \vec{x}) \frac{5}{2} \frac{1}{r^2} \left(\frac{R^3}{r^3} - \frac{R^5}{r^5} \right), \tag{7C.16}$$

$$P = -5\eta_{\mathrm{s}} R^3 \frac{\vec{x} \cdot \langle \underline{A} \rangle \vec{x}}{r^5}. \tag{7C.17}$$

These expressions give the surface force (purely viscous in this case) exerted by the fluid on the particle:

$$\underline{\sigma} \vec{v}|_{r=R} = 5\eta_{\mathrm{s}} \langle \underline{A} \rangle \vec{v}. \tag{7C.18}$$

yielding, in turn, the contribution of the particle to the average macroscopic stress:

$$\langle \underline{\sigma}^P \rangle / n = \int_{\text{particle}} (\underline{\sigma}\vec{v})\vec{x}\, dS = 5\eta_s \frac{4\pi}{3} R^3 \langle \underline{A} \rangle. \tag{7C.19}$$

This expression gives:

$$\langle \underline{\sigma}^P \rangle = 5\eta_s \phi \langle \underline{A} \rangle. \tag{7C.20}$$

Replacing $\langle \underline{\sigma}^P \rangle$ by its expression in Eq. (7C.2) finally yields:

$$\langle \underline{\sigma} \rangle = -\langle P \rangle \underline{I} + 2\eta_s \langle \underline{A} \rangle + 5\eta_s \phi \langle \underline{A} \rangle \tag{7C.21}$$

and thus the effective viscosity:

$$\eta = \eta_s (1 + 2.5\phi). \tag{7C.22}$$

We have recovered the Einstein Formula given in Section 7.5.1.5.

References

Allain, C., M. Cloitre, B. Lacoste and I. Marsone, *J. Chem. Phys.*, **100**, 4537 (1994).

Arrhenius, S., *Biochem. J.*, **11**, 112 (1917).

Ball, R. and P. Richmond, *Phys. Chem. Liq.*, **9**, 99 (1980).

Batchelor, G. K., *J. Fluid. Mech.*, **83**, 97 (1977).

Berret, J.-F., J. Appell and G. Porte, *Langmuir*, **9**, 2851 (1993).

Berret, J.-F., D. Roux and G. Porte, *J. Phys. II France*, **4**, 1261 (1994).

Berry, G. C. and T. G. Fox, *Adv. Polym. Sci.*, **5**, 261 (1970).

Candau, S. J., E. Hirsh and R. Zana, *J. Physique France*, **45**, 1263 (1984).

Cates, M. E., *J. Physique France*, **49**, 1593 (1988).

Cates, M. E. and S. J. Candau, *J. Phys. Condens. Matt.*, **2**, 6869 (1990).

Choi, S. J. and W. R. Schowalter, *Phys. Fluids*, **18**, 420 (1975).

Constantin, D., E. Freyssingeas, J.-F. Palierne and P. Oswald, *Langmuir*, **19**, 2554 (2003).

Cotton, J. P., *J. Physique Lett. France*, **41**, L231 (1980).

Edwards, S. F., *Proc. Phys. Soc. Lond.*, **92**, 9 (1967).

Edwards, S. F., *Br. Polym. J.*, **9**, 140 (1977).

Einstein, A., *Ann. Physik*, **19**, 289 (1906); *ibid.*, **34**, 591 (1911).

Fetters, L. J., D. J. Lohse, S. T. Milner and W. W. Graessley, *Macromolecules*, **32**, 6847 (1999).

Flory, P. J., *J. Chem. Phys.*, **17**, 303 (1949).

de Gennes, P. G., *J. Chem. Phys.*, **55**, 572 (1971).

Giesekus, H., *Rheol. Acta*, **5**, 29 (1966).

Grace, H. P., *Chem. Eng. Commun.*, **14**, 225 (1982).

Granek, R. and M. E. Cates, *J. Chem. Phys.*, **96**, 4758 (1992).

Hinch, E. J. and G. L. Leal, *J. Fluid. Mech.*, **52**, 683 (1972).

Israelachvili, J. N., *Intermolecular and Surface Forces*, London: Academic Press (1992).

Janeschitz–Kriegl, H., *Polymer Melt Rheology and Flow Birefringence*, New York: Springer-Verlag (1983).

Janssen, J., Ph.D. Thesis, *Dynamics of Liquid-Liquid Mixing*, Eindhoven University of Technology (1993).

Jeffery, G. B., *Proc. Roy. Soc. Lond. A*, **102**, 161 (1922).

Johnson, R. M., J. L. Schrag and J. D. Ferry, *Polym. J.*, **1**, 742 (1970).

Kitade, S., A. Ichikawa, N. Imura, Y. Takahashi and I. Noda, *J. Rheol.*, **41**, 1039 (1997).

Krieger, I. M., *Adv. Colloid Interface Sci.*, **3**, 111 (1972).

de Kruif, C. G., E. M. F. van Lersel, A. Vrij and W. B. Russel, *J. Chem. Phys.*, **83**, 4717 (1985).

Kuhn, W., *Kolloid Z.*, **68**, 2 (1934).

Massiera, G., Ph.D. Thesis, *Giant micelles decorated by copolymer amphiphiles: structural and dynamical properties* (in French), Université de Montpellier II (2002).

Mellema, J., C. G. de Kruif, C. Blom and A. Vrij, *Rheol. Acta*, **26**, 40 (1987).

Nawab, M. A. and S. G. Mason, *J. Colloid Sci.*, **13**, 179 (1958).

Oldroyd, J. G., *Proc. Roy. Soc. Lond. A*, **218**, 122 (1953).

Oldroyd, J. G., *Proc. Roy. Soc. Lond. A*, **232**, 567 (1955).

Onogi, S., T. Masuda and K. Kitagawa, *Macromolecules*, **3** (1970) 109.

Pal, R., *J. Rheol.*, **36**, 1245 (1992).

Palierne, J.-F., *Rheol. Acta*, **29**, 204 (1990).

Peterlin, A., *Makromol. Chem.*, **44**, 338 (1961).

Rouse, P. E., *J. Chem. Phys.*, **21**, 1272 (1953).

Rumscheidt, F. D. and S. G. Mason, *J. Colloid Sci.*, **16**, 238 (1961).

Taylor, G. I., *Proc. Roy. Soc. Lond. A*, **138**, 41 (1932).

Warner, H. R., *I&EC Fundam.*, **11**, 379 (1972).

Yang, J. T., *J. Am. Chem. Soc.*, **80**, 1783 (1958).

Zimm, B. H., *J. Chem. Phys.*, **24**, 269 (1956).

8

Rheology of liquid crystals

In the two previous chapters, we have considered viscoelastic isotropic liquids, which behave as Newtonian liquids at low frequency and as elastic solids at high frequency. The crossover between the two regimes is set by the viscoelastic relaxation time of the material. In simple liquids, such as water, this time is so short that the elastic effects can be neglected. On the other hand, this is not the case for liquid polymers of high molecular mass, where the elasticity plays a very important role.

In this chapter, we will study another class of materials, composed of *small molecules*, which are either *elongated* (and known as '*calamitic*') or *discoidal*. These materials, some examples of which were given in the first chapter, can form over certain temperature ranges *mesomorphic* or *liquid crystalline* phases, with *intermediate symmetries* between those of liquids and of solids. Best known among them are the *nematic phase*, with purely orientational order, the *smectic A phase*, possessing an additional positional order along one space direction, and the *columnar hexagonal phase*, where the positional order is two dimensional.

The main difference between a liquid and a liquid crystal is thus that the latter exhibits *orientational*, and sometimes even *positional, long-range order*, along the three space directions for the first type of order and in two or three space directions for the second type, whereas in a liquid positional and orientational order are *short range*. This crystalline character endows the phases with elasticity, acting down to *zero frequency*, as for three-dimensional crystals. This can be '*torque*' elasticity, given by the orientational degrees of freedom or '*stress*' elasticity in the directions along which the material possesses long-range positional order. However, a liquid crystal behaves as a liquid along those directions where it does not exhibit usual stress elasticity. For all these reasons, liquid crystals are *viscoelastic*. In their case, however, the viscoelasticity originates in the symmetries of the phase, and not in changes of the molecular conformation under flow, as in the case of the polymers described in the two previous chapters. There are also polymeric liquid crystals, where the viscoelasticity has a twofold origin, both structural (due to the symmetries) and conformational (because the molecules can change their conformation, and not only their orientation, under flow). In the following, we will neglect the conformational viscoelasticity and concentrate on structural viscoelasticity. This simplification is legitimate in the case of liquid crystals formed by small molecules, such as those described in the first chapter.

It is however less realistic for liquid crystalline polymers, where the two viscoelasticity sources can interfere, but we will not go into these details in the following. These aspects are discussed in [45].

Let us now present the plan of the chapter.

We will begin by examining the case of uniaxial nematics (Section 8.1). We will see that these materials exhibit '*torque*' elasticity, which can be demonstrated by specific experiments. The simplest such experiment is that of *Grupp*, directly revealing the existence of elastic torques, but the best known is that of *Frederiks*, since it is fundamental for the operation of displays based on nematic liquid crystals. We will describe these two experiments and build upon this base in order to develop a simplified theory of nematodynamics, only taking into account the in-place rotation of the director. Using this theory, we will explain the origin of the scintillation typical for the nematic phase. We will then tackle the more complicated problem of nematic flow. We will show that the viscosity of a nematic depends on its orientation with respect to the directions defining the flow, namely the velocity and its gradient, which will bring us to defining the three *fundamental Miesowicz viscosities*. We will then construct a hydrodynamic theory encompassing all of these phenomena (*the Leslie–Ericksen theory*), and show that it predicts novel effects, such as the presence of a transverse pressure gradient in Poiseuille flow, or the development of original instabilities under continuous shear. Finally, we will say a few words on defects of the *disclination* type and their coupling with the flow.

The chapter continues with a description of smectic phases (Section 8.2). We will see that these materials flow like liquids under shear parallel to the layers, but they behave like elastic solids under compression normal to the layers. In this respect, they are closer to solids than the nematics, which flow like liquids under compression, irrespective of their orientation. We will also see that *two types of sound* can propagate in smectics, and that smectic phases can exhibit elastic instabilities, such as the *layer undulation* instability, resembling the Euler instability of beams, discussed in Chapter 3. To explain these phenomena, we will develop a continuous theory, *smectodynamics*, which couples the elasticity of solids to the hydrodynamics of simple liquids. We will then describe some examples of *permeation flow*, and then an experiment of *creep under compression normal to the layers*. We will see that the data on the creep velocities can hardly be explained in the framework of this theory without invoking *dislocations*, linear defects which break the lamellar order locally. From this point of view, smectics behave as plastic solids. We will then try to visualise these defects and present some of their properties, before analysing their behaviour under stress. In particular, we will show that screw dislocations can develop *helical instabilities*, leading to faster sample creep. We will also analyse the nucleation, followed by growth or collapse of an *edge dislocation loop* in a *free-standing smectic film*. Finally, we will show that smectics make very good *lubricants*, which will lead us to completing the lubrication theory presented in Chapter 3 by accounting for the elasticity of the layers.

We will conclude this chapter with a few comments on the elasticity of columnar phases (Section 8.3). In particular, we will analyse two *column buckling instabilities*, one of them having no equivalent in smectics. We will highlight the need for *renormalising the elastic constants*, due to the presence of π-*type defects* (which are free column ends).

We should point out that this chapter owes a lot to Pawel Pieranski, our co-author on a two-volume book, *Liquid Crystals* [43], where the interested reader can find substantial complementary information. We should also note that almost all figures in this chapter are extracted from that book.

8.1 Nematodynamics

Our goal in this section is not to postulate a continuous theory of nematic hydrodynamics; instead, we will show how to obtain a theory starting from the relevant experimental observations. To this end, we will alternate between presenting the experiments and developing the theory, in order to render the account less abstract and (hopefully) more attractive. Let us start by describing a fundamental, but relatively recent, experiment, performed in 1983 (Grupp, 1983).

8.1.1 The Grupp experiment: evidence for elastic torques

The experimental setup is sketched in Fig. 8.1. It consists of two parallel glass discs, one of which is fixed (the lower one), while the upper one is suspended from a very thin quartz fibre (about 25 μm in diameter). The thickness of the nematic layer between the two discs can be adjusted between 0 and 200 μm, with micrometre precision. The surfaces of the two discs are treated so as to induce planar anchoring (with the molecules parallel to the plane of the discs); at rest, the orientation of the nematic is the same on both discs. The director (we recall that, according to the definition given in the first chapter, this is the unit vector describing the average orientation of the molecules) then has the same orientation everywhere (at the surface and in the bulk), that we take as parallel to the x axis. We assume that the anchoring is sufficient to prevent the director from slipping at the surface of the discs.

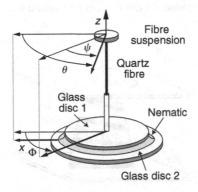

Fig. 8.1 Experimental setup used for measuring the twist constant of a nematic (Grupp, 1983).

The experiment consists in turning by an angle θ the suspension point of the quartz fibre. Experiment shows that the mobile disc also rotates in the same direction, through a smaller angle Φ. Thus, the quartz fibre is twisted by an angle $\Psi = \theta - \Phi$, resulting in an elastic torque:

$$\Gamma_{\text{fibre}} = D\,\Psi, \tag{8.1}$$

where D is the torsion constant of the fibre (here, its value is about 9.6×10^{-10} N m).

Equilibrium can only be reached if the nematic itself exerts on the mobile disc a torque Γ_{nem} exactly compensating that of the torsion fibre. One must then have:

$$\Gamma_{\text{fibre}} + \Gamma_{\text{nem}} = 0. \tag{8.2}$$

Grupp measured the angle Ψ as a function of the angle θ and showed that it was proportional to the twist angle of the nematic layer, $\Phi = \theta - \Psi$ as well as to the surface sample S. He also checked that Ψ is inversely proportional to the thickness d of the nematic layer. According to Eq. (8.1), it follows that the nematic exerts on the plate a surface torque with an amplitude:

$$C_{\text{nem}} = \frac{\Gamma_{\text{nem}}}{S} = -K_2 \frac{\Phi}{d}, \tag{8.3}$$

where K_2 is an elastic constant, with force units. Grupp measured $K_2 \approx 10^{-11}$ N in MBBA, at room temperature. This is a typical value for non-polymeric thermotropic liquid crystals.

From this experiment we can therefore conclude that nematics can sustain *elastic torques* without flowing. In this case, one speaks of a *twist torque*, since it is associated with a twist deformation of the director field (Fig. 8.2).

We emphasise that this experiment is particularly difficult to perform, requiring the measurement of extremely small torques (of the order of 10^{-9} N m, or 10 µgf cm). The main practical difficulties are due to the capillary forces acting at the edge of the sample, which can block its rotation, and to errors in the horizontality of the setup. To sidestep these problems, a similar experiment was performed a few years later by Faetti *et al.* (1985). This time, the mobile plate suspended on the torsion pendulum is completely immersed in the

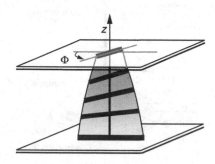

Fig. 8.2 Twist deformation of the director field.

nematic liquid crystal, which is oriented in the bulk by a strong external magnetic field. By changing the geometry (in particular, the nature of the anchoring on the suspended plate and its orientation with respect to the field), these authors were able to measure not only the twist constant, but also the other elastic constants, associated with the splay and bend deformations to be defined in the next section (Faetti *et al.*, 1986).

8.1.2 Nematic elasticity

8.1.2.1 The Frank–Oseen free energy

Clearly, distorting the director field leads to an increase in the free energy of the nematic phase. This increase, depending on the local orientation of the director $\vec{n}$ and its gradients $\vec{\nabla}\vec{n}$, is the *free energy of elastic distortion* $f(\vec{n}, \vec{\nabla}\vec{n})$. As in classical elasticity, one can develop this function as a power series of the distortion $\vec{\nabla}\vec{n}$. Keeping only terms up to second order one can show, taking into account the symmetries of the uniaxial nematic phase, that the most general expression for f is of the type (see Appendix 8.A):

$$f = \frac{1}{2}K_1 (\operatorname{div}\vec{n})^2 + \frac{1}{2}K_2 \left(\vec{n} \cdot \overrightarrow{\operatorname{curl}}\vec{n}\right)^2 + \frac{1}{2}K_3 \left(\vec{n} \times \overrightarrow{\operatorname{curl}}\vec{n}\right)^2. \tag{8.4}$$

This is the *Frank–Oseen free energy* (Oseen, 1933; Frank, 1958).

Each term of this expression accounts for a particular deformation of the director field.

Thus, the K_1 term describes a deformation of the '*splay*' type (Fig. 8.3a); the K_2 term represents a '*twist*' deformation (Fig. 8.3b) and the K_3 term accounts for a '*bend*' deformation (Fig. 8.3c).[1]

It is noteworthy that in ordinary (non-polymeric) nematics, the K_i ($i = 1, 2, 3$) constants are all of the same order of magnitude (up to a factor of 2 or 3), provided the system is not too close to the transition towards a smectic phase (the K_2 and K_3 constants diverge on approaching it). In these conditions, one usually has:

$$K \approx \frac{U}{a}, \tag{8.5}$$

with U the energy of molecular interaction (of the van der Waals type) and a the average distance between molecules. Taking $U \approx k_B T \approx 1/40$ eV and $a \approx 30$ Å yields $K \approx 10^{-11}$ N, which is exactly the order of magnitude determined experimentally.

Finally, let us give the expression for the energy in the case of isotropic elasticity, when all elastic constants are equal ($K_1 = K_2 = K_3 = K$):

$$f = K\nabla_i n_j \nabla_i n_j. \tag{8.6}$$

---

[1] We should point out here that we omitted from Eq. (8.4) the divergence term $(1/2)K_4\operatorname{div}(\vec{n}\operatorname{div}\vec{n} + \vec{n} \times \overrightarrow{\operatorname{curl}}\vec{n})$. This is a *surface term*, because its integral over the sample volume can always be converted into a surface integral. Hence, this term *does not contribute to the equilibrium equation for bulk torques*, but only to the equilibrium equation for the surface torques at the sample boundary, if the molecular anchoring energy is finite. In many experiments (in particular those described in this chapter), one can assume that the orientation of the molecules at the surfaces is fixed ('infinite' anchoring energy). The equilibrium equation for the surface torques is then reduced to a geometrical condition prescribing the molecular orientation at the surfaces, and K_4 drops out of the problem (see [43, volume 1] for a detailed discussion).

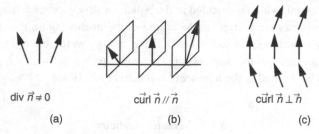

div $\vec{n} \neq 0$ $\text{curl}\,\vec{n} \,/\!/\, \vec{n}$ $\text{curl}\,\vec{n} \perp \vec{n}$

(a)   (b)   (c)

Fig. 8.3 The three elementary deformations of the nematic phase: (a) 'splay'; (b) 'twist'; (c) 'bend'.

The attentive reader will have noted that this expression differs from Eq. (8.4) by a div(...) term, which is generally irrelevant for the reasons invoked in Footnote 1.

8.1.2.2 The equilibrium equation for elastic torques

In the first chapter, we showed the existence of a tensor $\underline{C}$ such that, if the nematic medium is virtually cut into two parts (1) and (2) separated by an imaginary surface S, the side (2) exerts onto side (1) a surface torque $\vec{C}(M, \vec{v}) = \underline{C}\vec{v}$, where $\vec{v}$ is the unit vector perpendicular to S at point M, and pointing from (1) towards (2) (see Fig. 2.1).

In this section, we will express this tensor as a function of the Frank energy [39, 40, 41, 43].

The first step is calculating the variation δF in the total free energy F as the director fluctuates by $\delta \vec{n}$. By definition:

$$\delta F = \delta \int f \, dV = \int \delta f \, dV. \tag{8.7}$$

Since f is an explicit function of the n_i and their derivatives $n_{i,j}$, this yields:

$$
\begin{aligned}
\delta f &= \frac{\partial f}{\partial n_i}\delta n_i + \frac{\partial f}{\partial n_{i,j}}\delta n_{i,j} \\
&= \frac{\partial f}{\partial n_i}\delta n_i + \left(\frac{\partial f}{\partial n_{i,j}}\delta n_i\right)_{,j} - \left(\frac{\partial f}{\partial n_{i,j}}\right)_{,j}\delta n_i
\end{aligned}
\tag{8.8}
$$

and finally, after substituting into Eq. (8.7) and integrating by parts once:

$$\delta F = \int_V -h_i \delta n_i \, dV + \int_S \frac{\partial f}{\partial n_{i,j}}\delta n_i \, dS_j, \tag{8.9}$$

where:

$$h_i = -\frac{\partial f}{\partial n_i} + \frac{\partial}{\partial x_j}\left(\frac{\partial f}{\partial n_{i,j}}\right). \tag{8.10}$$

For reasons to be made apparent in the following, the vector $\vec{h}$ is known as the *molecular field*. Its expression is particularly simple in the case of isotropic elasticity where, from Eq. (8.6):

$$\vec{h} = K\Delta\vec{n}. \tag{8.11}$$

In the anisotropic case, the formulas are much more complicated (see, for instance, Eq. (B.2.36) in [43, volume 1].

Suppose now that the molecules locally turn in-place, with the rotation vector $\delta\vec{\omega}$, their centres of mass remaining fixed (the case of translation will be considered in Section 8.1.6.3). It follows that the director varies by:

$$\delta\vec{n} = \delta\vec{\omega} \times \vec{n}, \tag{8.12}$$

which can also be written, in index notation:

$$\delta n_i = e_{ijk}\delta\omega_j\, n_k. \tag{8.13}$$

Substituting in Eq. (8.9) gives:

$$\delta F = \int_V -h_i e_{ijk}\delta\omega_j\, n_k\, dV + \int_S \frac{\partial f}{\partial n_{i,l}} e_{ijk}\delta\omega_j\, n_k\, dS_l \tag{8.14}$$

or, in vector form:

$$\delta F = \int_V \delta\vec{\omega} \cdot (\vec{h} \times \vec{n}) + \int_S (\underline{C}\vec{v}) \cdot \delta\vec{\omega}\, dS, \tag{8.15}$$

where $\vec{v}$ is the normal unit vector, pointing out of the volume V, and $\underline{C}$ the second-rank tensor with components:

$$C_{ij} = \frac{\partial f}{\partial n_{k,j}} e_{ikl} n_l. \tag{8.16}$$

Let us now try to understand this result. According to the first principle of thermodynamics, δF is the work done by the external torques acting on the system as the director rotates by $\delta\vec{\omega}$. Hence, it is an elasticity of the '*torque*' type, $\vec{h} \times \vec{n}$ and $\underline{C}\vec{v}$ having units of torque per unit volume and torque per unit surface, respectively.

More precisely, $(\vec{h} \times \vec{n}) \cdot \delta\vec{\omega}$ represents the work done by the external bulk torque acting on the system. The equilibrium of bulk torques is written explicitly as:

$$\vec{h} \times \vec{n} = \vec{\Gamma}^V(\text{ext}). \tag{8.17}$$

This equation can be rewritten in the following form:

$$\vec{\Gamma}^E + \vec{\Gamma}^V(\text{ext}) = 0, \tag{8.18}$$

where the new quantity:

$$\vec{\Gamma}^E = \vec{n} \times \vec{h} \tag{8.19}$$

can be interpreted as the *bulk elastic torque exerted by the medium onto its own director $\vec{n}$*.

This equation also shows that, at equilibrium and in the absence of an external field, the director $\vec{n}$ aligns along the field $\vec{h}$, since $\vec{n} \times \vec{h} = 0$. That is why the field $\vec{h}$ bears the name '*molecular field*'.

In the same way, $\underline{C}\vec{v}$ must be equal to the surface torque exerted by the medium outside the volume V onto the surface S:

$$\underline{C}\vec{v} = \vec{C}(\text{ext}). \tag{8.20}$$

This is indeed the surface torque that we introduced in Chapter 1, and the existence of which we demonstrated using the tetrahedron construction.

Note that, in practice, $\vec{\Gamma}^V(\text{ext})$ can be a magnetic torque, given by:

$$\vec{\Gamma}^M = \vec{M} \times \vec{B} = \frac{\chi_a}{\mu_0}(\vec{n} \cdot \vec{B})\vec{n} \times \vec{B}. \tag{8.21}$$

Here, $\vec{M}$ is the magnetisation induced by the magnetic field $\vec{B}$, while χ_a represents the *anisotropy of the magnetic susceptibility* ($\chi_a = \chi_\parallel - \chi_\perp$, where $\chi_\parallel$ and $\chi_\perp$ are the magnetic susceptibilities parallel and perpendicular to the director, respectively; they are negative in the diamagnetic case).

The external field can also be an electric one (as in the case of displays, for instance). One must then replace $\vec{B}$ by the electric field $\vec{E}$ and χ_a/μ_0 by $\varepsilon_0\varepsilon_a$, where ε_a is the *dielectric anisotropy* of the material.

Note that, as an order of magnitude, $\chi_a \approx 10^{-6}$ (liquid crystals are most often diamagnetic) and $\varepsilon_a \approx 10$ (a typical value for 8CB, which has a strongly polarisable cyano- end group). With these values, a relatively modest electric field of 100 V/mm is equivalent to a rather strong 1 tesla magnetic field. Thus, the electric field is much more effective than the magnetic field. We recall that most materials have positive magnetic anisotropy, meaning that the director almost always tends to align along the magnetic field. On the other hand, dielectric anisotropy can be either positive or negative, depending on the materials. For instance, 8CB molecules, with $\varepsilon_a > 0$, will align along the electric field, while MBBA molecules, with $\varepsilon_a < 0$, will align perpendicular to the field. There are also some liquid crystal mixtures where the dielectric anisotropy varies strongly with the frequency of the applied electric field, changing sign at a well-defined frequency, in the range of a few kilohertz or tens of kilohertz.

8.1.2.3 Interpretation of the Grupp experiment

In this experiment, the director rotates by an angle Φ between the lower plate and the upper one, i.e. over a distance given by the sample thickness (Fig. 8.2). To show that the director turns uniformly, let us write the mechanical equilibrium for the liquid crystal slab between heights z and $z+dz$, see Fig. 8.4. For simplicity's sake, we assume that the director remains parallel to the (x, y) plane, making at height z an angle $\phi(z)$ with the x axis.

The nematic medium below the slab exerts upon it the surface torque:

$$C_z(z) = -K_2\frac{d\phi}{dz}(z), \tag{8.22}$$

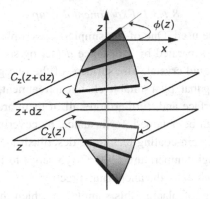

Fig. 8.4 The liquid crystal slab is acted upon by the torques $C_z(z)$ and $C_z(z + dz)$.

while the nematic above it exerts a torque:

$$C_z(z + dz) = K_2 \frac{d\phi}{dz}(z + dz). \tag{8.23}$$

At equilibrium, these two torques compensate exactly, yielding:

$$K_2 \frac{d\phi}{dz}(z + dz) - K_2 \frac{d\phi}{dz}(z) = 0 \tag{8.24}$$

or, to first order in dz:

$$K \frac{d^2\phi}{dz^2} = 0. \tag{8.25}$$

This equation is none other than the equilibrium equation of bulk torques ($\vec{\Gamma}^E = \vec{n} \times \vec{h} = 0$), since no external bulk torque acts on the sample ($\vec{\Gamma}^V(\text{ext}) = 0$).

It is obvious from Eq. (8.25) that the angle varies linearly in the thickness of the sample:

$$\phi = \Phi \frac{z}{d} \tag{8.26}$$

corresponding to uniform twist of the director field along the z direction, with a value $\vec{n} \cdot \overrightarrow{\text{curl}}\,\vec{n} = -\Phi/d$.

8.1.3 The Frederiks instability

This is a fundamental experiment, indispensable for the study of nematic liquid crystals, since it can be used to measure the elastic constants and to obtain optical valves, employed on a large scale in the production of liquid crystal displays.

8.1.3.1 Experimental setup

Various geometries can be used. One of the simplest uses a planar sample, consisting of two parallel glass plates, separated by a distance d (set by spacers) and filled by capillarity with the nematic liquid crystal. The surfaces of the two glass plates are treated in planar anchoring, meaning that, by a suitable chemical treatment, the molecules are forced to lie flat on the glass surface and, furthermore, all of them are aligned along the same direction x. This result can be obtained, for instance, by covering the surface with a thin polyimid layer (using the spin-coating technique described in Section 3.7.6). This layer is then polymerised at high temperature (300 °C typically) to harden it, then uniaxially rubbed with a felt-like cloth to fix the anchoring direction.

The experiment consists of placing this sample, in which the director is everywhere aligned along the x direction, in a magnetic field pointing along the y axis. The z axis is perpendicular to the glass plates. Experiment shows that, if the molecules have a positive magnetic anisotropy ($\chi_a > 0$), there exists a critical magnetic field B_c below which nothing changes (Fig. 8.5a), and above which the director starts to rotate in the centre of the sample. The result is a twisted configuration, shown in Fig. 8.5b. Using an appropriate optical method (conoscopy), one can measure the rotation angle ϕ of the director in the centre of the sample. Experiment shows that this angle can be either positive or negative, and that it increases as $(B - B_c)^{1/2}$ above the instability threshold. This result is a distinctive sign of a *supercritical bifurcation* (similar to a second-order phase transition in thermodynamics).

8.1.3.2 Formation of the equations

To interpret this experiment, one must minimise the free energy of the system. It contains two terms: a magnetic term, resulting from the coupling between the director and the external magnetic field, considered constant, and an elastic term due to the twist of the director field (assuming that, in the strong anchoring regime, the director cannot slip on the glass

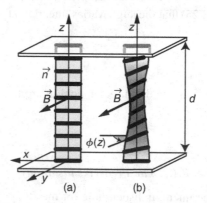

(a)   (b)

Fig. 8.5 Principle of the Frederiks instability. (a) Below the critical field, the director field remains unchanged; (b) above this value, the director field starts to twist.

plates). The energy per unit sample surface is:

$$F = \int_0^d \left(\frac{1}{2} K_2 \left(\vec{n} \cdot \overrightarrow{\text{curl}\,\vec{n}} \right)^2 - \frac{\chi_a}{2\mu_0} (\vec{B} \cdot \vec{n})^2 \right) dV, \tag{8.27}$$

where only the twist term in the Frank free energy was preserved, under the implicit assumption that the director remains in the (x, y) plane. With $\phi(z)$ the rotation angle of the director with respect to the x axis, this energy can also be written as:

$$F = \int_0^d \left[\frac{1}{2} K_2 \left(\frac{\partial \phi}{\partial z} \right)^2 - \frac{\chi_a}{2\mu_0} (B \cos \phi)^2 \right] dV. \tag{8.28}$$

Minimising the energy with respect to ϕ yields the following differential equation:

$$K_2 \frac{d^2 \phi}{dz^2} + \frac{\chi_a}{\mu_0} B^2 \sin \phi \cos \phi = 0. \tag{8.29}$$

This equation extends the equilibrium equation for bulk torques given above (Eq. (8.25)) to account for magnetic torques.

This equation must be supplemented by the two boundary conditions expressing the strong anchoring on the plates:

$$\phi(0) = \phi(d) = 0. \tag{8.30}$$

Before searching for the solution to these equations, note that the problem contains a typical length, given by:

$$\xi(B) = \sqrt{\frac{\mu_0 K_2}{\chi_a} \frac{1}{B}}. \tag{8.31}$$

This is the 'magnetic coherence length', quantifying the competition between the elastic and the magnetic torques; it gives the typical thickness of the twist boundary layer that would develop in the neighbourhood of each plate if the sample were 'infinitely thick.'

One would then expect two qualitatively different regimes, as the thickness d is larger or smaller than $\xi(B)$. More specifically, the elastic torque will dominate over the magnetic torque for $d < \xi(B)$, which explains why the sample remains planar. On the other hand, the director will start to turn as the magnetic torque overcomes the elastic torque, which is the case for $d > \xi(B)$.

Qualitatively, one can therefore predict that the critical magnetic field is reached when these two length scales are of the same order of magnitude.

We will now prove this result.

8.1.3.3 Calculating the Frederiks threshold

To solve Eq. (8.29) with the boundary conditions (8.30), let us assume that the angle ϕ remains much smaller than 1. It is then possible to expand Eq. (8.29) to third order in ϕ, yielding:

$$K_2 \frac{d^2\phi}{dz^2} + \frac{\chi_a}{\mu_0} B^2 \left(\phi - \frac{2}{3}\phi^3 \right) = 0. \tag{8.32}$$

Let us try solutions of the form:

$$\phi = \sum_m \phi_m \sin\left[(2m+1)\frac{\pi z}{d} \right]. \tag{8.33}$$

After substituting for ϕ in Eq. (8.32) and linearising the ϕ^3 term, setting to zero each coefficient of the $\sin(2m+1)\pi z/d$ terms gives a series of equations for ϕ_0, ϕ_1, etc. It is easily checked that the amplitude ϕ_0 of the fundamental mode is a solution of the following equation:

$$-K_2\phi_0 \frac{\pi^2}{d^2} + \frac{\chi_a}{\mu_0} B^2 \left(\phi_0 - \frac{\phi_0^3}{2} \right) = 0 \tag{8.34}$$

while $\phi_1 \propto \phi_0^3$, etc.

For all values of the field B, this equation has the trivial solution:

$$\phi_0 = 0, \tag{8.35}$$

corresponding to the undistorted planar sample, and possibly another solution, given by the equation:

$$1 - \frac{\phi_0^2}{2} = \frac{B_c^2}{B^2}, \tag{8.36}$$

where we introduced the field B_c given by:

$$B_c = \frac{\pi}{d} \sqrt{\frac{\mu_0 K_2}{\chi_a}} \tag{8.37}$$

with $\xi(B_c) = d/\pi$.

It is immediately apparent that Eq. (8.36) has a solution

$$\phi_0 = \pm 2 \sqrt{\frac{B - B_c}{B_c}} \tag{8.38}$$

if and only if $B > B_c$. In the opposite case ($B < B_c$), the problem only admits the solution $\phi_0 = 0$.

Thus, the field B_c is the *critical field* we had been looking for. It can be checked that, for $B > B_c$, the energy associated with the solution in Eq. (8.38) is lower than that of the trivial solution (so the latter is unstable in this regime). It is also noteworthy that *the bifurcation is supercritical*, from Eq. (8.38) (no jump in the angle ϕ_0 at the instability threshold).[2]

To fix the ideas, let us calculate the typical value of B_c. Using $d = 100$ µm, $\chi_a = 10^{-6}$ and $K_2 = 10^{-11}$ N (value measured in the Grupp experiment), one obtains $B_c = 0.1$ tesla. Such a field is easily obtained using a standard electromagnet.

It is then sufficient to measure the critical field to determine K_2, provided however the sample thickness and the magnetic susceptibility of the material are known. One can easily see that, by changing the experimental geometry, the other Frank constants also become accessible. This method was extensively used for measuring them (for more details, see [38, 39, 43]).

In the next section, we will analyse the dynamics of this instability.

8.1.3.4 Characteristic relaxation time: rotational viscosity

It so happens that the rotation of the director within the sample takes a certain time. Experiment shows that this time diverges on approaching the critical field as the reciprocal square root of the distance to the threshold. This divergence, reminiscent of the 'critical slowing down' phenomenon encountered in second-order phase transitions, is an additional hint that the bifurcation is supercritical.

To explain this finite tilt time, viscous dissipation must be considered. But where does it come from, since in the experiment we presented no flow occurs, the final distortion state being obtained simply by an in-place rotation of the molecules through an average angle $\phi(z)$.

To understand the origin of the dissipation, we show in Fig. 8.6 the changes in molecular configuration as all the molecules turn by a 90° angle around their centres of mass, taken as fixed. For comparison, we also show the molecular configuration obtained by a solid rotation of the entire set of molecules. Clearly, in the second case the statistical distribution of the molecules remains unchanged: the solid rotation is not accompanied by any dissipative effect; on the other hand, the distribution is significantly far from equilibrium when the molecules turn individually. Some energy is thus dissipated as the system returns to a local equilibrium configuration.

The physical phenomenon being understood, let us now see how it can be expressed mathematically as a function of macroscopic variables.

The simplest approach is to assume that the medium exerts on its director a viscous torque proportional to the local rotation rate $\vec{n} \times d\vec{n}/dt$, given by:

$$\vec{\Gamma}^v = -\gamma_1 \left(\vec{n} \times \frac{d\vec{n}}{dt} \right). \tag{8.39}$$

The coefficient γ_1 is the '*rotational viscosity*'. It plays an important role in the theory of nematodynamics, as we will see throughout this presentation.

---

[2] The Frederiks transition can however be subcritical in a few special cases, in particular when it occurs under electric field in the conductive regime (Deuling, 1974).

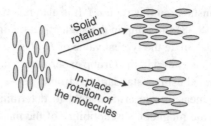

Fig. 8.6 Comparison between the final molecular configurations for a solid rotation of the entire nematic and for individual in-place rotation of the molecules by the same angle around their centres of mass, taken as fixed.

First of all, this coefficient (measured in Pa s, like the usual viscosities) must be positive since the dissipation, given by $\Phi = -\vec{\Gamma}^{\mathrm{v}} \cdot \left[\vec{n} \times (\mathrm{d}\vec{n}/\mathrm{d}t)\right]$ is always positive, according to the second principle of thermodynamics.

We are now equipped to calculate the typical tilt time for the Frederiks instability. It is sufficient to include the viscous torque in the torque equilibrium equation (8.18), which becomes (accounting for the magnetic field):

$$\vec{\Gamma}^{\mathrm{E}} + \vec{\Gamma}^{\mathrm{M}} + \vec{\Gamma}^{\mathrm{v}} = 0. \tag{8.40}$$

In the Frederiks problem under discussion, the most general form of this equation is:

$$K_2 \frac{\mathrm{d}^2\phi}{\mathrm{d}z^2} + \frac{\chi_a}{\mu_0} B^2 \sin\phi \cos\phi - \gamma_1 \frac{\partial\phi}{\partial t} = 0. \tag{8.41}$$

This formula, an extension of the static equation (8.29), can be solved in the small-angle limit by expanding ϕ in the Fourier series (Eq. (8.33)). In this limit, we have the following equation for ϕ_0:

$$\left(1 - \frac{B_c^2}{B^2}\right)\phi_0 - \frac{\phi_0^3}{2} = \frac{\mu_0 \gamma_1}{\chi_a B^2} \frac{\mathrm{d}\phi_0}{\mathrm{d}t} \tag{8.42}$$

with solution:

$$\phi_0^2 = \frac{\phi_0^2(\infty)}{1 + \left[\dfrac{\phi_0^2(\infty)}{\phi_0^2(0)} - 1\right]} \exp\left[-\frac{2\chi_a}{\mu_0 \gamma_1}\left(B^2 - B_c^2\right)t\right], \tag{8.43}$$

where $\phi_0(0)$ is the initial tilt angle, never strictly zero due to sample imperfections and to the thermal fluctuations, and $\phi_0(\infty)$ the long-time saturation angle, given by the static theory $\phi_0(\infty) \approx 2\left[(B - B_c)/B_c\right]^{1/2}$ (Eq. (8.38)).

This equation shows that the typical time for director tilt is:

$$\tau = \frac{\mu_0 \gamma_1}{\chi_a(B^2 - B_c^2)} \approx \frac{d^2}{2\pi^2} \frac{\gamma_1}{K_2} \frac{B_c}{B - B_c}. \tag{8.44}$$

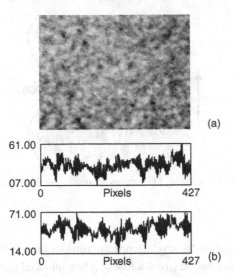

(a)

(b)

Fig. 8.7 (a) Planar sample observed in polarised light transmission; (b) two intensity profiles measured along the same horizontal line at a 0.04 s interval.

This formula predicts that the tilt time diverges on approaching the critical field as the reciprocal of the distance to the threshold (critical slowing down). It also predicts that it increases as the sample thickness squared. Both these results were confirmed experimentally. To estimate the time τ, take as the field value twice the critical field ($B = 2B_c$) and some typical parameters $\gamma_1 = 0.1$ Pa s, $d = 10$ μm and $K_2 = 10^{-11}$ N. We find $\tau(2B_c) = d^2\gamma_1/2\pi^2 K_2 = 0.05$ s. The tilt time is thus relatively short. Conversely, measuring this time gives access to the rotational viscosity γ_1. This method was extensively used by experimentalists.

We have just seen that the Frederiks instability is very useful for determining the elastic constants and the rotational viscosity. It can also be used for producing the optical valves employed in most liquid crystal displays. We refer the reader to [42] and [43, volume 1] for a detailed discussion.

8.1.4 On the scintillation of the nematic phase

When observing the nematic phase in polarised microscopy, one can immediately notice a very striking phenomenon: *the nematic phase scintillates*. This scintillation is particularly apparent in polarised light observation of planar samples. The photograph in Fig. 8.7 shows the sample in these conditions. The image here is inhomogeneous; it would be perfectly uniform for an isotropic liquid. The plot of the intensity profile recorded along a horizontal line shows the scintillation very clearly. If the profile is recorded an instant later, the plot changes completely. These fluctuations are at the origin of the scintillation phenomenon.

In the following, we will show that the phenomenon described above stems from the *fluctuation modes* of the director field (Fig. 8.8). The fact that they are perceptible to the

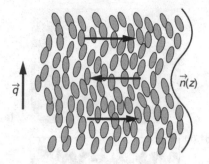

Fig. 8.8 Undulation mode in a nematic. The arrows represent the 'backflow' associated with rotation of the director.

naked eye means that they are 'slowly' damped in time, at least in the range of wavelengths observable under the microscope (from a few to a few hundred micrometres).

Let us show this by calculating their dispersion relation $\omega(k)$.

In the absence of any external field, and under the assumption of isotropic elasticity, the equilibrium equation for bulk torques (8.40) can be written, using Eqs. (8.11), (8.19) and (8.39), in the form:

$$-\gamma_1 \left(\vec{n} \times \frac{\partial \vec{n}}{\partial t} \right) + K(\vec{n} \times \Delta \vec{n}) = 0. \tag{8.45}$$

Since $\vec{n}$ is a unit vector, this equation can be rewritten in the simplified form:

$$\gamma_1 \frac{\partial \vec{n}}{\partial t} = K \Delta \vec{n}. \tag{8.46}$$

This is a diffusion equation, bringing into play the *diffusion coefficient for the molecular orientation*:

$$D_{\text{or}} = \frac{K}{\gamma_1}. \tag{8.47}$$

As an order of magnitude, $D_{\text{or}} \approx 10^{-10}$ m²/s, which is comparable to a molecular diffusion coefficient.

Specifying $\vec{n} = \vec{n}_0 \exp\left[i\vec{k} \cdot \vec{r} - \omega t \right]$ in Eq. (8.46) yields the dispersion relation:

$$\omega = D_{\text{or}} q^2. \tag{8.48}$$

For a wavelength of 10 μm, one has $\omega \approx 40$ s^{-1}, corresponding to a damping time $\tau = 2\pi/\omega \approx 0.2$ s. This time is longer than the reaction time of the human eye.

This calculation is approximate, insofar as it neglects the elastic anisotropy. In addition, we assumed here that the velocity is zero and the director turns in-place. In fact, a rotation of the director often sets the molecules in motion. This secondary flow (or 'backflow') is sketched in Fig. 8.8; it leads to a decrease in the global dissipation. However, it does not change the general form of the dispersion relation (8.48), where γ_1 is simply replaced by an effective viscosity γ_1^*, slightly lower than γ_1 (by about 10%).

Note that these damped modes are specific to nematic phases. They can be considered 'slow', since they are considerably less damped than the usual vorticity (or shear) 'fast' modes, for which $\omega = vq^2$, where v is a typical kinematic viscosity. Indeed, for the latter modes $v = \eta/\rho \approx 10^{-4}$ m²/s for a viscosity $\eta = 0.1$ Pa s (see below), proving that $v \gg D_{or}$.

Finally, it should be noted that, due to the orientation fluctuations of the director, *the nematic phase strongly scatters light*, about a million times more than the isotropic phase of the same material (de Gennes, 1969; [39, 40, 43]). By measuring the spectrum of the scattered light, one can even determine the Frank elastic coefficients (Langevin and Bouchiat, 1975). This technique is however less precise than that of Frederiks, which has the advantage of being more direct.

8.1.5 The Miesowicz experiment

So far we have only considered director dynamics, without taking into account the possibility of flow, since we always assumed the molecules were turning in-place, around their centres of mass.

We will now tackle a more difficult problem, the flow dynamics of a nematic phase.

In this area, the fundamental experiment was performed by M. Miesowicz in 1935 (Miesowicz, 1935). It is based on an oscillating pendulum immersed in a nematic phase where the director is fixed along a given direction by applying a strong magnetic field (Fig. 8.9).

Miesowicz's main result is that *the damping time of the pendulum, proportional to the viscosity of the phase, is strongly dependent on the orientation of the director with respect to the velocity and its gradient.*

In this way, Miesowicz defined three viscosities, which eventually came to bear his name, and which are given by:

- η_a, as $\vec{v} \perp \vec{n}$ and $\overrightarrow{\text{grad}}\, v \perp \vec{n}$ (diagram a);
- η_b, as $\vec{v} \parallel \vec{n}$ and $\overrightarrow{\text{grad}}\, v \perp \vec{n}$ (diagram b);
- η_c, as $\vec{v} \perp \vec{n}$ and $\overrightarrow{\text{grad}}\, v \parallel \vec{n}$ (diagram c).

The Miesowicz viscosities have been measured by various authors. Among the most precise measurements, we can cite those of Gähviller (1973), performed in a Poiseuille flow, using a strong magnetic field to impose the director orientation. The example of MBBA is shown in Fig. 8.10.

We now need to build a new theory to explain these results. This is the goal of the Leslie–Eriksen theory, to be presented in the next section.

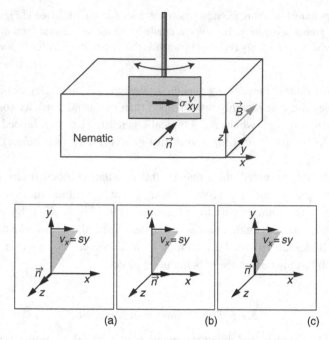

Fig. 8.9 Principle of the Miesowicz experiment: a pendulum oscillates in a nematic phase oriented by a strong magnetic field. The three diagrams represent the three Miesowicz geometries.

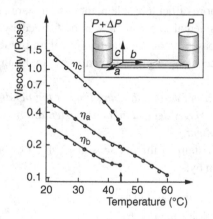

Fig. 8.10 The three Miesowicz viscosities for MBBA as a function of the temperature (Gähviller, 1973).

8.1.6 Constructing nematodynamics

Until now, we have concentrated on the dynamics of the director $\vec{n}$. We showed that it has a diffusive behaviour (see Eq. (8.46)), with a damping time that depends on the wave vector, diverging when this tends towards 0. *The director $\vec{n}$ is thus the hydrodynamic variable associated with the orientational order of the nematic phase.* The other two hydrodynamic variables, associated with the conservation laws, are the density ρ and the hydrodynamic velocity $\vec{v}$. We begin by repeating these laws.

8.1.6.1 The conservation equations

These equations were proven in Chapter 2.

The first one expresses *mass conservation*. It reads as for a simple fluid:

$$\frac{\partial \rho}{\partial t} + \operatorname{div}(\rho \vec{v}) = 0. \tag{8.49}$$

If the nematic phase is *incompressible*, as we will always assume in the following, the density is constant and div $\vec{v} = 0$.

The second equation is that of *momentum conservation*. In the Lagrangian form, it reads:

$$\rho \frac{D\vec{v}}{Dt} = \operatorname{div} \underline{\sigma} + \vec{f}. \tag{8.50}$$

Here, $\underline{\sigma}$ is the stress tensor (to be determined later) and $\vec{f}$ the density of bulk external forces (gravity, for instance). We recall that this equation translates Newton's law.

This set of two equations is at the core of classical hydrodynamics for simple fluids. Since nematic phases have an additional hydrodynamic variable, the director $\vec{n}$, another equation is needed, expressing the equilibrium of torques.

8.1.6.2 The torque equilibrium equation

This equation was also proved in Chapter 2. It is a direct consequence of the angular momentum theorem. Its expression is:

$$\vec{\Gamma}^{E} + \vec{\Gamma}^{v} + \vec{\Gamma}^{M} = 0, \tag{8.51}$$

where we assumed implicitly (as we will do throughout the discussion) that the external applied torque is the magnetic torque $\vec{\Gamma}^{M}$.

As to the two other torques, they are the elastic torque $\vec{\Gamma}^{E}$, already expressed as a function of the Frank energy in Eq. (8.19), and the viscous torque $\vec{\Gamma}^{v}$, for which we gave the simplified expression in the absence of flow (see Eq. (8.39)). Moreover, we showed in Chapter 2 that the internal torque exerted by the medium upon its director, of expression:

$$\vec{\Gamma} = \vec{\Gamma}^{E} + \vec{\Gamma}^{v}, \tag{8.52}$$

is related to the total stress tensor $\underline{\sigma}$ and to the $\underline{C}$ tensor (given by Eq. (8.16)) by the relation (see Eq. (2.23)):

$$\Gamma_i = -e_{ijk}\sigma_{jk} + C_{ij,j}. \tag{8.53}$$

We now need the expression for the stress tensor in order to calculate the viscous torque in the general case using the formula above.

8.1.6.3 Calculating the stress tensor

This tensor consists of three terms; a pressure term, $-P\underline{I}$, where P is the thermodynamic variable coupled to the density (or to the condition div $\vec{v} = 0$ in the incompressible case), an elastic term $\underline{\sigma}^E$ and a viscous term $\underline{\sigma}^v$. Together, they are written as:

$$\underline{\sigma} = -P\underline{I} + \underline{\sigma}^E + \underline{\sigma}^v. \tag{8.54}$$

We start by calculating the Ericksen elastic stress $\underline{\sigma}^E$. We showed in Chapter 4 that, during an isothermal process (as we will always assume in the following), the elastic stress is related to the free energy of elastic deformation f by the following general thermodynamic relation:

$$df = \sigma_{ij}^E \frac{\partial u_i}{\partial x_j}. \tag{8.55}$$

In this relation, the variable $\vec{u}(\vec{r})$ is the molecular displacement at *fixed orientation*, meaning that, as a molecule moves from position $\vec{r}$ to a position $\vec{r}'$ (with, by definition, $\vec{u}(\vec{r}) = \vec{r}' - \vec{r}$):

$$\vec{n}'(\vec{r}') = \vec{n}(\vec{r}). \tag{8.56}$$

Under these conditions, one can calculate the energy variation df:

$$df = \frac{\partial f}{\partial n_i} dn_i + \frac{\partial f}{\partial n_{i,j}} dn_{i,j} = \frac{\partial f}{\partial n_{i,j}} dn_{i,j}, \tag{8.57}$$

where we used $dn_i = 0$, since the change occurs at fixed director orientation. One then has:

$$dn_{i,j} = \frac{\partial n_i'}{\partial x_j'} - \frac{\partial n_i}{\partial x_j} = \frac{\partial n_i}{\partial x_k} \frac{\partial x_k}{\partial x_j'} - \frac{\partial n_i}{\partial x_j} = \frac{\partial n_i}{\partial x_k}\left(\delta_{kj} - \frac{\partial u_k}{\partial x_j'}\right) - \frac{\partial n_i}{\partial x_j} = -\frac{\partial n_i}{\partial x_k}\frac{\partial u_k}{\partial x_j} \tag{8.58}$$

leading to:

$$df = -\frac{\partial f}{\partial n_{i,j}} \frac{\partial n_i}{\partial x_k} \frac{\partial u_k}{\partial x_j} = -\frac{\partial f}{\partial n_{k,j}} \frac{\partial n_k}{\partial x_i} \frac{\partial u_i}{\partial x_j}, \tag{8.59}$$

which yields, by identification with the general relation (8.55):

$$\sigma_{ij}^E = -\frac{\partial f}{\partial n_{k,j}} \frac{\partial n_k}{\partial x_i}. \tag{8.60}$$

We still need to give the most general expression possible (respecting the symmetries of the uniaxial nematic phase) for the viscous stress tensor $\underline{\sigma}^v$. This is obtained by calculating the dissipation, as we did in the case of ordinary fluids at the beginning of Chapter 3, then, after having defined the forces and the fluxes, by writing the linear relations between these quantities. The demonstration is rather technical so we will not present it here, instead referring the reader to [39, 40, 43, volume 1]. It leads to the following result:

$$\sigma_{ij}^v = \alpha_1 (n_k A_{kl} n_l) n_i n_j + \alpha_2 n_j N_i + \alpha_3 n_i N_j + \alpha_4 A_{ij} + \alpha_5 n_j n_k A_{ki} + \alpha_6 n_i n_k A_{kj}, \quad (8.61)$$

where $A_{ij} = (\partial v_i / \partial x_j + \partial v_j / \partial x_i)/2$ is the symmetric deformation rate tensor and the vector $\vec{N}$ is the '*local rotation rate of the director*', defined by:

$$\vec{N} = \frac{D\vec{n}}{Dt} - \vec{\Omega} \times \vec{n}, \quad (8.62)$$

where $\vec{\Omega} = (1/2)\overrightarrow{\text{curl}}\,\vec{v}$ is the local rotation rate of the fluid.

The α_i coefficients are the six *Leslie viscosities*. O. Parodi showed, using Onsager's reciprocity relations, that they are not all independent, being connected by the following relation (Parodi, 1970):

$$\alpha_2 + \alpha_3 = \alpha_6 - \alpha_5. \quad (8.63)$$

Thus, there are only five independent viscosity coefficients.

It is noteworthy that the viscous stress tensor has the fundamental property of vanishing when the nematic undergoes solid rotation, as in this case $A_{ij} = 0$ and $\vec{N} = 0$.[3]

8.1.6.4 General expression of the viscous torque

We now have all the prerequisites for determining the complete expression of the bulk viscous torque. Indeed, from Eq. (8.53):[4]

$$\Gamma_i^v = -e_{ijk}\sigma_{ij}^v, \quad (8.64)$$

whence, using Eq. (8.61):

$$\vec{\Gamma}^v = -\vec{n} \times \left(\gamma_1 \vec{N} + \gamma_2 \underline{A}\vec{n} \right) \quad (8.65)$$

with:

$$\gamma_1 = \alpha_3 - \alpha_2, \quad (8.66)$$

$$\gamma_2 = \alpha_6 - \alpha_5. \quad (8.67)$$

Two particular combinations of the Leslie coefficients have appeared. The first one, γ_1, should be familiar as the rotational viscosity coming into play in the absence of hydrodynamic flow (see Section 8.1.3.4); the second one, γ_2, is new, and is only relevant when the velocity is finite.

---

[3] It is therefore independent of the chosen reference frame, in compliance with the objectivity principle.
[4] Note that the surface torque tensor $\underline{C}$ only contains elastic terms.

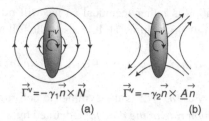

$$\vec{\Gamma}^{v} = -\gamma_1 \vec{n} \times \vec{N} \qquad \vec{\Gamma}^{v} = -\gamma_2 \vec{n} \times \underline{A}\vec{n}$$

(a)   (b)

Fig. 8.11 Viscous torque in rotational (a) and irrotational (b) flow.

To better understand the physical significance of each component of the viscous torque, let us decompose the local flow into its rotational and irrotational parts, as we have already done in Section 3.2.7.

It becomes apparent that *in the rotational flow* displayed in Fig. 8.11a *the viscous torque is proportional to* γ_1, *while it disappears for* $\vec{N} = 0$, *namely when the director rotates at the same velocity as the nematic.*

On the other hand, the viscous torque is proportional to γ_2 in the irrotational flow shown in Fig. 8.11b. This time, the torque vanishes when the director is aligned along the principal axes of the tensor A_{ij}, given by the asymptotes to four branches of hyperbola.

8.1.6.5 *Summary of the equations*

We find, in order:

- the incompressibility equation:

$$\operatorname{div} \vec{v} = 0, \tag{8.68}$$

- the equation of momentum conservation:

$$\rho \frac{D\vec{v}}{Dt} = \operatorname{div} \underline{\sigma} + \vec{f}, \tag{8.69}$$

- the equilibrium equation of bulk torques:

$$\vec{\Gamma}^{E} + \vec{\Gamma}^{v} + \vec{\Gamma}^{M} = 0. \tag{8.70}$$

In these equations, the expression for the stress tensor is:

$$\underline{\sigma} = -P\underline{I} + \underline{\sigma}^{E} + \underline{\sigma}^{v}, \tag{8.71}$$

where P is the hydrostatic pressure, $\underline{\sigma}^{E}$ the Ericksen elastic stress tensor, with components:

$$\sigma_{ij}^{E} = -\frac{\partial f}{\partial n_{k,j}} \frac{\partial n_k}{\partial x_i} \tag{8.72}$$

and $\underline{\sigma}^{v}$ the viscous stress tensor, with components:

$$\sigma_{ij}^{v} = \alpha_1 (n_k A_{kl} n_l) n_i n_j + \alpha_2 n_j N_i + \alpha_3 n_i N_j + \alpha_4 A_{ij} + \alpha_5 n_j n_k A_{ki} + \alpha_6 n_i n_k A_{kj}. \tag{8.73}$$

As to the bulk torques, they are given by the following formulas:

$$\vec{\Gamma}^{\mathrm{M}} = \frac{\chi_a}{\mu_0}(\vec{n} \cdot \vec{B})\vec{n} \times \vec{B},$$ (8.74)

$$\vec{\Gamma}^{\mathrm{v}} = -\vec{n} \times \left(\gamma_1 \vec{N} + \gamma_2 \underline{A}\vec{n}\right)$$ (8.75)

and

$$\Gamma_i^{\mathrm{E}} = -e_{ijk}\sigma_{jk}^{\mathrm{E}} + C_{ij,j}^{\mathrm{E}} = e_{ijk}\frac{\partial f}{\partial n_{l,k}}\frac{\partial n_l}{\partial x_j} + \frac{\partial}{\partial x_l}\left(\frac{\partial f}{\partial n_{k,l}}e_{ijk}\,n_j\right).$$ (8.76)

We emphasise that this last expression was deduced from Eqs. (8.16) and (8.60). We also could have used Eq. (8.19) to yield, using Eq. (8.10):

$$\Gamma_i^{\mathrm{E}} = e_{ijk}n_j h_k = e_{ijk}n_j \left(-\frac{\partial f}{\partial n_k} + \frac{\partial}{\partial x_l}\frac{\partial f}{\partial n_{k,l}}\right).$$ (8.77)

These two expressions of the elastic torque look quite different. They can however be shown to be equal, provided the free energy f exhibits rotational invariance. In other words, the energy must remain unchanged as the centres of mass of the molecules and the director turn by the same angle, i.e. under the transformation $\vec{u}(\vec{r}) = \vec{\omega} \times \vec{r}$ and $\delta\vec{n}(\vec{r}) = \vec{\omega} \times \vec{n}$. This condition is indeed fulfilled by the Frank–Oseen free energy.

We recall that the Leslie viscosities satisfy the Parodi relation:

$$\alpha_2 + \alpha_3 = \alpha_6 - \alpha_5.$$ (8.78)

Finally, here is the formula for the bulk dissipation (in the isothermal case):

$$\rho T \mathring{s} = \sigma_{ij}^{\mathrm{v}} A_{ij} + (\gamma_1 \vec{N} + \gamma_2 \underline{A}\vec{n}) \cdot \frac{D\vec{n}}{Dt} = \sigma_{ij}^{\mathrm{v}} A_{ij} - \vec{\Gamma}^{\mathrm{v}} \cdot \left(\vec{n} \times \frac{D\vec{n}}{Dt}\right)$$ (8.79)

with $\gamma_1 = \alpha_3 - \alpha_2$ and $\gamma_2 = \alpha_6 - \alpha_5$.

In this formula, the first term is due to the viscous stress (as for ordinary liquids) and the second one to viscous torques.

We will now use this theory to study some particular examples of flow.

8.1.7 Couette flow

This is the most common flow in rotating rheometers and in many applications. We will show that nematic phases exhibit novel features, not encountered in simple or viscoelastic liquids. One example is the appearance of a transverse shear stress when the flow symmetry is broken due to the director pointing out of the shear plane. Some instabilities can also develop due to the coupling between the director and the flow field. Before addressing these questions, let us return to Miesowicz's classical experiment and calculate the corresponding viscosities as a function of the Leslie viscosities.

8.1.7.1 Calculating the Miesowicz viscosities

We start by recalling that in the Miesowicz experiment, described in Fig. 8.9, the molecular orientation during flow is fixed by applying a magnetic field. This condition is only fulfilled if the imposed magnetic torque is much stronger than the hydrodynamic torque, leading to the following constraint:

$$\chi_a B^2 / \mu_0 \gg \alpha \dot{\gamma}, \tag{8.80}$$

where $\dot{\gamma}$ is the applied shear rate and α a typical viscosity. As an order of magnitude, taking $\alpha = 0.1\,\mathrm{Pa\,s}$ and $\chi_a = 10^{-6}$ yields:

$$B \gg \sqrt{\frac{\alpha \mu_0 \dot{\gamma}}{\chi_a}} \approx \frac{1}{3}\sqrt{\dot{\gamma}}, \tag{8.81}$$

where $\dot{\gamma}$ is expressed in s^{-1} and B in tesla. This equation shows that a strong field (of the order of one tesla) is needed to fix the director orientation under flow, even for relatively moderate shear rates of the order of $0.1\,\mathrm{s}^{-1}$. In the following, we consider this condition to be fulfilled.

As the viscosity is defined by the general relation:

$$\eta = \frac{\sigma_{ij}^{\mathrm{v}}}{2A_{ij}}, \tag{8.82}$$

one can determine the viscosity of the nematic phase in each of the Miesowicz geometries described in Fig. 8.9.

Thus, in the (a) geometry, where $\vec{v} \perp \vec{n}$ and $\overrightarrow{\mathrm{grad}\,v} \perp \vec{n}$:

$$A_{xy} = \frac{1}{2}\frac{\partial v}{\partial y}, \tag{8.83}$$

$$\sigma_{xy}^{\mathrm{v}} = \alpha_4 A_{xy}, \tag{8.84}$$

yielding, from Eq. (8.82):

$$\eta_a = \frac{1}{2}\alpha_4. \tag{8.85}$$

In the (b) case, $\vec{v} \parallel \vec{n}$ and $\overrightarrow{\mathrm{grad}\,v} \perp \vec{n}$, so that:

$$A_{xy} = \frac{1}{2}\frac{\partial v}{\partial y} = N_y, \tag{8.86}$$

$$\sigma_{xy}^{\mathrm{v}} = \alpha_3 N_y + (\alpha_4 + \alpha_6)A_{xy}, \tag{8.87}$$

whence:

$$\eta_b = \frac{1}{2}(\alpha_3 + \alpha_4 + \alpha_6). \tag{8.88}$$

Finally, in the (c) geometry, with $\vec{v} \perp \vec{n}$ and $\overrightarrow{\text{grad } v} \parallel \vec{n}$, one has:

$$A_{xy} = \frac{1}{2}\frac{\partial v}{\partial y} = -N_x, \tag{8.89}$$

$$\sigma_{xy}^v = \alpha_2 N_x + (\alpha_4 + \alpha_5)A_{xy} \tag{8.90}$$

leading to:

$$\eta_c = \frac{1}{2}(-\alpha_2 + \alpha_4 + \alpha_5). \tag{8.91}$$

One can also calculate the viscosity for the case when the director makes an angle θ with the velocity, provided it remains contained in the shear plane, defined by the velocity and its gradient. This calculation yields:

$$\eta(\theta) = \eta_b \cos^2 \theta + \eta_c \sin^2 \theta + 2\alpha_1 \sin^2 \theta \cos^2 \theta. \tag{8.92}$$

By measuring the three Miesowicz viscosities, as well as γ_1 and $\eta(45°)$, one can thus find the values of the six Leslie viscosities, assuming that the Parodi relation holds.

In MBBA, one of the most thoroughly studied materials, the room-temperature values of these viscosities are [39]:

$$\alpha_1 = 6.5 \text{ cP}, \quad \alpha_2 = -77.5 \text{ cP}, \quad \alpha_3 = -1.2 \text{ cP},$$
$$\alpha_4 = 83.2 \text{ cP}, \quad \alpha_5 = 46.3 \text{ cP}, \quad \alpha_6 = -34.4 \text{ cP}.$$

These values are given in centipoise (1 cP = 10^{-3} Pa s).

One can notice that, in MBBA, the α_2 and α_3 coefficients are negative, which is not a universal feature; if α_2 is always negative, α_3 can sometimes become positive. In particular, this occurs close to the transition to a smectic phase. This phenomenon was interpreted as being due to the formation of cybotactic groups, or molecular 'bunches' with a locally lamellar structure, in the vicinity of the transition (Helfrich, 1972; [43, volume 1]).

The next question concerns the stability of these flows. Indeed, what happens if the external magnetic field is suppressed?

8.1.7.2 Calculation of the viscous torque and flow stability

To find out whether the director will tend to turn under shear one must determine the viscous torque acting upon it.

This torque depends on the flow geometry, as we will show in a few simple cases (Fig. 8.12).

In the Miesowicz geometry (a), one has:

$$\vec{N} = A\vec{n} = 0 \quad \text{and} \quad \vec{\Gamma}_{(a)}^v = 0. \tag{8.93}$$

The torque being zero in this geometry (Fig. 8.12a), one would expect *the director to maintain its orientation under flow*, even in the absence of the magnetic field. This is indeed

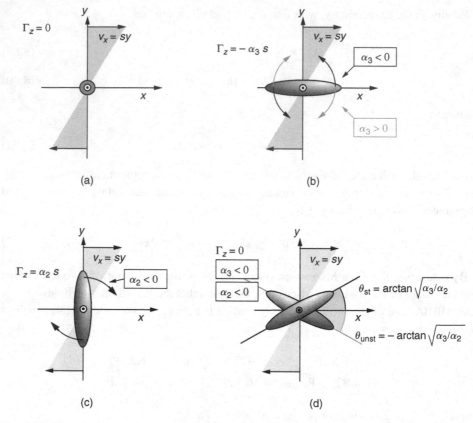

Fig. 8.12 Viscous torque exerted on the director by a simple shear flow in the three Miesowicz geometries (a, b and c). As the director makes with the velocity an angle $\pm \arctan(\alpha_3/\alpha_2)^{1/2}$, the torque vanishes (diagram d). In the figure, $s = \dot{\gamma}$.

the case at low shear rate. However, *above a critical shear rate the flow becomes unstable*. We will give later a qualitative description of this phenomenon.

In the Miesowicz geometry (b), on the other hand:

$$\vec{N} = \underline{A}\vec{n} = (\dot{\gamma}/2)\vec{y} \quad \text{and} \quad \vec{\Gamma}^v_{(b)} = -(\dot{\gamma}/2)(\gamma_1 + \gamma_2)\vec{z} = -\alpha_3 \dot{\gamma}\vec{z}, \qquad (8.94)$$

showing that *the director tends to turn perpendicular to its axis in the shear plane* (Fig. 8.12b). Note that the rotation direction depends on the sign of the viscosity α_3, since we know that this coefficient can be either positive or negative.

In the Miesowicz geometry (c), the calculation yields:

$$\vec{N} = -\underline{A}\vec{n} = -(\dot{\gamma}/2)\vec{x} \quad \text{and} \quad \vec{\Gamma}^v_{(c)} = (\dot{\gamma}/2)(\gamma_2 - \gamma_1)\vec{z} = \alpha_2 \dot{\gamma}\vec{z}, \qquad (8.95)$$

signifying that *the director starts to turn as soon as the nematic is sheared* (Fig. 8.12c).

One can legitimately wonder whether an intermediate situation between the Miesowicz geometries (b) and (c) exists, such that *the torque vanishes*. To find out, let us calculate the viscous torque when the director makes an angle θ with the velocity, while contained in the shear plane. The result is:

$$\vec{\Gamma}^{v} = (\alpha_2 \sin^2 \theta - \alpha_3 \cos^2 \theta)\dot{\gamma}\vec{z}. \tag{8.96}$$

The formula shows that, for $\alpha_2/\alpha_3 > 0$ (or, equivalently, for $\alpha_3 < 0$, since α_2 is always negative in nematic phases of elongated rods), the viscous torque vanishes for two values of the angle θ, given by:

$$\theta_{st} = +\arctan\sqrt{\alpha_3/\alpha_2} \quad \text{and} \quad \theta_{unst} = -\arctan\sqrt{\alpha_3/\alpha_2}. \tag{8.97}$$

One can show that the first solution corresponds to a stable orientation under flow, while the second one is completely unstable (Fig. 8.12d). Such materials are aligned by the flow ('flow-aligning nematics').

On the contrary, for $\alpha_3 > 0$ the torque does not vanish, whatever the angle θ; in this case, the nematic is unstable under flow and the director keeps on turning ('tumbling nematics').

The following dimensionless *'tumbling parameter'*:

$$\lambda = \frac{1 + \alpha_3/\alpha_2}{1 - \alpha_3/\alpha_2}, \tag{8.98}$$

quantifies this tendency. For $\lambda > 1$, the nematic is of the 'flow-aligning' type. One can check that, in this case, the angle θ_{st} given by Eq. (8.97) can also be written in the form:

$$\theta_{st} = \frac{1}{2} \arccos\left(\frac{1}{\lambda}\right). \tag{8.99}$$

If, however, $\lambda < 1$, the nematic is of the 'tumbling' type. Experiment shows that this is usually the case on approaching the transition towards a smectic phase. Thus, $\lambda \approx 0.5$ at $34\,°C$ in 8CB (this material goes to the smectic phase at about $33.5\,°C$), while in 5CB, which does not have a smectic phase, $\lambda \approx 1.15$ at the same temperature (Ternet *et al.*, 1999).

8.1.7.3 Non-Newtonian behaviour under simple shear, Ericksen number and stability

The results of the section above lead to some conclusions on the rheological behaviour of a nematic under shear, in the absence of a stabilising field.

To fix the ideas, consider the case of a *planar sample*, sandwiched between two parallel plates. If the anchoring is strong, the molecules cannot slip on the surfaces. Two limiting cases can occur:

- either the anchoring direction is perpendicular to the velocity, corresponding to the Miesowicz geometry (a) (Fig. 8.12a). The viscous torque vanishes, so the director preserves everywhere the orientation imposed by the plates and the measured viscosity is constant and independent of the shear rate: $\eta_a = \alpha_4/2$. The nematic exhibits *Newtonian behaviour*, at least at low shear rate, when no instability develops (for a discussion of the instabilities in this geometry, see (Pieranski and Guyon, 1973; [43, volume 1]).

- or the anchoring direction is parallel to the velocity, corresponding to the Miesowicz geometry (b) (Fig. 8.12b). In this case, the director feels a torque tending to make it turn within the shear plane. The planar sample will therefore lose its orientation under the action of the flow. It can be shown that its subsequent evolution depends on the sign of α_3 (Pieranski and Guyon, 1976; [43, volume 1]).

To show this qualitatively, assume that the director stays in the shear plane (this is indeed the case at low velocity). Splay deformation of the director field follows, since the director is anchored at the surfaces. This leads to a restoring torque acting on the director and opposing its rotation. Clearly, in the stationary regime (if present) the equilibrium of these two torques will determine the tilt angle of the director with respect to the x axis (parallel to the velocity). Let θ_0 be its value at the centre of the sample (in $y = d/2$, with the y axis perpendicular to the plates situated in $y = 0$ and $y = d$). To estimate this angle, let us write the equilibrium of torques. At the centre of the sample, the viscous torque is, from Eq. (8.96), $(\alpha_2 \sin^2 \theta - \alpha_3 \cos^2 \theta)\dot{\gamma}$. The elastic torque is given by $K_1 d^2\theta/dz^2 \approx K_1\pi^2\theta_0/d^2$, assuming a distortion of the type $\theta = \theta_0 \sin(\pi z/d)$. In the stationary regime, one must have:

$$(\alpha_2 \sin^2 \theta_0 - \alpha_3 \cos^2 \theta_0)\dot{\gamma} + K_1\pi^2\theta_0/d^2 = 0. \tag{8.100}$$

If $|\alpha_2| \gg |\alpha_3|$ (which is very often the case) and for a small angle θ_0 (a condition to be checked *ex post facto*), expanding to second order in θ_0 yields:

$$\dot{\gamma}\alpha_2\theta_0^2 + \frac{K_1\pi^2}{d^2}\theta_0 - \alpha_3\dot{\gamma} = 0. \tag{8.101}$$

This equation always has solutions for $\alpha_3 < 0$, since the discriminant is always positive, for all values of $\dot{\gamma}$. In this case, one finds two solutions to the problem:

$$\theta_0^{\pm} = \frac{-\dfrac{\pi^2 K_1}{d^2} \pm \sqrt{\dfrac{\pi^4 K_1^2}{d^4} + 4\alpha_2\alpha_3\dot{\gamma}^2}}{2\dot{\gamma}\alpha_2} \tag{8.102}$$

of which only the θ_0^+ solution is stable (as shown in Fig. 8.12d).

This formula states that, at low shear rates, the tilt angle is proportional to the shear rate:

$$\theta_0^+ = \frac{d^2}{\pi^2}\frac{\alpha_3}{K_1}\dot{\gamma}, \tag{8.103}$$

while at higher shear rates the angle reaches a saturation value:

$$\theta_0^+(\infty) = \sqrt{\frac{\alpha_3}{\alpha_2}}. \tag{8.104}$$

As expected intuitively, this is the value at which the viscous torque vanishes, already given in the previous section (noting that $\arctan\sqrt{\alpha_3/\alpha_2} \approx \sqrt{\alpha_3/\alpha_2}$ in the $|\alpha_2| \gg |\alpha_3|$ limit).

An important rheological consequence of this calculation is that the nematic now exhibits *non-Newtonian behaviour*, its viscosity changing with the shear rate. Indeed, its apparent viscosity is:

$$\eta_{app}(\dot{\gamma} \to 0) = \eta_b = \frac{1}{2}(\alpha_3 + \alpha_4 + \alpha_6) \tag{8.105}$$

at low shear rate, while at high shear rate it tends towards the viscosity corresponding to the entire sample being oriented at an angle $\theta_0^+(\infty)$. This viscosity is given by the formula (8.92) as:

$$\eta_{app}(\dot{\gamma} \to \infty) = \eta_b \cos^2\left[\theta_0^+(\infty)\right] + \eta_c \sin^2\left[\theta_0^+(\infty)\right] + 2\alpha_1 \sin^2\left[\theta_0^+(\infty)\right]\cos^2\left[\theta_0^+(\infty)\right] \tag{8.106}$$

yielding:

$$\eta_{app}(\dot{\gamma} \to \infty) = \eta_b + (2\alpha_1 + \eta_c - \eta_b)\frac{\alpha_3}{\alpha_2} \tag{8.107}$$

in the $|\alpha_2| \gg |\alpha_3|$ limit.

To fix the ideas, consider the example of MBBA, for which all the Leslie coefficients have been measured accurately. For a sample of this material, the viscosity varies between 23.8 cP at low shear rate and 25.2 cP at high shear rate.

It becomes apparent that, in the Miesowicz geometry (b), the nematic behaves as a *shear-thickening fluid*, similar to most concentrated emulsions.[5]

Let us now try to estimate the shear rate value corresponding to the transition from one regime to the other. This can be done by comparing the two terms under the square root in Eq. (8.102). These two terms represent the square of the elastic torque and the viscous torque, respectively; their ratio defines the *Ericksen number* of the flow:

$$Er = \frac{\text{Viscous torque}}{\text{Elastic torque}}. \tag{8.108}$$

In this experiment, it is given by:

$$Er = \frac{2}{\pi^2}\frac{\sqrt{\alpha_2\alpha_3}\dot{\gamma}d^2}{K_1}. \tag{8.109}$$

The crossover from one regime to the other typically occurs when this number is of unit order, which fixes a value $\dot{\gamma}_{cross}$ for the shear rate. For a 50 μm thick MBBA sample, $\dot{\gamma}_{cross} \approx 1\ \text{s}^{-1}$, which is not too high; it follows that the director tilts fairly easily under the action of shear. It is very important to note that the Reynolds number is extremely low for this shear rate value. Indeed, for simple shear flow, by definition:

$$Re = \frac{\rho\dot{\gamma}d^2}{\eta}, \tag{8.110}$$

---

[5] Note that this shear-thickening behaviour is not systematic; the same nematic material will behave as a shear-thinning fluid in the Miesowicz geometry (c); the proof is left as an exercise for the reader.

where η is the viscosity of the sample. The values given above yield $Re \approx 10^{-4}$, which is indeed much smaller than 1.

This comparison shows that *the relevant number in nematodynamics is the Ericksen number, rather than the Reynolds number.*

This conclusion is confirmed by the behaviour of the sample for $\alpha_3 > 0$. In this case, Eq. (8.101) has no solution when the discriminant becomes negative, i.e. when the Ericksen number is equal to or larger than 1, namely for:

$$Er = \frac{2}{\pi^2} \frac{\sqrt{-\alpha_2 \alpha_3}\, \dot\gamma d^2}{K_1} \geq 1. \tag{8.111}$$

This inequality defines a critical shear rate (corresponding here to $Er = 1$), above which the flow loses its stationary character and develops instabilities (the formation of convection rolls, or of disordered thread-like textures in the 'director turbulence' regime (Manneville, 1981)). We will not study these complicated flow phenomena in the present work. The important conclusion is that, in this experiment, *the stability limit of the flow is set by its Ericksen number.*

All these results have been checked by experiments. In particular, the tilt angle θ_0 of the director was measured experimentally as a function of the shear velocity, in the HBAB material, where the α_3 viscosity changes sign at a certain temperature. Some data are plotted in Fig. 8.13. They show that θ_0 is indeed positive and reaches saturation at high velocity for $\alpha_3 < 0$, while for $\alpha_3 > 0$, θ_0 is negative and diverges when the velocity approaches its critical value, close to 80 μm/s in a 80 μm thick sample.

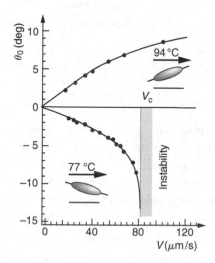

Fig. 8.13 Rotation angle θ_0 as a function of the shear velocity, at two different temperatures (Pieranski, 1976).

8.1.7.4 First and second normal stress differences

It is also extremely instructive to calculate the first normal stress difference. Two cases must be distinguished, depending on whether the director orientation under shear is maintained fixed or not.

The first case is obviously the simplest one. It corresponds to applying a very strong magnetic field to the sample. As before, we assume that the director is contained in the shear plane, making an angle θ with the velocity. Keeping in mind that the elastic stress is exactly zero in this case, since the director field is not distorted, one can calculate using the formula (8.73) the three normal stress values, given respectively by:

$$\sigma_{xx} = \sigma_{xx}^v = \frac{1}{2}\dot{\gamma} \sin\theta \cos\theta \left[2\alpha_1 \cos^2\theta + (-\alpha_2 - \alpha_3 + \alpha_5 + \alpha_6)\right], \tag{8.112}$$

$$\sigma_{yy} = \sigma_{yy}^v = \frac{1}{2}\dot{\gamma} \sin\theta \cos\theta \left[2\alpha_1 \sin^2\theta + (\alpha_2 + \alpha_3 + \alpha_5 + \alpha_6)\right], \tag{8.113}$$

$$\sigma_{zz} = \sigma_{zz}^v = 0. \tag{8.114}$$

Hence, *when the director is blocked along the direction θ*, the two normal stress differences (defined in Section 6.3.2) take values:

$$N_1 \equiv \sigma_{xx} - \sigma_{yy} = \frac{1}{2}\dot{\gamma} \sin 2\theta \left[\alpha_1 \cos 2\theta - \alpha_2 - \alpha_3\right], \tag{8.115}$$

$$N_2 \equiv \sigma_{yy} - \sigma_{zz} = \frac{1}{2}\dot{\gamma} \sin 2\theta \left[\alpha_1 \sin^2\theta + \frac{1}{2}(\alpha_2 + \alpha_3 + \alpha_5 + \alpha_6)\right]. \tag{8.116}$$

These formulas show that *the normal stress differences are of viscous (hydrodynamic) origin and related to the symmetries of the nematic phase*. Here, they are *linear in $\dot{\gamma}$, and not quadratic* as for isotropic viscoelastic liquids at low shear rate. This is due to the difference between shearing 'with or against the pile' a nematic phase where the molecules are tilted with respect to the velocity (which is made possible by blocking the director orientation using the magnetic field). Obviously, the normal stress differences are zero for $\theta = 0$ or $\pi/2$, on symmetry grounds.

In the second case, *the director orientation is not blocked*. Under these conditions, the director orientation changes due to the viscous torque acting upon it. If the anchoring energy on the plates is zero (experimentally, this is almost feasible), the director is oriented so as to nullify the viscous torque (we assume here implicitly that $\alpha_3 < 0$). This condition sets the value of the angle θ at $\arctan(\alpha_3/\alpha_2)^{1/2}$ when $\dot{\gamma} > 0$ and $-\arctan(\alpha_3/\alpha_2)^{1/2}$ for $\dot{\gamma} < 0$. The formulas above are still valid, since the director has the same orientation everywhere, and yield the following relations:

$$N_1 = |\dot{\gamma}|\sqrt{\alpha_3\alpha_2}\left[1 - \alpha_1 \frac{\alpha_2 - \alpha_3}{(\alpha_2 + \alpha_3)^2}\right], \tag{8.117}$$

$$N_2 = -|\dot{\gamma}|\frac{\sqrt{\alpha_3\alpha_2}}{2}\left[1 + \frac{\alpha_5 + \alpha_6}{\alpha_2 + \alpha_3} + \frac{2\alpha_1\alpha_3}{(\alpha_2 + \alpha_3)^2}\right]. \tag{8.118}$$

Here, the important result is that *the normal stresses are no longer analytic in $\dot{\gamma}$, since they vary as $|\dot{\gamma}|$.*

Note that these formulas still apply at strong shear rates in the limit of strong anchoring on the plates, since we know that, in this limit, the angle θ tends towards $\pm \arctan(\alpha_3/\alpha_2)^{1/2}$ for $\dot{\gamma} \to \pm\infty$.

Another important observation is that the predictions of this model also apply to polymeric liquid crystals, provided the shear rate is sufficiently low. The explanation is that the nematic contribution, linear in $\dot{\gamma}$, will always dominate at low shear rate over the elastic contribution (due to the deformation of the chains), which is proportional to $\dot{\gamma}^2$.

Until now, we have only considered flows that are symmetric with respect to the shear plane. Let us now analyse the situation when this symmetry is broken.

8.1.7.5 Transverse shear stress

Once again, we assume that the orientation of the director is fixed by applying a strong magnetic field. This time, however, the director makes with the shear plane an angle that is neither 0 nor $\pi/2$.

The mirror symmetry of the flow with respect to the shear plane is now broken, so that the viscous stress tensor, written in the reference frame (x, y, z) (defined in Fig. 8.14), need not be diagonal.

We can show this explicitly, assuming that the director is parallel to the (x, z) plane and makes an angle ϕ with the z axis, perpendicular to the shear plane. In this case, calculating the σ^v_{xy} component of the Leslie–Ericksen viscous stress tensor gives:

$$\sigma^v_{xy} = \eta_{\text{eff}}\,\dot{\gamma} \tag{8.119}$$

with

$$\eta_{\text{eff}} = \eta_a \cos^2\phi + \eta_b \sin^2\phi. \tag{8.120}$$

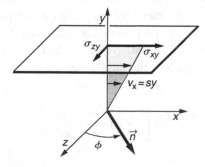

Fig. 8.14 Appearance of a transverse stress when the symmetry of the flow with respect to the shear plane is broken.

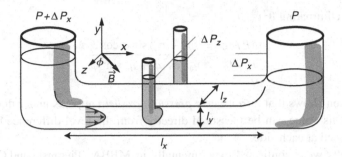

Fig. 8.15 Setup used for demonstrating the transverse pressure gradient in Poiseuille flow.

This calculation shows that the effective viscosity of the nematic is between η_a and η_b, which was to be expected, since this geometry is intermediate between the two Miesowicz geometries (a) and (b).

Another, more interesting, result is that *the viscous stress tensor has an off-diagonal component*:

$$\sigma_{zy}^{v} = (\eta_b - \eta_a)\sin\phi\cos\phi\dot{\gamma}, \tag{8.121}$$

which is finite for $\phi \neq 0$ and $\pi/2$.

Consequently, *a transverse tangent force* σ_{zy}^{v} must be applied to the upper plate, aside from the σ_{xy}^{v} force, in order to shear the nematic along the desired direction (Fig. 8.14). To our knowledge, no direct measurement of this force has been performed yet.

We will show that this effect is also manifested by *the appearance of a transverse pressure in a Poiseuille flow*; this phenomenon was demonstrated experimentally by P. Pieranski and E. Guyon in 1976.

8.1.8 Poiseuille flow

This experiment is described in Fig. 8.15. The Poiseuille flow occurs in a capillary of rectangular cross-section, with inner dimensions $l_y \ll l_z \ll l_x$. Under these conditions, the velocity field is of the type:

$$v_x = v_x(y), \qquad v_y = v_z = 0. \tag{8.122}$$

Equations (8.69), (8.71) and (8.73) give:

$$0 = -\frac{\partial P}{\partial x} + \eta_{\text{eff}}(\phi)\frac{\partial^2 v_x}{\partial y^2}, \tag{8.123}$$

$$0 = -\frac{\partial P}{\partial z} + (\eta_b - \eta_a)\sin\phi\cos\phi\frac{\partial^2 v_x}{\partial y^2} \tag{8.124}$$

and yield, by eliminating $\partial^2 v_x / \partial y^2$:

$$\frac{\partial P / \partial z}{\partial P / \partial x} = \frac{(\eta_b - \eta_a) \sin \phi \cos \phi}{\eta_a \cos^2 \phi + \eta_b \sin^2 \phi}. \tag{8.125}$$

This equation shows that a *transverse pressure gradient* appears under flow for $\phi \neq 0$ and $\pi/2$, and its value can be measured directly from the level difference between two capillaries placed at each side.

These results were confirmed experimentally in MBBA (Pieranski and Guyon, 1974; [43, volume 1]). In this material, $\eta_a \approx 40$ cP and $\eta_b \approx 30$ cP, and the transverse gradient reaches a maximum, of about 0.14 times the longitudinal gradient, for an angle $\phi \approx 50°$. This is exactly what Eq. (8.125) predicts.

8.1.9 Disclination lines

So far, we have assumed that the nematic was well oriented, in particular that its director field was regular, exhibiting no topological singularities. This result can be achieved experimentally by an adapted treatment of the glass plates containing the sample, or by applying a magnetic field.

There are however situations where the director field exhibits singularities. In nematics, these singularities can be point-like or linear. They locally break the orientational order of the phase, like dislocations break the positional order in crystals. Note that the concept of dislocations is not relevant in nematics insofar as these defects, which one might try to create using the Volterra construction used in solids, will disappear by viscous relaxation in microscopic times.

In the following, we will be concerned with disclinations, defects that form thread-like textures as shown in Fig. 1.29.

8.1.9.1 De Gennes–Friedel construction

We have already shown how to construct a disclination using the Volterra method (see Section 5.1.2.1). This construction is not the only one possible, nor is it the most intuitive one. That is why we will use an alternative one, due to P. G. de Gennes, and whose equivalence with the Volterra construction in nematics was shown by J. Friedel [39].

Consider a line L in a perfect nematic and a cut surface Σ limited by this line. Virtually cut the nematic open along this surface. This defines the two lips of the cut, Σ_1 and Σ_2. It is then sufficient to rotate in place the molecules on Σ_2 (by an angle Ω around the $\vec{v}$ axis) relative to the molecules on Σ_1. If the $(\vec{v}, \Omega)$ rotation belongs to the symmetry elements of the phase, the two lips Σ_1 and Σ_2 can then be perfectly resealed. Finally, letting the system relax creates a disclination line of angle Ω.

This method is illustrated in Fig. 8.16, showing a circular disclination loop. In this example, the nematic orientation starts by being everywhere perpendicular to the $\vec{v}$ vector and parallel to the plane of the loop. The disclination line is constructed by creating a 180° difference in the director orientation across the cut surface Σ contained by the loop. Since

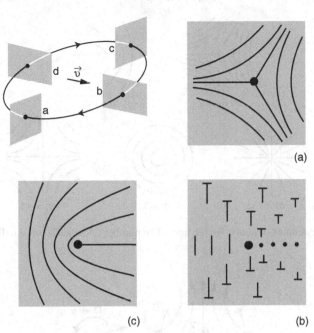

(a)

(c)   (b)

Fig. 8.16 Disclination loop in a nematic phase. Insets (a) and (c) depict the director field lines in the vicinity of the defect line in the normal sections (a) and (c). In inset (b), corresponding to the section at point (b), the nails represent molecules tilted with respect to the plane of the drawing. The length of each nail is proportional to the projection of the director onto this plane. By convention, the point of the nail represents the end of the molecule closer to the observer.

$\vec{n}$ and $-\vec{n}$ are equivalent, the director field can be resealed on this surface with no discontinuity. Relaxing the system yields the line shown in Fig. 8.16. Note that, according to the definitions given in Chapter 5, the disclination line is purely of *wedge* type ($\vec{v}$ parallel to the line) in sections (a) and (c), and of *twist* type ($\vec{v}$ perpendicular to the line) in sections (b) and (d). Such lines give rise to the *thin threads* in thread-like nematic textures (Fig. 1.29). One can show these defects to be *topologically stable*, since they cannot be removed by a continuous transformation of the director field (they have the same topology as a Moebius strip).

A different kind of defect is obtained by turning the director through 360° across the cut surface. The resulting line resembles the *thick threads* in nematic textures (Fig. 1.29). One can show that *these defects are not topologically stable*, since they can be eliminated by a continuous transformation of the director field. Unlike thin threads, these defects do not have a singular core, explaining their different appearance in microscope observation. Thus, the former scatter light strongly and exhibit sharp contrast, while the latter have a diffuse aspect, making them more difficult to observe.

We will try to understand these issues by concentrating on the case of planar wedge lines, where the director is confined within the plane perpendicular to the line.

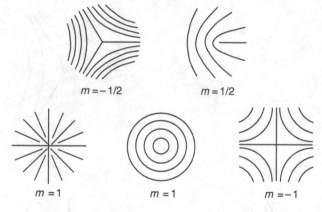

Fig. 8.17 Some examples of planar wedge lines. The number under each curve indicates its topological rank.

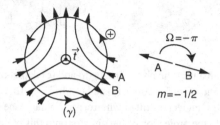

Fig. 8.18 Hodograph of a wedge line (from [41]).

8.1.9.2 Topological rank of a planar wedge line

We start by noting that a wedge line is necessarily straight since (by definition) it is parallel to the $\vec{v}$ vector (this vector is conserved along the line). If, moreover, the director remains in the plane perpendicular to the line, the latter is said to be *planar*. We will make this simplifying assumption in the following.

In Fig. 8.17 we show the field lines of the director field around some planar wedge lines. It is easy to check that in the two examples at the top the rotation angle of the director is π (in absolute value), while in bottom ones this angle is 2π (in absolute value, once again). Another remark is that the shape of the director field depends on the rotation direction, as shown by the top images. Hence the need to introduce a sign convention.

To this end, let us construct *the hodograph of a wedge line* (Fig. 8.18). The first step is to orient this line by choosing a unit vector $\vec{t}$ tangent to the line. One then constructs a *Burgers circuit* around this line (as we did for dislocations, see Section 5.1.2.2), and orients it according to Maxwell's 'corkscrew' rule. This fixes the positive direction of rotation and, at the same time, defines the sign of the angle Ω, describing the rotation of the director on the cut surface. It is then sufficient to start from a point A of the Burgers circuit, to make one turn in the positive direction to point B, and to note the angle Ω by which the director $\vec{n}$

has turned. It is easily checked that the sign of Ω does not depend on the (arbitrarily chosen) orientation. This angle is therefore *a topological parameter, intrinsic to the line*.

The line is more commonly described by the number m, its *topological rank*, defined by the relation:

$$\Omega = 2\pi m. \tag{8.126}$$

The value of m is indicated under each disclination line in Fig. 8.17.

In the following, we will show that the topological rank m of a wedge line plays the same role as the Burgers vector b for a dislocation, or the circulation Γ for a vortex.

Note that, in the case of nematics, m is necessarily a (positive or negative) multiple of $1/2$: $m = \pm 1/2, \pm 1, \pm 3/2$, etc. The topological rank of a disclination is therefore *quantified*, like the Burgers vector of a perfect dislocation in a solid ($b = \pm ka$, $k \in N$, a being the lattice spacing), or the circulation of a vortex in superfluid helium ($\Gamma = \pm k \frac{h}{m}$, $k \in N$, where h is the Planck constant and m the mass of a helium atom).

8.1.9.3 Energy of a planar line

For ease of calculation, we will assume all Frank constants to be equal. In this case, the distortion free energy is reduced to formula (8.6). If the director field is planar (contained in the (x, y) plane), a single angle is needed to describe its orientation. Let φ be the angle between the director and the x axis. In this case, the distortion energy has a particularly simple form:

$$f = \frac{1}{2} K \left(\overrightarrow{\text{grad}\,\varphi} \right)^2. \tag{8.127}$$

At equilibrium, the director field must minimise the total energy. This condition leads to the following equation:

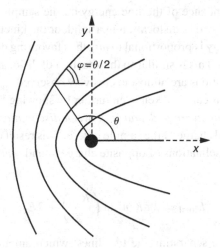

Fig. 8.19 Reference frame and disclination line of rank $m = +1/2$.

$$\Delta\varphi = 0 \tag{8.128}$$

or, in polar coordinates (r, θ):

$$\frac{1}{r}\frac{\partial}{\partial r}\left(r\frac{\partial\varphi}{\partial r}\right) + \frac{1}{r^2}\left(\frac{\partial^2\varphi}{\partial\theta^2}\right) = 0. \tag{8.129}$$

This equation has a completely obvious solution, of the type:

$$\varphi = m\theta \tag{8.130}$$

representing a disclination of rank m (Fig. 8.19).

The energy of the line is then calculated by integrating the energy density over the surface of the plane:

$$E_1 = \int \frac{1}{2}K\frac{m^2}{r^2}2\pi r\,dr. \tag{8.131}$$

Clearly, the integral diverges in 0 and at infinity. One must then introduce cutoff distances, the core radius r_c (microscopic in size, of the order of 100 Å) and the sample size R. Finally, this leads to:

$$E_1 = \pi Km^2\ln\left(\frac{R}{r_c}\right) + E_c, \tag{8.132}$$

where we added the core energy E_c (which can be shown to be of the order of $\pi Km^2/2$, and thus generally negligible in comparison with the other term).

This logarithmic dependence of the line energy on the sample size is the same as that found for the elastic energy of a dislocation in a solid, or the kinetic energy of a vortex.

Also note that the energy is proportional to m^2, thus favouring disclinations of low topological rank ($m = \pm 1/2$). This is similar to the situation of dislocations in solids ($E \propto b^2$), where elementary dislocations are almost exclusively observed.

The analogy with solids can be taken even further by showing that *the cutoff distance R must be taken equal to the average distance between disclinations of opposite sign*, provided the total topological charge of the sample is zero. This result can be proved rigorously in the presence of two disclinations of opposite charge m and $-m$. Indeed, in this case the energy is exactly:

$$E_{\text{total}} = 2\pi Km^2\ln\left(\frac{R_{12}}{r_c}\right) + 2E_c, \tag{8.133}$$

where R_{12} is the distance separating the two lines, which amounts to assigning to each disclination the energy (8.132), with $R = R_{12}$.

8.1.9.4 Dynamics of a planar line

Experiment shows that it is very easy to generate disclination lines under flow. It suffices, for instance, to place a nematic droplet on a glass plate treated in homeotropic anchoring (molecules perpendicular to the surface) and to stir it vigorously using a spatula. This yields a beautiful thread-like texture, with a high density of defects. These threads, which are disclinations, are very mobile when they are not attached to the surfaces, and are able to disappear quickly once the flow that produced them has stopped.

This is illustrated by the sequence of photographs in Fig. 8.20, where the threads recombine and then form loops that shrink and finally collapse. Clearly, the force driving defect healing is a global decrease in the elastic energy of the system, which is far from equilibrium at the beginning of the process. The same phenomenon occurs at high temperature in solids, where it is known as 'recrystallisation'.

The analysis of the dynamics of a disclination line is in general very complicated, so we will simplify the problem as much as possible and start by considering the case of a planar wedge line moving at a constant velocity (under the action of a force that we will not yet discuss) in the nematic phase, considered at rest at infinity.

To find the effective friction force applied by the medium to the line, one must first calculate the dissipation. This is not a simple procedure because, strictly speaking, one would have to solve the nematodynamics equations around the line. This problem does not have an analytical solution, and a numerical solution was given only recently

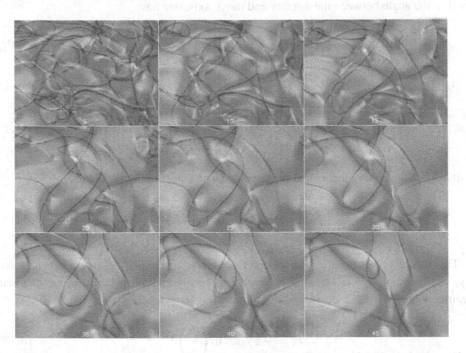

Fig. 8.20 Sequence of photographs, taken every 5 s, showing the recombination and progressive collapse of the loops formed by disclinations in a thread-like texture.

(Tóth *et al.*, 2002). This friction force can nevertheless be estimated by assuming that (Imura and Okano, 1973):

(1) the line moves only by in-place rotation of the molecules, without any backflow ($\vec{v} = 0$);
(2) the director field around the moving line is the same as for the static case (in the reference frame of the line).

Under these conditions, the total dissipation takes the following simple form:

$$\Phi_{\text{tot}} = \gamma_1 \int \left(\vec{n} \times \frac{\partial \vec{n}}{\partial t} \right)^2 dx\, dy, \tag{8.134}$$

where $\vec{n}(x, y, t) = \vec{n}(x - Vt, y)$, assuming that the line moves at a constant velocity V in the x direction.

Using assumption 2 and the fact that the director is a unit vector, and setting $X = x - VT$ and $Y = y$ yields:

$$\Phi_{\text{tot}} = \gamma_1 V^2 \int \left(\frac{\partial \vec{n}}{\partial X} \right)^2 dX\, dY, \tag{8.135}$$

where $\vec{n}(X, Y)$ is the static configuration of the director, in the reference frame of the line. With φ the angle between the director and the X axis, one has:

$$\Phi_{\text{tot}} = \gamma_1 V^2 \int \left(\frac{\partial \varphi}{\partial X} \right)^2 dX\, dY, \tag{8.136}$$

yielding for a disclination with rank m, after switching to polar coordinates:

$$\Phi_{\text{tot}} = \gamma_1 V^2 m^2 \int \left(\frac{\partial \theta}{\partial X} \right)^2 dX\, dY = \gamma_1 V^2 m^2 \int_{r_c}^{R} \int_0^{2\pi} \left(\frac{\sin \theta}{r} \right)^2 r\, dr\, d\theta, \tag{8.137}$$

using again the core radius r_c and the large distance cutoff radius R. Finally:

$$\Phi_{\text{tot}} = \pi \gamma_1 V^2 m^2 \ln \left(\frac{R}{r_c} \right). \tag{8.138}$$

This formula gives the effective friction force acting on the line $\vec{F}_v$. It opposes the velocity, and its amplitude is obtained by applying the dissipation theorem in the stationary regime $\Phi_{\text{tot}} = -\vec{F} \cdot \vec{V}$ (see Section 3.2.6), resulting in:

$$\vec{F}_v = -\pi \gamma_1 m^2 \ln \left(\frac{R}{r_c} \right) \vec{V}. \tag{8.139}$$

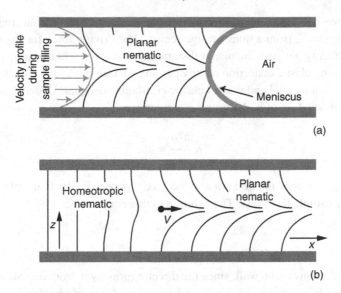

(a)

(b)

Fig. 8.21 (a) Inversion wall (where the director turns by π from one plate to the other) formed upon filling a homeotropic sample; (b) disclination line separating the planar area (to the right) from the homeotropic one (to the left).

We should point out that, even in an infinite sample, R remains finite, being typically given by $3.6K/\gamma_1 V$ (Ryskin and Kremenetsky, 1991; [43, volume 1]).[6]

To conclude this section, let us analyse two experimental situations where these theoretical predictions can be tested.

The first one concerns the dynamics of a disclination line isolated in a sample contained between two glass plates treated for homeotropic anchoring (with the molecules perpendicular to the glass surface). Experiment shows that an *inversion wall* forms in the thickness of the sample while the nematic fills the gap between the plates by capillarity, starting at the edges. This wall, represented in Fig. 8.21a, is due to the homeotropic anchoring of the molecules on the meniscus of the air interface. It is also favoured by the Poiseuille flow between the two plates (since $\alpha_2 < 0$). Note that the sample can be filled by capillarity because the liquid crystal almost always wets the glass surface, meaning that the surface energy between air and glass is higher than between glass and the liquid crystal. Obviously, immediately after filling, the sample is not in its minimal energy configuration, the director field being distorted.

Experiment shows that homeotropic areas (where the director is everywhere perpendicular to the glass plates) nucleate fairly rapidly. This nucleation is heterogeneous, since it always occurs on dust particles or on defects at the surface of the plates. These homeotropic areas progressively invade the entire sample, which is perfectly oriented at the end of the

---

[6] The formula (8.139) predicts a friction force independent of the sign of m. This result no longer holds when the elastic anisotropy and, above all, the 'backflow' effects are taken into account, as shown by several authors, using both numerical simulations (Tóth *et al.*, 2002) and analytical methods (Kats *et al.*, 2002). Recent experiments bear out these predictions (Blanc *et al.*, 2005; Oswald and Ignés-Mullol, 2005).

process. As shown in Fig. 8.21b, a disclination line separates each 'planar' region, containing an inversion wall, from a homeotropic region. The driving force for the motion is the difference in energy between the planar distorted region and the homeotropic one.

To calculate the elastic distortion energy per unit surface of sample E_P, assume that the director field is not twisted and that all elastic constants are equal. Under these conditions, the equilibrium equation of bulk torques is written:

$$K\frac{\partial^2 \theta}{\partial z^2} = 0, \tag{8.140}$$

with θ the angle between the director and the z axis, perpendicular to the plates.

Let d be the sample thickness. The solution to this equation is:

$$\theta = \pi \frac{z}{d}, \tag{8.141}$$

which is indeed an inversion wall, since the director turns by π from one plate to the other. Assume the line is parallel to the x axis and denote by L the width (along y) of the planar region. The energy of the planar strip is, per unit length along y:

$$E_P = L \int_0^d \frac{1}{2} K \left(\frac{d\theta}{dz}\right)^2 dz = \frac{\pi^2}{2} \frac{K}{d} L. \tag{8.142}$$

Since no elastic energy is stored in the homeotropic region ($E_H = 0$), the total energy of the sample is:

$$E = E_P + E_H + E_1 = \frac{\pi^2}{2}\frac{K}{d}L + E_1, \tag{8.143}$$

with E_1 the line energy. Thus, the elastic configuration force acting on the line is:

$$F = \frac{dE}{dL} = \frac{\pi^2}{2}\frac{K}{d}. \tag{8.144}$$

In the stationary regime, the absolute value of this force is equal to the viscous force acting on the line (here, of rank $m = -1/2$). This latter force is given by Eq. (8.139), where one must take $R = d$, since the dissipation occurs in a region of size d around the line. This yields the velocity of the disclination line:

$$V = \frac{2\pi}{\ln(d/r_c)}\frac{K}{\gamma_1 d}. \tag{8.145}$$

Thus, the line velocity is typically given by D_{or}/d, where $D_{or} = K/\gamma_1$ is the diffusion coefficient of director orientation (see Eq. (8.47)).

As an order of magnitude, $V \approx 10$ μm/s for a sample thickness $d = 10$ μm, using $D_{or} = 10^{-10}$ m²/s. This value agrees reasonably well with the experimental results. The $1/d$ dependence of the velocity is also borne out by the experiments.

Another interesting problem is that of the *annihilation dynamics of two parallel wedge lines with opposite topological rank m and −m*. Let ξ_0 be the distance between them at time $t = 0$. Since the energy of the system decreases as the lines come closer, they attract each other and finally annihilate.

To calculate the annihilation time, one must solve the dynamics of the system. Let $\xi(t)$ be the distance between the two lines at time t. From Eq. (8.133), each line exerts upon the other an elastic force with an amplitude:

$$F_e = 2\pi K \frac{m^2}{\xi}. \tag{8.146}$$

This force opposes the viscous force (assuming that the inertial forces are completely negligible). The viscous force is given by Eq. (8.139), and its amplitude is:

$$F_v = \pi \gamma_1 m^2 V \ln\left(\frac{R}{r_c}\right), \tag{8.147}$$

where the cutoff radius R is not easy to determine; it depends on ξ and V, where $V = -(1/2)d\xi/dt$. Note that, within the approximations made in this chapter, the dynamics is symmetric, the viscous force being independent of the sign of m.

Comparing these forces yields the following differential equation for ξ:

$$2\pi K \frac{m^2}{\xi} + \frac{\pi}{2}\gamma_1 m^2 \frac{d\xi}{dt} \ln\left(\frac{R}{r_c}\right) = 0. \tag{8.148}$$

This equation (with R constant) has a solution:

$$8\frac{K}{\gamma_1}t = \left(\xi_0^2 - \xi^2\right) \ln\left(\frac{R}{r_c}\right), \tag{8.149}$$

resulting in the annihilation time:

$$t_a = \frac{\gamma_1}{8K}\xi_0^2 \ln\left(\frac{R}{r_c}\right) \approx \frac{\gamma_1}{K}\xi_0^2. \tag{8.150}$$

This time increases as the square of the initial distance ξ_0 between the defects.

In practice, to reproduce the configuration used in the calculation, one would need to produce planar samples where the director is perfectly free to turn on the surfaces. The experimentalists are now able to obtain such *sliding* anchoring conditions (Oswald and Ignés-Mullol, 2008), but the corresponding experiments have not yet been performed systematically.

However, A. Bogi *et al.* (2002) were able to study the annihilation dynamics of two lines with rank 1/2 and −1/2 in a planar sample, as the anchoring direction is not degenerate (Fig. 8.22).

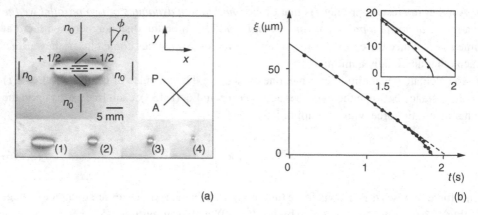

(a)   (b)

Fig. 8.22 (a) Annihilation of two disclination lines, of rank +1/2 and −1/2, in a planar sample. On both plates, the preferred alignment direction is along *y*. Photographs taken between crossed polarisers. Photos 1 to 4 show the two defects at the moments $t = 1.5$ s, 1.75 s, 1.78 s and 1.82 s; (b) the distance ξ as a function of time (Bogi *et al.*, 2002).

In this case, the problem is much more complicated, since one must account for the azimuthal anchoring energy of the director at the surface with respect to the preferred alignment direction.

In this experiment, the authors show that a distinction must be made between two regimes:

- a regime where the defects are far away from each other; in this case, they are connected by an inversion wall, of angle π (Fig. 8.22a), whose width and energy depend on the Frank constants and on the *anchoring energy* of the director on the plates, W. Note that introducing W leads to defining an additional *length scale*, known as the *anchoring penetration length*, $L_p = K/W$. The width of the wall is then explicitly dependent on this length and on the sample thickness (since the director field is no longer invariant along the z direction, perpendicular to the plates). An important consequence of the presence of the wall is that the attractive interaction force between the defects is constant and equal to the wall energy (while for $W = 0$, or $L_p = \infty$, the force is inversely proportional to the distance ξ between the defects). Therefore, the two lines must approach at a constant velocity, such that ξ decreases linearly with time;

- in the other asymptotic regime, the defects are very close together, at a distance ξ much smaller than the typical width of the inversion wall; the problem reduces to the one we treated analytically immediately before describing this experiment. In this case, the defects approach at a velocity that grows as $(t_0 - t)^{-1/2}$ (up to a logarithmic correction). Consequently, ξ decreases as $(t_0 - t)^{1/2}$, where t_0 is the moment when the two defects annihilate.

These predictions are borne out by the experiments, as shown in Fig. 8.22b.
In the next section we will study the smectic A phase.

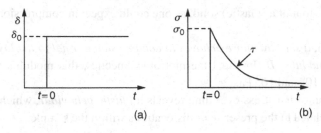

Fig. 8.23 Viscoelastic response of a smectic under compression normal to the layers.

8.2 Smectodynamics

As we mentioned in the first chapter, a smectic A phase resembles a nematic phase crystallised along one dimension. More specifically, the molecules, aligned on the average along the director $\vec{n}$ (with $\vec{n}$ equivalent to $-\vec{n}$, as always), gather in layers, themselves perpendicular to $\vec{n}$. The appearance of this long-range positional order has profound rheological consequences, as illustrated by the following experiment.

8.2.1 The experiment of Bartolino and Durand: evidence for the elastic stress

This experiment was performed using a piezoelectric rheometer identical to that described in Fig. 3.49. The glass plates containing the sample are planar and parallel; they are treated for homeotropic anchoring, i.e. the smectic layers are parallel to the glass surface. In practice, the sample is aligned in the nematic phase, then slowly cooled down to the smectic phase, yielding almost perfect monodomains. The experiment (Bartolino and Durand, 1977) consists of suddenly compressing the smectic phase using one of the ceramics, by a constant amount δ_0 starting at a time $t = 0$, and of measuring the normal stress $\sigma(t)$ using the other ceramic.

Experiment shows that this stress jumps suddenly from zero to a value σ_0, then relaxes slowly to 0 following a mono-exponential law (Fig. 8.23). This can be formally stated as:

$$\delta(t) = \begin{cases} 0 & \text{for} \quad t < 0 \\ \delta_0 & \text{for} \quad t > 0 \end{cases} \tag{8.151}$$

and

$$\sigma(t) = \begin{cases} 0 & \text{for} \quad t < 0 \\ \sigma_0 \exp(-t/\tau) & \text{for} \quad t > 0 \end{cases} \tag{8.152}$$

where τ is the relaxation time, varying from a fraction of a second to several minutes, depending on the sample. On the other hand, the initial stress σ_0 is almost insensitive to the quality of the sample; it is proportional to the initial deformation δ_0/d, where d is the sample thickness:

$$\sigma_0 = \bar{B} \frac{\delta_0}{d}. \tag{8.153}$$

This is the behaviour of a (plastic) solid, as one could expect in compression normal to the layers.

In conclusion, *a smectic phase behaves in compression normal to the layers as a solid, with a Young modulus $\overline{B}$*. In usual thermotropic smectics, this modulus typically takes values between 10^6 and 10^7 Pa.

The relaxation of the stress over time reveals a *plastic behaviour*, which we will show later to be correlated to the presence of dislocations within the sample.

8.2.2 Smectic elasticity

Since smectic phases are partially crystallised nematics, they exhibit mixed elastic properties, intermediate between those of solids and those of the nematic phase.

8.2.2.1 The free energy of elastic deformation

Before giving the mathematical expression of the elastic free energy, let us list the thermodynamic variables to be used for describing a smectic phase [39–41, 43].

In a compressible liquid, the relevant variable is the density ρ, or the bulk dilation θ, defined by:

$$\theta = -\frac{\delta\rho}{\rho}, \tag{8.154}$$

where $\delta\rho$ is the difference with respect to the equilibrium density.

In a three-dimensional solid, the relevant variable is the displacement $\vec{u}$ of the sites of the crystal lattice (often identified with the displacement of the atoms or molecules since the diffusion is negligible), with components u_x, u_y and u_z along the three space directions x, y and z along which the material exhibits long-range order. We recall that in a solid $\theta = \operatorname{div}\vec{u}$.

In smectics, the material is only ordered along the direction perpendicular to the layers (conventionally taken as the z axis) and behaves as a liquid in the (x, y) plane. The relevant variables for its description are therefore the layer displacement u_z (denoted simply by u in the following) and the dilation of the liquid within the layers $\theta_\perp$, where the $\perp$ sign refers to the plane perpendicular to the director. Since $\theta = \theta_\perp + \partial u/\partial z$, one can also choose as independent variables u and θ, as we will do in the following. We are left with the director $\vec{n}$, which is an additional thermodynamic variable, associated with the nematic orientational order of the smectic phase. Note that this variable is not independent of the displacement u, as long as *the molecules remain perpendicular to the layers*. We will make this assumption in the following, keeping in mind that it is only valid far away from the transition to the smectic C phase (where the molecules are tilted with respect to the layer normal). Under these conditions, $\vec{n} = \overrightarrow{\operatorname{grad}}\,\Phi/|\overrightarrow{\operatorname{grad}}\,\Phi|$ where $\Phi = z - u(x, y, z)$.

Once we have made our choice of variables, we can search for the general expression of the elastic free energy.

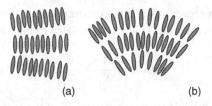

(a)   (b)

Fig. 8.24 (a) In a bend deformation of the director field, the layer thickness must necessarily change; (b) for a splay deformation, the layer thickness remains constant.

We start by considering the situation where the layers remain planar. The elastic deformation energy reads, to second order in the strain:

$$f_e = \frac{1}{2}A\theta^2 + \frac{1}{2}B\left(\frac{\partial u}{\partial z}\right)^2 + C\theta\frac{\partial u}{\partial z}, \tag{8.155}$$

where the first term stands for the energy associated with a change in the density of the phase, the second one for the energy of layer compression at constant density, and the last term describes the coupling between the first two. This quadratic form must be positive definite, leading to the following inequalities:

$$A > 0, \qquad B > 0 \qquad \text{and} \qquad AB > C^2. \tag{8.156}$$

During layer deformation, the director field can also be distorted. Under these conditions, the elastic deformation energy (8.155) must be supplemented by the energy due to the deformation of the director field. This contribution has the same expression as in nematics, namely:

$$f_c = \frac{1}{2}K_1(\text{div}\,\vec{n})^2 + \frac{1}{2}K_2(\vec{n}\cdot\overrightarrow{\text{curl}}\,\vec{n})^2$$
$$+ \frac{1}{2}K_3(\vec{n}\times\overrightarrow{\text{curl}}\,\vec{n})^2 + \frac{1}{2}K_4\text{div}(\vec{n}\,\text{div}\,\vec{n} + \vec{n}\times\overrightarrow{\text{curl}}\,\vec{n}), \tag{8.157}$$

where we also added the div(...) term, more significant in smectics than in nematics.

Some terms in this expression can be suppressed.

This is the case for the twist term, which is strictly *zero* in smectics due to their layered structure (indeed, $\vec{n}\cdot\overrightarrow{\text{curl}}\,\vec{n} = 0$ since $\vec{n} = \text{grad}\,\Phi/|\text{grad}\,\Phi|$).

The bend term in $(\vec{n}\times\overrightarrow{\text{curl}}\,\vec{n})^2$ can also be ignored, since it is necessarily associated with a variation of the layer thickness (see Fig. 8.24a). Indeed, changing the layer thickness bears a much higher energy cost than that of bending the director field. The bend term is therefore hidden by the layer elasticity term, so the former can be safely neglected.

The situation is different for the last two terms, which can come into play even at constant layer thickness.

Thus, the $(\operatorname{div}\vec{n})^2$ term represents a splay deformation of the director field or, equivalently, a layer bending deformation. This term is related to their total curvature, since:

$$\operatorname{div}\vec{n} = \frac{1}{R_1} + \frac{1}{R_2},\tag{8.158}$$

where R_1 and R_2 are the two principal curvature radii of the layers at the current point.

As to the div(...) term, one can show that it is related to the Gaussian curvature of the layers:

$$\operatorname{div}\left(\vec{n}\operatorname{div}\vec{n} + \vec{n}\times\overrightarrow{\operatorname{curl}}\vec{n}\right) = \frac{1}{R_1 R_2}.\tag{8.159}$$

In conclusion, one can write that, for a smectic:

$$f_c = \frac{1}{2}K\left(\frac{1}{R_1} + \frac{1}{R_2}\right)^2 + \frac{1}{2}\overline{K}\frac{1}{R_1 R_2},\tag{8.160}$$

where we set $K_1 = K$ and $\overline{K} = K_4$.[7]

In practice, the expressions (8.155) and (8.160) can often be simplified even further.

Indeed, let us assume that the smectic is not too far from its equilibrium configuration, where all layers are planar and equidistant, and consider the case of *static* deformation.

If the sample boundaries are free (which is almost always the case in experiments), the pressure within the smectic phase is constant and equal to the ambient pressure P_a. It follows that:

$$P - P_a = -\frac{\partial f_e}{\partial\theta} = -A\theta - C\gamma = 0,\tag{8.161}$$

where $\gamma = \partial u/\partial z$. Thus $\theta = -(C/A)\gamma$, which leads, by substitution in expression (8.155), to:

$$f_e = \frac{1}{2}\overline{B}\left(\frac{\partial u}{\partial z}\right)^2,\tag{8.162}$$

where we set:

$$\overline{B} = B - \frac{C^2}{A}.\tag{8.163}$$

The $\overline{B}$ modulus (positive, from the inequalities (8.156)) is *the experimentally accessible compression modulus of the layers* (see for instance the Bartolino and Durand experiment described above).

The other simplification is that *the Gaussian curvature term can be neglected as long as the topology of the layers remains that of a plane*. In this case, the Gauss–Bonnet theorem

---

[7] Note that $K > 0$, while $\overline{K}$ can be of either sign, depending on the material (in practice, $\overline{K} > 0$ in most thermotropic smectics, while it can be either positive or negative in the lamellar phases of lyotropic systems). We recall that in the classical elasticity of plates $\overline{K}$ is always negative (see Eq. 4.151).

ensures that the integral over the entire sample of the Gaussian curvature term is strictly zero (obviously, this is not the case if the layers have the topology of a sphere or a torus, as they do inside the focal domains described in Chapter 1). As to the total curvature term, it has a simple expression as a function of the displacement u in the limit of weakly deformed layers. Indeed, in this case $n_x = -\partial u/\partial x$, $n_y = -\partial u/\partial y$ and $n_z = 1$, yielding:

$$\text{div}\,\vec{n} = \frac{1}{R_1} + \frac{1}{R_2} \approx -\left(\frac{\partial^2 u}{\partial x^2} + \frac{\partial^2 u}{\partial y^2}\right). \tag{8.164}$$

Finally, the free energy of elastic distortion takes, *in the static case* and *in the small deformation limit*, the following simple form:

$$f = \frac{1}{2}\overline{B}\left(\frac{\partial u}{\partial z}\right)^2 + \frac{1}{2}K\left(\frac{\partial^2 u}{\partial x^2} + \frac{\partial^2 u}{\partial y^2}\right)^2, \tag{8.165}$$

where K is the *bending modulus of the layers* (of the same order of magnitude as the Frank moduli for nematics, namely 10^{-11}N) and $\overline{B}$ their *compression modulus* (of the order of 10^7 Pa, as we have seen).

8.2.2.2 Penetration length

Since the compression modulus of the layers $\overline{B}$ is measured in Pa (or N/m^2) and the bending modulus in N, starting from these two quantities one can define a new length scale, intrinsic to the material:

$$\lambda = \sqrt{\frac{K}{\overline{B}}}. \tag{8.166}$$

This is the *smectic penetration length* (Durand, 1972). Before justifying this name, let us point out that this length must be of the order of the layer thickness, which is the only length scale of the problem (since there is no interlayer correlation in smectics A). The reader who is not convinced by this argument should remember the relation demonstrated in Chapter 4 stating that the bending modulus (per unit surface) of a plate with thickness a is of the order of Ba^3, where B is the Young modulus of the material (see Eq. (4.151)). This result leads to a bending modulus K per unit volume of the order of Ba^3/a, showing that λ must be of the order of a.[8]

To justify the name given to λ, let us analyse the experimental situation described in Fig. 8.25. The smectic phase is spread over a plate with a sinusoidal profile, described by a wave vector $q = 2\pi/L$ and an amplitude u_0. If the layers perfectly follow the form of the surface, the smectic phase is distorted over a certain thickness ξ, which we intend to calculate.

---

[8] Note that this line of reasoning applies essentially to thermotropic smectics, and not to the lamellar phases of lyotropic liquid crystals, where the layers can be swollen by a solvent and the discussion becomes more complicated.

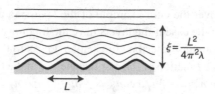

Fig. 8.25 Smectic phase in contact with a corrugated surface. The perturbation propagates over the finite distance ξ.

To do this, one must minimise the free energy of elastic distortion (8.165). This calculation leads to the following differential equation, describing the equilibrium of elastic stress in the bulk of the smectic phase:

$$K\frac{\partial^4 u}{\partial x^4} - \overline{B}\frac{\partial^2 u}{\partial z^2} = 0. \tag{8.167}$$

This equation is supplemented by the boundary condition of layer anchoring at the solid surface, which imposes the displacement at the $z = 0$ plane:

$$u(z = 0) = u_0 \cos(qz). \tag{8.168}$$

At infinity, on the other hand, the layers are not deformed, so that the displacement u and all its derivatives must vanish.

We can search for solutions to this problem in the following simple form:

$$u = u_0 \cos(qx) \exp\left(-\frac{z}{\xi}\right). \tag{8.169}$$

Substitution in Eq. (8.167) yields:

$$\xi = \frac{1}{\lambda q^2} = \frac{L^2}{4\pi^2 \lambda}, \tag{8.170}$$

showing that λ is indeed connected to the penetration distance of a perturbation within the smectic phase (hence its name).

Note that the perturbations propagate over a distance of the order of $L \times (L/\lambda)$, which can be much larger than L, if $L/\lambda \gg 1$. This peculiar behaviour is typical for smectics, being related to their mixed-type elasticity. Indeed, in solids or in nematics, where the elasticity is of stress type or torque type, respectively, a perturbation necessarily propagates over a distance of the order of L, since there is no other length one can use for constructing ξ.

Let us now see how one can measure λ experimentally.

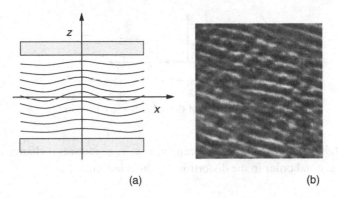

(a)   (b)

Fig. 8.26 Layer undulation instability under dilation. (a) Principle of the experiment; (b) photograph of the undulation taken between crossed polarisers (Ribotta, 1975).

8.2.3 *The layer undulation instability*

The best way of determining λ is to perform an experiment on layer buckling, somewhat similar to the Euler instability of beams, described in Chapter 4.

8.2.3.1 *The experiment of Delaye, Ribotta, Durand, Clark and Meyer*

In this experiment, a homeotropic sample is suddenly dilated by a constant amount δ using a piezoelectric ceramic. Observing it in polarised microscopy shows that it becomes unstable above a certain critical thickness variation δ_c, independent of the thickness of the sample, of the order of a few nanometres. The instability leads to the formation of a striped structure (with a very low optical contrast) revealing a periodic undulation of the layers (Fig. 8.26). The wavelength of this undulation can be measured precisely in light scattering by shining a laser beam onto the sample. This measurement shows that the wave vector $\vec{q}$ of the undulation varies as the square root of the sample thickness.

To explain these experimental results, anharmonic terms must be taken into account in the expression of the free energy of elastic deformation (Clark and Meyer, 1973; Delaye et al., 1973).

8.2.3.2 *The anharmonic correction*

It is readily apparent that the elastic deformation energy (8.165) is not invariant under solid rotation of the smectic phase. This shortcoming, often encountered in linear elasticity, can be corrected by adding an anharmonic correction to the layer dilation term $\partial u / \partial z$. To find this correction, assume that the displacement is of the form $u = \beta x$ (Fig. 8.27). To first order in the distortion, the layer thickness does not change, as $\partial u / \partial z = 0$. It does however change to second order, by an amount:

$$\frac{\delta a}{a_0} = \cos \beta - 1 \approx -\frac{\beta^2}{2} = -\frac{1}{2}\left(\frac{\partial u}{\partial x}\right)^2,\tag{8.171}$$

where a stands for the layer thickness after the displacement and a_0 for the equilibrium thickness.

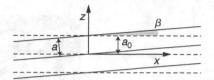

Fig. 8.27 Origin of the anharmonic correction.

More generally, an unspecified displacement $u(x, y, z)$ induces a change in layer thickness given, to second order in the distortion, by the relation:

$$\frac{\delta a}{a_0} = \frac{\partial u}{\partial z} - \frac{1}{2}\left(\frac{\partial u}{\partial x}\right)^2 - \frac{1}{2}\left(\frac{\partial u}{\partial y}\right)^2. \qquad (8.172)$$

Finally, the free energy of elastic deformation can be written in the following form:

$$f = \frac{1}{2}\bar{B}\left[\frac{\partial u}{\partial z} - \frac{1}{2}\left(\frac{\partial u}{\partial x}\right)^2 - \frac{1}{2}\left(\frac{\partial u}{\partial y}\right)^2\right]^2 + \frac{1}{2}K\left(\frac{\partial^2 u}{\partial x^2} + \frac{\partial^2 u}{\partial y^2}\right)^2. \qquad (8.173)$$

This expression is more accurate than Eq. (8.165), but it is not rigorously exact, since it only accounts for the first anharmonic corrections (for an exact formulation, see Holyst and Oswald, 1995). This degree of approximation is nevertheless sufficient for predicting the instability threshold, as we will see now.

8.2.3.3 Calculation of the threshold by the energy method

Before tackling this problem, let us point out that the undulation of the layers contributes to the local relaxation of the dilation stress since, wherever they are tilted, the layer compression is reduced, because their total number over the sample thickness must remain constant. On the other hand, the undulating layers build up bending energy, contributing to the energy cost. Thus, it is the competition between these two mechanisms that sets the instability threshold.

To determine this threshold, suppose that a simple layer undulation develops within the sample, with an amplitude u_0 and a wave vector q, their values unknown for the time being. Under these conditions, the displacement reads:

$$u = \alpha z + u_0 \sin\left(\frac{\pi z}{d}\right)\cos(qx), \qquad (8.174)$$

where we accounted for the term of uniform layer dilation, $\alpha = \delta d/d$, which must be added to the undulation. Note that the general form chosen above obeys the condition of planar anchoring of the layers (corresponding to homeotropic anchoring of the molecules) on the two glass plates, at heights $z = 0$ and $z = d$, respectively.

Direct calculation of the elastic deformation energy, per unit of sample surface, using the modified expression (8.173) yields, to second order in the amplitude:

$$F = \frac{1}{2} d \overline{B} \alpha^2 + \frac{d}{8} u_0^2 \left[\overline{B} \left(\frac{\pi^2}{d^2} - \alpha q^2 \right) + K q^4 \right] + O(u_0^4). \tag{8.175}$$

Denote the coefficient of the u_0^2 term by $g(\alpha, q)$. The undulation instability can only develop if it becomes negative. This can only occur under layer dilation, i.e. for $\alpha > 0$. We will assume that this is the case. The $g(\alpha, q)$ term goes through a minimum for a q^2 value of:

$$q^2 = q_{min}^2 = \frac{\alpha \overline{B}}{2K} = \frac{\alpha}{2\lambda^2} \tag{8.176}$$

and is given by (for this particular value of the wave vector):

$$g_{min} = g(q_{min}, \alpha) = \frac{d}{8} \overline{B} \left(\frac{\pi^2}{d^2} - \frac{\alpha^2}{4\lambda^2} \right). \tag{8.177}$$

This term becomes negative if the dilation exceeds the *critical dilation value*:

$$\alpha_c = 2\pi \frac{\lambda}{d}, \tag{8.178}$$

corresponding to the *critical thickness variation*:

$$\delta d_c = 2\pi \lambda. \tag{8.179}$$

Hence, at the threshold of the instability, the *critical wave vector of the undulation* is, from Eq. (8.176):

$$q_c^2 = \frac{2\pi}{d\lambda}. \tag{8.180}$$

Thus, measuring q_c gives access to λ, provided the sample thickness is known. This method typically yields $\lambda = 20$ Å in 8CB at room temperature and, in general, the values of λ are comparable to the layer thickness, or even lower.

In the next section, we will show how the Navier–Stokes equations for simple fluids can be generalised to the smectic case.

8.2.4 Smectodynamics equations

The procedure used in constructing the dynamic theory of smectics is the same as that employed in the hydrodynamics of simple liquids or in nematodynamics. It consists of first recalling the conservation equations applicable to all continuous media and then, as a second step, of deducing the constitutive laws of the material starting from the expression of irreversible entropy production [39, 40, 43, volume 2].

8.2.4.1 Recalling the conservation laws

The first is the *mass conservation* law. It can be expressed as a function of the bulk dilation in the form:

$$\frac{D\theta}{Dt} - \text{div}\,\vec{v} = 0, \tag{8.181}$$

where $\vec{v}$ is the hydrodynamic velocity of the molecules.

The second law expresses *momentum conservation* and is most generally expressed as:

$$\rho\frac{D\vec{v}}{Dt} = \text{div}\,\underline{\sigma} + \vec{f}, \tag{8.182}$$

where $\underline{\sigma}$ is the total stress tensor, containing a hydrostatic pressure term $-P\,\underline{I}$, coupled to the density or to the dilation θ, an elastic term $\underline{\sigma}^E$ conjugated to the layer displacement u and a viscous term $\underline{\sigma}^v$ depending on the velocity gradients:

$$\underline{\sigma} = -P\,\underline{I} + \underline{\sigma}^E + \underline{\sigma}^v. \tag{8.183}$$

To find the mathematical expressions for the pressure and the elastic stress, let us determine the variation in the elastic free energy as the density varies by $d\theta$ and the layer displacement by du. One has:

$$df = \frac{\partial f}{\partial\theta}d\theta + \frac{\partial f}{\partial u_{,i}}du_{,i} + \frac{\partial f}{\partial u_{,ij}}du_{,ij}, \tag{8.184}$$

which can also be written in the form:

$$df = \frac{\partial f}{\partial\theta}d\theta + \frac{\partial f}{\partial u_{,i}}du_{,i} - \frac{\partial}{\partial x_{,j}}\frac{\partial f}{\partial u_{,ij}}du_{,i} \tag{8.185}$$

up to an additive divergence term, which we will neglect since it is a surface term corresponding to the work of the surface elastic torques.

Three terms appear in this expression:

- the first one represents the work of the pressure $-Pd\theta$, so that simple identification gives:

$$P = -\frac{\partial f}{\partial\theta}; \tag{8.186}$$

- the last two correspond to the work $\sigma^E_{zi}\,du_{,i}$ of the elastic stress associated with the layer displacement u, so that:

$$\sigma^E_{zi} = \frac{\partial f}{\partial u_{,i}} - \frac{\partial}{\partial x_{,j}}\frac{\partial f}{\partial u_{,ij}}. \tag{8.187}$$

Note that the other components of the elastic stress tensor, in σ^E_{xi} and σ^E_{yi}, are zero.

Using the expression of the elastic free energy (8.165), to which one must add the magnetic term $-\frac{\chi_a}{2\mu_0}(\vec{B}\cdot\vec{n})^2$ if the sample is exposed to a constant magnetic field $\vec{B}$, one obtains explicitly (neglecting the anharmonic terms):

$$P = -A\theta - C\frac{\partial u}{\partial z}, \tag{8.188}$$

$$\sigma^{E}_{zz} = B\frac{\partial u}{\partial z} + C\theta, \tag{8.189}$$

$$\sigma^{E}_{zx} = -K\frac{\partial}{\partial x}(\Delta_\perp u) - \frac{\chi_a}{\mu_0}B_x\left(B_x u_{,x} + B_y u_{,y} - B_z\right), \tag{8.190}$$

$$\sigma^{E}_{zy} = -K\frac{\partial}{\partial y}(\Delta_\perp u) - \frac{\chi_a}{\mu_0}B_y\left(B_x u_{,x} + B_y u_{,y} - B_z\right), \tag{8.191}$$

where we set $\Delta_\perp = \frac{\partial^2}{\partial x^2} + \frac{\partial^2}{\partial y^2}$.

We can see that here *the elastic stress tensor is not symmetric* due to the layer bending term and the presence of elastic torques, in agreement with the general results of Chapter 2.

It is also noteworthy that, unlike the case of nematics, the elastic stress σ^E is not determined for a fixed director orientation, but assuming that the director is always perpendicular to the layers. For this reason, the equilibrium equation for the bulk torques does not appear explicitly among the equations of motion. One can indeed show that it is automatically satisfied, which is a significant simplification with respect to a more complete theory, considering both the director and the layer displacement as independent variables (Kléman and Parodi, 1975).

8.2.4.2 Energy dissipation and constitutive laws

In order to find the constitutive laws of the material, one needs to calculate the energy dissipation. This calculation, presented in detail in [43, volume 2], leads to the following result (without neglecting the surface terms, but assuming an isothermal transformation):

$$\rho T\overset{\circ}{s} = \sigma^{v}_{ij}\frac{\partial v_i}{\partial x_j} + \left(\frac{Du}{Dt} - v_z\right)G + \mathrm{div}\left[\sigma^{E}_{zi}(\dot{u} - v_z)\right], \tag{8.192}$$

where the div(...) term gives the dissipation at the surface of the smectic. In this formula, G is the z component of the bulk elastic force $\vec{G} = \mathrm{div}(\underline{\sigma}^E)$ associated with the layer displacement u. From Eqs. (8.189) to (8.191), its expression is:

$$\vec{G} = (0, 0, G) \quad \text{with} \quad G = B\frac{\partial^2 u}{\partial z^2} + C\frac{\partial\theta}{\partial z} - K\Delta_\perp(\Delta_\perp u). \tag{8.193}$$

Note that formula (8.192) is less general than that provided in Chapter 2 (Eq. (2.36)), which contained an additional term related to the viscous torque. This results from our assumption that the director remains perpendicular to the layers, for any deformation (the director $\vec{n}$ is fixed by the displacement u) suppressing the rotational degrees of freedom of the director with respect to the normal to the layers.

The constitutive laws are obtained by writing, in the bulk and at the surface, linear relations between the thermodynamic forces σ_{ij}^V and $\text{div}(\underline{\sigma}^E)$ and the associated fluxes $\partial v_i/\partial x_j$ and $Du/Dt - v_z$. This procedure, which must take into account the uniaxial symmetry of the smectic A phase and the fact that the dissipation must remain null in a 'solid'-type rotation, leads to the following constitutive laws in the bulk (Martin *et al.*, 1972):

$$\sigma_{ij}^V = 2\eta_2 A_{ij} + 2(\eta_3 - \eta_2)(\delta_{iz}A_{zj} + \delta_{jz}A_{zi}) + (\eta_4 - \eta_2)\delta_{ij}A_{kk}$$

$$+ (\eta_1 + \eta_2 - 4\eta_3 - 2\eta_5 + \eta_4)\delta_{iz}\delta_{jz}A_{zz}$$

$$+ (\eta_5 - \eta_4 + \eta_2)(\delta_{ij}A_{zz} + \delta_{iz}\delta_{jz}A_{kk}) \tag{8.194}$$

and

$$\dot{u} - v_z = \lambda_p G, \tag{8.195}$$

where $A_{ij} = (1/2)(\partial v_i/\partial x_j + \partial v_j/\partial x_i)$ is the deformation rate tensor and $\dot{u} = Du/Dt$ the material derivative of the displacement with respect to time. Note that $\sigma_{ij}^V = \sigma_{ji}^V$.

The law (8.194) generalises the constitutive law for ordinary liquids, involving however five independent viscosity coefficients (instead of two). We recall that these coefficients are in units of Pa s.

The second law, known as the *permeation equation*, is new; it describes the flow of molecules across the layers. It involves the permeation coefficient λ_p, measured in $\text{m}^2\text{Pa}^{-1}\text{s}^{-1}$.

Note that, in the case of incompressible smectics, the viscous stress tensor only contains four viscosity coefficients, since in this case $A_{kk} = 0$, yielding:

$$\sigma_{xy}^V = 2\eta_2 A_{xy},$$

$$\sigma_{xz}^V = 2\eta_3 A_{xz}, \qquad \sigma_{yz}^V = 2\eta_3 A_{yz},$$

$$\sigma_{xx}^V = 2\eta_2 A_{xx} + (\eta_5 - \eta_4 + \eta_2)A_{zz}, \qquad \sigma_{yy}^V = 2\eta_2 A_{yy} + (\eta_5 - \eta_4 + \eta_2)A_{zz}, \tag{8.196}$$

$$\sigma_{zz}^V = (\eta_1 - \eta_5)A_{zz}.$$

In the following we will further simplify the problem by making the assumption that $\eta_4 = \eta_5 + \eta_2$ and $\eta_2 = \eta_3 = (\eta_1 - \eta_5)/2 = \eta$. Under these conditions, Eq. (8.182), generalising the Navier–Stokes equation for simple fluids, is written:

$$\rho \frac{D\vec{v}}{Dt} = -\overrightarrow{\text{grad}}\, P + \vec{G} + \eta\, \Delta \vec{v}, \tag{8.197}$$

where the only difference with respect to ordinary fluids is the presence of the elastic force $\vec{G}$ normal to the layers.

For completeness, one must add a surface constitutive law resulting from the div(...)
term in Eq. (8.192):

$$\dot{u} - v_z = -\zeta \, \sigma_{zi}^E \, \nu_i, \qquad (8.198)$$

where ν_i is the unit vector perpendicular to the surface and pointing out of the smectic
phase. In this formula, ζ is another phenomenological coefficient, describing the surface
permeation process.

Note that, since the dissipation must also be positive, the same condition holds for the
coefficients η, λ_p and ζ.

8.2.5 Elastic waves: first and second sound

We know that liquids only sustain longitudinal waves (i.e. where the displacement is parallel to the propagation direction), representing two counter-propagating longitudinally
polarised modes. The same holds for nematics (which are essentially fluid materials).

However, solids can also sustain transverse waves, propagating at a different velocity
from longitudinal waves. In all, there are six modes, two of them longitudinally and the
other four transversely polarised.

The behaviour of smectics is intermediate between that of liquids and solids, insofar as
they exhibit a total of four modes. Indeed, we will show that two types of sound exist, each
accounting for two modes: *first sound* (longitudinally polarised), the same as in liquids or
solids, and *second sound* (with a mixed polarisation) with no equivalent in other materials
(except for superfluid helium, which represents a very special case).

To calculate the propagation velocity of these two sounds as simply as possible, we will
ignore all viscosities (which amounts to neglecting the damping, $\eta = 0$) and will assume
negligible permeation ($\lambda_p = 0$) and very flexible layers ($K = 0$).

In this very simplified framework, and furthermore neglecting the nonlinear terms when
determining the material derivatives, the equations of motion become:

$$\rho \frac{\partial \vec{v}}{\partial t} = -\overrightarrow{\text{grad}}\, P + \vec{G}, \qquad (8.199)$$

$$\text{div}\,\vec{v} - \frac{\partial \theta}{\partial t} = 0,$$

$$v_z = \frac{\partial u}{\partial t}, \qquad (8.200)$$

with

$$P = -\frac{\partial f}{\partial \theta} = -A\theta - C\gamma, \qquad (8.201)$$

where $\gamma = \partial u / \partial z$.

Eliminating the pressure between the equations above yields the following equation system:

$$\rho \frac{\partial v_x}{\partial t} = A \frac{\partial \theta}{\partial x} + C \frac{\partial \gamma}{\partial x},$$

$$\rho \frac{\partial v_y}{\partial t} = A \frac{\partial \theta}{\partial y} + C \frac{\partial \gamma}{\partial y},$$

$$\rho \frac{\partial v_z}{\partial t} = (A + C) \frac{\partial \theta}{\partial z} + (B + C) \frac{\partial \gamma}{\partial z}, \qquad (8.202)$$

$$\frac{\partial v_x}{\partial x} + \frac{\partial v_y}{\partial y} + \frac{\partial v_z}{\partial z} = \frac{\partial \theta}{\partial x},$$

$$\frac{\partial v_z}{\partial z} = \frac{\partial \gamma}{\partial x}.$$

Searching for solutions of the type $v_i = v_i^0 \exp\left[i(\vec{q} \cdot \vec{r} - \omega t)\right]$, $\theta = \theta^0 \exp\left[i(\vec{q} \cdot \vec{r} - \omega t)\right]$ and $\gamma = \gamma^0 \exp\left[i(\vec{q} \cdot \vec{r} - \omega t)\right]$ leads, after substituting in Eqs. (8.202), to a homogeneous equation system in the unknowns v_i^0, θ^0 and γ^0, which only admits non-trivial solutions if its determinant vanishes. This condition leads to the following secular equation:

$$\omega \left\{ \left[\rho \omega^2 - (B + C)q_z^2\right]\left[\rho \omega^2 - A q_\perp^2\right] - (A + C)\left(\rho \omega^2 + C q_z^2\right) q_z^2 \right\} = 0, \qquad (8.203)$$

where we set $q_\perp^2 = q_x^2 + q_y^2$.

This fifth-order equation has five distinct solutions:

- the first one corresponds to the neutral mode $\omega = 0$. For this solution, $u = \theta = 0$ and the velocity is perpendicular to $\vec{n}$ and $\vec{q}$. It is the transverse shear mode, encountered in liquids and nematics. Accounting for the viscosity shows that this mode is in fact diffusively damped (a more complete calculation would give $\rho \omega = -i \eta q^2$);
- the other solutions are of the form:

$$\omega_1 = \pm c_1(\Phi) q, \qquad \omega_2 = \pm c_2(\Phi) q, \qquad (8.204)$$

where Φ is the angle $(O\vec{z}, \vec{q})$. Defining $x = \rho c^2$ leads to ρc_1^2 and ρc_2^2 as solutions to the following equation:

$$x^2 - x \left[A \cos^2 \Phi + (A + B + 2C) \sin^2 \Phi\right] + \sin^2 \Phi \cos^2 \Phi (AB - C^2) = 0. \quad (8.205)$$

Note that one of these velocities goes to zero for $\Phi \to 0$ ($\vec{q} \parallel \vec{n}$) and for $\Phi \to \pi/2$ ($\vec{q} \perp \vec{n}$). We exclude here these two limiting cases, discussed in [43, volume 2], and assume that $A \gg B$ and C, which is generally the case in smectics. The previous equation reduces to:

$$x^2 - Ax + \sin^2 \Phi \cos^2 \Phi \left(AB - C^2\right) = 0, \qquad (8.206)$$

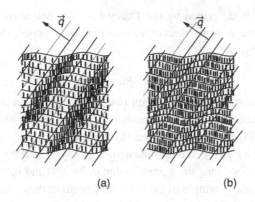

(a)   (b)

Fig. 8.28 Representation of first (a) and second (b) sound in smectics A (drawing by P. Pieranski).

with the approximate solutions:

$$x = A \quad \text{and} \quad x = \frac{AB - C^2}{A} \sin^2\phi \cos^2\phi = \overline{B} \sin^2\phi \cos^2\phi. \quad (8.207)$$

The first of these solutions should be familiar, since it corresponds to ordinary sound (consisting of two counter-propagating modes). This is *first sound*, isotropic in the first approximation, as it propagates at the constant speed:

$$c_1 = \sqrt{\frac{A}{\rho}}. \quad (8.208)$$

It corresponds to a longitudinal compression–dilation wave (Fig. 8.28a).

The second solution is much more unusual. This is *second sound*, also accounting for two modes and propagating at a speed:

$$c_2 = \sqrt{\frac{B}{\rho}} \sin \Phi \cos \Phi. \quad (8.209)$$

In this mode, the layer thickness varies, but the density remains constant. Its name comes from superfluid helium, which also exhibits two phonon branches (in helium, second sound is associated with the phase fluctuations of the complex order parameter, analogous to the layer displacement u in smectics). Second sound is depicted in Fig. 8.28b.

These two sound modes have been demonstrated experimentally by Brillouin light scattering (Liao *et al.*, 1973), confirming the theory proposed above.

8.2.6 *Permeation flow*

The permeation process (passage of molecules from one layer to the next) is a distinctive feature of smectics. This process resembles fluid flow across a porous medium which, as we know by now, is described by the Darcy law. We will show that this analogy is legitimate by solving the flow problem of a smectic phase experiencing a pressure gradient perpendicular to the layers.

8.2.6.1 'Plug flow'

Consider a smectic phase confined between two parallel plates, a distance $2d$ apart. We assume that the molecules are aligned in planar anchoring on the two plates (along z, for instance) and that the layers cannot slip over the surfaces.

A pressure gradient $\partial P/\partial z \neq 0$ is applied across the layers and the system is considered invariant under translation along the y axis, so that $\partial/\partial y = 0$ and $v_y = 0$. The equations of motion are then particularly simple in the stationary regime; they can be written explicitly (knowing that $v_x = 0$ and $v_z = v(x, z)$):

$$\frac{\partial v}{\partial z} = 0 \qquad \text{(incompressibility)}, \tag{8.210}$$

$$-\frac{\partial P}{\partial z} + \eta_3 \frac{\partial^2 v}{\partial x^2} + G = 0 \qquad \text{(generalised Navier–Stokes)}, \tag{8.211}$$

$$v = -\lambda_p G \quad \text{(permeation)}. \tag{8.212}$$

Eliminating G between the last two equations yields the following equation for v:

$$-\frac{\partial P}{\partial z} + \eta_3 \frac{\partial^2 v}{\partial x^2} - \frac{v}{\lambda_p} = 0, \tag{8.213}$$

to be solved subject to the condition $v = 0$ at $z = \pm d$. The solution is:

$$v = -\lambda_p \frac{\mathrm{d}P}{\mathrm{d}z} \left[1 - \frac{\cosh(x/l_p)}{\cosh(d/l_p)} \right], \tag{8.214}$$

where we introduced a new characteristic length, intrinsic to the material, named the *permeation length*:

$$l_p = \sqrt{\eta_3 \lambda_p}. \tag{8.215}$$

As shown in Fig. 8.29, the flow (8.214) is of the *'plug'* type, the velocity in the centre of the sample being constant, given by the equivalent of a Darcy law:

$$v = -\lambda_p \frac{\mathrm{d}P}{\mathrm{d}z} \qquad \text{(in the centre of the sample)} \tag{8.216}$$

and going to zero on approaching either wall over a thin layer with thickness l_p (which must be of the order of a layer thickness).

In principle, this experiment should thus provide a direct measurement of the permeation coefficient λ_p. Unfortunately, it is not experimentally feasible, since the layers break (by

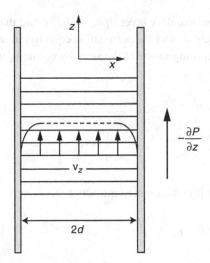

Fig. 8.29 Permeation flow in a smectic phase.

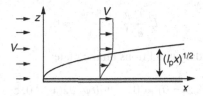

Fig. 8.30 Permeation boundary layer close to a thin semi-infinite plate.

the nucleation of edge dislocations) and slip at the walls, invalidating the assumption that $\dot{u} = 0$. In the next section we propose an alternative method for measuring λ_p; it has not yet been attempted experimentally.

8.2.6.2 Permeation boundary layer and Poiseuille flow

In Chapter 3 we analysed the flow of a fluid close to a semi-infinite planar wall, showing that a boundary layer developed at the wall, with thickness growing as the square root of the distance from the leading edge (Blasius problem).

We will solve the same problem in the smectic case, with the layers parallel to the semi-infinite plane (de Gennes, 1974), and then analyse the consequences of the result on Poiseuille flow.

The flow geometry is represented in Fig. 8.30. The layers are parallel to the x axis and the system is taken as invariant under translation along the y direction. The velocity is constant far away from the plate, equal to V. We also make the assumption (to be checked *ex post facto*) that:

$$l_p \ll \delta \ll x. \tag{8.217}$$

It follows that within the boundary layer $\partial/\partial z \gg \partial/\partial x$ and that $v_z/\lambda_p \gg \eta \, \partial^2 v_z/\partial z^2 \approx \eta v_z/\delta^2$. Using these inequalities and the permeation equation in the stationary regime $v_z = -\lambda_p G$, we find that the two components of the velocity v_x and v_z must satisfy the following equations:

$$\frac{\partial P}{\partial z} = -\frac{1}{\lambda_p} v_z, \tag{8.218}$$

$$\frac{\partial P}{\partial x} = \eta_3 \frac{\partial^2 v_x}{\partial z^2}. \tag{8.219}$$

To solve these equations, it is convenient to introduce the stream function $\psi(x, z)$ such that

$$v_x = \frac{\partial \psi}{\partial z}, \qquad v_z = -\frac{\partial \psi}{\partial x}. \tag{8.220}$$

Eliminating the pressure between Eqs. (8.218) and (8.219) yields the following equation for ψ:

$$l_p^2 \frac{\partial^4 \psi}{\partial z^4} - \frac{\partial^2 \psi}{\partial x^2} = 0, \tag{8.221}$$

to be solved with the boundary conditions on the plate:[9]

$$\psi(x > 0, z = 0) = 0, \qquad \partial \psi/\partial z(x > 0, z = 0) = 0 \tag{8.222}$$

and at infinity:

$$\psi(x, z \to \infty) \to Vz. \tag{8.223}$$

It is easily checked that the following integral form:

$$\psi(x, z) = \frac{1}{2} \int_0^z du \int_0^u dv \, h(x, v), \tag{8.224}$$

where

$$h(x, v) = \frac{2V}{\sqrt{\pi l_p x}} \exp\left(-\frac{v^2}{4 l_p x}\right) \qquad (x > 0) \tag{8.225}$$

is the solution to this problem. It yields the velocity profile in the vicinity of the plate:

$$v_x = V \operatorname{erf}\left(-\frac{z}{2\sqrt{l_p x}}\right), \tag{8.226}$$

where erf is the error function, $\operatorname{erf}(t) = \frac{2}{\sqrt{\pi}} \int_0^t \exp(-u^2) du$.

---

[9] We recall that the streamlines are given by the equation $\psi = \mathrm{Cst}$ and that ψ is only defined up to an additive constant.

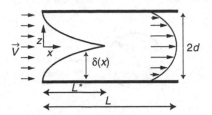

Fig. 8.31 Poiseuille flow in a flat capillary (its width along y is much larger than its thickness d) when the layers are parallel to the flat walls of the capillary (and thus to the velocity). The asymptotic regime is reached as the two boundary layers merge.

This expression shows that the permeation boundary layer has a thickness:

$$\delta(x) = 2\sqrt{l_p x}. \tag{8.227}$$

This result is very different from the Blasius profile calculated in Chapter 3 (notwithstanding the same $x^{1/2}$ dependence), as it involves the (microscopic) permeation length instead of the macroscopic length v/V (where v is the kinematic viscosity of the fluid).

One predicts thus that the boundary layer is much thinner in a smectic phase than in a simple liquid of identical viscosity since, as a general rule, $l_p \ll v/V$.

This result has an important consequence in the set-in regime leading to the parabolic velocity profile in Poiseuille flow (Fig. 8.31). In this case, the asymptotic regime is reached when the permeation boundary layers formed on either side of the capillary (treated in homeotropic anchoring, so that the velocity is parallel to the layers) merge in the centre of the sample. This occurs at a distance:

$$L^* = \frac{d^2}{l_p} \tag{8.228}$$

from the inlet of the capillary. This distance is generally very large, since l_p is a microscopic length. Indeed, consider $l_p = 3$ nm and $d = 100$ μm. This leads to $L^* \approx 3$m, conclusively showing that, in practice, the velocity profile is never parabolic, remaining of 'plug' type within the capillary.

To conclude this discussion, let us give the expression for the effective viscosity of a smectic when the length L of the capillary containing it is much smaller than L^* [43, volume 2]. This viscosity is that of a Newtonian fluid which, submitted to the same pressure gradient, would give the same total flow rate. Writing that the work of the pressure is completely dissipated in the boundary layers yields:

$$\eta_{\text{eff}} = \frac{4\sqrt{2}}{3\sqrt{\pi}} \frac{d}{\sqrt{l_p L}} \eta_3. \tag{8.229}$$

This relation provides a method for measuring l_p, as the viscosity can be measured directly in the Couette geometry using a rotating rheometer (note, however, that this latter

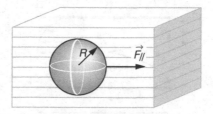

Fig. 8.32 Sphere moving in a smectic phase.

measurement is not easily performed due to subtle parasitic effects, such as lubrication, to be described later).

8.2.7 Force on a moving sphere

We will not treat this problem in detail, giving here only a few indications.

We will consider that the sphere surface is treated such that the anchoring energy of the molecules is zero. We should point out straightaway that this anchoring is almost impossible to achieve experimentally. Assuming, however, that this is the case, the smectic layers remain planar and are not deformed in the presence of the sphere.

Imagine now that the sphere moves parallel to the layers (Fig. 8.32). In this case the flow remains essentially localised within the plane of the layers, the permeation being very difficult. Calculating the drag force (de Gennes, 1974), gives (neglecting permeation and assuming that $\eta_2 = \eta_3 = \eta$):

$$F_\parallel = 8\pi \eta R V, \tag{8.230}$$

where R is the radius of the sphere. This formula is very close to that of Stokes (although the 6π factor is replaced by 8π).

Obviously, the motion of the sphere along the layer normal is much more difficult. A dimensional argument shows that the drag force must vary in the following way [43, volume 2]:

$$F_\perp \approx \frac{R^3}{\lambda_p} V \tag{8.231}$$

up to a numerical factor.

This force, necessarily involving permeation across the layers, is much larger than the previous one. In practice, the sphere can hardly move from one layer to the next when its size exceeds the molecular dimension.

8.2.8 Creep under compression normal to the layers

In this section we will describe an experiment that shows very clearly the shortcomings of the continuous model in describing the rheological behaviour of smectic phases. This experiment concerns the creep law for a smectic undergoing compression perpendicular to

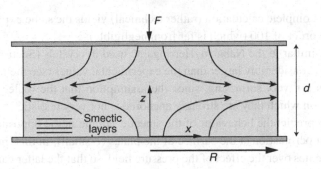

Fig. 8.33 Creep of a smectic phase in compression normal to the layers.

the layers. Before presenting the experimental results, let us briefly recall the predictions of the continuous model.

8.2.8.1 The continuous model of the Orsay group on liquid crystals

The experimental geometry is shown in Fig. 8.33. The smectic phase is confined between two horizontal plates treated to ensure homeotropic anchoring of the molecules.

Let d be the sample thickness and R its radius. We are interested in the deformation rate of the sample $\dot{\varepsilon}$ as a function of the applied stress σ.

The Orsay group on liquid crystals was the first to perform this calculation (Orsay Group, 1975). They assumed that the molecules cannot slip at the two plates (where their horizontal velocity is therefore zero) but can 'easily' leave them (so that the layers can be eliminated through the plates) and diffuse towards the edge of the sample, considered stress-free. In this context, 'easily' means that the surface permeation coefficient ζ is infinite, imposing that the stress σ_{zz}^E vanishes at the plates. Under these conditions, when determining the force exerted by the smectic phase on the plates, only the pressure field comes into play, the viscous stress in $\eta \partial v_z / \partial z$ being completely negligible. We can find the creep law without calculation, by noting that two boundary layers develop in the vicinity of the surfaces, with a thickness going from zero at the centre of the sample to about $(l_p R)^{1/2}$ at the edges. Assume a sample thickness much larger than this value $(d > (l_p R)^{1/2})$, of the order of a micrometre for a sample radius $R = 1$ cm. Since, according to the permeation equation and the generalised Navier–Stokes equation:

$$v_z \approx -\lambda_p \frac{\partial P}{\partial z}, \tag{8.232}$$

one has, knowing that on the upper plate $v_z = \partial d / \partial t = d\dot{\varepsilon}$, $-P \approx F/\pi R^2$ and $\partial/\partial z \approx 1/\delta \approx 1/(l_p R)^{1/2}$ (because $d > (l_p R)^{1/2}$):

$$\dot{\varepsilon} \approx \frac{\lambda_p}{d(l_p R)^{1/2}} \sigma, \tag{8.233}$$

where $\sigma = F/\pi R^2$ is the applied stress (negative in compression).

Note that the complete calculation (rather technical) yields the same expression, up to a prefactor of the order of 100 (which is far from negligible).

This model, similar to the Nabarro–Herring one used in crystals (Section 5.3.4.1), predicts creep rates considerably larger than the experimental values (see the next section).

This result is not very surprising, since the assumption that the molecules can easily leave the plates on which they are strongly anchored is not very realistic.

One can also predict the behaviour of the smectic sample in the opposite case, defined by very difficult permeation at the surface of the plates (ζ small). In this limit, the surface stress σ_{zz}^{E} dominates over the effect of the pressure field, so that the latter can be neglected. This yields immediately, using the surface permeation law:

$$\sigma \approx \sigma_{zz}^{E} = \frac{v_z}{\zeta} = \frac{d\dot{\varepsilon}}{\zeta}. \tag{8.234}$$

The resulting creep law is given by:

$$\dot{\varepsilon} \approx \frac{\zeta}{d}\sigma. \tag{8.235}$$

Clearly, the creep process is completely blocked if the molecules cannot leave the surface ($\zeta = 0$).

8.2.8.2 *Experimental results: the need to account for defects*

The experiments were performed in 8CB, at constant creep stress (Oswald, 1983a) and at constant creep strain (Oswald, 1984). They show that, as long as the deformation and the applied stress remain small (typically below 2×10^{-2} for the first parameter, and a few hundred Pa for the second corresponding to an elastic compression of the sample by three to five layers, depending on the temperature), the absolute value of the creep rate is well described by a law of the type:

$$\dot{\varepsilon} = \frac{m\alpha}{d}(\sigma - \sigma_{c}), \tag{8.236}$$

where m is a thermally activated coefficient (its meaning will be discussed later), given in 8CB by:

$$m = m_0 \exp\left(-\frac{E}{k_B T}\right), \tag{8.237}$$

with $m_0 \approx 7 \times 10^{22} \text{ m}^2 \text{ s kg}^{-1}$ and $E \approx 1.8$ eV.

Note that, for ease of calculation, the stress σ is taken as positive in compression.

The parameter α represents the angle between the two surfaces containing the sample (experimentally, they are never rigorously parallel). In the experiments under discussion, the angle was varied between 10^{-2} and 10^{-4} rad and the thickness between 40 and 400 μm.

Experiment also shows the presence of a yield stress (or elastic limit) σ_c, below which the smectic does not creep (within experimental precision). This stress corresponds experimentally to a sample compression by a half-layer:[10]

$$\sigma_c \approx \frac{1}{2} \bar{B} \frac{a_0}{d},$$ (8.238)

where a_0 is the equilibrium layer thickness.

The results also show that the smectic suddenly departs from the creep law (8.236) and starts to creep much more rapidly when the (absolute value of the) stress exceeds a critical value corresponding to elastic compression of the sample by three to five layers, depending on the temperature (Oswald, 1984).

This body of experimental results is clearly incompatible with the predictions of the continuous model presented in the section above. On the other hand, the linear variation of the creep rate with the angle α between the plates hints at the direct intervention of dislocations. Before establishing a convincing creep model, let us start with a quick reminder on dislocations and show that their density is fixed at equilibrium (and in the low-stress regime) by the angle α.

8.2.9 Dislocations and smectic plasticity

Note that F. C. Frank was the first to suggest the existence of dislocations in smectics, in 1958 (Frank, 1958). As in solids, these defects locally break the positional order of the molecules, and hence the stacking of the layers. Smectic phases exhibit two types of dislocations, with fairly different properties. We will start by describing them, and then show how they interact with the flow and how they are nucleated, first in the case of creep under compression normal to the layers, then during spontaneous thinning of a free-standing smectic film, and finally under shear parallel to the layers (lubrication theory).

8.2.9.1 Topological construction

The Volterra construction, described in Section 5.1.2.1 for solids, is directly applicable to smectics. It can be used to construct dislocations with a Burgers vector $\vec{b}$, which must be a translation vector of the crystal lattice.

Since the system exhibits one-dimensional order along the layer normal z, in a smectic one has necessarily:

$$\vec{b} = na_0\vec{z}$$ (8.239)

with n integer. Let $\vec{t}$ be the unit vector tangent to the line.

For $\vec{b} \perp \vec{t}$, one has an *edge dislocation* (Fig. 8.34a).

If $\vec{b} \parallel \vec{t}$, one speaks of a *screw dislocation*. In this case, the smectic layers take the shape of a helicoid (Fig. 8.34b).

---

[10] Considerably higher values for σ_c were reported recently in the smectic A phase of other materials close to their smectic C phase (smectic fluid phase where the molecules are tilted with respect to the layer normal). These results are not yet well understood (Blanc *et al.*, 2004).

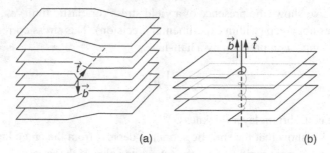

(a)   (b)

Fig. 8.34 (a) Edge dislocation; (b) screw dislocation.

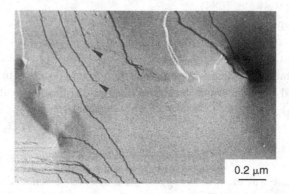

Fig. 8.35 Screw dislocations revealed by freeze-fracture and electron microscopy. Wherever a screw dislocation pierces the surface, it produces a step, advancing in the direction of fracture propagation (Allain, 1985).

8.2.9.2 Experimental illustration

Several techniques have been used experimentally to demonstrate the existence of dislocations.

One can cite electron microscopy which, coupled to freeze-fracture, can be used to observe replicas of the fracture surface of frozen samples. This method, used in particular for the study of lamellar lyotropic phases (Kléman *et al.*, 1977), revealed the existence of screw dislocations, sometimes at a very high density (Fig. 8.35).

Edge dislocations are much more difficult to observe using this method, since the sample must be broken perpendicular to the layers, and their structure resolved in electron microscopy.

Another – more ingenious – way to observe them consists in approaching the transition towards a smectic C phase (Meyer *et al.*, 1978). This method renders the dislocations visible directly in polarised microscopy, between crossed polarisers, since the transition towards the smectic C phase first occurs close to the cores of the dislocations. These then become bright on a dark background due to the local molecular tilt (Fig. 8.36).

This decoration technique showed that a good homeotropic sample, after annealing, only contains the dislocations geometrically necessary to accommodate the variations in sample

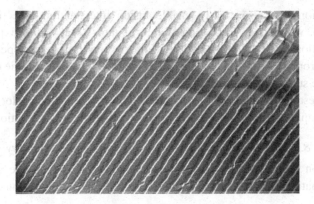

Fig. 8.36 Edge dislocations observed between crossed polarisers in polarised microscopy, close to a smectic A–smectic C transition (photograph by S. Lagerwall).

Fig. 8.37 Arch texture in a free-standing smectic film observed in reflection microscopy. There is a difference in thickness of the order of one layer (sometimes even more) between two areas with different shades of grey. Each circular separation line reveals the presence of a dislocation, centred in the thickness of the film (photograph by F. Picano, 2000).

thickness related to the imperfect plate parallelism. Experiment also shows that the dislocations are elementary ($b = a_0$ when the angle is small: below 10^{-3} rad, typically), but tend to bunch into dislocations with larger Burgers vectors at higher angles. This is an unusual observation as we know that in solids the dislocations are almost always elementary.

The dislocations are also clearly visible inside free-standing smectic films (to be described later), where they form typical arch textures (Fig. 8.37). Experiment shows that the dislocations take up position in the centre of the film, being repelled by the free surfaces. This property is once again unusual, if we think that in solids the dislocations are attracted by free surfaces.

To explain these observations, we will outline a few properties of dislocations in smec-tics, trying to emphasise the essential differences with respect to dislocations in solids.

8.2.9.3 *Properties: line energy, line tension and mobility*

The deformation field around a dislocation line must satisfy, in the static case, the equilibrium equation for the elastic stress $G = 0$, found by minimising the elastic free energy (8.165):

$$\frac{\partial^2 u}{\partial z^2} + \lambda^2 \left(\frac{\partial^4 u}{\partial x^4} + \frac{\partial^4 u}{\partial y^4} + 2\frac{\partial^4 u}{\partial x^2 \partial y^2} \right) = 0, \tag{8.240}$$

where λ is the smectic penetration length.

We start by considering the case of an edge dislocation parallel to the y axis (such that $\partial/\partial y = 0$).

The deformation field must obey the boundary conditions:

$$u(z = 0, x \leq 0) = 0, \tag{8.241}$$

$$u(z = 0, x > 0) = \text{sgn}(z)\frac{b}{2}, \tag{8.242}$$

where sgn stands for the 'signum' function.

It is easy to verify that the following deformation field is the solution to the linearised problem (de Gennes, 1972):[11]

$$u(x, z) = \frac{b}{4}\text{sgn}(z) \left[1 + \text{erf}\left(\frac{x}{2\sqrt{\lambda|z|}} \right) \right], \tag{8.243}$$

with erf the error function $\text{erf}(t) = \frac{2}{\sqrt{\pi}} \int_0^t \exp(-u^2)\, du$.

This equation yields the tilt angle of the layers $\partial u/\partial x$ and the layer dilation $\partial u/\partial z$:

$$\frac{\partial u}{\partial x} = \frac{\text{sgn}(z)b}{4\sqrt{\pi\lambda|z|}} \exp\left(-\frac{x^2}{4\lambda|z|} \right), \tag{8.244}$$

$$\frac{\partial u}{\partial z} = \frac{-b}{8\sqrt{\pi\lambda}} \frac{x}{|z|^{3/2}} \exp\left(-\frac{x^2}{4\lambda|z|} \right). \tag{8.245}$$

The last formula shows that *the compression stress propagates over long distances (fol-lowing a $1/|z|^{3/2}$ law) in the glide plane of the dislocation, but is quickly damped (as*

---

[11] This problem has been solved analytically, taking into account the anharmonic terms in Eq. (8.240) (Brener and Marchenko, 1999). The results are qualitatively unchanged.

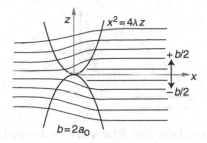

Fig. 8.38 Shape of the layers around an edge dislocation and deformation parabolas.

$\exp(-x^2))$ *in the plane of the layers.* This behaviour is in contrast with that of solids, where the stress decreases as $1/r$ around a dislocation.

This very strong anisotropy of the stress field is due to layer fluidity in smectics. Note that, due to the exponential term in formulas (8.244) and (8.245), the stress only takes significant values within two parabolas, with equations:

$$x^2 = 4\lambda|z| \tag{8.246}$$

known as the '*deformation parabolas*' (Fig. 8.38).

These formulas also give the energy of an edge dislocation. The calculation leads to the following result (Kléman, 1974):

$$E_{\text{edge}} = \frac{Kb^2}{2\lambda r_c} + E_c, \tag{8.247}$$

where r_c is a cutoff radius in the glide plane, corresponding to the radius of the dislocation core, and E_c is a core energy, often written as $E_c = 2\gamma_c r_c$, where γ_c is a cutoff energy of the layers per unit surface. Minimising the dislocation energy with respect to r_c yields:

$$E_{\text{edge}} \approx 2\,\overline{B}^{1/4} K^{1/4} \gamma_c^{1/2} b. \tag{8.248}$$

This formula shows that *the energy of an edge dislocation in smectics is proportional to its Burgers vector* (and not to b^2, as in solids). *This dependence explains the ease of dislocation bunching into dislocations with large Burgers vectors in smectic phases*; sometimes, they even contain '*giant*' dislocations, with Burgers vectors of several hundred layers.

Another important observation concerning the edge dislocations refers to their interaction with the free surfaces. To predict the type of this interaction, let us compare the energy of a bulk dislocation to that of a step at the free surface of the smectic, with a height equal to the Burgers vector of the dislocation. In the first case the energy is of the order of $(K\overline{B})^{1/2}b$, while in the second it is about $\gamma_{\text{air}}b$, with γ_{air} the surface tension at the interface with air. Thus, if

$$\gamma_{\text{air}} > \sqrt{K\overline{B}}, \tag{8.249}$$

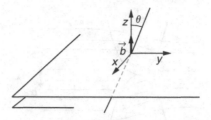

Fig. 8.39 Mixed dislocation, tilted with respect to the layer normal.

a bulk dislocation is preferable to a surface step. This condition is almost always fulfilled in practice, meaning that *the edge dislocations are repelled by the free surfaces in smectics*.[12] Once again, this behaviour is in stark contrast with that of solids, where the dislocations, incurring an energy cost much larger than that of surface steps, are attracted by the free surfaces.

We will now say a few words on screw dislocations. They are aligned along the z axis and their (multivalued) deformation field is of the type:

$$u = \frac{b\varphi}{2\pi} = \frac{b}{2\pi}\arctan\left(\frac{y}{x}\right) \quad (\mathrm{mod}\,\pi), \tag{8.250}$$

with φ the polar angle in the plane of the layers. It is easily checked that this deformation obeys the bulk equation of layer equilibrium (8.240). One can also see that, to lowest order in the distortion, the elastic energy is zero, since $\partial u/\partial z = 0$ and $\Delta_\perp u = 0$. On the other hand, the anharmonic term does not vanish, and neither does the K_3 term in the Frank energy (8.157); close to the dislocation core, the latter is of the same order of magnitude as the former (so it cannot be neglected). Taking these two contributions into account and using the deformation field (8.250) yields the energy of a screw dislocation (Kléman, 1976):

$$E_{\mathrm{screw}} = \frac{\overline{B}b^4}{128\pi^3 r_c^2} + \frac{K_3 b^4}{64\pi^3 r_c^4} + E_c, \tag{8.251}$$

here r_c is a core radius and E_c a core energy.

This formula shows that the elastic energy of a screw dislocation is very low (about 1000 times lower than for an edge dislocation), and varies as b^2 (since $b \approx r_c$). These results explain why elementary screw dislocations proliferate in the samples.

Finally, we would like to emphasise a fundamental feature of screw dislocations, concerning their *line tension*. Indeed, in Section 5.1.2.6 we pointed out that the line tension of a dislocation is different from its line energy in anisotropic materials. This is the case for smectics, which present a huge anisotropy. A direct consequence is that *the line tension of a screw dislocation is much larger than its line energy*.

To show this, let us imagine that the screw dislocation, perpendicular to the layers, becomes tilted by a small-angle θ with respect to the layer normal, acquiring a small edge

---

[12] Note that the condition (8.249), obtained here in a qualitative manner, can be proven rigorously (Lejcek and Oswald, 1991).

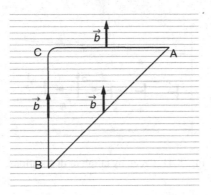

Fig. 8.40 Polygonisation of a mixed dislocation in a smectic A.

component (Fig. 8.39): it is now a mixed dislocation. Under these conditions, the deformation field around the line is given, in the first approximation, by:

$$u = \frac{b}{2\pi} \arctan\left(\frac{y - \theta x}{x}\right),$$ (8.252)

allowing us to calculate the energy of the slightly tilted line:

$$E(\theta) = E_{\text{screw}}(\theta = 0) + \frac{\overline{B}b^2}{4\pi}\theta^2 \ln\left(\frac{R}{r_0}\right),$$ (8.253)

where R is an outer cutoff radius. The line tension of a screw dislocation is obtained using Eq. (5.48):

$$\tau_{\text{screw}} = \left[E(\theta) + \frac{d^2 E(\theta)}{d\theta^2}\right]_{\theta=0} \approx E_{\text{screw}} + \frac{\overline{B}b^2}{2\pi}\ln\left(\frac{R}{r_c}\right) \gg E_{\text{screw}}.$$ (8.254)

An important consequence of this very strong line tension is that *dislocations in smectics are strongly polygonised*. More specifically, a mixed dislocation AB, tilted with respect to the layers, will have a strong tendency to decompose into a purely screw segment and a purely edge segment, as shown in the diagram of Fig. 8.40.

Note that the line tension of edge dislocations is equal to their energy, the smectic phase being isotropic in the plane of the layers:

$$\tau_{\text{edge}} = E_{\text{edge}}.$$ (8.255)

In the next section, we will return to the experiment of creep under compression, described in Section 8.2.8.1.

8.2.9.4 Interpretation of the experiment of creep under compression

The experimental results suggest that, contrary to the assumptions of the Orsay group on liquid crystals, the smectic layers are not eliminated at the plates, but rather in the centre

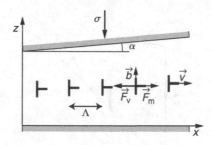

Fig. 8.41 Creep by climb of dislocations.

of the sample, at the position of the dislocation lattice revealed by the decoration technique of Meyer and Lagerwall (Meyer *et al.*, 1978) (Fig. 8.41). Indeed, the applied stress exerts on each dislocation a climb force tending to move it parallel to the layers, in order to remove layers from the sample. Note that the Peach and Koehler formula, proven in the case of solids (Eq. (5A.9) in Appendix 5.A), is still valid in smectics, provided one uses the stress field given by the formula (8.187) (Kléman and Williams, 1974). In the experiment mentioned above, this force is:

$$F_\text{m} = \sigma b, \tag{8.256}$$

with b the Burgers vector and σ the applied stress (positive in compression, by convention). The climb motion induced by the stress is not conservative, being accompanied by molecular flow in the vicinity of the dislocation, within the layers and, above all, across the layers (permeation). By studying the flow around the core of the dislocation, the dislocation can be shown to experience a drag force, proportional to its velocity and Burgers vector:

$$F_\text{v} = -b\frac{v}{m}, \tag{8.257}$$

where m stands for the *mobility* of the dislocation, given by (Orsay Group, 1975):

$$m \approx \sqrt{\frac{\lambda_\text{p}}{\eta}}. \tag{8.258}$$

In the stationary regime, the elastic force exactly compensates the viscous force, yielding:

$$v = m\sigma. \tag{8.259}$$

It suffices now to use the Orowan relation to find the creep rate of the smectic material. This relation is applicable to smectics, being of a purely geometrical nature. It states that the macroscopic creep rate is proportional to the velocity v of the mobile dislocations, to their Burgers vector b and their density ρ_m (see Eq. (5.116)). In the experiment under consideration, the only dislocations are those in the central grain boundary (Fig. 8.41).

Since the distance Λ between two successive dislocations is fixed by the angle of the plates α ($\tan \alpha = b/\Lambda \approx \alpha$), then:

$$\rho_m = \frac{1}{\Lambda d} \approx \frac{\alpha}{bd},$$

(8.260)

yielding the creep rate:

$$\dot{\varepsilon} = \rho_m b v = \frac{m\alpha}{d}\sigma.$$

(8.261)

Up to the presence of the σ_c constant, this law is the same as that describing the experimental behaviour (Eq. (8.236)). In particular, we can see that the coefficient m can be interpreted as the mobility of edge dislocations in climb motion parallel to the layers.

One still needs to explain the origin of σ_c. This stress value represents the elastic limit of the sample. It could be attributed to the edge dislocations becoming pinned on the screw dislocations crossing the layers.[13] This interpretation is not satisfactory for the experiments performed on 8CB, as it leads to a value of σ_c independent of the sample thickness, in contradiction with the observations. Moreover, the energy of the kinks formed on the edge dislocation on crossing a screw dislocation (see Section 5.2.5) is (as one can easily check) much smaller than $k_B T$. Hence, the crossings are in principle very easy in the smectic A phase (for 8CB, at any rate) and their role should therefore be negligible. Another possibility is that the plates containing the sample are not smooth at the scale of the layer thickness. In this case, edge dislocations must appear at the surface, in order to 'take up' its irregularities. It is easy to verify, using Eq. (8.245), that each surface dislocation creates at the centre of the sample an internal stress presenting two extrema in $x = \pm(\lambda d)^{1/2}$, with an amplitude (Oswald, 1985)

$$\sigma_M = \sqrt{\frac{2}{\pi e}} \overline{B} \frac{a_0}{d} \approx 0.5 \overline{B} \frac{a_0}{d}.$$

(8.262)

The mobile dislocations cannot cross these extrema unless the (absolute value of the) applied stress exceeds σ_M. Hence, this value is indeed the yield stress, and it also has the correct form ($\sigma_c = \sigma_M$).

This calculation concludes the explanation of smectic creep under compression normal to the layers, in the case of moderate stress.

We have mentioned that the smectic phase reaches a much higher creep rate when the stress exceeds a threshold corresponding to compression by a few layers. This acceleration suggests the presence of a *threshold for the spontaneous nucleation of new edge dislocations*.

In the next section we will propose an original mechanism that accounts for this observation and which involves the screw dislocations.

---

[13] This interpretation was proposed on the basis of polarised microscopy observations to explain the high values of σ_c observed in certain materials close to the SmA–SmC transition temperature (Blanc *et al.*, 2004).

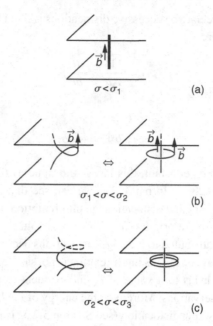

Fig. 8.42 Helical instabilities of a screw dislocation under stress. (a) Stable dislocation below the instability threshold; (b) first instability mode ($p = 1$); (c) second instability mode ($p = 2$). For clarity's sake, we do not show the kink of height b formed on an edge dislocation loop.

8.2.9.5 The helical instability of screw dislocations

The samples presumably contain important numbers of screw dislocations. Due to their very high line tension, they are perpendicular to the layers and cross the sample from top to bottom.

We will show that a screw dislocation spanning the distance between the two plates can become unstable under stress (Bourdon *et al.*, 1981; Oswald, 1984).

More precisely, let us assume that the dislocation takes a helical form, with a pitch $P = d/p$. If the line is anchored at the surfaces the number p is necessarily an integer, and we will make this assumption in the following. Topologically speaking, a helical mixed dislocation with a Burgers vector $\vec{b} = a_0\vec{z}$ is equivalent to a screw dislocation encircled by p edge dislocation loops with the same Burgers vector (Fig. 8.42). This amounts to saying that p layers were removed from within the cylinder with radius r bearing the helix, whose equation in polar coordinates is $z = d\theta/2\pi p$. This process speeds up the creep, since it increases the density of mobile dislocations. Note that in smectics work hardening is completely negligible, since the dislocations can cross very easily and their elastic interaction is very weak, as one can check using the Peach and Koehler formula and the expression for the stress field around a dislocation (in particular, the interaction between two edge dislocations with the same climb plane is strictly zero).

To show that above a critical stress value the screw dislocation becomes unstable, let us write down the radial forces acting upon it as it winds into a helix.

We find:

- the Peach and Koehler force:

$$F_1 = \sigma b \cos \Phi = \sigma b \frac{r}{\sqrt{r^2 + d_0^2}}, \qquad (8.263)$$

where Φ is the tilt angle of the helix with respect to the horizontal plane and $d_0 = d/(2\pi p)$;
- the elastic restoring force associated with the line tension τ:

$$F_2 = -\frac{\tau}{R} = -\tau \frac{r}{r^2 + d_0^2}, \qquad (8.264)$$

with R the radius of curvature of the line;
- a viscous drag force:

$$F_3 = -\frac{b}{M} \frac{dr}{dt}, \qquad (8.265)$$

where M is a mobility that can be calculated; for a screw dislocation, it is given by (Oswald, 1985):

$$M = \frac{8\pi^2 r_c^2}{\eta b}. \qquad (8.266)$$

The inertia of a dislocation line being negligible (see Section 5.2.1) entails that $F_1 + F_2 + F_3 = 0$ or, more explicitly:

$$\sigma b \frac{r}{\sqrt{r^2 + d_0^2}} - \tau \frac{r}{r^2 + d_0^2} = \frac{b}{M} \frac{dr}{dt}. \qquad (8.267)$$

This dynamical equation defines the stress value σ_p above which a dislocation line becomes unstable. We can see it by writing that $dr/dt > 0$ for $r \to 0$, showing that a p-order helix develops spontaneously if:

$$\sigma \geq \sigma_p = 2\pi p \frac{\tau}{bd}. \qquad (8.268)$$

This stress corresponds to sample compression by a number of layers N_p given by:

$$N_p = p \frac{2\pi \tau}{B a_0^2}, \qquad (8.269)$$

where we took $b = a_0$. The first instability corresponds to $p = 1$ and sets the stress threshold σ_1 (corresponding to elastic compression of the sample by N_1 layers) below which the screw dislocations remain stable. Under these conditions ($\sigma < \sigma_1$) the model of creep by climb of the edge dislocations in the grain boundary is valid, in agreement with the experimental observations.

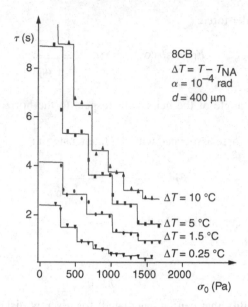

Fig. 8.43 Relaxation time as a function of the initial compression stress, measured in 8CB at different temperatures (Oswald, 1984).

For an applied stress above σ_1, the screw dislocations wind into helices and the creep rate increases. The model predicts that this acceleration occurs for a compression:

$$N_1 = \frac{2\pi\,\tau}{\overline{B}a_0^2},\tag{8.270}$$

which gives, using the exact result for the line tension of a helical screw dislocation (Bourdon *et al.*, 1981):

$$N_1 = \frac{2\pi}{\overline{B}a_0^2}\left[E_c + 0.04\overline{B}a_0^2 \ln\left(\frac{2\pi\,d\lambda}{r_c^2}\right)\right].\tag{8.271}$$

In this formula, all parameters are known except for the core energy of the dislocation (and the core radius, which we can take to be of the order of λ). Using the measured value for N_1 (of about 3–4) and the experimental values of $\overline{B}$ and λ, one finds $E_c \approx 0.1\overline{B}a_0^2$. Consequently, this method provides an estimate for the core energy of a screw dislocation (which, as we can see here, is relatively large compared to its elastic energy, given by the formula (8.251)).

Other experiments, more precise than those performed at constant creep stress, have proven the existence of the helical instabilities of screw dislocations. One of them (Oswald, 1984), performed using a piezoelectric setup, consists of measuring the relaxation time of a sample suddenly compressed by N layers, following the method described in Section 8.2.1. This experiment shows that the relaxation time varies by successive jumps as a function of the initial applied stress ($\sigma_0 = Na_0\overline{B}/d$), as shown in Fig. 8.43.

The first jump signals the first instability mode ($p = 1$), and the pth jump, the instability mode of order p. This experiment can be used to follow the instabilities up to the order $p = 7$ or so. The instability was also observed while compressing a smectic phase in a surface force apparatus similar to that described in Fig. 3.50 (Herke et al., 1995).

We have seen that the equivalent of edge dislocation loops can be nucleated by destabilising the screw dislocations present in the samples. One can however wonder whether edge dislocation loops can be nucleated *ex nihilo*, without the help of screw dislocations.

This problem will be analysed in the next section.

8.2.9.6 'Ex-nihilo' nucleation of an edge dislocation loop

Imagine a perfect sample, limited by two rigorously parallel planar surfaces. Apply to this sample the compression stress σ (once again taken as positive). Assume that an edge dislocation loop appears, with radius R. The energy of the sample varies by:

$$\Delta E = -\pi R^2 \sigma b + 2\pi R E_{\text{edge}}, \tag{8.272}$$

where the first term corresponds to the work of the applied stress, and the second one to the excess line energy associated with the creation of the loop.

This energy grows as its radius increases from zero, goes through a maximum:

$$\Delta E_n = \frac{\pi E_{\text{edge}}^2}{\sigma b} \tag{8.273}$$

at a radius value:

$$R_n = \frac{E_{\text{edge}}}{\sigma b}, \tag{8.274}$$

then decreases as the radius keeps increasing.

Consequently, the radius R_n corresponds to the *ex nihilo nucleation radius* for an edge dislocation loop. The value ΔE_n sets the energy barrier to be crossed by thermal activation at the moment of nucleation.

In the experiments described above, the stress does not exceed 2000 Pa. Using this value for σ and taking as the typical energy of an edge dislocation line $E_{\text{edge}} \approx K \approx 10^{-11}$ N, one has $\Delta E_n = 5 \times 10^{-17}$ J, representing an activation energy of $10\,000\,k_B T$! Such a considerable value excludes this nucleation mechanism for the creep experiments we have just discussed.

Nevertheless, we will show that one can artificially induce the nucleation of a loop in a free-standing smectic film, and subsequently study its growth or collapse dynamics.

8.2.9.7 Nucleation and growth of a dislocation loop in a free-standing film

As G. Friedel had already noticed at the beginning of the twentieth century, one can stretch free-standing films of the smectic A phase (Friedel, 1922). To this end, one only needs to place a small amount of liquid crystal on a thin plate pierced by a hole, and then spread

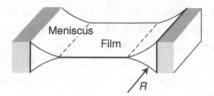

Fig. 8.44 Smectic film and its meniscus at equilibrium.

it slowly over the hole using a spatula. The film thus obtained initially looks as in the photograph in Fig. 8.37. It is composed of a multitude of areas with different thicknesses, separated by bulk edge dislocations. This structure is out of equilibrium and it changes with time such that, at equilibrium, the film has a homogeneous thickness over its entire surface. It is noteworthy that the final film thickness is always given by that of the thinnest area in the initial texture. This observation shows that the films have a systematic tendency to thinning, which can be explained by matter being drained into the meniscus around the film. The presence of the meniscus is indispensable for connecting the film to its supporting frame (Fig. 8.44). It also plays the role of the matter reservoir, its volume being much larger than that of the film (Pieranski *et al.*, 1993).

Experiment also shows that smectic films are very stable, irrespective of their thickness – which can vary between two layers and several thousand layers. This behaviour is very different from that of soap films, which thin by draining the liquid trapped between the surface monolayers, leaving behind a Newton black film, where the two monolayers are almost in contact (note however that, in general, there are two kinds of Newton black films, with different thicknesses).

Before analysing the dynamics of dislocations in smectic films, let us explain the origin of their exceptional stability. As we have already pointed out, a film is inseparable from its meniscus. Thus, the analysis must be global. For more details, see Geminard *et al.* (1997) and Picano *et al.* (2000, 2001).

Throughout the film, the number of layers N is constant.

In the meniscus, on the other hand, the thickness increases away from the edge of the film, so that the meniscus surface is curved. Interferometric measurements in reflected light show that the meniscus has a perfectly circular profile, with a radius R ranging from a few tens of micrometres to several millimetres in experiments. The structure of the meniscus is sketched in Fig. 8.45, where the edge dislocations are placed halfway between the free surfaces, since they are repelled by these surfaces (Lejcek and Oswald, 1991).

A theory of the meniscus shows that the average value of the elastic stress σ_{zz}^{E} inside it vanishes, since the edge dislocations allow it to relax; on the other hand, the hydrostatic pressure is fixed by the curvature of the surfaces *via* the Laplace law:

$$\Delta P = P_{a} - P = \frac{\gamma}{R} \qquad \text{(in the meniscus)}, \qquad (8.275)$$

where P_{a} is the ambient pressure and γ the surface tension with air.

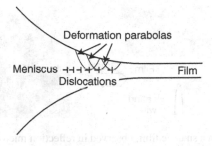

Fig. 8.45 Cross-section view of a film and its meniscus. The meniscus contains edge dislocations that control its thickness increase. The deformation parabolas of the dislocations overlap, so that the free surface of the meniscus is smooth over the typical distance between dislocations (well below the micrometre range).

In the film, on the other hand, the surface is flat. Nevertheless, in an equilibrium state the pressure inside it must be the same as in the meniscus. The conclusion is that the smectic layers are uniformly compressed to obtain mechanical equilibrium:

$$\sigma_{zz}^{E} = -\Delta P \qquad \text{(in the film)}. \qquad (8.276)$$

Hence, it is the elasticity of the smectic layers that keeps the film from collapsing. However, the film must also be stable with respect to the nucleation of edge dislocation loops. This can be verified by determining the nucleation energy associated with this mechanism. From Eqs. (8.273), (8.275) and (8.276) we find:

$$\Delta E_{n} = \frac{\pi R E_{\text{edge}}^{2}}{\gamma b}. \qquad (8.277)$$

This formula leads to extremely high activation energies in normal conditions (of the order of $10^5 k_B T$ using $R = 100$ μm, $\gamma = 25 \times 10^{-3}$ J/m^2 and $E_{\text{edge}} = K = 10^{-11}$ Pa). Consequently, spontaneous nucleation of dislocation loops in films is impossible, hence their excellent stability.

It is however possible to induce artificially the nucleation of an edge dislocation loop in a smectic film. The technique relies on the energy of an edge dislocation varying as the compression modulus $\overline{B}$; this modulus vanishes on approaching the transition towards the nematic phase when this transition is second order (as for 8CB). In principle, it is then sufficient to approach the transition temperature T_{NA} towards the nematic phase in order to reduce the dislocation energy to the point where their nucleation energy (8.277) becomes comparable to $k_B T$. This condition can be achieved by placing under the film a heating micro-tip, heated by a brief current pulse. In this way, the film temperature can be increased locally up to T_{NA}. Experiment shows that heating for a few milliseconds is enough to nucleate within the film a dislocation loop (Fig. 8.46), with an initial radius that can be tuned by controlling the heating time. Note that the initial radius is defined as the loop radius at the moment when the film temperature returns to that of the oven containing it, i.e. a few seconds after having sent the current pulse into the heating tip.

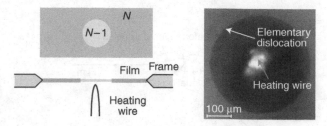

Fig. 8.46 Dislocation loop in a smectic film, observed in reflection microscopy. The extremity of the heating tip situated under the film shows through (Géminard *et al.*, 1997).

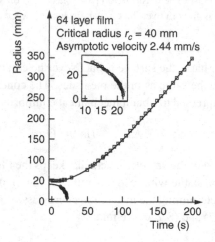

Fig. 8.47 Radius of a loop as a function of time, for different initial radii. The solid curves are fits with the law (8.279) (Géminard *et al.*, 1997).

Experiment shows that the subsequent evolution of the loop, at the temperature T of the oven (lower than T_{NA}), depends on its initial radius r_0. More precisely, there is a critical radius r_c such that:

- for $r_0 > r_c$, the loop keeps on growing, accelerating up to a limiting constant speed v_∞, independent of its radius;

- for $r_0 < r_c$, the loop collapses, first slowly, and then faster and faster, until it disappears completely.

Two examples of the measured radius r as a function of time are shown in Fig. 8.47, illustrating the two cases described above.

To explain these observations, let us write the balance of forces acting on the dislocation. First comes the driving Peach and Koehler force, associated with the lower pressure in the film, pointing out of the loop (the film tends to thin) and with an absolute value ΔPb, opposed by the viscous force and by the line tension force, both pointing inwards. Writing the viscous force as in the bulk samples (in the form bv/m, where m is the mobility of the dislocation and $v = dr/dt$, its velocity) and knowing that the absolute value of the line tension force is E_{edge}/r yields the dynamic equation for the loop:

$$\Delta Pb - \frac{E_{edge}}{r} = \frac{b}{m}\frac{dr}{dt}. \tag{8.278}$$

The solution to this equation reads:

$$r - r_0 + r_c \ln\left|\frac{r - r_c}{r_0 - r_c}\right| = m\Delta Pt, \tag{8.279}$$

where r_c is the critical radius (already calculated, see Eq. (8.274)) at which the loop is in unstable equilibrium ($v = 0$):

$$r_c = \frac{E_{edge}}{\Delta Pb}. \tag{8.280}$$

This formula shows that the loop grows if its initial radius exceeds r_c; its velocity increases, and then saturates at a constant value $v_\infty = m\Delta P$, which is set by the curvature radius of the meniscus through the Laplace law (8.275):

$$v_\infty = m\frac{\gamma}{R}. \tag{8.281}$$

These theoretical predictions have been verified experimentally in thick films, containing more than 100 layers (Picano *et al.*, 2000a). For instance, direct microscopy measurements of the asymptotic velocity v_∞ confirmed its $1/R$ variation and yielded the mobility m of the dislocations with very good precision (note that such a measurement is very difficult in solids). The result is in excellent agreement with that obtained by the creep experiments described above. Measuring the critical radius of the loops also yielded the line tension of the edge dislocations. For instance, $E_{edge} \approx 7 \times 10^{-12}$ N in 8CB at room temperature. On the other hand, this method was used to check that the energy of edge dislocations is indeed proportional to their Burgers vector, as predicted by Eq. (8.248) (Géminard *et al.*, 1997).

The situation becomes more complicated in very thin films (containing fewer than ten layers), where the interactions between the free surfaces induce additional forces (van der Waals, induced order, etc.) (Picano *et al.*, 2001; Holyst *et al.*, 2002). There are also subtle effects due to the fact that the meniscus is not necessarily a good pressure reservoir, as we assumed in our calculations (Oswald *et al.*, 2002). These effects become important in thin films, of fewer than 100 layers, and can lead to surprising phenomena, such as a notable increase in the critical nucleation radius or a marked slowing down of the loop growth dynamics (Oswald *et al.*, 2003; Caillier and Oswald, 2004).

To conclude our discussion of dislocations, we will show that, surprisingly, they play an important role under shear parallel to the layers in lubrication geometry.

8.2.9.8 *Theory of smectic lubrication*

What happens when a smectic phase is sheared parallel to the layers in lubrication geometry (Fig. 8.48)? In this case, the layers slide viscously on top of each other with the

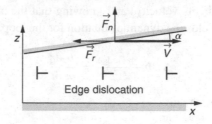

Fig. 8.48 Classical lubrication geometry for a smectic phase. The lower plate is fixed; the upper one undergoes horizontal translation at a velocity V.

viscosity η_3. However, they are also compressed since, at the position of a given point on the lower plate (considered fixed), the sample thickness decreases with time. This layer compression generates an important force, opposing sample squeezing. This force adds to the usual lubrication force of simple liquids, and can be much more important than the latter. This is why a smectic phase is a much more effective lubricant than a simple liquid with the same shear viscosity (Oswald and Kléman, 1982). This feature is well known in practice. Thus, custard, soap or the very elaborate lyotropic mixtures used in metallurgy for drawing or pulling metals have a lamellar structure that gives them excellent lubricant power.

To show this, let us compare the viscous lubrication force to the elastic force due to compression of the layers.

To do that, one must solve the dynamics of edge dislocations under flow. The complete calculation can be found in [43, volume 2]. A quicker way of reaching the result simply consists of considering that shearing a smectic amounts to shearing a simple liquid of the same viscosity η_3 and compressing the smectic phase at the same compression rate $\dot{\varepsilon} = \alpha V /2d$, where V is the velocity of the upper plate and d the average sample thickness. Note that the factor of 2 in the denominator is due to the fact that the dislocations are carried by the average flow, at a velocity $V/2$ in the stationary regime.

Denoting by L the horizontal sample size, Eq. (3.277) then yields:

$$F_N(\text{viscosity}) = \frac{1}{2}\eta_3 \left(\frac{L}{d}\right)^3 V\alpha \tag{8.282}$$

and, from Eq. (8.261):

$$F_N(\text{elasticity}) = L\frac{d\dot{\varepsilon}}{m\alpha} = L\frac{V}{2m}. \tag{8.283}$$

In practice, these two forces add and contribute together to the lubrication effect. Comparing them is very interesting. In 8CB at room temperature, $\eta_3 \approx 0.4$ Pa s and $m \approx 10^{-8}$ m^{2}s kg^{-1} (from Eq. (8.237)) yielding, as an aside, a permeation length $l_p = m\eta_3 \approx 40$ Å. With $L = 1$ cm, $d = 100$ μm and $\alpha = 10^{-3}$ rd, one has:

$$\frac{F_N(\text{elasticity})}{F_N(\text{viscosity})} = \frac{1}{\alpha}\frac{d^3}{m\eta_3 L^2} = \frac{1}{\alpha}\frac{d^3}{l_p L^2} \approx 2500. \tag{8.284}$$

As expected, *the elastic contribution completely dominates the viscous one*, hence the interest in using lamellar materials when a strong lubrication effect is desired.

It is also interesting to calculate the horizontal drag force. It also contains two terms: the usual viscous term given by Eq. (3.278), supplemented by a term resulting from the lubrication force F_N (knowing that this force acts perpendicularly to the surface of the mobile plate). One finds, neglecting the viscous term in α^2 (which would be the only contribution in a viscous fluid):

$$F_T = -\eta_3 \frac{VL}{d} - \frac{\alpha VL}{2m}. \tag{8.285}$$

Comparing these two contributions is once again interesting. Using the values above leads to:

$$\frac{\alpha VL/(2m)}{\eta_3 VL/d} = \frac{\alpha d}{2l_p} \approx 10. \tag{8.286}$$

This result shows that the *lubrication effect leads to a notable increase in the apparent shear viscosity of the smectic phase*:

$$\eta_{app} = \eta_3 + \frac{\alpha d}{2m}. \tag{8.287}$$

Due to this effect, η_3 is difficult to determine experimentally.

We should point out that other complications can occur as smectics are sheared parallel to the layers and contribute to the increase in their apparent viscosity. One of them is due to the coupling between the flow and the screw dislocations (Oswald, 1983b, 1986). At low shear rate this coupling is strong, since the dislocations refuse to bend, due to their strong line tension. This leads to a secondary flow around each screw dislocation and thus to a notable increase in the apparent viscosity. This effect decreases at high shear rates, when the dislocations align along the velocity. The resulting behaviour is shear-thinning, with a viscosity that saturates at high shear rate. The measured viscosity is then equal to η_3, provided the lubrication effect is eliminated using, for instance, a 'floating plate' setup (Oswald, 1986; [43, volume 2]).

8.2.10 *Towards more complex dynamics*

In the experiments we have just described, the organisation of the layers is well defined and it remains unchanged under shear. This requires rather specific conditions: thin samples, very good anchoring of the layers at the surfaces containing the sample, low shear rates.

In practice, one can also perform less controlled experiments, and they do not necessarily lead to inextricable situations.

Such an example was provided by a team at the Centre de Recherche Paul Pascal in Bordeaux, led by D. Roux (Panizza *et al.*, 1995).

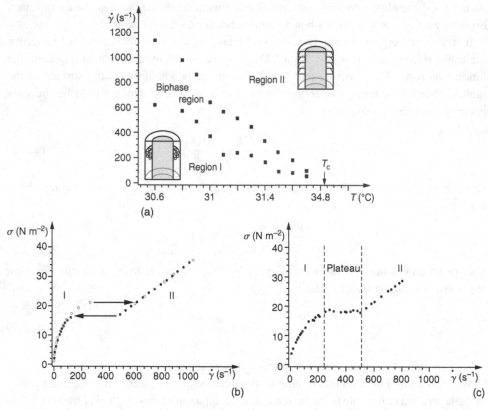

Fig. 8.49 (a) Orientation diagram for 8CB, the *y*-axis represents the shear rate and the *x*-axis the temperature; (b) and (c) stress as a function of the shear rate at 31° C, measured at imposed stress (b), or at imposed shear rate (c) (Panizza *et al.*, 1995).

Their experiment consists in placing a smectic phase in a rotating rheometer and shearing it. In this experiment, the surfaces of the cylinders are not specially treated. The sample is observed under shear by a variety of techniques (optical microscopy as well as light, X-ray or neutron scattering), thus following the evolution of the average orientation of the layers at increasing shear rate. These experiments helped establish *characteristic orientation diagrams*, such as that in Fig. 8.49a.

Three regions are clearly distinguishable in this diagram:

- region I, where the smectic layers roll up forming 'leek'-shaped structures, shown schematically in the lower left corner of the figure. These leeks align along the velocity;

- in region II, on the other hand, the layers tend to align perpendicular to the cylinders (in a planar orientation, as shown in the top right corner of the figure). The flow takes place in the layers (with a viscosity η_2);

- in the intermediate region, the system is *biphasic*, meaning that leek domains coexist with domains in planar orientation. The sample often exhibits a *banded structure*.

The evolution of the stress as a function of the shear rate is also very interesting, since the behaviour of the smectic phase depends crucially on the configuration of the rheometer.

Indeed, *under imposed stress*, the behaviour is as shown in Fig. 8.49b, where the open dots are experimental values taken as the stress increases, and the solid ones are taken on the decrease. One can see that, in this case, the transition from region I to region II is sharp, and no biphase regions are observed within the sample. Note that the jump from one region to the other occurs at different values of the applied stress, depending on whether this parameter increases or decreases: *hysteresis* occurs.

If, on the other hand, the shear rate is imposed, the stress increases continuously in region I, displaying shear-thinning behaviour, then exhibits a *plateau* in the coexistence region, before once again increasing linearly in region II, showing Newtonian behaviour (with the viscosity η_2).

This rather complex behaviour is similar to a *first-order phase transition*.

The same type of behaviour was observed in the lamellar phases of lyotropic liquid crystals. The main difference with respect to 8CB is that the 'leeks' are replaced by '*onions*', or '*spherulites*', where the lamellas are stacked in concentric spheres (Diat and Roux, 1993). A very interesting result, since it is experimentally robust, although it has not yet been explained convincingly, is that the radius of the onions decreases as the reciprocal square root of the shear rate.

Lamellar phases are not the only systems that exhibit dynamic phase transitions under shear. Complex systems, such as concentrated giant micelle phases (already discussed in Section 7.4) or polymer mixtures, exhibit similar behaviour that starts to be actively investigated by physicists.

In the next section we will discuss a few questions related to columnar hexagonal phases.

8.3 Canodynamics

In 1977, the team of S. Chandrashekhar in Bangalore discovered that, in a narrow temperature range, the molecules of a substituted benzene derivative can assemble in fluid columns that form a two-dimensional lattice (Chandrashekhar *et al.*, 1977). A few years later, chemists (H. T. Nguyen, J. Malthête, and others) synthesised a large number of new molecules exhibiting the same type of phases, but over much wider temperature ranges. These columnar phases were termed 'canonic' phases by Sir Charles Frank (from the Greek $\kappa\alpha\nu\acute{\omega}\nu$, meaning 'reed stalk'). Hence the term 'canodynamics', by analogy with smectodynamics or nematodynamics.

In this section, we will be concerned with the elasticity and the mechanical instabilities of these phases, and show that they possess specific features, with no counterpart in smectics. We will restrict ourselves to hexagonal phases, where the columns form – by definition – a hexagonal lattice.

8.3.1 The elasticity of columnar phases

The elasticity can be constructed as for smectics, taking as thermodynamic variables the density (or bulk dilation θ) and the column displacement in the plane of the hexagonal lattice. Let u and v be the displacement components along the x and y axes (perpendicular to the columns), the z axis being parallel to the columns, by convention.

By analogy with smectics, the elastic deformation energy reads:

$$f_e = \frac{1}{2}A\theta^2 + \frac{1}{2}B\left(u_{xx} + u_{yy}\right)^2 + \frac{1}{2}C\left(u_{xy}^2 - u_{xx}u_{yy}\right) - D\theta\left(u_{xx} + u_{yy}\right), \quad (8.288)$$

to which must be added the bending energy of the columns, given by the K_3 term in the Frank energy:

$$f_c = \frac{1}{2}K_3\left[\left(\frac{\partial^2 u}{\partial z^2}\right)^2 + \left(\frac{\partial^2 v}{\partial z^2}\right)^2\right]. \quad (8.289)$$

In the following, we will set $K_3 = K$. Note that the K_1 and K_2 terms do exist, but they cannot be observed since the deformations they describe do not conserve the distances between the columns.

The strain u_{ij} is calculated as in solids. It contains anharmonic corrections (Kléman and Oswald, 1982) that must be taken into account to study the buckling instabilities to be considered in the next section. Explicitly:

$$u_{xx} = \frac{\partial u}{\partial x} - \frac{1}{2}\left(\frac{\partial u}{\partial z}\right)^2 + \frac{1}{2}\left(\frac{\partial v}{\partial x}\right)^2,$$

$$u_{yy} = \frac{\partial v}{\partial y} - \frac{1}{2}\left(\frac{\partial v}{\partial z}\right)^2 + \frac{1}{2}\left(\frac{\partial u}{\partial y}\right)^2, \quad (8.290)$$

$$u_{xy} = \frac{1}{2}\left(\frac{\partial u}{\partial y} + \frac{\partial v}{\partial x}\right) - \frac{1}{2}\left(\frac{\partial u}{\partial x}\frac{\partial v}{\partial x} + \frac{\partial u}{\partial y}\frac{\partial v}{\partial y} + \frac{\partial u}{\partial z}\frac{\partial v}{\partial z}\right).$$

As in smectics, the expression (8.288) can be simplified in the static case, assuming that the extremities of the molecular cylinders are free (at ambient pressure). Under these conditions:

$$P = -\frac{\partial f_e}{\partial \theta} = 0, \quad (8.291)$$

yielding the relation:

$$\theta = \frac{D}{A}\left(u_{xx} + u_{yy}\right). \quad (8.292)$$

Substituting in Eq. (8.288) and taking into account the bending term yields the following expression for the elastic deformation energy (valid in the static case or, for a finite wave vector, at a much lower frequency than the frequency of sound corresponding to that particular wave vector):

$$f = \frac{1}{2}\bar{B}\left(u_{xx} + u_{yy}\right) + \frac{1}{2}C\left(u_{xy}^2 - u_{xx}u_{yy}\right) + \frac{1}{2}K_3\left[\left(\frac{\partial^2 u}{\partial z^2}\right)^2 + \left(\frac{\partial^2 v}{\partial z^2}\right)^2\right], \quad (8.293)$$

where

$$\overline{B} = B - \frac{1}{2}\frac{D^2}{A}. \tag{8.294}$$

As for smectics, one can define a penetration length, intrinsic to the material, with the same interpretation as in smectics since it is clearly related to the penetration distance of a perturbation perpendicular to the columns:

$$\lambda = \sqrt{\frac{K}{\overline{B}}}. \tag{8.295}$$

Another parameter characterises the elastic anisotropy of the hexagonal lattice:

$$\mathcal{A} = \frac{1}{4}\frac{C}{\overline{B}}. \tag{8.296}$$

Analysis of the diffuse X-ray scattering around the Bragg peaks suggests that $\mathcal{A}$ is rather small, meaning that it is easier to shear the lattice than to compress it.

In conclusion, we emphasise that the energy must be positive, imposing $A > 0$, $B > 0$, $C > 0$ and $\overline{B} > 0$ (or $D^2 > 2BA$).

Let us now apply this formalism to the study of instabilities.

8.3.2 Two mechanical instabilities

The first instability, corresponding to a column undulation (or zigzag), is immediately apparent when a planar sample of a hexagonal phase is observed under the microscope. We recall that the term 'planar' refers to the orientation of the director; in a hexagonal phase, this is the unit vector tangent to the columns. The second instability corresponds to the buckling of the columns under uniaxial compression and resembles the Euler instability of beams in solids. We will now describe these two instabilities in detail.

8.3.2.1 Zigzag of the columns

When the hexagonal phase nucleates from the isotropic phase, it often forms a fan-shaped texture, as shown in Fig. 1.35. Microscope analysis of the extinction bands between crossed polarisers shows that the columns are parallel to the glass plates but wind inside the sample, forming circular arcs centred upon singular lines perpendicular to the plates, which are wedge disclinations of rank 1 or 1/2 (Bouligand, 1980; Kléman, 1980; Oswald, 1981; Oswald and Kléman, 1981).

Experiment shows that this texture is very quickly destabilised by a temperature change, and the columns locally start to zigzag around their average orientation. The photograph in Fig. 8.50 shows such a zigzag texture. This instability exists in both thermotropic and lyotropic liquid crystals, occurring either on heating or on cooling, depending on the material.

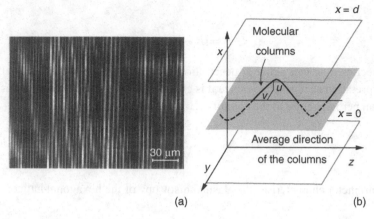

(a)   (b)

Fig. 8.50 (a) Zigzag of the columns in a lyotropic hexagonal phase of a non-ionic surfactant; (b) puta-
tive structure of the column undulation close to the instability threshold.

The simplest way of explaining this instability relies on the assumption that the spacing
of the hexagonal lattice changes with the temperature, as one can confirm by precise X-ray
scattering measurements.

Consider now a temperature change such that the spacing decreases. If the thickness d of
the sample remains constant, and if the columns (taken as aligned along the z direction) do
not slide on the surfaces, this leads to a homogeneous dilation γ of the hexagonal lattice,
equivalent to the displacement field:

$$u = \gamma x \qquad \text{and} \qquad v = \gamma y. \tag{8.297}$$

To find out whether this state of homogeneous deformation is stable, let us envisage the
development of a column undulation of the type:

$$u = \gamma x + u_0 \sin(q_z z) \sin(q_x x), \tag{8.298}$$
$$v = \gamma y + v_0 \sin(q_z z) \sin(q_x x), \tag{8.299}$$

with $q_x = \pi/d$ and $q_z = \pi/\Lambda$. Plugging this displacement field into the energy equation (8.293) and integrating over the entire sample volume yields (per unit sample surface),
to second order in the deformation amplitude:

$$\frac{F}{Bd} = 2(1 - A)\gamma^2 + \frac{1}{8}\left\{\lambda^2 q_z^4 + q_x^2 - 2\gamma(1 - A)\right\} u_0^2$$
$$+ \frac{1}{8}\left\{\lambda^2 q_z^4 + q_x^2 \left[2\gamma(1 - A) + A(1 - \gamma)^2\right] - 2\gamma(1 - A)q_z^2\right\} v_0^2. \tag{8.300}$$

Note the absence of cross-terms in $u_0 v_0$ from the energy, showing that the undulations in
the horizontal and vertical planes are decoupled.

Reasoning as in the case of smectics for each component gives for the vertical plane instability:

$$\gamma_{uc} = \frac{\pi\lambda}{d(1-\mathcal{A})} \quad \text{and} \quad \Lambda_{uc} = 2\sqrt{\pi\lambda d}, \tag{8.301}$$

and for that in the horizontal plane:

$$\gamma_{vc} = \frac{\pi\lambda}{d(1-\mathcal{A})}\sqrt{\mathcal{A}} \quad \text{and} \quad \Lambda_{uc} = \frac{2\sqrt{\pi\lambda d}}{\sqrt[4]{\mathcal{A}}}. \tag{8.302}$$

Since in columnar phases $\mathcal{A} \ll 1$, it follows that the threshold for the development of undulations in the horizontal plane is lower than in the vertical plane.

Hence, this calculation predicts that the *columns start to undulate in the horizontal plane*. Its applicability is however quite limited, since in practice the thermal dilation is of the order of 10^{-3} K^{-1}, much larger than the theoretical threshold calculated using the typical values $\lambda = 15$ Å, $d = 100$ μm and $\mathcal{A} = 0.05$: $\gamma_{vc} \approx 10^{-5}$.

Consequently, the observed textures will have formed very far from the instability threshold. One can legitimately wonder whether the formula (8.301), giving the wavelength at the threshold, is still experimentally relevant.

It turns out that only a calculation *in the strongly nonlinear regime* can provide an answer to this question. This calculation, which we will not detail here (Oswald *et al.*, 1996), shows that the undulation very quickly transforms into a *zigzag structure* with a wavelength Λ very close to that calculated at the threshold. *Formula (8.301) therefore remains valid far from the threshold.*

On the other hand, the calculation shows that *the columns zigzag in a plane parallel to the plates* ($v = 0$ far from the threshold), a prediction confirmed by X-ray experiments (Impéror-Clerc and Davidson, 1999). Another result of the nonlinear analysis is that the zigzag angle with respect to the average direction of the columns is $\theta = 1.9\sqrt{\gamma}$. On the photograph in Fig. 8.50, this angle is of a few degrees, resulting in a γ value close to 10^{-3}, in good agreement with the experimental value inferred from the temperature dependence of the lattice parameter.

The $\sqrt{d}$ dependence of the wavelength of the stripes has also been verified experimentally in several materials, leading to λ values comparable to the distance between the columns (a few tens of angstroms), as expected when there is no correlation between the columns.

8.3.2.2 *Buckling instability of the columns under uniaxial compression*

This instability is all the more interesting since it has no equivalent in smectic phases. First predicted theoretically (Kléman and Oswald, 1982), it was observed experimentally two years later in a columnar phase of discoidal molecules (Cagnon *et al.*, 1984).

The principle of the experiment is described in Fig. 8.51. The homeotropic sample (with the columns perpendicular to the glass plates), of thickness d, is suddenly compressed by a constant amount δd. We will show that, above a certain (absolute value of the) critical compression δd_c, the columns buckle spontaneously, as shown in Fig. 8.51.

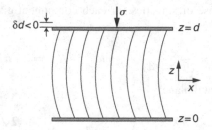

Fig. 8.51 Buckling instability of the columns under compression.

For simplicity's sake, we assume that the sample is very thin compared to its lateral size. The edge effects are therefore negligible, so that $\partial/\partial x = \partial/\partial y = 0$. We also admit that the columns do not slip on the glass surface and that permeation flow is negligible over the duration of the experiment.

Under these conditions, the fundamental solution reads:

$$\theta = \frac{\delta d}{d} \tag{8.303}$$

with

$$u = v = 0. \tag{8.304}$$

To test the stability of this solution, assume that the columns bend in the vertical plane according to the equation:

$$u = u_0 \sin\left(\frac{\pi z}{d}\right). \tag{8.305}$$

From Eqs. (8.288)–(8.290), the distortion energy can be written as:

$$\rho f = \frac{1}{2}A\theta^2 + \frac{1}{2}Bu_{xx}^2 - D\theta u_{xx} + \frac{1}{2}K\left(\frac{\partial^2 u}{\partial z^2}\right)^2, \tag{8.306}$$

where $u_{xx} = -(1/2)(\partial u/\partial z)^2$. Calculating the elastic deformation energy per unit sample surface yields, to second order in u_0:

$$F = \frac{1}{2}A\frac{(\delta d)^2}{d} + \frac{\pi^2}{4}\frac{u_0^2}{d}\left[D\frac{\delta d}{d} + \pi^2\frac{K}{d^2}\right]. \tag{8.307}$$

This calculation shows that the instability develops in compression when:

$$\frac{\delta d}{d} < \alpha_c = \left(\frac{\delta d}{d}\right)_c = -\pi^2\frac{K}{Dd^2}. \tag{8.308}$$

The experiments of M. Cagnon *et al.* confirmed the existence of a buckling threshold in compression, varying as the reciprocal sample thickness squared, in agreement with the expression above.

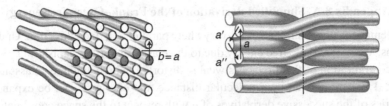

Fig. 8.52 Examples of dislocations in the columnar phases. (a) Transverse edge dislocation; (b) screw dislocation (Bouligand, 1980).

However, the value of the penetration length λ obtained in this way (admitting that D, which has not been measured, is of the same order of magnitude as B) is considerably larger than that obtained by measuring the wavelength of the thermal stripes described above, since we have here $\lambda \approx 1000$ Å instead of 50 Å.

This profound disagreement between the two measurements remains somewhat enigmatic for now.

One explication, proposed by J. Prost in 1990, and which deserves careful consideration, is that the thermotropic columnar phases contain *a high density of column ends* (sometimes termed 'π *defects*' due to their shape) (Prost, 1990). The presence of these defects has been confirmed recently by X-ray scattering in some columnar phases of substituted aromatic diamides (Albouy *et al.*, 1995). But, topologically speaking, a column end is equivalent to a segment of a transverse edge dislocation (Fig. 8.52a), with a length given by the spacing of the hexagonal lattice. Since a dislocation cannot stop suddenly within the material, a column end is necessarily connected to another defect of the same type by a screw dislocation dipole (Fig. 8.52b). Thus, one can imagine that the sample is filled with column ends, connected to one another by screw dislocation dipoles. What does this change with respect to column bending? If the column ends are 'frozen in' along the columns, bending the columns leads to a deformation of the lattice formed by the screw dislocation dipoles, which requires much more energy than bending each column individually. One could say that the bending constant measured in the experiment of buckling under uniaxial compression is not that of the individual columns, but rather an effective constant associated with the defect lattice connecting the isolated column ends. Thus, the 1000 Å found in the column buckling experiment would be related to the average distance between column ends.

We emphasise that this model assumes the permeation to be negligible during the buckling process, because otherwise the column ends can move along the columns and relax the stress. This assumption is certainly satisfied, since the experimental time scale of the buckling is very fast (a few milliseconds).

On the other hand, the formation of the thermal stripes described in the previous section is much slower (quasi-static). Under these conditions, only the bending modulus of the individual columns is relevant, as the elastic stress associated with the defect lattice can relax by permeation. It is then natural that the experimental value for λ measured via this method should be close to the lattice spacing, as in smectics, since the columns are not correlated in the investigated materials.

Appendix 8.A Simplified derivation of the Frank–Oseen free energy

When a nematic is unconstrained, $\vec{n}$ is everywhere parallel to a fixed direction. Under these conditions, $\vec{\nabla}\vec{n} = 0$ and the free energy due to the elastic distortion vanishes: $f = 0$.

If the configuration of the director is weakly distorted, which amounts to assuming that $a|\vec{\nabla}\vec{n}| \ll 1$, where a is an intermolecular distance, the energy f can be expanded in a power series of the successive derivatives of $\vec{n}$ with respect to the space coordinates.

Several kinds of terms can be distinguished in this development.

First come the terms *linear* in $\vec{\nabla}\vec{n}$. Among them, only terms of the form $\mathrm{div}\,\vec{n}$ and $\vec{n} \cdot \overrightarrow{\mathrm{curl}}\,\vec{n}$ are invariant under changes in the reference frame. However, the first term must be discarded, since the energy has to remain invariant as $\vec{n}$ is changed to $-\vec{n}$ (nematics are not ferroelectric). The second term is also forbidden, because it changes sign under the transformation $x \to -x$, $y \to -y$ and $z \to -z$, which is incompatible with nematics because they count an inversion centre among their symmetry elements.

One must then consider the *quadratic terms*, in $(\vec{\nabla}\vec{n})^2$. These terms are somewhat more difficult to find and classify. The most apparent second order invariants are the squares of the first-order scalars given above:

$$(\mathrm{div}\,\vec{n})^2, \qquad \left(\vec{n} \cdot \overrightarrow{\mathrm{curl}}\,\vec{n}\right)^2. \tag{8A.1}$$

One must also consider the scalar products generated using $\vec{n}$ and its first derivatives $\vec{\nabla}\vec{n}$. There are three independent vectors of this type:

$$\overrightarrow{\mathrm{curl}}\,\vec{n}, \qquad \vec{n}\,\mathrm{div}\,\vec{n}, \qquad \vec{n} \times \overrightarrow{\mathrm{curl}}\,\vec{n}, \tag{8A.2}$$

yielding three additional scalar invariants:

$$\left(\overrightarrow{\mathrm{curl}}\,\vec{n}\right)^2, \qquad \left(\vec{n} \times \overrightarrow{\mathrm{curl}}\,\vec{n}\right)^2, \qquad \left(\vec{n} \cdot \overrightarrow{\mathrm{curl}}\,\vec{n}\right)\mathrm{div}\,\vec{n}. \tag{8A.3}$$

It is immediately apparent that the combination $\left(\vec{n} \cdot \overrightarrow{\mathrm{curl}}\,\vec{n}\right)\mathrm{div}\,\vec{n}$ must be eliminated, as it changes sign under $\vec{n} \to -\vec{n}$. Furthermore, the $\left(\overrightarrow{\mathrm{curl}}\,\vec{n}\right)^2$ term is not independent from the others, since it can be decomposed as $\left(\overrightarrow{\mathrm{curl}}\,\vec{n}\right)^2 = \left(\vec{n} \cdot \overrightarrow{\mathrm{curl}}\,\vec{n}\right)^2 + \left(\vec{n} \times \overrightarrow{\mathrm{curl}}\,\vec{n}\right)^2$ (this relation only holds for a unit vector, which is indeed the case for the $\vec{n}$ vector). Thus, one is left with the three fundamental invariants:

$$(\mathrm{div}\,\vec{n})^2, \qquad \left(\vec{n} \cdot \overrightarrow{\mathrm{curl}}\,\vec{n}\right)^2, \qquad \left(\vec{n} \times \overrightarrow{\mathrm{curl}}\,\vec{n}\right)^2. \tag{8A.4}$$

Finally, two more invariants can be constructed from the vectors (8A.2) above:

$$\mathrm{div}\left(\vec{n}\,\mathrm{div}\,\vec{n}\right), \qquad \mathrm{div}\left(\vec{n} \times \overrightarrow{\mathrm{curl}}\,\vec{n}\right). \tag{8A.5}$$

Note that these quantities, containing second derivatives of the form $\partial^2 n_i/\partial x_j \partial x_k$, also vary as the square k^2 of the perturbation wave vector: there is no a-priori reason to discard them, especially since they cannot be expressed as a function of the invariants defined in (8A.4).

Finally, the most general form of the distortion free energy f (to second order in the perturbation) is the following:

$$f = \frac{1}{2}K_1 (\operatorname{div}\vec{n})^2 + \frac{1}{2}K_2 \left(\vec{n} \cdot \overrightarrow{\operatorname{curl}}\,\vec{n}\right)^2 + \frac{1}{2}K_3 \left(\vec{n} \times \overrightarrow{\operatorname{curl}}\,\vec{n}\right)^2$$
$$+ \frac{1}{2}K_4 \operatorname{div}\left(\vec{n}\operatorname{div}\vec{n} + \vec{n} \times \overrightarrow{\operatorname{curl}}\,\vec{n}\right) + K_{13}\operatorname{div}\left(\vec{n}\operatorname{div}\vec{n}\right), \qquad (8A.6)$$

where the two invariants defined in (8A.5) were regrouped in the K_4 term, in order to eliminate the terms containing the second derivatives $\partial^2 n_i/\partial x_j \partial x_k$ (Frank, 1958). Finally, the K_{13} term is of second order in the derivation. In the framework of a microscopic density functional theory, one can however show that this elastic constant vanishes: $K_{13} = 0$ (Yokoyama, 1997). This is a very fortunate result, because otherwise severe mathematical contradictions can arise in the elasticity calculations (Oldano–Barbero paradox).

Finally, the distortion free energy of a nematic can be written in the following general form:

$$f = \frac{1}{2}K_1 (\operatorname{div}\vec{n})^2 + \frac{1}{2}K_2 \left(\vec{n} \cdot \overrightarrow{\operatorname{curl}}\,\vec{n}\right)^2 + \frac{1}{2}K_3 \left(\vec{n} \times \overrightarrow{\operatorname{curl}}\,\vec{n}\right)^2$$
$$+ \frac{1}{2}K_4 \operatorname{div}\left(\vec{n}\operatorname{div}\vec{n} + \vec{n} \times \overrightarrow{\operatorname{curl}}\,\vec{n}\right). \qquad (8A.7)$$

This expression is known as the *Frank–Oseen free energy*.

References

Albouy, P. A., D. Guillon, B. Heinrich, A.-M. Levelut and J. Malthête, *J. Phys. II France*, **5**, 1617 (1995).

Allain, M., D.Sci. Thesis, *Thermodynamic defects in a lamellar phase of a nonionic system* (in French), Paris-Sud University, Orsay (1985).

Bartolino, R. and G. Durand, *Phys. Rev. Lett.*, **39**, 1346 (1977).

Blanc, C., N. Zuodar, I. Lelidis, M. Kléman and J.-L. Martin, *Phys. Rev. E*, **69**, 011705 (2004).

Blanc, C., D. Svenšek, S. Žumer and M. Nobili, *Phys. Rev. Lett.*, **95**, 097802 (2005).

Bogi, A., P. Martinot-Lagarde, I. Dozov and M. Nobili, *Phys. Rev. Lett.*, **89**, 225501 (2002).

Bouligand, Y., *J. Physique France*, **41**, 1297 (1980).

Bourdon, L., M. Kléman, L. Lejcek and D. Taupin, *J. Physique France*, **42**, 261 (1981).

Brener, E. A. and V. I. Marchenko, *Phys. Rev. E*, **59**, R4752 (1999).

Cagnon, M., M. Gharbia and G. Durand, *Phys. Rev. Lett.*, **53**, 938 (1984).

Caillier, F. and P. Oswald, *Phys. Rev. E*, **70**, 031704 (2004).

Chandrashekhar, S., B. K. Sadashiva and K. A. Suresh, *Pramana*, **9**, 471 (1977).

Clark, N. A. and R. B. Meyer, *Appl. Phys. Lett.*, **22**, 493 (1973).

Delaye, M., R. Ribotta and G. Durand, *Phys. Lett. A*, **44**, 139 (1973).

Deuling, H. J., *Solid State Physics, Supp. 14*, Academic Press, L. Liebert Ed. (1974), p. 77.

Diat, O. and D. Roux, *J. Phys. II France*, **3**, 9 (1993).

Durand, G., *C. R. Acad. Sci. Paris*, **275B**, 629 (1972).

Faetti, S., M. Gatti and V. Palleschi, *J. Phys. Lettres*, **46**, L881 (1985).

Faetti, S., M. Gatti and V. Palleschi, *Rev. Physique Appl.*, **21**, 451 (1986).

Frank, F. C., *Disc. Faraday Soc.*, **25**, 19 (1958).

Friedel, G., *Ann. Phys.* **18**, 273 (1922).

Gähviller, C., *Mol. Cryst. Liq. Cryst.*, **20**, 301 (1973).

Géminard, J.-C., R. Holyst and P. Oswald, *Phys. Rev. Lett.*, **78**, 1924 (1997).

de Gennes, P. G., *Mol. Cryst. Liq. Cryst.*, **7**, 325 (1969).

de Gennes, P. G., *C. R. Acad. Sci. Paris*, **275B**, 939 (1972).

de Gennes, P. G., *Phys. of Fluids*, **17**, 1645 (1974).

Grupp, J., *Phys. Lett. A*, **99**, 373 (1983).

Helfrich, W., *J. Chem. Phys.*, **56**, 3187 (1972).

Herke, R. A., N. A. Clark and M. A. Handschy, *Science*, **267**, 651 (1995).

Holyst, R. and P. Oswald, *Intern. J. Mod. Phys. B*, **9**, 1515 (1995).

Holyst, R., P. Oswald and A. Poniewierski, *Langmuir*, **18**, 1511 (2002).

Impéror-Clerc, M. and P. Davidson, *Eur. Phys. J. B*, **9**, 93 (1999).

Imura, H. and K. Okano, *Phys. Lett. A*, **42**, 405 (1973).

Kats, E. I., V. V. Lebedev and S. V. Malinin, *JETP*, **95**, 714 (2002).

Kléman, M., *J. Physique France*, **35**, 595 (1974).

Kléman, M., *Phil. Mag.*, **34**, 79 (1976).

Kléman, M., *J. Physique France*, **41**, 737 (1980).

Kléman, M. and P. Oswald, *J. Physique France*, **43**, 655 (1982).

Kléman, M. and O. Parodi, *J. Physique France*, **36**, 671 (1975).

Kléman, M. and C. E. Williams, *J. Physique Lett. France*, **35**, L49 (1974).

Kléman, M., C. E. Williams, J. M. Costello and T. Gulik-Krzywicki, *Phil. Mag.*, **35**, 33 (1977).

Langevin, D. and M. J. Bouchiat, *J. Physique Coll. France*, **36**, C1-197 (1975).

Lejcek, L. and P. Oswald, *J. Phys. II France*, **1**, 931 (1991).

Liao, Y., N. A. Clark and P. S. Pershan, *Phys. Rev. Lett.*, **30**, 639 (1973).

Manneville, P., *Mol. Cryst. Liq. Cryst.*, **70**, 1501 (1981).

Martin, P. C., O. Parodi and P. S. Pershan, *Phys. Rev. A*, **6**, 2401 (1972).

Meyer, R. B., B. Stebler and S. T. Lagerwall, *Phys. Rev. Lett.*, **41**, 1393 (1978).

Miesowicz, M., *Nature*, **136**, 261 (1935).

Orsay Group on Liquid Crystals, *J. Physique Coll. France*, **36**, C1-305 (1975).

Oseen, C. W., *Trano. Faraday soc.*, **29**, 883(1933).

Oswald, P., *J. Physique Lett. France*, **42**, L171 (1981).

Oswald, P., *C. R. Acad. Sci. Paris*, **296**, 1385 (1983a).

Oswald, P., *J. Physique Lett. France*, **44**, L303 (1983b).

Oswald, P., D.Sci. Thesis, *Dislocation dynamics in Smectics A and B* (in French), Paris-Sud University, Orsay (1985).

Oswald, P. *J. Physique France*, **47**, 1091 (1986).

Oswald, P., and J. Ignés-Mullol, *Phys. Rev. Lett.*, **95**, 027801 (2005).

Oswald, P. and M. Kléman, *J. Physique France*, **42**, 1461 (1981).

Oswald, P. and M. Kléman, *J. Physique Lett. France*, **43**, L411 (1982).

Oswald, P. and M. Kléman, *J. Physique Lett. France*, **45**, L319 (1984).

Oswald, P., A. Dequidt and A. Żywociński, *Phys. Rev. E*, **77**, 061703 (2008).

Oswald, P., J.-C. Géminard, L. Lejcek and L. Sallen, *J. Phys. II France*, **6**, 281 (1996).

Oswald, P., F. Picano and F. Caillier, *Phys. Rev. E*, **68**, 061701 (2003).

Oswald, P., P. Pieranski, R. Holyst and F. Picano, *Phys. Rev. Lett.*, **88**, 015503 (2002).

Panizza, P., P. Archambault and D. Roux, *J. Phys. II France*, **5**, 303 (1995).

Parodi, O., *J. Physique France*, **31**, 581 (1970).

Picano, F., R. Holyst and P. Oswald, *Phys. Rev. E*, **62**, 3747 (2000).

Picano, F., P. Oswald and E. Kats, *Phys. Rev. E.*, **63**, 021705 (2001).

Pieranski, P., D.Sci. Thesis, *Hydrodynamic instabilities in nematics. Experimental study* (in French), Paris-Sud University, Orsay (1976).

Pieranski, P. and E. Guyon, *Solid St. Comm.*, **13**, 435 (1973).

Pieranski, P. and E. Guyon, *Phys. Lett. A*, **49**, 237 (1974).

Pieranski, P. and E. Guyon, *Comm. Phys.*, **1**, 45 (1976).

Pieranski, P. *et al.*, *Physica A*, **194**, 364 (1993).

Prost, J., *Liq. Cryst.*, **8**, 123 (1990).

Ribotta, R., D.Sci. Thesis, *Experimental study of smectic phase elasticity* (in French), Paris-Sud University, Orsay (1975).

Ryskin G. and M. Kremenetsky, *Phys. Rev. Lett.*, **67**, 1574 (1991).

Ternet, D. J., R. G. Larson and L. G. Leal, *Rheol. Acta*, **38**, 183 (1999).

Tóth, G., C. Denniston and J. M. Yeomans, *Phys. Rev. Lett.*, **88**, 105504 (2002).

Yokoyama, H., *Phys. Rev. E*, **55**, 2938 (1997).

General references

On continuum mechanics

[1] L. A. Segel and G. H. Handelman, *Mathematics Applied to Continuum Mechanics*, Philadelphia: Society For Industrial & Applied Mathematics (2007).

On the hydrodynamics of simple liquids

[2] É. Guyon, J.-P. Hulin, L. Petit and C. D. Mitescu, *Physical Hydrodynamics*, Oxford: Oxford University Press (2001).
[3] L. Landau and E. Lifshitz, *Fluid Mechanics*, Oxford: Pergamon Press (1987).
[4] D. J. Tritton, *Physical Fluid Dynamics*, Cambridge: Cambridge University Press (1982).
[5] G. K. Batchelor, *An Introduction to Fluid Dynamics*, Cambridge: Cambridge University Press (1981).
[6] H. Schlichting, *Boundary-Layer Theory*, New York: McGraw-Hill (1968).
[7] J. Happel and H. Brenner, *Low Reynolds Number Hydrodynamics*, Dordrecht: Martinus Nijhoff Publishers (1983).
[8] M. Van Dyke, *An Album of Fluid Motion*, Stanford, California: The Parabolic Press (1982).

On elasticity

[9] L. Landau and E. M. Lifshitz, *Theory of Elasticity*, Oxford: Pergamon Press (1986).
[10] J. F. Nye, *Physical Properties of Crystals*, Oxford: Clarendon Press (1957).
[11] A. E. H. Love, *A Treatise on the Mathematical Theory of Elasticity*, New York: Dover Publications (1944).

On defects and the plasticity of solids

[12] Y. Quéré, *Physics of Materials*, Amsterdam: Gordon and Breach Science Publishers (1998).
[13] *Éléments de métallurgie physique*, edited by Y. Adda, J.-M. Dupouy, J. Philibert and Y. Quéré, Paris: La documentation Française (1976).
Tome 1: rappels
Tome 2: physique du métal

Tome 3: alliages défauts
Tome 4: diffusion – transformation
Tome 5: déformation plastique

[14] J.-L. Martin, *Dislocation et plasticité des cristaux*, Lausanne: Presses polytechniques et universitaires romandes (2000).

[15] D. Hull, *Introduction to Dislocations*, Oxford: Pergamon Press (1979).

[16] J. Friedel, *Dislocations*, Oxford: Pergamon Press (1964).

[17] W. Dekeyser and S. Amelincks, *Les dislocations et la croissance des cristaux*, Paris: Masson (1955).

[18] F. R. N. Nabarro, *Theory of Crystal Dislocations*, New York: Dover (1987).

[19] J. P. Poirier, *Creep of Crystals*, Cambridge: Cambridge University Press (1985).

[20] J. P. Hirth and J. Lothe, *Theory of Dislocations*, 2nd edition, Melbourne: Krieger Publication (1992).

[21] J. Philibert, *Atom Movements, Diffusion and Mass Transport in Solids*, Les Ulis: EDP Sciences (1991).

[22] Y. Quéré, *Défauts ponctuels dans les métaux*, Paris: Masson (1967).

[23] B. R. Lawn, *Fracture of Brittle Solids*, Cambridge: Cambridge University Press (1993).

On isotropic viscoelastic liquids and solids

[24] H. A. Barnes, J. F. Hutton and K. Walters, *An Introduction to Rheology*, Amsterdam: Elsevier (1989).

[25] G. Couarraze and J.-L. Grossiord, *Initiation à la rhéologie*, 3rd edition, Paris: Édition TEC & DOC Lavoisier (2000).

[26] R. G. Larson, *Constitutive Equations for Polymer Melts and Solutions*, Boston: Butterworths (1988).

[27] R. G. Larson, *The Structure and Rheology of Complex Fluids*, Oxford: Oxford University Press (1999).

[28] R. B. Bird, R. C. Armstrong and O. Hassager, *Dynamics of Polymeric Liquids*, New York: Wiley-Interscience (1981).
Volume 1: Fluid Mechanics
Volume 2: Kinetic Theory

[29] J. D. Ferry, *Viscoelastic Properties of Polymers*, New York: Wiley (1980).

[30] G. Astarita and G. Marrucci, *Principle of Non-Newtonian Fluid Mechanics*, New York: McGraw-Hill (1974).

[31] P.-G. de Gennes, *Scaling Concepts in Polymer Physics*, Ithaca: Cornell University Press (1979).

[32] M. Doi and S. F. Edwards, *The Theory of Polymer Dynamics*, Oxford: Clarendon Press (2001).

[33] P. J. Flory, *Principles of Polymer Chemistry*, Ithaca: Cornell University Press (1979).

[34] R. H. Boyd and P. J. Phillips, *The Science of Polymer Molecules*, Cambridge: Cambridge University Press (1996).

[35] Ph. Coussot, *Rheometry of Pastes, Suspensions and Granular Materials: Applications in Industry and Environment*, Weinheim: Wiley (2005).

[36] *Comprendre la rhéologie: de la circulation du sang à la prise du béton*, Editors Ph. Coussot and J.-L. Grossiord, Les Ulis: EDP Sciences (2001).

[37] D. H. Everett, *Basic Principles of Colloid Science*, London: Royal Society of Chemistry (1988).

[38] W. B. Russel, D. A. Saville and W. R. Schowalter, *Colloidal Dispersion*, Cambridge: Cambridge University Press (1989).

On liquid crystals

[39] P.-G. de Gennes and J. Prost, *The Physics of Liquid Crystals*, Oxford: Clarendon Press (1995).

[40] S. Chandrasekhar, *Liquid Crystals*, Cambridge: Cambridge University Press (1992).

[41] M. Kléman, *Points, Lines and Walls*, New York: Wiley (1983).

[42] P. J. Collings and M. Hird, *Introduction to Liquid Crystals*, London: Taylor & Francis, (1997).

[43] P. Oswald and P. Pieranski, *Nematic and Cholesteric Liquid Crystals: Concepts and Physical Properties Illustrated by Experiments*, Volume 1, Boca Raton: Taylor & Francis (2005). P. Oswald and P. Pieranski, *Smectic and Columnar Liquid Crystals: Concepts and Physical Properties Illustrated by Experiments*, Volume 2, Boca Raton: Taylor & Francis (2006).

[44] M. Kléman and O. D. Lavrentowich, *Soft Matter Physics: An Introduction*, New York: Springer (2003).

[45] A. Donald, A. Windle and S. Hanna, *Liquid Crystalline Polymers*, 2nd edition, Cambridge: Cambridge University Press (2006).

Notation

$\underline{A}$	strain rate tensor with components $A_{ij} = (1/2)(v_{i,j} + v_{j,i})$ p. 54
$\alpha_{1,...,6}$	Leslie viscosities (nematics) p. 531
$\overline{B}$	compression modulus at constant pressure
	(1) of the layers (smectics) p. 558
	(2) of the hexagonal lattice (columnar phases) p. 599
c	speed of sound (simple liquids) p. 76
c_1	speed of first sound (smectics) p. 569
c_2	speed of second sound (smectics) p. 569
C_1	speed of longitudinal waves (solids) p. 214
C_t	speed of transverse waves (solids) p. 214
C_R	speed of Rayleigh waves (solids) p. 218
$\vec{C}(M, \vec{\nu})$	surface torque at point M for an orientation $\vec{\nu}$ (liquid crystals) p. 46
$\underline{C}(M)$	surface torque tensor at point M (liquid crystals) p. 49
D/Dt	material derivative with respect to time p. 50
$\overset{\triangledown}{}$	upper-convected derivative (or contravariant) p. 383
$\vartriangle$	lower-convected derivative (or covariant) p. 383
E	Young modulus (solids) p. 187
Er	Ericksen number (dimensionless) (liquid crystals) p. 539
ε	plastic strain (solids) p. 295
$\dot{\varepsilon}$	(1) plastic strain rate (solids) p. 296
	(2) elongation rate (viscoelastic fluids) p. 373
$\underline{\varepsilon}$	strain tensor $\varepsilon_{ij} = (1/2)(u_{i,j} + u_{j,i})$ (usual definition for solids) p. 172
η	dynamic shear rate (simple fluids) p. 4
$\eta(\dot{\gamma})$	steady-state dynamic shear viscosity (complex fluids) p. 9, p. 384
η^*	complex viscosity (viscoelastic fluids) p. 344
η_e	elongational viscosity (solids) p. 305 (viscoelastic fluids) p. 373
η_i	viscosity of the ith Maxwell element (viscoelastic fluids) p. 332
η_0	dynamic shear viscosity at 'vanishing' shear rate (non-Newtonian fluids) p. 8 (viscoelastic fluids) p. 332
G	elastic shear modulus (general definition) p. 3
$G(t)$	relaxation modulus of shear stresses (viscoelastic fluids) p. 333
G^*	complex modulus (viscoelastic fluids) p. 341
G'	elastic modulus (or storage modulus) (real part of G^*) p. 341
G''	loss modulus (imaginary part of G^*) p. 341

611

G_i	elastic modulus of the ith Maxwell element (viscoelastic fluids) p. 332
G_p	elastic modulus on the rubber plateau (or plateau modulus) (entangled polymers) p. 416
γ	simple shear strain p. 3
$\dot\gamma$	shear rate p. 4
$\underline{\gamma}$	strain tensor (alternative definition, mainly used in the rheology of complex fluids) ($\underline\gamma = 2\underline\varepsilon$) p. 347
γ_1	rotational viscosity (nematics) p. 523
γ_c	intrinsic deformation of the 'particles' at the crossover between the linear regime and the nonlinear regime (viscoelastic fluids) p. 21
$\dot\gamma_c$	critical shear rate determining the crossover from the linear to the nonlinear regime (viscoelastic fluids) p. 21
$\vec\Gamma^E$	bulk elastic torque (liquid crystals) p. 52 and p. 533
$\vec\Gamma^M$	bulk magnetic torque (liquid crystals) p. 52 and p. 533
$\vec\Gamma^v$	bulk viscous torque (liquid crystals) p. 52 and p. 533
$\underline{H}$	Finger strain tensor (viscoelastic fluids) p. 379
K	(1) compression modulus (solids) p. 185 (2) layer bending modulus (smectics) p. 558
$\overline{K}$	Gaussian curvature modulus of the layers (smectics) p. 558
$K_{1,2,3,4}$	Frank constants (liquid crystals) p. 515 and p. 605
λ	first Lamé coefficient (isotropic solids) p. 184
λ_p	permeation coefficient (smectics) p. 566
Ma	Mach number (dimensionless) p. 77
μ	second Lamé coefficient or shear modulus (isotropic solids) p. 184
$\vec n$	director (liquid crystals) p. 38
$N_1(\dot\gamma)$	first normal stress difference (viscoelastic fluids and liquid crystals) p. 358
$N_2(\dot\gamma)$	second normal stress difference (viscoelastic fluids and liquid crystals) p. 358
$\vec N$	local (vector) rotation rate of the director (liquid crystals) p. 54 and p. 531
$\vec v$	unit vector normal to the surface and pointing outwards
ω	angular frequency (in general)
$\vec\omega$	vorticity ($\vec\omega = \overrightarrow{\mathrm{curl}\vec v}$) p. 79
$\vec\Omega$	local rotation rate p. 79
P	pressure p. 49
Pe	Péclet number (dimensionless) (suspensions) p. 472
Re	Reynolds number (dimensionless) p. 81
$\rho T\overset{\circ}{s}$	irreversible entropy production per unit volume p. 53
σ	stress p. 3
$\sigma^* = \sigma - \sigma_i$	effective stress in plasticity (solids) p. 299
σ_c	elasticity limit (or critical stress) (solids) p. 7 and p. 178
σ_i	internal stress (solids) p. 277
σ_r	rupture stress (solids) (definition p. 312; Griffith p. 315)
σ_y	yield stress (yield stress fluids) p. 12
σ_{FR}	Frank–Read stress (solids) p. 283
σ_{PN}	Peierls–Nabarro stress (solids) p. 278
$\underline\sigma(M)$	total stress tensor at point M p. 49

$\underline{\sigma}^{\mathrm{v}}$	viscous part of the stress tensor p. 49
$\underline{\sigma}^{\mathrm{E}}$	elastic part of the stress tensor p. 49
t	present time (in general)
t'	past time (in viscoelasticity, Chapters 6 and 7)
$\vec{T}(\mathrm{M}, \vec{v})$	stress at point M for an orientation $\vec{v}$ p. 47
τ	viscoelastic relaxation time p. 19
τ_i	relaxation time associated with the ith Maxwell element p. 332
τ_{life}	lifetime of a giant micelle p. 469
τ_{rep}	reptation time (entangled polymers) p. 442
τ_{R}	Rouse time (polymers) p. 431
u	layer displacement (smectics) p. 556
$\vec{u}$	deformation vector (solids) p. 172
$\vec{v}$	velocity vector p. 49
$\vec{\nabla}\vec{v}$	velocity gradient tensor $(\vec{\nabla}\vec{v})_{ij} = \partial v_{\mathrm{i}}/\partial x_{\mathrm{j}}$ p. 382

Physical constants

Designation	Symbol	Value (SI)
Proton charge	e	1.602×10^{-19} C
Planck constant	h	6.626×10^{-34} J s
Boltzmann constant	k_{B}	1.381×10^{-23} J deg^{-1}
Avogadro's number	N_{A}	6.022×10^{23} mol^{-1}
Permittivity of vacuum	ε_0	$\dfrac{1}{36\pi \times 10^9}$
Permeability of vacuum	μ_0	$4\pi \times 10^{-7}$
1 electronvolt	eV	1.602×10^{-19} J

Unit conversions

Quantity	SI	CGS	Conversion factor
Length	m	cm	1 m = 10^2 cm
Mass	kg	g	1 kg = 10^3 g
Time	s	s	
Force	N	dyn	1 N = 10^5 dyn
Torque	N m	dyn cm	1 N m = 10^7 dyn cm
Pressure (or stress)	Pa (N/m^2)	dyn/cm^2	1 Pa = 10 dyn/cm^2
Surface torque	N/m	dyn/cm	1 N/m = 10^3 dyn/cm
Energy	J	erg	1 J = 10^7 erg
Elastic moduli (G, λ, μ, E, etc.)	Pa	erg/cm^3	1 Pa = 10 erg/cm^3
Bending moduli (K_{1-4})	N	dyn	1 N = 10^5 dyn
Dynamic viscosity	Pa s (poiseuille)	P (poise)	1 Pa s = 10 P
Kinematic viscosity	m^2/s	cm^2/s	1 m^2/s = 10^4 cm^2/s
Surface tension	J/m^2	erg/cm^2	1 J/m^2 = 10^3 erg/cm^2

Index

Printed in the United States
By Bookmasters